DISCOVERING

DIS

Harvard University Press
Cambridge, Massachusetts
London, England

COVERING

Robert Scott
Root-Bernstein

Printed in the United States of America

10 9 8 7 6 5 4 3 2

First Harvard University Press paperback edition, 1991

Library of Congress Cataloging-in-Publication Data

Root-Bernstein, Robert Scott.
Discovering.
Bibliography: p.
Includes index.
1. Science—Methodology. 2. Science. I. Title.
Q175.R546 1989 502.8 88-35768
ISBN 0-674-21175-8 (alk. paper) (cloth)
ISBN 0-674-21176-6 (paper)

The progress of science is threatened today not only by loss of financial support and of good people—which is bad enough; not only by diversion of too much of its energy to applied or engineering work that may not yet be bolstered by enough basic knowledge . . . No, what seems to me to be the most sensitive, the most fragile part of the total intellectual ecology of science is the understanding, on the part of scientists themselves, of the nature of the scientific enterprise, and in particular the hardly begun study of the nature of scientific discovery. In this pursuit, our own day-by-day experience as scientists will help us if we set it into the historic framework provided by those who went before us.

—Gerald Holton,
physicist and historian of science

Contents

PROLOGUE: On Fact and Fiction ▷ **xi**

PREPARATIONS: Toward a Science of Science ▷ **1**

Jenny's Notebook: Impish Ideals 3
Imp's Journal: Dissatisfactions 5
Jenny's Notebook: The Problem Area Defined 7
Imp's Journal: Who Cares? 8
Jenny's Notebook: Dinner with Hunter 9
Transcript: Courting Novelty (Berthollet) 10
Jenny's Notebook: The Arts of Scientists 24

DAY ONE: The Problem of Problems ▷ **27**

Imp's Journal: Discovering a Problem 29
Transcript: How Does Science Grow? 31
Jenny's Notebook: Implications, Contradictions 48
Transcript: What's Worth Investigating? 50
Imp's Journal: Dogma Denied 69

DAY TWO: Planning or Chance? ▷ **73**

Imp's Journal: Alternative Hypotheses 75
Transcript: Planning (Pasteur) 80
Jenny's Notebook: Personal Knowledge 96
Transcript: Chance (Pasteur) 99
Jenny's Notebook: The Eye of the Mind 117
Imp's Journal: Surprises 123

DAY THREE: Logic of Research, Surprise of Discovery ▷ **127**

Imp's Journal: Competition 129
Jenny's Notebook: The Body Is Part of the Mind 130
Transcript: The Probability of Discovering (Fleming) 133

Jenny's Notebook: Patterns in Mind Space 159
Transcript: The Fun of Discovering (Fleming) 166
Imp's Journal: Unexpected Connections 190

DAY FOUR: Creating Unity from Diversity ▷ **193**
Imp's Journal: Thematum 195
Transcript: Modeling the Process (van't Hoff) 196
Jenny's Notebook: Falsification 223
Transcript: Global Thinking (van't Hoff) 231
Jenny's Notebook: Renewing Old Knowledge 260

DAY FIVE: Insight and Oversight ▷ **267**
Imp's Journal: Illumination 269
Jenny's Notebook: Patterns on Paper 272
Transcript: Insight and Oversight (Arrhenius) 273
Jenny's Notebook: Modeling Illumination 291

DAY SIX: Complementary Perspectives ▷ **309**
Jenny's Notebook: Asymmetry and Fault Tolerance 311
Transcript: Reports 312
Ariana's Report: Who Discovers, How? Personality and the Art of Science 312
Transcript: Discussion 340
Constance's Report: History and Philosophy of Science in Science 342
Transcript: Discussion 353
Richter's Report: The Evolution of Science 354
Transcript: Discussion 378
Hunter's Report: Obstacles and Inducements to Exploratory Research 382
Transcript: Discussion 405
Imp's Report: A Manual of Strategies for Discovering 407
Transcript: Discussion 420

POSTSCRIPT ▷ **423**

Imp's Journal: Ambiguities 425

Jenny's Notebook: Conclusions Are Questions 426

Notes ▷ **431**

References ▷ **455**

Acknowledgments ▷ **493**

Index ▷ **495**

Prologue: On Fact and Fiction

The design of a book is the pattern of reality controlled and shaped by the mind of the writer. This is completely understood about poetry or fiction, but is too seldom realized about books of fact.

—John Steinbeck, novelist, in collaboration with Edward Ricketts, marine biologist, The Sea of Cortez *(1941)*

The thought processes involved in scientific discovering are, by their nature, hidden—from observers and practitioners alike. Lack of knowledge concerning scientific creativity affects not only the broadest social perceptions of what science is and how it functions, but also patterns of recruitment, training, and funding. Until we know more, we can do none of these things well.

What makes the process of scientific discovering so mysterious? Scientists themselves are partly to blame. Few take the time or make the effort to be self-conscious about how they work. Some even argue that to do so would interfere with the doing of their science, just as trying to hit the ball while playing tennis can interfere with one's stroke. Few, in any case, believe that the methodological problem of how science is done is itself amenable to scientific research. So they ignore it. In consequence, each new generation of scientists must relearn the nuances of its craft by trial and error, much as if each new tennis player had to learn by watching other players, without benefit of a knowledgeable coach.

The process of scientific discovery is obscured in a second, less innocent way. As the crystallographer J. D. Bernal notes, "it is not possible in any published book to speak freely and precisely about the way science is run. The law of libel, reasons of State, and still more the unwritten code of the scientific fraternity itself forbid particular examples being held up alike for praise or blame."[1] Indeed, I have repeatedly encountered scientists who would not discuss how they arrived at an insight or what motivated their research because something embarrassing was lurking beneath the placid exterior presented in their published reports.[2] Others have demurred because their comments might offend some member of a peer-review board or reveal another colleague's feet of clay. Criticism, even of the constructive variety, is rare in science except when priority for a discovery is at stake. Scientists have an image to maintain; to maintain it, they resort, just as often as politicians, to convoluted rhetoric, convenient back doors, and cover-ups. The only way past these obstacles is access to original research notes, which reveal that many scientists have not told the truth about how they made their discoveries—not out of any intention to defraud, but simply because the code of the scientific community forbade them to speak.

The philosophical and methodological conventions of science present the third obstacle to understanding the process of discovering. From the time that modern science emerged in response to medieval skepticism, its spokesmen have portrayed it as an objective enterprise carried out by objective individuals. Scientific education has been, and often still is, promulgated as the finest way to train unbiased minds. As part of that training, the scientist learns to convey his or her results in such a way that anything unique in the history, sociology, or psychology of the research is irrelevant. The object of the scientific presentation is to convince, not to reveal. What scientists generally fail to admit to others, or often to themselves, is that this stripping away of the unique aspects of the process of discovering may make the enterprise appear to be uniformly objective, but only by inventing a fiction. This objective fiction is just as much a product of the human imagination as a novel or a building. Scientists are as subjective as everyone else.

The best scientists recognize these problems. Einstein, for example, wrote that "science as something existing and complete is the most objective thing known to man. But, science in the making, science as an end to be pursued is as subjective and psychologically conditioned as any other branch of human endeavor."[3] "There are exactly as many ways of approaching the scientific world as there are individuals in science," asserts the scientist and novelist C. P. Snow. "It is only because the results are expressed in the same language, are subject to the same control, that science seems more uniform than, say, original literature. In effect, in the end, it *is* more uniform; but if we could follow the process of scientific thought through many minds, as it actually happens and not as it is conventionally expressed after the event, we should see every conceivable variety of mental texture."[4]

This is my purpose in writing *Discovering*: to bring to light the "process of scientific thought through many minds, as it actually happens" and to investigate how these subjective and fallible human minds can nonetheless produce something as powerful as science. But how to do this, given the obstacles of ignorance, public images, and objectivism?

I found advice in the writings of various people. The novelist Leo Tolstoy: Play games. "One may tell the truth in games. But in real life . . . I should never dare to tell the truth."[5] The poet Stephen Spender: Abandon realism. It is folly to discuss "in terms of knowledge what would only be real in terms of imagination."[6] The physiologist L. J. Henderson: Put yourself in the discoverer's place. "The discoverer . . . can only be *known* through an effort of imagination."[7] The physicist

Victor Weisskopf: Turn to fiction. Especially when we are trying to understand human beings, "a piece of art or a well-written novel could be much more revealing than any scientific study."[8]

The result is the book you hold in your hand: an imaginary reconstruction of the arguments, reflections, and games of six fictional characters attempting to understand how scientists invent. They talk science, think science, do science. They examine historical figures, one another, and themselves. They praise, compare, condemn; poke fun, have fun, examine fun; contemplate, expiate, prevaricate; consent, dissent, cajole, and condescend. In short, they act more like real people than the cardboard pastiches and cosmetically treated advertisements that too often appear in accounts of science.

Yet, despite its fictional form, this is a book of fact. It differs from other books of fact only in that its fictional structure is explicit rather than implicit. The medium, in this case, truly is the message. For it is my contention that what scientists do rests upon the stuff of subjectivity—personality, experience, self-revelation. Therefore, the only way in which the process of discovering can be understood is through imaginative recreation of mental dialogues, nonverbal images and feelings, and unexpected revelations such as I have attempted here. Indeed, one of the surprises that emerged as I wrote the book was that just such mental recreations characterize the manner in which the best scientists always learn to do and to understand science. *Discovering* is, in this sense, simply a public airing of what happens in the privacy of individual minds.

Let the reader be duly warned: This is not a common adventure.

> *These tales concern the doing of things recognised as impossible to do; impossible to believe; and, as the weary reader may well cry aloud, impossible to read about. Did the narrator merely say that they happened, without saying how they happened, they could easily be classified with the cow who jumped over the moon and the more introspective individual who jumped down his own throat. In short, they are tall stories; and though tall stories may also be true stories, there is something in the very phrase appropriate to such topsy-turveydom; for the logician will presumably class a tall story with a corpulent epigram or a long legged essay.*[9]

Let him. It is not logic that creates, but imagination. This is as true in science as elsewhere.

Let the reader be further warned that there are various features of this book that will irk some as much as they please others. The most salient is that the arguments evolve according to the very process being elucidated by the participants. Thus, one thing leads to another not in the post hoc, linear, objectivist form of academic writing, but through the interweavings, digressions, detours, and surprises that attend exploration in any new field. The book exemplifies what it describes. Rest assured that the vast majority of these meanderings turn out to be essential forays and that, in the end, the various elements of the book do tie together into coherent theory.

Those who must know my conclusions before they can agree to accompany me on this journey, and those for whom it is enough to know my conclusions without understanding how they are reached, are advised to begin with "Day Six: Complementary Perspectives." This section contains the final reports submitted by each of the characters.

One further caveat may also prevent undue disappointment. This is not a definitive treatise on discovering. *Discovering* is intended to be a new way of questioning the subject. It is a rethinking and repatterning of what is known that yields new possibilities. My individual conclusions must therefore be considered hypotheses to be tested by future research. The book as a whole is a program statement. Nothing would give me greater pleasure than if the book were to provoke new studies and create controversy. If so, it will do its job.

PREPARATIONS

Toward a Science of Science

It is one of the hopes of the science of science that, by careful analysis of past discovery, we shall find a way of separating the effects of good organization from those of pure luck, and enabling us to operate on calculated risks rather than on blind chance.

—J. D. Bernal, crystallographer and historian of science

There is only one way to approach this study: We must look at the conduct of past searches in order to discern trends that can lead to assessments of the future.

—Martin Harwit, astronomer and historian of science

▼ JENNY'S NOTEBOOK: Impish Ideals

Tonight Imp looked up from his book and asked, "What is *the* most important problem in science?" Imp loves riddles, especially when they disturb someone else's reading. But he was wearing his boyish grin, so I played along. "Developing a unified field theory? Finding a cure for cancer? Discovering life in outer space? I don't know." My heart wasn't in it.

"No, no, no! You're thinking too narrowly," he said, brushing these trite suggestions aside with a wave of his hand. "Come on. *Think*!" He was obviously enjoying himself. Why he insists on asking me questions he already has the answers to, I don't know. I used to think he was just asserting his intellectual superiority, but I long ago absolved him of such arrogance. In matters intellectual he's like a child with a newfound toy. The only difference is that Imp sees all of nature as his sandbox. When we first met in college, I realized that he was either a nut or a genius, because he was constantly bearding authority and inventing his own grand schemes to explain the universe. His "efforts," as one professor sneeringly called them, offended many of his superiors as well as his peers. Yet, unlike so many of them, he wasn't afraid to think for himself or to say what he thought. And he did it with such zest and enjoyment that I soon developed considerable pleasure from watching him. I even composed a limerick in his honor (this was after we'd fallen in love, of course):

There once was a man most impious
Who oft did his most to defy us.
We all tried converting him
From fantastical flirting in
Ideas that could only belie us.

Between his impious, impulsive, and impish nature, and the profound dislike I felt for his given name (Ernest), he soon became, for better or worse, my dear Imp.

Which makes him no easier to live with. There is no denying he is a difficult man. Even after ten years of marriage, I can no more read his mind than prevent his exuberant interruptions. I shrugged my shoulders. "So tell me."

He beamed. "Why, the most important question in science is how to discover! Just imagine the possibilities if we could teach people how to do that, instead of teaching them a laundry list of what other people have already discovered!"

I should have guessed. We'd been through things like this before. "Why can't pedagogy be future-oriented?" Imp likes to ask. "Why can't we teach what needs to be done, instead of congratulating ourselves on the little we've already learned? God! If you could weigh knowledge versus ignorance, knowledge wouldn't even budge the scale!"

"Aren't you being just a bit unreasonable?" I ventured. "Isn't teaching how to discover the same as asking that literature courses teach how to write great novels? Or that art courses teach how to paint masterpieces?"

"Ah! The old Bronowski line! When the Greek departments produce a Sophocles, or the English departments produce a Shakespeare, then I'll start looking in my laboratory for a Newton.[1] But I'm not asking for a Newton. I'm simply asking for people who can *create* instead of copy. Should literature courses teach how to write great novels? Sure!" Imp has a disconcerting way of turning my arguments against me. "They should at least try! Better aim for excellence than be satisfied with mediocrity. After all, what major college—or for that matter, high school—doesn't offer courses in creative writing, or sculpture, or music composition? Okay, so most students never *do* turn out a masterpiece. At least they learn to appreciate how difficult it really is. Yet find me a textbook that even discusses the subject of discovering! And I don't mean the two paragraphs of crap that appear under the heading 'Scientific Method.' I'm talking about how to generate problems and figure out how to solve them; or how to interpret data. What to do if your data don't fit the popular theory. How to invent a totally new theory. And how to get the skeptics to pay attention to results that upset their preconceived little notions of how the universe *ought* to be but isn't!" He was getting serious now.

"Is it any wonder that George Kester can't find any reasonably creative chemical engineers to hire, or that I can't find any biologists? We teach science like art appreciation courses. No, more like paint-by-number kits! 'Here are all the colors and here's where each one goes. Fill in the blanks.' Lab exercises are rigged so that only a total incompetent fails to get the expected results. Cookbook science! When do we teach students how to find answers their own way, using their own little box of tools? Because that's what science requires in real life! There aren't any guidebooks to the unknown. You've got to invent your own!"

"Okay. But this idea isn't new, Imp. One of Snow's characters in *The Search* said much the same thing: If you want your students to learn how to discover, put them in the position of the discoverer. But another character responds: Don't be ridiculous! Scientists stumble onto their discoveries, through unpredictable mistakes or flashes of insight. How are you going to teach that?"[2]

"I say the opposite is no better," countered Imp. "I say that students who never get the slightest inkling of the stumbles and mistakes will never be able to invent new science. They don't need recipes. They need stratagems and tactics for dealing with ambiguity and contradiction. They need to know how to increase the probability of discovery by minimizing the effects of pure chance. Moreover, it's our duty as scientists to search for such stratagems and tactics."[3]

He wasn't finished. "Look, I have no objection to teaching what we know already about science in whatever way is most efficient. I do object to people who have no training in the discovering aspect of science being given degrees in science. They're really just glorified technicians! All they can do is reproduce well what we already know how to do. Or add a further decimal point of accuracy. If they were producing art or music, we'd rate them no better than forgers or mimics. In literature they'd be your hack writers of Gothic romances and murder mysteries. Epigones—inferior imitators—that's what we train! Yet just because they persevere sufficiently to earn a doctorate in—oh, we *must* kowtow!—Science with a capital S, we treat them as if they were automatically scientists! Discoverers! *B.S.*! Ninety-nine percent of these so-called scientists have never had an original thought in their heads."[4]

"Is that their fault, or the fault of the system?" I asked, thinking how unoriginal my own work was.

"Some of both, I suspect," replied Imp, beginning to settle down a little. "You can't expect anybody to be creative in science or anywhere else if the only thing you reward is answering pat questions with predetermined answers. If you're going to discover anything, you have to practice discovering, for chrissakes! And it had damn well better be before you start spending millions of everyone else's tax dollars!

"And that's another point. It isn't just an education problem. It affects every level of science. I mean, think about it—we train scientists, allocate resources, set science policy, all without having the slightest idea of how discoveries are made, or who makes them, or under what educational or economic or institutional conditions! It's a travesty!"

There was only one thing to say. "So what are you going to do about it?"

"I'll write a manual of discovering, or something," he replied.

"Good luck." I smiled and thankfully returned to deciphering the almost impenetrable jargon (dare I say nonsense?) of Michel Foucault and Jacques Derrida. Amazing what French philosophy has spawned! Imp remained surprisingly quiet for the rest of the evening, scribbling notes to himself.

I wonder what's up.

▶ IMP'S JOURNAL: Dissatisfactions

The problem of fostering science is one of the greatest unsolved problems of our day. T. H. Huxley once remarked that the new truths of

> *science begin as heresy, advance to orthodoxy, and end up as superstition. It is not science in its last two phases that we are interested in promoting: such kinds of science can take care of themselves only too well. It is young science, new science, science that is heretical that is our problem. How do we encourage that?*
>
> *—Garrett Hardin, biologist and historian of science (1959)*

Back from running. Realized science isn't fun any more. Too regimented, controlled, specialized. Must have imbibed too much Galileo, Darwin, Pasteur, and Einstein as a kid. *Their* kind of science excites me. Pioneering science! Exploratory science! Antidogmatic science!

Instead, what do we have? Thousands of little minds shaping tiny bricks to add to the Great Edifice of Science. Shape your block of mud, desiccate it, fit into predetermined spot, and voilà! You've made your contribution!

Crap! All you get's a pile![1] A heap of meaningless facts. What is it Truesdell wrote? Petty wits raised aloft by the crooked legs of pygmies distort the history of science into capsules of indoctrination for pygmylets.[2] Amen. Everyone a part in the enterprise, regardless of talent. Oh sure, somebody's got to work out every last detail of enzyme *x* or substrate *y*—we've got to have the facts—but dammit, somebody's also got to design the edifice that makes *sense* of facts. How they relate. Well, that's what *I* want: to be the architect who foresees the shape of the building beforehand, one who knows where the little bricks will fit. Even better, an architect-engineer, designing beautiful buildings and evaluating them for structural integrity. At heart, I'm an integrator, a synthesizer, an idea man. I want to know how it all fits together. I want to know what it *means*.

No, I take that back. Not just to know—to *understand*. Distinction fundamental. To know about something, passive; to understand, active. To understand is to be able to act on it or with it, even to create it.[3] That's what I want from science: *understanding*. Understanding not only of nature but of the nature of science itself.

How to *do* that? How to make a lasting conceptual contribution? Get the freedom, time, money to try the Big Thing? The thing that may not—probably won't—work? How do you explore, when system only rewards doing what we already know can be done? How do you cross disciplinary boundaries? Push beyond them when the only thing that's valued is specialized expertise? Where do you learn the skills to carry out truly original work? How do you gain the integrity and self-assurance to turn your back on the rest of the scientific community to search for unknown frontiers? Because that's what I damn well intend to do. Somehow.

Szent-Györgyi once said it's better to use a big hook and not catch a big fish than to use a little hook and not catch a little fish.[4] Damn right!

▼ JENNY'S NOTEBOOK: The Problem Area Defined

Imp handed me the following memo today. Now I begin to see what's bothering him.

MEMO: *Re: Discovering*
FROM: *Dr. Imp*
TO: ?

Dear Friends and Colleagues:

Recently I came upon four passages that I found pleasantly diverting and all too true. The first was written some four hundred years ago by Sir Francis Bacon: "Invention is of two kinds much differing; the one, of arts and sciences; and the other, of speech and arguments. The former of these I do report deficient; which seemeth to me to be such a difficience as if in the making of an inventory touching the estate of a defunct it should be set down *that there is no ready money*. For as money will fetch all other commodities, this knowledge is that which should purchase all the rest . . . so it cannot be found strange if sciences be no further discovered if the art itself of invention and discovery hath been passed over."[1]

Yet four hundred years later, I say that there is still no ready money! The art of invention *hath* been passed over! To quote J. V. McConnell, "Most of what is wrong with science these days can be traced to the fact that scientists are willing to make objective and dispassionate studies of any natural phenomenon at all—except their own scientific behavior. We know considerably more about flatworms than we do about people who study flatworms."[2] Similarly, the Nobel laureate Sir Peter Medawar recently wrote that "what scientists *do* has never been the subject of scientific . . . inquiry."[3] And finally, the doyen of historians of science, Thomas Kuhn, has just announced that not only have historians of science "failed utterly to bring an understanding of science to nonscientists," but "we simply no longer have any useful notion of how science evolves, or what scientific process is."[4]

Now I ask you: If scientists have no idea how they do science, and historians of science and related types don't know how scientists do science, then who does? Are we all wandering blindly? Must we feel our way forward, tapping randomly with our sticks before us? Or can we enlighten our dark world by placing ourselves in the glare of our own scientific instruments? "Would not," to quote Linus Pauling, "instruction in the art of having ideas about good experiments be just as valuable to the young experimental scientist as instruction in laboratory technique?"[5] "Would not," as Martin Harwit suggests, "any reliable study concerned with progress in science—with procedures, attitudes, or working conditions under which great advances are made . . . be useful in bringing about further advances and . . . ultimately influence how we conduct science?"[6] I say yes! I say, let us invent a "science of science!" Let us coin some ready money!

I, like you, am constantly faced with a myriad of problems ranging from what to study and who to hire to study it, to what conditions are most conducive to problem raising and problem solving, to how much money and time I should allocate toward finding solutions. In short, who discovers, how, when, where, why? Who does not? Why not? Do patterns exist—educational, psychological, institutional, economic; is there a cranial bump for inventiveness?—that can help me match people, problems, and resources so as to increase the probability of making discoveries? Is there any way I can increase my own inventiveness? Are there *a priori* means of determining which problems are susceptible of solution and which are not, so that I can minimize time lost chasing down dead ends? And how can I recognize the best solution when it appears, if it appears?

As you can see, these are simultaneously very theoretical questions and very practical ones: Theoretical in that they concern questions such as "What is discovery?" and "Do discoveries have common patterns?" Practical in that if answers are forthcoming, one might expect to be able to increase the rate of scientific and technological development by means such as more efficient use of time, money, and equipment, or by new forms of scientific education more appropriate to producing discoverers rather than technicians. We might even figure out how to make a better mousetrap!

Toward these ends, I hereby invite you to participate in a colloquium on "Discovering"—note that I use the gerund, as I am interested in the ongoing process rather than in specific products (discoveries) or individuals—to begin meeting next month on Saturday mornings at my home. Please bring lots of problems, anecdotes, case studies, and whatever knowledge and understanding you possess. Most important, bring an active intellect (preferably your own)!

One caveat, however, is in order: "If you believe . . . that science is no more than the accumulation of facts, a social process susceptible of computerization, whose future will unfold unerringly without imaginativeness and a capacity to dream, an ant-hill kind of enterprise that can be bureaucratized, then turn away."[7] My purpose is not to repeat old rhetoric or kneel before ancient idols such as money or "organization." It is not to examine what scientists say they do or what they say science is, but to grasp what scientists *actually* do and how they actually do it. I repeat, therefore, stay away unless you enjoy looking at things in new ways from unusual perspectives.

But forgive me, my friends, for this caveat. I would not invite you to share my journey of discovering if I doubted you. RSVP ASAP.

"What do you think?" Imp asked when I'd finished.

"A bit silly, but fine as far as it goes. May I make some suggestions?"

"On one condition."

"Which is?"

"That you'll help me edit the conference proceedings as well."

"What makes you think there'll be any?" I asked.

"Faith. As the Chinese say, a long journey starts with a small step . . ."

► IMP'S JOURNAL: Who Cares?

Damn! How can people be so stupid and self-centered? I talked to Oliphant about my Discovering Project today. His response: "I tell you frankly, you're wasting your time. If you want to make a discovery, do science, don't talk about how to do it." Davidson was even more complacent! "Sorry old man. No time. As my Prof at Cambridge used to say, 'Time is data!' To succeed at science, you only need one quality: persistence. Work fourteen hours a day, seven days a week and something's bound to pop up. Law of averages and all that. If I stopped to ask questions like 'What's a discovery?'—that's philosophy, not science! 'Who discovers?' Leave that to psychologists. Better still, let time winnow the men from the boys. Look. I simply haven't the time."

I could see what he was really thinking: "Look at me, old boy. How could I be more successful than I already am? Full professor at age forty-two. Over a hundred publications. A lab crawling with postdocs, grad students, and lab techs! Couple million dollars a year in grants! Editor of my own journal! Chairman of review panels for NIH and half-a-dozen other foundations. What more could I want?"

"To make a discovery," I thought as I walked out. He's never once *initiated* a line of research! Only does safe research. Sure things. He lets some poor guy nobody's ever heard of make the breakthrough and then hires him or overwhelms the journals with papers by his dozens of flunkies. Either way he gets credit. And funding. But no surprises. No unexpected results out of *his* lab! Just more of the same. He's no better than that guy at the university who makes a pact with his grad students: in return for their good ideas and results, he'll get 'em a good job. But he's the one whose reputation grows!

What do people like Davidson really think of scientists like Mendel, who wrote only a handful of papers with no funding? Banting, who did his insulin work while unemployed? Einstein, who spent his seven most creative years employed in a patent office? Losers. "Men who didn't know how to play the game, old boy." Yet they'll be remembered when Davidson no longer warrants a footnote. I wonder if he ever thinks of that.

And Saunders! My god! "Don't do it," he warned, "You'll ruin your scientific reputation.[1] Look at the ruckus raised by *The Double Helix*! You can't go peeking into scientists' closets and attics and expect to get away with it. Nobody in their right mind will participate anyway. Really: don't do it! You'll only regret it later." Well, as C. P. Snow once wrote, there are two kinds of advice. Only one is meant to help the listener. I wonder what Saunders is hiding.

At least Hunter's positive.

▼ JENNY'S NOTEBOOK: Dinner with Hunter

Imp's friend Hunter Smithson came to dinner tonight. What a tonic for Imp! And what a surprise for me. Imp always insisted Hunter was not your usual chemist, but his knowledge of history quite took me by surprise.

As part of Imp's efforts at documenting what he's now calling the Discovering Project, he asked Hunter if we could tape-record the dinner conversation. "Isn't this going a little too far?" I asked, somewhat embarrassed, but Hunter seemed to be perfectly agreeable. "My students do it all the time. Go right ahead."

◀ TRANSCRIPT: Courting Novelty

▷ **Hunter:** Look. Let's not waste any more of your tape with idle chitchat. I don't know if Imp's told you, Jenny, but I've been working on something like your Discovering Project myself—looking at a series of related discoveries in physical chemistry, to figure out how they came about. You know—is there a structure of discovery similar to the structure of scientific revolutions? I've turned up some pretty fascinating material, if I do say so myself.

▷ **Imp:** Such as?

▷ **Hunter:** Well, let me see. I could tell you a little story to whet your appetite. Actually, Jenny, you should find this interesting for its French connection. Besides, it's an intrinsically *fun* discovery—at least to my mind.

▷ **Imp:** Go to it.

▷ **Hunter:** Okay. Perhaps I should begin by trying to put you in the position of the discoverer. Imagine you're a doctor turned chemist named Claude Louis Berthollet. You're, let me see, fifty years old.

▷ **Imp:** Hold it! You're going to tell us about a discovery made by a fifty-year-old chemist?

▷ **Hunter:** Actually, Berthollet worked out the details of his theory between the ages of fifty-one and fifty-four.

▷ **Imp:** But it's common knowledge that scientists don't do anything worth talking about after they reach thirty—especially not physical scientists![1]

▷ **Hunter:** That's the myth, all right. Berthollet's an exception. And I know of a few others, too. Pasteur, Pauling, and Chandrasekhar, for example. That's one of the things I'm trying to figure out, how they maintain their creativity over many decades. That's one reason Berthollet's interesting. He was practically forced to see new things because he was literally displaced from his usual activities by the French Revolution and the Napoleonic Wars. You could even make a case that he wouldn't have discovered anything had he followed the normal routine of a middle-aged chemist. I find that thought suggestive.

▷ **Imp:** To say the least!

▷ **Hunter:** So, you're Berthollet. A big, energetic man; gray eyes, blond hair; eager for adventures. Your passions are the theater and gambling, and your encyclopedic mind ranges over everything from Newtonian physics to the fine arts. But you're a skeptic tending toward analytical and abstract thought, you're distinctly materialistic, rationalistic, and antireligious. A true son of the Enlightenment.[2] The year's 1798, and you've

Claude Louis Berthollet at the time of the French Revolution. (Académie des Sciences, Paris)

been chosen by Napoleon Bonaparte to lead a 167-man Commission of Arts and Sciences to Egypt as part of a military expedition. Now, if you're up on the latest gossip, you've probably heard that Napoleon is getting a little too popular in Paris, so the Directory wants him someplace far away. That suits Napoleon's plans, too. So he turns to you, both as a fellow member of the Institut de France—yep, Napoleon acquired that scientific distinction even before he became emperor—and as a former leader of a similar arts and science commission that he had taken to Italy when he conquered its constituent principalities two years earlier.

▷ **Jenny:** So having plundered Italy's treasures for the Louvre, Napoleon now proposes to bring Enlightenment back to the cradle of civilization.

▷ **Hunter:** Right. The commission includes artists, a theater troupe, economists, physicists, chemists, doctors, and the necessary equipment for them to do their jobs.[3]

▷ **Jenny:** Old Puss-in-Boots had rather a swelled head, didn't he?

▷ **Imp:** How else does one rise from common soldier to emperor? Something to think about with regard to scientific leaders, too.

▷ **Hunter:** Self-assurance doesn't hurt. You arrive in Egypt. What do you do? If you're Berthollet, you organize a counterpart to the Parisian Institut, called the Institut d'Egypte. Your major tasks as a physician and chemist consist of three basic sorts. The first is to deal with medical problems, because the army is besieged by dysentery, plague, and eye infections. The second is water analysis—you've just invaded a desert nation, so you need to locate all sources of water and determine whether they're potable. The third is dyes. Why dyes? Because Egypt is a rich source of cochineal insects and various plants used to make dyes, and this is your area of expertise.[4] In 1784, you succeeded your teacher Pierre Joseph Macquer as director of the Manufacture Nationale des Gobelins and as inspector general of Dyeing Industries for the Bureau du Commerce, both government appointments.

▷ **Jenny:** Is that the same Gobelins that made—make—those beautiful tapestries?

▷ **Hunter:** The very same. Colbert, one of Louis XIV's ministers, had undertaken to modernize French industry by introducing so-called "scientific methods" as early as the seventeenth century. Soda production, the Sèvres porcelain works, and various paper mills were also involved, for example.[5] This gave the French industrialists a distinct advantage over the English and Germans until around 1800. And this science-industry link also had purely scientific benefits as well, which then yielded back principles applicable in industry.

▷ **Imp:** Sounds like the Bell Labs story all over again!

▷ **Jenny:** Excuse me?

▷ **Imp:** The research arm of AT&T—Bell Telephone. Bell Labs has probably produced more Nobel Prize winners than any university in the world. I saw a poll recently that said they employed more of the so-called "One Hundred Brightest Scientists under Forty" than any other institution, academic or industrial![6] Their select researchers have (or had—I hear the breakup of AT&T has threatened the existence of the lab) almost total freedom to explore anything they want to, related to communications in the broadest possible sense, on the theory that even the purest research will have some practical application and even the most mundane research can advance theory.

▷ **Hunter:** IBM does the same sort of thing for its research fellows. So does Du Pont. Interesting comparison.

Anyway, Berthollet was best known for two things. The first was a book on the theory of dyeing fabric—a classic.[7] The second was the invention of chlorine bleach. C. W. Scheele, in 1785, isolated a new gaseous element he called chlorine, which he also noted had the property of bleaching vegetable matter. Being Berthollet, you immediately see the potential of the new substance. Cotton and linen are vegetable stuffs, so you wonder if chlorine might bleach them. And perhaps even wool? At present (we're still talking 1785), you have to wash fabric over and over in animal urine and hang it out in the sun. It takes six months to get the fabric white, not to mention odor-free! So you try chlorine, and the gas whitens the same cloth in a day. Unfortunately, it's poisonous and difficult to control.

▷ **Imp:** As became all too clear in World War I.

▷ **Hunter:** Yes. However, you find that the chlorine loses its poisonous properties and retains its bleaching power when absorbed by potash (that's potassium carbonate, or sometimes potassium hydroxide). The combination is soluble, so you invent your famous bleaching solution, *Eau de Javelle*. A short while later, an even better and more easily used bleach is developed by mixing the chlorine with soda (sodium carbonate, Na_2CO_3) to form "bleaching powder." Your friend James Watt, who watches the initial experiments in 1785, introduces the process to England, and by the time you go to Egypt in 1798 it's standardized in both countries.[8] So standardized, we still use it in a modified form today.

Now, your duties at the Gobelins factory and the Bureau du Commerce require you to standardize and improve the production not only of bleaches, but of soaps, mordants—

▷ **Jenny:** What's a mordant?

▷ **Hunter:** A mordant's a chemical used to prepare fabric to take a dye, or to fix the dye afterward. Not all dyes will bond to a given fabric without a mordant. How mordants worked was quite a mystery in the eighteenth and nineteenth centuries. Of course we know a lot more today, but there's still more to some mordants than fits our current theories. Not to mention catalysts and lots of other commonly used chemicals. Lots of practice in need of theory!

So you concern yourself with the manufacture of mordants, soaps, bleaches, dyes, cloth preparation, and so on. You're particularly interested in soda production. Soda is used in the manufacture of bleach, soap, glass, and other necessities, and France imports millions of pounds of soda yearly from Spain. This suppy is cut off during the Revolution. You're asked to act as patent examiner for proposed soda processes.[9]

▷ **Jenny:** Ah! Another genius in a patent office.

▷ **Hunter:** I have a vague recollection that Dorothy Hodgkin—one of my mentors—also worked in a patent office. Keeps you up on the latest chemical advances. Anyway, these concerns come together for you—Berthollet—in Egypt. You prepare extensive reports to the Institut d'Egypte on such indigenous dyes as indigo, henna, and cochineal, which are of economic importance to the French dye industry. You also make periodic expeditions into the regions surrounding Cairo as a scientific adviser to the military. In January 1799 you accompany General Andreossi to the Natron Lakes—one of the most spectacular areas of Egypt. These lakes are actually carved by water out of limestone. You perform a standard analysis of the water and find a good deal of common salt as well as a high proportion of soda (hence the name Natron, which means soda). You've just made the discovery that will set your name indelibly in the history of chemistry!

▷ **Imp:** I don't get it.

▷ **Jenny:** He's discovered a cheap source of soda. Right? France desperately needs soda, so . . .

The expedition to the Natron Lakes. (From Description de l'Egypte, *courtesy of the University of Chicago Library)*

▷ **Hunter:** Well, there is certainly an economic side to the discovery, which Berthollet reports immediately to Paris, but that's not the answer.[10]

▷ **Imp:** Enlighten us, oh Hunter, as Napoleon enlightened Egypt!

▷ **Hunter:** Napoleon failed! But never mind. Have you ever heard of the principle of mass action?

▷ **Jenny:** Only as it applies to riots.

▷ **Hunter:** Imp?

▷ **Imp:** Afraid not. Bonds and structures I know a bit about, but—

▷ **Hunter:** What are universities coming to when biologists don't know even the rudiments of physical chemistry? Well, I'll educate you. Let's start with Berthollet's perspective. As far as he's concerned, the salt in the lake is not supposed to be there at the same time as the soda.

▷ **Jenny:** According to whom?

▷ **Hunter:** According to Berthollet and all the chemical experts of his day.

▷ **Jenny:** Everybody? Surely science isn't *that* monolithic. History certainly isn't.

▷ **Hunter:** You have a point. Most chemists wouldn't have known enough about soda reactions to have cared. And of those who did, Berthollet was practically the only one with the proper background to realize the significance of the Natron Lakes as a huge natural chemical experiment. Now that's something worth thinking about in regard to your Discovering Project, Imp.

▷ **Jenny:** You mean, to what extent does the discovery choose the discoverer, rather than the other way around?

▷ **Hunter:** That's one way to put it. Have you ever noticed in your work how you tend to make use of just about anything and everything you know sooner or later? Pierre Duhem once said something about that—that the scientist's education, hobbies, philosophy, personality all influence the form of the theory he invents.[11] Actually Poincaré and a lot of other guys said it, too. In Duhem's case, he was not only a physical chemist but a philosopher, historian of science, and painter, and he used all that in his research. Same with Berthollet. So maybe Berthollet invents mass action, rather than someone else, because he's the first to combine the prerequisite knowledge, experience, and tools: physical, chemical, industrial, and—what shall we say?—geographical? geological?

▷ **Jenny:** Personal and even political. He's got to get to Egypt first. And to do that, he's got to be friendly with Napoleon. But what about this business that the salt and soda shouldn't have been present simultaneously?

▷ **Hunter:** Yes, I'm sorry. I do tend to digress. But I often find the view to the side too interesting to ignore. And I've made a few discoveries that way. But, as you say, back to the salt and soda.

Why should the coexistence of the two catch your eye, if you're Berthollet? It's because you've been taught that chemicals combine according to their "elective affinities."

▷ **Imp:** Like political action committees, right?

▷ **Jenny:** Or lovers! Didn't Goethe write a novel comparing love to chemical affinity? Some people attract and others repel.

▷ **Imp:** "Where the hormone's, there moan I"?

▷ **Hunter:** Ahem. To take an example, you mix oil and water and what happens? They separate out again. Why? Because the molecules of oil have a greater affinity—or attraction—for themselves than for water; and molecules of water have a greater affinity for themselves than for oil. During the eighteenth century, these kinds of affinities were said to be "elective." If one chemical was placed in the presence of two others, it "elected" to combine *exclusively* with the chemical with which it had the greatest affinity. So, if C had a greater affinity for A than B had, C would replace B in the compound AB. There were lists of rules of this sort made by, among others, Berthollet's teacher, Macquer.[12] And before you start thinking of contradictory examples, Imp, let me say that the concept of elective affinities was abandoned a very long time ago, in part due to Berthollet's ideas.

The basic problem with affinity theory seems to have been that most chemists studied precipitation reactions almost exclusively. They're the easiest chemical reactions to study because you can see the product form as it falls out of solution. Precipitation reactions also have another useful characteristic: they go to completion.

▷ **Jenny:** Imp, what *are* you doing?

▷ **Imp:** I'm gonna produce a precipitation reaction! An acid here (lemon juice); soluble proteins here (milk); and hot water here (not absolutely necessary, but the heat does speed up the reaction). Add milk to hot water (or simply heat the milk). Then add lemon juice, stir, et voilà! Gunk! Or, to use Hunter's fancy terminology, a precipitate.

▷ **Jenny:** I get the picture. You clean up tonight!

▷ **Imp:** No experimental proclivities!

▷ **Hunter:** Nonetheless, a useful demonstration. Now, consider this. If there are more milk proteins than acid molecules (that's what's in the lemon juice), then *all* of the acid molecules will combine with the milk molecules to create the precipitate. No acid will be left in solution, in the water. Adding more milk doesn't create more "gunk" because there's no acid left to combine with it. So let's imagine you just showed there's no acid left in the solution. Okay?

It turns out that chemists habitually used precipitation reactions to determine affinities because the reaction was always complete. You got one product or another, not some confusing mixture. Presumably, the situation eventually became tautological, with affinities thought to be elective due to the existence of precipitation reactions, and precipitation reactions being used to measure the elective nature of affinity. But the fact that the affinities appeared to be elective turns out to be an artifact of the method. The method could yield no other results. Chemists ignored most equilibrium reactions because they yielded little data about the relative strength of affinity. It was too difficult to determine how much of what had formed.

▷ **Imp:** So you only work on problems that can be solved by techniques that fit the current paradigm, as Kuhn might say. You know, I've often wondered how much of genetics is still a mystery simply because we keep trying to fit everything into Mendel's laws. What if there are other modes of inheritance, and we're just "proving" Mendelian inheritance by limiting ourselves to cases that fit it. We could be brushing aside the other modes as being too complicated—things like penetrance and expressivity—when they're really different *kinds* of inheritance. They may be no more complicated than equilibrium reactions! Just different properties that require different concepts and tools for analysis.

▷ **Jenny:** One of your typically impious ideas?

▷ **Hunter:** It certainly never hurts to try a different perspective. Especially since you only see what you look for. I remember when I was in college we were taught that all of the elements in the helium series—helium, neon, argon, and so on—were incapable of forming compounds. There was a whole theory as to why these elements were nonreactive. And then Neil Bartlett demonstrated that these "noble gases" do indeed form compounds under appropriate conditions.[13] The work isn't even that terribly complex. Mainly, it was a matter of knowing how to look and what to look for. So, yeah: sometimes you have to imagine the impossible to make it possible.

But back to Berthollet. In this case, the reaction involving soda was considered, correctly, to be a precipitation reaction by Berthollet and his contemporaries. Soda at that time usually designated what we call sodium carbonate, but included sodium bicarbonate and sodium hydroxide as contaminants. When mixed with calcium chloride, ordinary table salt (sodium chloride) and lime (calcium carbonate or calcium hydroxide) were formed:

$$Na_2CO_3 + CaCl_2 \rightarrow 2NaCl + CaCO_3 \downarrow$$

The same reaction occurs with NaOH (sodium hydroxide, or Drano for you home plumbers), giving $Ca(OH)_2$ (calcium hydroxide) as a precipitate.

How's your chemistry, Jenny? The arrow pointing from left to right indicates the direction of the reaction. In other words, it points to the side on which the affinities are stronger. The arrow pointing down indicates that the calcium carbonate precipitates. Okay?

The point is that you—Berthollet—are intimately familiar with this and related reactions from your patent and bleach work. You know Guyton de Morveau's theories and patents for soda production. You've reviewed the as-yet unworkable Leblanc soda process, which will eventually revolutionize soda manufacture. And so on. So when you analyze the water from the Natron Lakes, you're instantly surprised. You've seen lots of soda reactions and you've analyzed water from all over Egypt, and these results just don't make sense. Along with soda and salt there's also lime and calcium chloride in the water. In other words, the reaction has not gone to completion as you expect. It represents an anomaly: an observation that contradicts accepted theory.[14]

▷ **Imp:** Kuhn says all important shifts in theory begin with anomalies.[15] But let's get back to Jenny's point that maybe the discovery picks the discoverer. Who else would have noticed that the results were anomalous?

▷ **Hunter:** It's hard to say for certain. Probably Guyton de Morveau. Jean Darcet, who had recently reviewed all known soda processes. Maybe Watt. Macquer and Scheele, had they been alive. Probably not Leblanc, for all his subsequent fame—he seems to have been a pretty sloppy chemist. In fact, he was a doctor. But that's a point in itself. Why does a man described as being a third-rate scientist invent the only truly original and useful contribution to soda manufacture?[16] I think it's because Leblanc was willing to invest his time and energy in ideas that better-trained scientists thought were useless or crazy. He was willing to try things, in other words. But the same explanation won't work for Berthollet. He *is* one of the first-rate scientists of his time. So, here's another puzzle for your Discovering Project . . .

▷ **Imp:** Yes—why don't the experts make all the discoveries?

▷ **Hunter:** But you have to realize that perceiving the anomaly is only the first step. Berthollet brings lots of other unique knowledge into play as he tries to resolve the anomaly.

▷ **Jenny:** Like what?

▷ **Hunter:** Well, for instance, one of Berthollet's earliest projects was to analyze the composition of steel.[17] Steel is an alloy rather than a chemical compound, so it can be composed of a wide range of proportions of iron to carbon. But eighteenth-century chemists hadn't yet distinguished clearly between mixtures, alloys, and compounds. So, as far as Berthollet was concerned, affinities in metal "compounds" were not elective. Any proportion of constituent elements was possible. Even his friend Lavoisier was coming to that point of view. He suggested during the 1780s that when several elements with mutual affinities are present together, an equilibrium will be set up among the various affinities so that more than one product is formed.[18] Lavoisier was executed by the revolutionaries before he could develop the idea.

▷ **Jenny:** Ah, yes! "The Revolution has no need of savants," or some such nonsense.

▷ **Hunter:** Except that Lavoisier and his friends were the ones who provided the expertise necessary to make gunpowder, forge cannons, and so on. In any case, Lavoisier left Berthollet a legacy of doubt about elective affinity.

Berthollet's experience at Gobelins provided other unsolved problems to ponder, too. Early in his book on dyeing, Berthollet noted that experimental processes that work on a small scale often fail during large-scale production.[19] This is, of course, *the* classic problem for all chemical engineers. Yes, it works in a test tube, but will it work in thousand-gallon mixing tanks? Why should the scale of the reaction, the actual physical amount of the reactants, change the outcome of the reaction?

▷ **Jenny:** But why didn't these industrial anomalies serve to overthrow the elective affinities theory?

▷ **Hunter:** Ah! Good question. Because no one knew what was going on inside a metal, how a dye attached itself to fabric, or even what the composition of any dye was.[20] Ignorance is a lousy replacement for knowledge, even limited knowledge. A scientist—heck, a human being—always prefers an inadequate explanation to no explanation at all.

The point is that, as Berthollet, you have a great deal of experience with reactions that clearly don't obey elective affinities, but you don't know why, so all you can do is doubt the theory. You can't overthrow it because all you've got is a vague problem (not a concrete one) and no solution.

So, you're reviewing your analysis of the Natron Lakes water. What do you see? You see this beautifully simple, graphic illustration of a well-known reaction in which the precipitation doesn't occur. And the outcome of the reaction seems to have been changed by its scale. No question as to what the reaction is this time. Elective affinities have not been obeyed. Instead, you see an equilibrium between reactants and products, much as Lavoisier had suggested. The problem is now clear-cut.

▷ **Imp:** So what you're saying is this. You have this anomaly—not just your garden-variety anomaly, but a very special sort: an anomaly that clearly falls into a *class* of anomalies (all those industrial reactions and alloys that don't obey elective affinities), and an anomaly whose characteristics are simple and clear enough to promise a solution. So, by solving your simple anomaly, there's hope of getting insight into the whole class of more complicated—and more vague—ones as well.

▷ **Hunter:** Precisely. That's how you know you're onto an important problem, not just some freak occurrence.

▷ **Jenny:** So how do you solve your problem? What's the discovery?

▷ **Hunter:** We're almost there! Just one more bit of preparation. You've got to remember: Berthollet's fifty-one years old now, so he's got lots of experience to draw on.

Okay. The key to solving your problem comes from your earlier association with Lavoisier and Pierre Simon de Laplace, the "French Newton." Lavoisier and Laplace had begun collaborating on studies of heat and affinity in 1782. Almost the first thing they discovered was that acids and bases react with water to give off heat. Pure sodium, for example, explodes on contact with water! In consequence, these chemicals must have an affinity for water—which until that time had been considered to be a totally passive, nonreactive substance. This observation raised a whole set of new problems for affinity studies because many affinities were measured in water solutions and now one had to take into account the affinity of the water for the compounds. Moreover, many acids were totally unreactive in their dry or crystalline forms, which indicated that their reactivity was not due to intrinsic affinity but depended on solvation. Another anomaly.[21]

An accumulation of such anomalies between 1783 and 1792 finally caused Lavoisier to abandon totally the concept of elective affinities.[22] As far as he could see, temperature and concentration and who knew what else affected whether a reaction occurred or not under any given circumstances, so no absolute affinities could be measured.

Now Laplace, for his part, suggested as early as 1784 that physics would soon explain the laws governing chemical interactions.[23] And the conceptual tool would be the same as the one that had explained the mechanics of the universe, namely, gravity. He maintained that all the physical effects of chemical combinations could ultimately be explained if chemical affinity were simply considered to be a form of gravitational attraction.[24]

There were complications, of course. Contemporaries believed that the atoms in every element had a different size and shape that would affect their gravitational field at the close distances required for combining. These elemental shapes were unknown, for the obvious reason that you couldn't see atoms, and consequently the application of gravitational theory to chemical reactions became rather complex, if not impossible, in practice.[25]

▷ **Jenny:** But let me get something straight. Did Laplace actually intend to use the mathematics of gravitational theory to analyze chemical reactions, or was gravity just an analogy?

▷ **Hunter:** You have to be careful when you say "just an analogy." We scientists tend to distrust analogies, but they can be very powerful tools when properly used. But, yes, I think Laplace really did expect to be able to apply the mathematical theory of gravitation to chemistry, if only the proper data were forthcoming.

Now, all of this is important because, although in 1796 Berthollet doubted the gravitational attraction analogy for affinity, in 1799 he used it to solve the Natron Lakes anomaly.[26] That's another topic I'm particularly interested in: what you might call resistance to knowledge. Berthollet's actions suggest that the acceptance of theories may be dependent on whether one can see how to apply them.

▷ **Imp:** Or whether one's faced with a problem unsolvable in any other way.

▷ **Hunter:** Sure. There is, by the way, no doubt that Berthollet was aware of the ideas of Lavoisier and Laplace. He had been actively collaborating with both men since 1785.[27] Again, vis-à-vis the problem choosing the investigator, that puts Berthollet in a privileged position, shared only by a few chemists within France and perhaps nowhere else.

▷ **Jenny:** Fine. But when does he *do* it?

▷ **Hunter:** Right away. Within days of making his analysis at the Natron Lakes, Berthollet writes a report putting together the following pieces of the puzzle. The reaction usually observed is soda plus calcium chloride resulting in salt plus lime precipitate. In the presence of water, Berthollet observes the opposite reaction: lime plus salt forming soda plus calcium chloride. Why? Just looking about suggests an answer. The Natron Lakes are salt lakes carved into limestone. In other words, exactly the opposite situation that you see in the laboratory. In the laboratory, you start with soda and calcium chloride and form salt and lime. Here at the Natron Lakes you start with salt and limestone—

▷ **Jenny:** And create soda and calcium chloride.

▷ **Hunter:** There's so much of what have been classically considered products (salt and lime) present that they start converting back to what were classically considered reactants (soda and calcium chloride). A textbook case of elective affinities unexpectedly being compromised to yield an equilibrium situation.

Berthollet's most important insight is *how* the elective affinities are being overcome. He suggests that the "mass" of the chemical affects the direction the reaction takes—toward products or reactants. More on that in a second.

Finally, he hints that his ideas may revolutionize the concept of affinity.[28]

Now, the crucial question is why Berthollet uses the word "mass" in discussing the Natron Lakes reaction, rather than, say, "quantity," or "amount."[29]

▷ **Imp:** Isn't that nit-picking?

▷ **Jenny:** Napoleon liked to say that in affairs of magnitude, everything turns on a trifle. Just think of the limes fed to the "Limeys" of the British navy to prevent scurvy. Or the invention of canning during Napoleon's campaigns. Better health, stronger soldiers—little things can have big consequences.

▷ **Hunter:** In this case, we can be sure that Berthollet used the word "mass" with forethought, for in subsequent publications and lectures he reiterates its use, explaining in detail that "mass" refers, as in physics, to "gravitational mass." Chemical combination is determined not only by the attraction of affinity but by the mass of chemicals present. A small affinity can be compensated for by a large mass of chemical.

What Berthollet is saying is that, under normal conditions, there are two competing sets of affinities in a reaction. In Imp's "gunk" experiment, one set of affinities is that between water and milk, and water and lemon juice (acid). This set of affinities keeps the milk and lemon juice in solution. The other set of affinities is between the lemon juice and the milk. This latter affinity is the stronger. If the affinity is purely elective, all of the milk and acid that can combine will combine. If the product has little affinity for water, it will precipitate. If, on the other hand, the amount of reactants or products affects the reaction, one should be able to produce an equilibrium between the sets of affinities.

Ah, Imp's at it again!

▷ **Imp:** Sure! What that means in our case can be demonstrated as follows: I pour out all of the liquid (solution) from our cup full of "gunk," leaving the "gunk" in place. Then I add more water (again, preferably hot) and what happens? The "gunk" begins to dissolve! Not much, but enough to make the hot water murky! Pour out the water. Do it again. And so forth ad infinitum. Why? Because a large amount of water has a small affinity for even the precipitated milk, and causes a fraction of the milk and lemon juice to go back into solution. If we used a whole bathtub full of hot water, we could probably get all of the "gunk" back into solution!

▷ **Jenny:** So my reference to mass action as a riot wasn't all that far off, right? What even a few strong men can't rearrange, a crowd of weaklings can pull down with ease. The French Revolution in a nutshell!

▷ **Hunter:** Well, certainly, what a little water can't do, a million gallons can accomplish easily. Which is precisely what Berthollet observes at the Natron Lakes—lots of water dissolving just enough lime to allow the soda to form. By 1801, he maintains that all chemical reactions are reversible if a sufficient mass of a compound is present to make up for its weak affinity, and by 1803 he is actually equating affinity with gravitational attraction.[30]

▷ **Imp:** A simple restatement of Laplace's position!

▷ **Hunter:** Not quite. It sounds the same, but think about the application for a moment. Laplace and his predecessors used the gravitational attraction analogy to try to explain the forces acting between individual atoms. They were preoccupied with how the shapes of atoms would alter the gravitational field at close range. This is the microscopic case, if you will. Berthollet, in invoking mass action, is utilizing gravitational attraction in a macroscopic sense. He refers not to attractions between individual atoms but to the attraction that the entire mass of one type of compound has for another. This is the only way that the amount of chemical can overcome a weak individual affinity. So, he couldn't care less about what shape the atoms are, nor their individual affinities.

▷ **Jenny:** Oh, I see it. So it's sort of like a gestalt shift! Same picture, but two different faces.

▷ **Hunter:** Precisely the mental model Thomas Kuhn has suggested to describe scientific revolutions.[31]

▷ **Imp:** Something's wrong here.

▷ **Jenny:** What?

▷ **Imp:** I don't know. Maybe it's just my antipsychologism, but something doesn't fit. I mean, the gestalt pictures I know, the vase-made-up-of-two-profiles sort of thing, aren't analogous to Berthollet's case at all. Look: with the vase-profiles switch, you have exactly the same data points defining both. That's not Berthollet's situation. He's got new information—he's added new elements to the picture, if you will. And these elements don't just change his perception—they change his *perspective.* He's paying attention to new features of the picture.

▷ **Jenny:** Not to sound too trite, but couldn't you say he's in a position now to see the forest, not just the individual trees? Mass action instead of individual attraction?

▷ **Hunter:** Right. But a lot of people take Kuhn very seriously, and they aren't going to be interested in criticism till there's a new and better model.

▷ **Imp:** Great! Something to work on!

▷ **Hunter:** Why not? But look, lots of other elements of Berthollet's discovery are suggestive for your Discovering Project, too. For instance, consider the equipment needed to make the discovery. Minimal. He could carry it in a box on a camel. The techniques used for analyzing water were just as simple. Total cost of discovering: negligible. All he really needed was to observe the right kind of anomaly, a simple, well-understood one. How typical is this of important breakthroughs in science?

▷ **Imp:** And there's the more general question Jenny mentioned of whether scientists *make* discoveries or are somehow "selected for" by the structure of the discovery. A selection mechanism suggests that the more broadly trained you are, the greater your probability of having the knowledge to solve an important problem. Or recognize it in the first place.

▷ **Hunter:** Alternatively, perhaps discovery requires having an unusual background or different experiences than your colleagues. Berthollet's a dye and metal chemist. His colleagues, in general, aren't, especially outside France.

▷ **Jenny:** And, along the same lines, you could argue for a strong economic side to discovery and invention. Berthollet wouldn't have invented his theory if the French government hadn't been so interested in its own economic welfare.

▷ **Hunter:** And he wouldn't have seen the Natron Lakes except for his friendship with Napoleon; and Napoleon wouldn't have been in Egypt if not for the Revolution; and the Revolution wouldn't have occurred but for various social forces. And then there's the institutional side, too. Napoleon would never have been admitted to the old Académie des Sciences as he was to the Institut de France after the Revolution overturned all the old institutions. And without Napoleon, there would have been no scientific commission to Egypt, and so on.

▷ **Jenny:** You're pushing my point too far. I mean, if we extrapolate, France itself wouldn't have existed as a nation but for all of its history; the Natron Lakes wouldn't have existed but for geology and the history of the earth; which wouldn't have existed but for the history of the universe; which all leads to the conclusion that Berthollet's discovery was predetermined by the Big Bang! One of the problems with doing history is where to set your explanatory boundaries.

▷ **Imp:** Well, I suppose everything is relevant at *some* level. The question is, what level of relevance are we looking for? But let's go a step further. Did Berthollet really have to see the Natron Lakes, or would some other simple anomaly have done as well? If Berthollet hadn't invented mass action, would someone else have done it eventually anyway? If so, under what circumstances? And if the circumstances differ, what becomes of social and economic determination of discovery?

▷ **Jenny:** That's counterfactual history. What if Napoleon had died of scarlet fever as a child; would history be different? How can you answer such a question?

▷ **Hunter:** In science we do it all the time. Set different parameters and see what happens to the equations. That's what experimenting is all about.

▷ **Jenny:** I don't see how you can experiment with history. What happened happened.

▷ **Imp:** But what happened? Darnton or Mandrou say this; you say that. Same data, different interpretations. Aren't you experimenting as much as we are with possible explanations and reinterpretations?

▷ **Jenny:** All right, have it your way. But then why can't you just say that Berthollet was the right person at the right place at the right time?

▷ **Imp:** I don't know. You're the historian. Why can't you?

▷ **Jenny:** Well, it depends on what you mean by an explanation. Every historical event is unique—

▷ **Imp:** So is every scientific one!

▷ **Jenny:** Let me finish! So the question becomes whether you focus on the unique aspects of the event or the general ones. Take Napoleon, for example. Why does he win any particular battle, or come to power when he does? You can treat these events as being due to unique, unrepeatable circumstances (which is how most historians look at history), or you can take the perspective of a military strategist like Patton (who studied Napoleon's battles), or a political strategist like Lenin (who tried to model the Russian Revolution on the French). Presumably Patton and Lenin looked for patterns that characterize a successful battle or a successful revolution. Then they tried to recognize opportunities or find ways to recreate or encourage similar patterns in their own unique historical setting. There's no *a priori* reason you couldn't do the same in trying to explain discovering. You just have to find the patterns first, and try to figure out what controls them.

▷ **Imp:** So you're saying that, essentially, it's a matter of whether you believe that history contains lessons or not. And that, I suppose, depends on why you're interested in history. Is it just a useless humanistic diversion, a mirror in which to examine the present, or is it a source of Machiavellian knowledge?

▷ **Jenny:** Faustian, too!

▷ **Hunter:** Well, then, here's my answer to your question. I don't think Berthollet was the right person at the right time because that's not a useful hypothesis. It gets us nowhere. It's no better than saying he invented mass action by chance, or that God willed it. Science is a search for patterns, rules, causative agents. It advances only when we eschew first causes and noncausal explanations alike. And I see no reason why the history of science should be less amenable to analysis than, say, the history of the earth—geology—or the history of life—evolution. Perhaps that's crazy, but that's my belief, and thus I can accept only certain kinds of answers about Berthollet. Like your suggestion, Jenny, of an overlap between the experience of the investigator and the problem he recognizes and solves. That's something we can test by examining other cases.

▷ **Imp:** Good. Because I'm a dyed-in-the-wool Machiavellian, too. I assume you've got other cases.

▷ **Hunter:** Many. I've looked at physicochemical discoveries all the way up to the 1920s. But I wouldn't depend entirely on me. I'm no authority on the history of science, you know. This is just a hobby.

▷ **Jenny:** That's probably why you have a refreshing perspective. You weren't brainwashed in grad school. Why don't we invite some more people to participate in our tête-à-têtes? Get some other perspectives.

▷ **Imp:** Easier said than done.

▷ **Jenny:** Is it? Just because a couple of your colleagues said you're wasting your time? When did that ever stop you?

▷ **Imp:** Never. Quite the opposite—it makes me more stubborn.

▷ **Hunter:** So let's give it a try. I can think of a couple people who might be willing.

▷ **Imp:** Why the hell not!

▼ JENNY'S NOTEBOOK: The Arts of Scientists

We've found three other participants for the Discovering Project. Saturday I met Ariana for lunch at the County Museum of Art restaurant. After that we took in the new Armand Hammer wing. Impressive, but I knew what Imp would have said: "Imagine the number of aspiring artists who could have been trained and supported with half that money. Instead, they build another shrine to memorialize the same guys who are hung in every other museum in the world. Meanwhile what's happening to the future of art? Who's fostering tomorrow's masters?"

Nonetheless, the inaugural exhibit, on "The Spiritual in Art," was stunning.[1] Never realized how much cultural trends affected the content of art. And that includes science. As we passed one table of artifacts, I noticed an Annie Besant book opened to a page illustrating various phantasmagoric renderings of the periodic table of elements.[2] Well, "table" isn't really the right word. The periodicity was displayed by repeating figure-eights and crazy loops. Certainly nothing like what I remember from college chemistry! Apparently artists of the 1920s were intrigued by such "natural patterns."

I drew Ariana over for a look. Reminded of chemistry, I told her about our dinner with Hunter last week, and about our Discovering Project.

"Oh! Then there're a couple of other things you've got to see." First she led me to the back of the gallery where a series of "Rotoreliefs" by Marcel Duchamp were hung.[3] These were black and white spiral designs on round paper meant to be viewed on a rotating record turntable. "Duchamp developed these into a true artform, creating a set of twelve optical illusions around 1935," Ariana explained. "I know some visual psychologists who use these rotoreliefs for studying the perception of motion. In some of them, you actually get the impression of two spirals moving in opposite directions. Like a barbershop pole, but spiraling two ways at once. Fascinating psychological problem. And I think it's interesting that it takes an artist to create the experimental device for studying it in the first place."[4] I confessed that I'd never thought of art as being useful to science. Although, when I thought about it, Hunter had said that Berthollet's knowledge of tapestry design and manufacture was an important source of scientific problems.

I mentioned Berthollet to Ariana. "Oh, sure," she responded. "There's a very famous chemist, Chevreul—who I think followed Berthollet as director of Gobelins—anyway, among the problems he faced were complaints from the dyers and weavers that they couldn't get consistency of color. Sometimes a blue or a yellow was bright, sometimes it was not. Couldn't he do something about it? Well, after many years of work, Chevreul discovered that it wasn't the dyes that were the problem. The controlling factor was which colors of thread were woven next to one another. Contrasting colors, such as orange and blue, appeared very bright together, whereas closely related colors, such as purple and dark blue, weakened each other's visual effect. As a result, Chevreul invented the concept of complementary and contrasting colors, and a whole color theory to match.[5] Probably as important as the work of Helmholtz, Rood, and Ostwald in terms of the impact on art and design, and equally important for understanding the physics and physiology of perception. Amazing what you can learn in art museums, isn't it?" I made a mental note to include some of this material in my next "Revolution and Reaction" course. Students should eat it up.

But there was more. Next thing I knew, Ariana was hurrying me across the street to the Craft and Folk Art Museum. "Puzzles, Old and New: Head Crackers, Patience Provers and Other Tactile Teasers," said the poster in the window. My own look of puzzlement provoked a laugh from Ariana. "Inside," she directed. "You'll understand in a moment."

Sure enough, as we wandered past the exhibits, we found references to various scientists and mathematicians. Many of the puzzles were from the collection of Solomon Golomb, an important mathematician who invented a puzzle game called Polyominoes. Sir William Hamilton, another mathematician, had invented the Icosian Game to illustrate a new form of calculus. The Tower of Hanoi was invented by a French mathematician named Edouard Lucas. Richard Feynman, Roger Penrose, and P. A. M. Dirac had invented science-related puzzles as well.[6]

I even came across a few tidbits of personal interest, such as the fact that Napoleon liked peg solitaire and related games of strategy. That set me thinking. Might the games people play be a good reflection of personality and interests? Could it be true of scientists? After all Hunter said that Berthollet was a gambler both in games and in his science. And certainly Imp's penchant for turning things upside down is as obvious in his research as it is in our personal life. And then, of course, there's Ariana with her incredibly eclectic mind. I wonder what style her medical research takes?

As if reading my mind, Ariana said, "You see, if you're going to study scientific creativity, you can't ignore the way scientists play. While we like to think that once a person dons a lab coat, she suddenly becomes objective and serious, it isn't true, you know. People are no different inside the lab than out. Gamblers are gamblers; strategists, strategists; pranksters, pranksters; drudges, drudges. I think there's something significant—and fun—about the fact that a Golomb, a Hamilton, or a Lucas illustrates his mathematical insight by inventing a game. Suggests that science itself is a game to them. And we know how seriously little boys take their games, don't we!"

It didn't take much persuasion to get Ariana to agree to make room in her busy schedule for our Saturday-morning sessions.

Our other two participants agreed to come yesterday. Hunter called to ask if he could bring an interested acquaintance who's a historian of science. She's helped him with his research. And Imp invited his colleague Richter Zweifel. I'm not too pleased about that. I find Richter loud, tactless, arrogant, and occasionally offensive. But he knows that, and so does Imp. "He is the epitome of the maverick scientist!" responded Imp. "Seriously," he continued, "we need his skeptical Germanic mind. And his knowledge of methodology and of philosophy of science. You've got to have *some* hard-nosed, devilish personality around to keep me in line, don't you?" He's got a point.

So, as A. A. Milne might have said, "Now We Are Six."

DAY ONE

The Problem of Problems

For Carruthers, the dawn of excitement came while he was a graduate student at Cornell. He had been studying quantum field theory, a subject that confused him because it seemed riddled with dogma rather than equations that were simple or elegant. "The nightmare of it always made me uneasy in my stomach," he recalls. "I'd go to class and the students would sit there nodding their heads in rhythm to these incantations from the lectern. And I'd be sitting there thinking, 'They all understand it and I don't.' I was sure they were all much brighter than I was." Put off by the confusion, he was about to flee particle physics when Dr. Richard P. Feynman, a Nobel Laureate, came to Cornell and taught a course on the subject. "He made complete fun of the ridiculous problems of field theory. And I thought, 'My God, maybe I'm right. Maybe there's a reason I don't understand this stuff.' That experience became the turning point in my development."

—Interview of Peter Carruthers, theoretical physicist and violinist, by William Broad (1984)

Camille Jordan wrote of Henri Poincaré's work: "It is beyond ordinary praise, and forcefully recalls what Jacobi wrote of Abel: that he solved problems which before him nobody would even have dared to pose."

—René Taton, historian of science (1957)

►

IMP'S JOURNAL: Discovering a Problem

So what *is* process of discovering? How to get at it? Where to start?

Scratch that. Too general. Too theoretical. Won't get anywhere thinking like that. Start with something I understand: *my* science. Where has *my* research started?

Problems. Always with problems. Things that don't make sense. Puzzles. Anomalies. Inconsistencies. Discrepancies. Like amino acid pairing. Now *that* was an experience!

Lecture to introductory bio students about Landsteiner's classic work on ABO blood types. Landsteiner trying to demonstrate every person unique by analyzing proteins—not too different from our modern concept of tissue typing. But at the time proteins, not nucleic acids, were thought to carry the hereditary principle. Blood easily obtained in reasonably large quantities from many individuals; has high percentage of proteins, so convenient material for protein analysis.

Simple technique: person's blood is injected into an animal; the animal develops antibodies specific to the proteins on that blood. Next time animal is injected with the same kind of blood, antibodies immediately cause the blood to clump together in a clot. But if blood a different type, no clotting. Landsteiner apparently hoping to show each person's blood creates a unique immunological reaction.

Surprise! Blood proteins not unique to individuals. Rather, fall into major groups: A, B, and O; M and N; Rh, etc. (Interesting, that: Berthollet wasn't trying to solve the mass action problem either. Typical?) If person with Type O blood injected with Type A or Type B blood, his antibodies recognize the blood as foreign, clump it, and thereby cause heart attacks, strokes, hemorrhaging. But if Type O blood injected into Type O person, no immunological reaction. The blood is same type: looks like "self." In fact, no reaction if Type O blood is injected into anyone. Hence, people with Type O blood "universal donors." A to A, and B to B, also safe. But injecting Type A blood into a Type B person or vice versa deadly. All of this is subsequently found due to special chemical groups attached to a particular blood protein. Type O has no side chains. Types A and B have different ones. The immune system can determine whether those side chains are present, and if so whether they are the same as the body's or not. In vitro tests used to determine blood type for transfusions. Okay so far?

Then the shocker. Phyllis raises her hand. Oh, no. Phyllis my best student. Sharp, witty, inventive. If she didn't understand, *nobody* did.

This may be stupid question, she says, but if the Central Dogma of molecular biology is correct—I'm already lost, because I can't even imagine what the Central Dogma has to do with immunology—then how do you react immunologically to proteins? I don't follow. Okay, let me put it this way, says Phyllis. You taught us that the Central Dogma says that information can be transferred from DNA to RNA to protein (as in protein translation), and sometimes even from RNA to DNA (as in viral replication), but never from protein to protein or from protein to RNA or DNA. Right? Exactly, I say. I believe Crick summarized it by saying that once information gets into a protein, it can't get out again. Okay, says Phyllis. Then how does a foreign protein induce the immune system to produce antibodies—also proteins—that are specific to it? Doesn't that imply the existence of either protein-to-protein information transfer or protein-to-DNA information transfer? I mean, isn't one protein causing the specific production of another? And doesn't that mean that it sent information about itself to an immunological cell which utilizes that information either to turn on protein synthesis of antibodies (if the protein is foreign) or not to (if the protein is "self")? And isn't "selfness" determined by your genes—that is to say, DNA? So how does the protein sequence get compared to the DNA sequence?

First reaction: relief. She understood the lesson. "Good question," I say. Then helplessness. She understood too well. She's obviously thinking like an instructionist, but arguments concerning info transfer at level of clonal diversity and recognition just as problematic for clonal selection. [Editor's note: Instructionist theories of immunology assumed that the antigen acts as a template for making the antibody; these were replaced by clonal selection theories positing that all possible antibodies are generated shortly after birth, and the antigen "selects" or activates the cell producing the appropriate antibody.] I don't know how to answer her. Her logic is unassailable, which means it's probably based on a faulty assumption. But what? Definition of information? Central Dogma itself? Misunderstanding of immunology? What can I say? "I'll look into it and try to answer you next week." Then third reaction: This is *great*! So *fundamental*! It's got to be important, even if wrong. Because somewhere in the confusion something's not clear. This is payoff for hammering into my students' heads that there's no such thing as a stupid question! Thank you, Phyllis!

So—that was the beginning of amino acid pairing research. A "stupid" question that turned out not to be stupid at all. Crick never defined "information," so conditions bounding the application of the Central Dogma unclear.[1] Problem of how body differentiates between "self" and "nonself" also unresolved. "Clonal selection theory" still has big gaps—such as how original clones are generated; what turns them on; how? Lots of hypotheses, none based on anything but wishful thinking. So how *is* information concerning "self" and "nonself" encoded and processed by the immune system? And does processing obey Central Dogma? Nobody knows, least of all me, and I've been trying to get at the problem for ten years now. And, like Berthollet and Landsteiner, solving other problems along the way.

But back to the problem generation. Isn't there some story about Fermi agreeing to join the University of Chicago graduate faculty only if allowed to teach introductory physics to the freshmen? Now why would he have done that unless *he* got something out of it? Didn't Feynman write that students' questions are a fertile source of research ideas? (Look up.) George Wald and Walter B. Cannon, too.[2] Something about forcing one to pay attention to the basic weave of the fabric of science, instead of the individual threads. Sometimes details obscure fact that the overall pattern doesn't hang together. You can't see that until you try to weave it together for someone else, who isn't sophisticated enough to overlook the holes.

So what's the lesson here? Think basic? Ask how every theory in your field fits with every other? Teach it to know it? Wrap it all into one: the most important advances in science occur at levels comprehensible even to the introductory student. Which is why the most important insights are in the intro texts, the classic experiments in the intro labs. So, new rule for myself: never work on any problem that can't be explained to college freshmen. If that doesn't keep my work fundamental, nothing will!

◀ TRANSCRIPT: How Does Science Grow?

▷ **Imp:** Ah, Richter! Better late than never.

▷ **Richter:** Had business to attend to . . . So.

▷ **Imp:** So, let's get started. You all got a transcript of our dinner with Hunter? And I take it we're all agreed that these sessions can be tape-recorded? Good. Allow me to thank you all in advance for participating in our colloquium on discovering. We're here to practice what the physiologist Walter B. Cannon aptly termed "the fecundity of aggregation."[1] All you have to do is gather together a group of people, he said, and something new will emerge. May we fertilize one another's minds!

First, let me make introductions. You all know my wife, Geneviève—Jenny for those of us whose French is *comme çi comme ça*. As always, she has agreed to accompany me on this intellectual adventure.

▷ **Ariana:** And do you ever accompany her on hers?

▷ **Imp:** Hey! Ask me about popular culture during the French Revolution. Or Napoleon! Or Diderot!

▷ **Jenny:** It's true. I have more trouble keeping him *out* of my work than getting him interested in it. Some of our best arguments have been about how susceptible history is to scientific methodology. And that's one reason I'm here. I'd like to find out how people other than Imp think scientists work. My first impression is that what most of us nonscientists glean about the scientific process is pretty inaccurate.

▷ **Imp:** And you can blame the scientists for that. But let's get on with the introductions.

To Jenny's right is Ariana Parergon. She's our feisty scientific humanist par excellence: artist, photographer, amateur cellist—she performs with the Doctors' Symphony—and, incidentally, she's an endocrinologist. In short, the embodiment of the modern "renaissance woman." We have invited her not only because she's an old friend and would never forgive us if we hadn't but because she has a deep interest in creativity, and I feel certain we'll want to discuss discovering as a creative activity. Besides, we need someone to bring a little harmony—dare I add beauty?—to our business.

▷ **Ariana:** As usual, you are being Imp-possible! But yes: I'm here because I believe that creativity is creativity, whether expressed in art or science. Understanding one provides an understanding of the other. I would even go so far as to say that the arts provide tools of thought for the sciences.

▷ **Imp:** Good! I'm sure we'll get around to testing that hypothesis before these sessions are over.

Beside Ariana sits Richter Zweifel, whose black scowl proclaims his role among us as professional skeptic. No doubt he's already questioning what possible relevance the fine arts can have for the sciences and has concluded there can be none. Not to worry. Richter's skepticism will be invaluable to us. You might say he is an epicure of the mind, an intellectual gastronome whose mental tastebuds flinch at the slightest inconsistency, revolt at bland platitudes, and shudder at the hint of a specious argument. It is, one might say, a "rare bit" that gets past him!

▷ **Richter:** Enough, enough. Imp misportrays me. I am not a skeptic, but a devil's advocate whose foe is complacency. I accept nothing till I have satisfied myself; and, I might add, I'm damn hard to satisfy.

But Imp is right on one count. I do not like this art-science comparison. I've read Kuhn, Popper, Gombrich.[2] They don't see any useful similarities, nor do I.

▷ **Ariana:** Yet perfectly good scientists such as Jacob Bronowski and MIT's Cyril Stanley Smith—even Jonathan Miller—do.[3]

▷ **Hunter:** You could add Nobel laureates such as Roald Hoffmann, Victor Weisskopf, and Sabrumanian Chandrasekhar, too.[4]

▷ **Imp:** As you see, the last thing we'll have to worry about is unanimity of opinion! Let me just add that Richter's field is biomedical research, especially its theoretical aspects. He's also dabbled a bit in philosophy of science.

Next to Richter sits our chemist, Hunter Smithson. He tells me that he's tired of making stinks in sinks and wants to find out why some people make discoveries and others don't. So, in addition to his chemical research and teaching, he's been writing a book about how physical chemistry was invented as a discipline. Who did it, how, and why—that sort of thing, eh, Hunter?

▷ **Hunter:** More or less. Though as you know full well, physical chemists don't make "stinks in sinks." We turn chemicals into numbers. A more fascinating transformation, and much cleaner!

▷ **Imp:** I stand corrected. Beside Hunter sits our final guest, Constance Delaney. But perhaps you'll do the honors, Hunter.

▷ **Hunter:** Constance is a patent attorney by profession and historian of science by training. We met when I had a piece of apparatus made in the lab and wanted to know whether it was patentable. One day I asked her about her background, because she seemed to know quite a bit about chemistry as well as law. She informed me that she had majored in chemistry in college, taken a Ph.D. in history of science, and then found that there were no jobs to be had. So, to make the best use of her talents and knowledge, she turned to patent law. And a good thing for me, too, because she's been just a gold mine of information, both about patents and about the history of chemistry. But I'm sure you'll find that out for yourselves. Constance?

▷ **Constance:** Well, as you might imagine, I've always been interested in science. I was the kid who read scientific biographies for fun—still do, to tell the truth. I guess you can tell from all the boxes of notes I brought. Anyway, I got interested in how science progressed, and that led to graduate school, and you know the rest.

There is one thing I ought to point out. Maybe everyone knows this already, but the idea of a science of science isn't new. I mean, no offense to Hunter or Imp, but the term "science of science" originated in 1935 in a paper by Stanislaw Ossowski and Maria Ossowska.[5] There's a vibrant Polish tradition of research on the subject, and it would, I think, be fair to see Derek de Solla Price's books, like *Little Science, Big Science,* and much of J. D. Bernal's work as relevant. There's even a journal, called *Minerva,* dedicated to the "science of science."

▷ **Imp:** Yes, but little of that literature addresses what scientists actually do in the laboratory—let alone what goes on in their minds.

▷ **Hunter:** Not much about how to train scientists either. I mean, it seems to me that one of my jobs is not only to do good research, but to train other people to do it well, too. How can I do that without knowing what differentiates good science from bad?

▷ **Richter:** You don't have to. Great scientists are born, not made.[6] All one must do is weed out the mediocre.

▷ **Ariana:** Einstein was almost weeded out as being mediocre.

▷ **Richter:** Myth!

▷ **Hunter:** Well it's no myth that van't Hoff, Ostwald, and Arrhenius, who subsequently received Nobel Prizes in chemistry, were almost weeded out. So even if you're right about geniuses being born, we still have a definite problem in figuring out how to recognize them. However, I don't believe you *are* right. I say we just haven't figured out what to teach or how to teach it!

▷ **Imp:** Besides, Richter, if scientists are born, doesn't that imply the existence of genes for scientific genius? Where's your evidence?

▷ **Richter:** Look who walks into your classroom. They have it or they don't. Besides, if E. O. Wilson can postulate a gene for altruism, I see no problem postulating one for scientific genius.

▷ **Constance:** Actually, there *is* evidence of a sort. Zuckerman's study of Nobel Prize winners, for instance.[7] She lists Christian Bohr, a distinguished physiologist, whose son was Niels Bohr (Nobel Prize, physics, 1922), whose son was Aage Bohr (Nobel Prize, 1975). The Braggs, Curies, Darwins, Herschels, Becquerels—they're a good one. Four generations, all physicists: Antoine, Edmond, Henri, who won a Nobel Prize in 1903, and Jean. And look at the eminent contributions to biology and physiology made by J. S. Haldane, his son J. B. S., and his daughter Naomi's family, the Mitchisons! Maybe there *is* a gene for scientific genius.

▷ **Imp:** Great! So to increase scientific progress we must employ eugenics: Great scientific families must be interbred!

▷ **Hunter:** You're joking, of course.

▷ **Imp:** Am I?

▷ **Hunter:** The point, Constance, is: How do you distinguish between the transmission of genes and the transmission of knowledge or habits of thought and action?

▷ **Richter:** How do we know that a gene for cancer runs in families rather than a habit of eating unusually carcinogenic foods? We correlate. We guess. We test.

▷ **Hunter:** Then consider the importance of master-apprentice training as a possible mechanism. Unrelated scientists cluster in eminent groups, too, which you wouldn't expect on a random genetical basis.[8] Right, Constance? You're the one who told me about the laboratories of William Thomson and Justus Liebig being training grounds for half the great physicists and chemists of the late nineteenth century. And I've found in my research that one can link virtually every great physical chemist between 1890 and 1920 to the Leipzig laboratory of Nobel laureate Wilhelm Ostwald—even those who worked in the United States, England, and Japan!

▷ **Richter:** Making Thomson, Liebig, and Ostwald good talent scouts.

▷ **Hunter:** Yet you can construct "family trees" of scientists related by their training. Von Baeyer, who won a Nobel Prize, trained four Nobel laureates, these another six, and so on. Seventeen altogether. Same thing can be done with Nobelists in physics. So I believe, along with Hans Krebs, that given basic intelligence, scientists are not so much born as made by those who teach them research.[9] If that's true, then it isn't just a matter of selecting the best minds—we also have to train them properly.

▷ **Imp:** I agree—which means that what we need is what Steven Dedijer calls a "summation of practical knowledge." He says—allow me to quote him because this is very much what I'd like to get out of our sessions:

> *What do we know? Fourteen generations of scientists have been trained. Every one of those scientists had a teacher. That teacher had* know-how. *But very few have summed up that* know-how. *I like to*

> *cite the example of J. J. Thomson, who had nine Nobel Prize winners, thirty-two fellows of the Royal Society, and eighty-three professors of physics among his pupils . . . Yet, when you look—what were his rules of thumb? How did he teach?—you find practically nothing.*[10]

Well, that's what I'm after. All those rules of thumb, that tacit knowledge great scientists pass on without ever recording it.

▷ **Richter:** Assuming it exists. Assuming it *can* be known.

▷ **Ariana:** But is science any different in this respect than art? Art can only be learned by direct experience—by being apprenticed, if you like. Supposedly, scientific knowledge is explicit and publicly accessible, yet what Hunter says about Krebs suggests that the *doing* of science is still an art. You still have to be apprenticed.

▷ **Imp:** Well, I for one don't see why the doing of science has to remain an art. I don't see why only a few people should have access to the rules of thumb that one great investigator learns from another. It seems to me that if we want to improve scientific research in this country, it's not more money or equipment that needs to be infused into the system, but the accumulated experience of the best minds who've ever done science. You want better science? Train everyone with the likes of J. J. Thomson

Hunter's diagram of the apprenticeship lineage of some eminent physical chemists and biochemists.

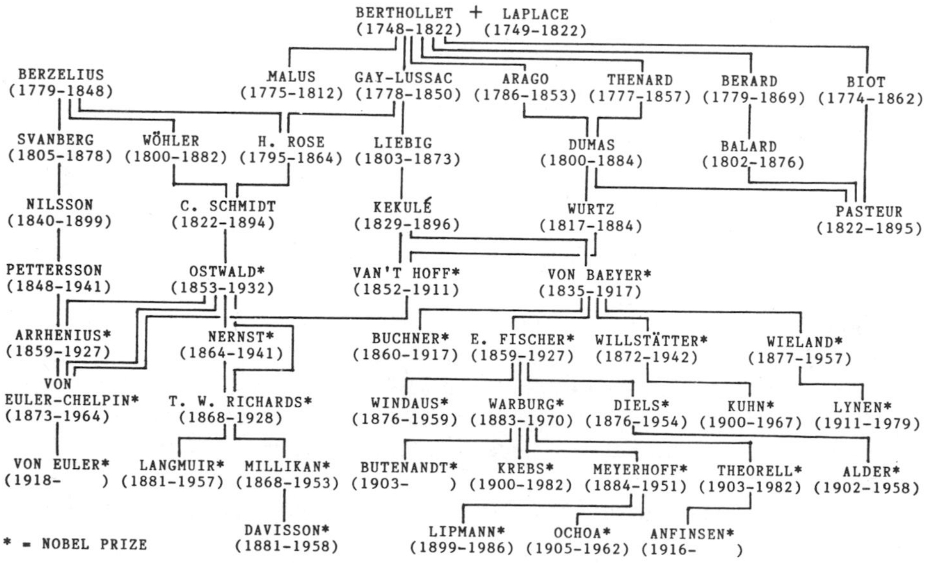

or Lord Rutherford or von Baeyer. Or if you can't do that, find out how these men did their science and teach that!

Look at it this way. If we adopt Richter's point of view—that scientists are born, not made—

▷ **Richter:** Not just *my* point of view.[11]

▷ **Imp:** Whoever's. The point is that with the genetic model, we're stuck with a very small group of potentially great investigators. The talent available is proportional to the population size.[12] If the population stops growing, as it's done in most developed nations, the fund of geniuses stops growing, too. In fact, increasing the number of scientists we train, or creating more scientific positions, would just increase the number of incompetent (or less competent) scientists!

▷ **Ariana:** Adding to the bottom of the hierarchy of excellence rather than to the top, as it were.

▷ **Imp:** Which may be precisely what we've been doing for the last fifty years or so.

▷ **Hunter:** Right. But if scientists *are* educable, then the factor limiting the growth of science is how many first-rate men—

▷ **Ariana:** *And* women!

▷ **Hunter:** —and women we can train and how we train them.

▷ **Ariana:** Or the discovery of more efficient and creative ways to use the existing pool of brainpower.[13]

▷ **Constance:** Hunter, have you seen the report by Brinkman for the National Academy of Science? It's called *Physics through the 1990s*, and it points out that although most research money goes to large labs with a hundred or more team members—

▷ **Imp:** Big science!

▷ **Constance:** —yet little training of scientists is possible in such large labs. The report therefore recommends stronger support for the small labs that actually train the majority of physicists.[14]

▷ **Jenny:** But if the best scientists run big labs, then won't further support of small labs just result in the training of more second-rate scientists?

▷ **Imp:** Is it true the best scientists run big labs? Francis Crick doesn't, for example.

▷ **Hunter:** Neither did Richard Feynman.

▷ **Ariana:** Then we're faced with a tripartite conundrum: either, one, the fund of first-rate scientists is limited by nature; or, two, the nature of first-rate scientists is limited by funds; or, three, the nature of the fund (of scientists) is limited by first-rate scientists.

▷ **Jenny:** So how do we tell which formula is correct?

▷ **Imp:** Ah—my cue! By playing games! By generating paradoxes! Contradictions! Anomalies! As Oscar Wilde says in *Dorian Gray*: "The way of paradoxes is the way of truth. To test Reality we must see it on the tightrope. When the Verities become acrobats we can judge them."[15]

▷ **Ariana:** Lovely!

▷ **Constance:** Niels Bohr said much the same thing: "How wonderful that we have met with paradox. Now we have some hope of making progress."[16]

▷ **Imp:** And the more important the idea or the observation to which you can find a paradox, contradiction, or anomaly, the greater your chance of having identified an important problem. So let's play some games designed to try to find out how science is growing.

Having invented these games, I will, of course, begin. The first game is "Implications." The rules are simple. Take any well-established observation or theory and extrapolate the most extreme case you can think of. Things either simplify or fall apart. If the case still appears reasonable, then the original observation or theory is probably accurate. If not, you've generated a paradox, anomaly, or contradiction, and you've found the limits of your idea.

Take this slide—please! You've probably all seen it or something like it, showing the exponential growth of science journals from 1665 to the present. The same sort of curve describes the growth in the number of scientists over time. Derek de Solla Price, who collected these data, demonstrated that due to this exponential growth, almost 90 percent of all the scientists who have ever lived are alive today, and 90 percent of all the literature of science is being written by the scientists now alive.

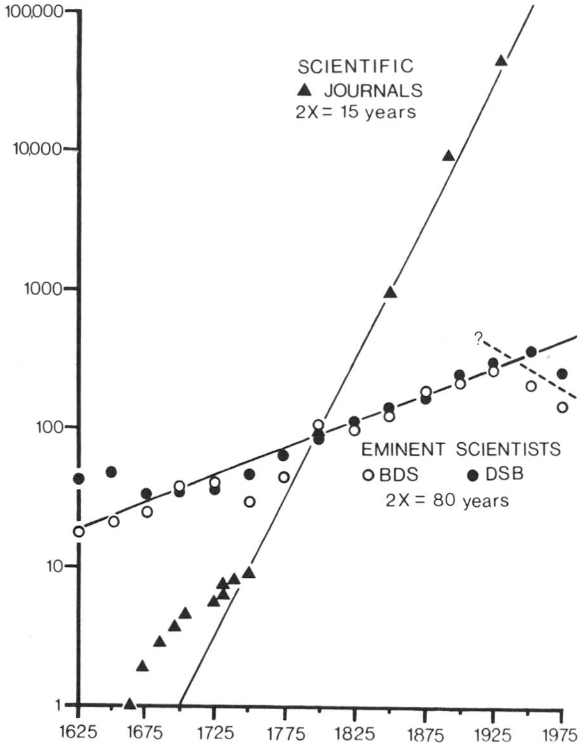

Number of scientists active between 1625 and 1975 included in the Biographical Dictionary of Science *(T. Williams, editor) and the* Dictionary of Scientific Biography *(C. C. Gillispie, editor) compared with the number of journals published during the same period (after Price, 1963). Note that the scale is logarithmic.*

Similarly, 80 percent of all the science professorships have been founded in the last generation, and over 90 percent of all the money for research has been spent. Price concludes that since 1660, when the Royal Society was founded in England, the "size" of science, anyway you measure it, has increased about a millionfold.[17]

Now we play "Implications." The question is whether we are witnessing today a millionfold or even a thousandfold increase in the number of fundamental discoveries that, say, Galileo or Newton would have witnessed three or four hundred years ago. Price says yes. He claims that the number of important discoveries doubles every twenty years. How he determines this, he doesn't say. But, if true, then for every major discovery made in the year 1665, a *hundred thousand* should be occurring in 1985! You will have to excuse me if I express my doubts! We do not have a million Galileos or even a thousand Newtons alive today. We do not have a million books being published in our lifetimes that are the caliber of *The Two New Sciences* or even a thousand that rival the *Principia*. I doubt we have even two dozen. In short, common sense says that the progress of scientific discovery is not keeping up with the growth in the size of science. That's my anomaly. The problem is to demonstrate it. Yes, Richter—you agree?

▷ **Richter:** To this extent, yes. Hundreds of years of research in biology and medicine have yielded no useful understanding of embryology, development, morphology, homeostasis, how most germs cause disease, how virtually any drug works, how healing occurs, what sleep is, how we feel pain or pleasure, the bases of cognition—I could go on for half an hour. If you're interested in this sort of thing, see the *Encyclopedia of Ignorance*, or some of the recent volumes of the *Journal of Theoretical Biology* devoted to unsolved problems. I realy envy the physicists, who seem much closer to wrapping up their understanding of the inanimate world.

▷ **Hunter:** Some of my colleagues would disagree. They believe another cosmological revolution is in the making.

▷ **Richter:** Fine. Then why not in biology, too? That's why I'm here. To find out why we don't have any—or at least not many—modern Galileos and Newtons and Einsteins in biology and medicine. We don't even seem to have a modern Darwin or Pasteur. Why not? Certainly not because the genetic combinations for such individuals don't exist. As Imp says, there should be hundreds or thousands of geniuses today for every one in the sixteenth century. So where are they? Are they all designing microchips or recombining DNA? Or is there something more to it? Because God knows we've tons of data and precious few ideas.

▷ **Imp:** So, let's generate some ideas. Play a second game, called "Contradictions"—you'll like this, Richter; it's right up your alley. Take whatever statement confronts you and assert the opposite. If you can't find any supporting data, your contradictory hypothesis is incorrect and your faith in the original hypothesis is strengthened. If, however, supporting data for the contradiction are forthcoming, then serious doubt will be cast upon the validity of the original observations or ideas.

▷ **Ariana:** Stock in trade for a devil's advocate.

▷ **Imp:** Now, in this case, Price says significant scientific discoveries are increasing exponentially with time. I contradict his observation by asserting that they aren't. For the sake of argument, I'll assert that the number of significant discoveries is constant with time (though I frankly find Parkinson's law of scientific growth attractive, too: "Actual progress must vary inversely with the number of journals published").[18] The exact hypothesis is irrelevant as long as it forces you to search for new sources and types of data.

Et voilà! Here are my data, taken from I. Bernard Cohen's book *Revolution in Science.*[19] Despite the exponential growth in the number of scientists and publications, Cohen's analysis suggests that the number of scientific revolutions in the twentieth century is about the same as the number in the seventeenth century. The seventeenth century had Galileo in mechanics, Newton in cosmology, and Harvey in physiology. Set that off against Einstein and relativity theory, Planck and the quantum, Wegener and plate tectonics, and, if you're generous, Watson-Crick and DNA—though what they revolutionized I fail to see. There wasn't a prior theory of genes to overthrow, was there?

▷ **Constance:** But what about the computer revolution, and all the revolutions in communications, and space travel, and even warfare?

▷ **Richter:** That's just technology. Computers only allow us to do faster what we already know how to do. I object more to the paucity of data upon which Imp bases his conclusion.

▷ **Imp:** It's limited, true, but remember, the smaller your sample size, the better your chance of finding what you're looking for. A sample size of one is optimal!

But seriously, there have *not* been many great revolutions in twentieth-century science, which is really the point I'm trying to make. And other measures show the same thing. The American Chemical Society made a list of the most significant advances in chemistry—pure and applied—a few years ago, and two photography journals have likewise evaluated the history of breakthroughs in photography. All three of these studies show the same thing: Over as much as a hundred and fifty years, there has been no increase in the rate of discovery, despite a several-hundredfold increase in the number of investigators in each field.[20]

▷ **Richter:** Again, I'm skeptical. You're relying on some subjective evaluation of importance.

▷ **Imp:** Why the hell not? You do that every time you read a scientific article, every time you recommend a research proposal for funding or a student for a professorship. Surely you aren't saying that every scientific paper is of equal value, or that each one represents a discovery?

▷ **Constance:** Actually, there is other evidence that tends to support Ernest—um, Imp's point. Most striking is recent evidence that the number of

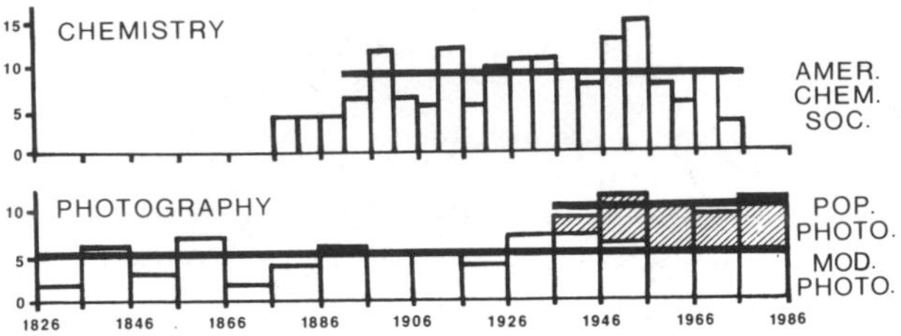

Number of important inventions or innovations introduced in chemistry according to the editors of Chemical and Engineering News *(published by the American Chemical Society), and in photography according to the editors of* Popular Photography *and* Modern Photography.

patents being filed by U.S. citizens has been declining steadily over the last few decades. U.S. citizens were issued about fifty thousand patents per year during the 1960s. It's only about forty thousand today. Almost half of all U.S. patents now go to foreigners, whereas that share was only 20 percent just two decades ago.[21] So increasing the pool of scientists and engineers doesn't necessarily lead to more breakthroughs or inventions.

Harriet Zuckerman has reported data compatible with the same interpretation. She says that the number of scientists nominated for Nobel Prizes has increased only two- or threefold since 1900, although there are several hundred times as many scientists working today.[22] So the number of fundamental discoveries certainly isn't keeping up with other growth indicators. William Beaver found the same thing concerning eponyms, laws and principles considered significant enough to bear the inventor's name. They've been invented at a rate of about two per year for several hundred years.[23] And I once did a study of entries in biographical dictionaries and found that despite the almost millionfold increase in numbers of scientists, publications, and so forth between the Renaissance and today, the number of scientists deemed worthy of an entry has only increased by a factor of about twenty-five.

▷ **Jenny:** So what are all the other scientists doing? Wasting their time and our money?

▷ **Hunter:** No, they're doing paperwork and directing other people's work (or lack of it). According to one study I saw, 90 percent of British chemists (I ssume the same is true here in the U.S.)—and this study included everyone from postdocs to full professors—spend less than 10 percent of their time doing experiments or writing up results.[24] The majority of their time's devoted to raising money, administration, teaching, and travel. In consequence, it may take—this is conjecture, mind you—five chemists to do as much research today as a hundred years ago, when the average chemist had only a handful of students whom he taught by direct example at the lab bench.

▷ **Imp:** Reminds me of a joke: How many scientists does it take to change a light bulb? Answer: Twenty-one. A peer review committee of ten to decide who gets the job; another committee of five to make sure the work satisfies human-use guidelines; a health and safety inspector to ensure that all NIH and OSHA guidelines are obeyed; three grad students and postdocs for whom this will be a valuable laboratory experience; and the laboratory director, who will, as his title indicates, direct the operation (heaven forbid he actually *do* it!). That's twenty. And last and least (salary- and educationwise, at any rate), a lab tech—the only experienced member of the team—who will actually change the light bulb!

▷ **Richter:** You know, of course, Leo Szilard's parody of the current system—"The Mark Gable Foundation"? Szilard asks how may science best be retarded. His answer: by just the sort of bureaucracy that runs science today, in which we spend all our time justifying, planning, and evaluating, and none of it simply trying things.[25] You get what you pay for.

▷ **Hunter:** Yes. J. J. Thomson used to say that if government patronage of science and technology had existed in the Stone Age, we'd all have wonderful stone tools today, and no metals.[26] But that's only part of the problem. The other answer to Jenny's question requires a look at what scientists do when they actually get into the lab. Most don't attempt to make major discoveries, or at least they don't succeed. You can see this historically. Constance once showed me a graph of the name entries in a major chemistry textbook plotted according to the date of the person's contribution. It differentiated between citations referring to data and those referring to concepts and novel techniques. Overall, the number of citations grew exponentially with time between the Renaissance and 1975. However, the number of scientists contributing fundamental concepts and techniques formed only a small fraction of the number cited, and showed a very slow, linear growth. The overall exponential growth was completely accounted for by references to data citations. New determinations usually supplant old ones.

In other words, it appears that the rate at which major inventions and conceptual breakthroughs occur is independent of the number of investigators in the field (assuming, of course, a minimum number), and depends instead upon the body of concepts and techniques already in existence. These, as Imp says, seem to grow slowly, linearly. On the other hand, the more scientists one trains, the more data will be generated, and the more accurate and extensive the data base will become. Hence, the data base supporting or applying basic concepts and techniques can increase exponentially.

▷ **Constance:** And the faster it grows, the more applications will be found for it. Which is really Kuhn's point, isn't it, in drawing a distinction between "revolutionary" and "normal" scientists? He says most scientists spend their lives articulating existing discoveries, inventions, and theories.

▷ **Imp:** But not inventing or making them. Hence the search for basic knowledge becomes the rate-limiting step in the scientific process. Unless the rate of revolutionary breakthroughs is maintained or increased, the number of new applications will eventually diminish. Right?

▷ **Richter:** Okay, look. Let us accept your position for the sake of argument and get on with this. Certainly it would be nice to do better what we, in my opinion, already do well. But you seem to want to argue that there is some way to turn every scientist into a revolutionary.

▷ **Imp:** Even I wouldn't go *that* far.

▷ **Richter:** Fine. Then according to the rule of your own "Implications" game, there is something wrong. So let me employ "Contradictions" to suggest what it is. Let me assert that nothing can be done to increase the rate of invention or discovery. For example, Freeman Dyson maintains that, in physics at least, the rate of discovery is relatively constant. Thirty years to properly define and solve a problem; another thirty years to work out its consequences, recognize exceptions, and raise a new fundamental question. That's fixed.[27]

▷ **Constance:** Sure, and Mitchell Wilson says one of the lessons of the history of technology is that nothing can be done to alter the rate of invention, either by individual inventors or by huge nationally funded programs. Inventions are made when they're ready to be made.[28]

▷ **Jenny:** The discovery picks the discoverer again.

▷ **Richter:** The point is that there are only a finite number of things that *can* be discovered or invented at any given time, and these must be discovered in some order.[29] Therefore, nothing can be done to increase the rate of discovery.

▷ **Imp:** By god! I like it! Certainly it makes sense that only certain discoveries are possible at any given time, and that their possibility is a function of the order in which previous concepts and information accumulate. Berthollet again—right, Hunter?

▷ **Hunter:** Absolutely.

▷ **Imp:** So it doesn't matter what sort of data you've got! They don't lead to a discovery unless you have the right problem, the right anomaly hooked up with the right theory, and the right data. Otherwise, you don't know what the data mean.

But for all that, you're wrong, Richter! Ninety-nine percent of scientists work on the same subset of problems using the same techniques and see the same thing. Go to any professional meeting and just look at how uniform the topics of the papers are. And how repetitious from year to year! The presentations cluster into "hot" topics while huge areas of interest are totally ignored. There're fads in science just as in every other human activity, like it or not.[30]

▷ **Richter:** I don't like it. You know very well what I think of all these people running after cancer money or AIDS money or Star Wars money.

If the problem were solvable, it would have been solved. Money does not buy ideas—only applications of existing ideas.

▷ **Ariana:** But it *can* cause other things that might be investigated to be ignored. I gather that when the DNA double helix was discovered, most of the funding and researchers left protein chemistry.[31] Biochemistry today is virtually synonymous with recombinant DNA techniques, as if that were some kind of panacea for ignorance. What really happens is that unsolved problems remain unsolved. Take staphylococcus infections, for example. When penicillin was introduced in the 1940s, doctors promptly lost interest in the question of how staph causes diseases. As far as I know, it's never been answered. Of course, the advent of antibiotic-resistant strains may once again push the problem to our attention.[32] So you might say that the course of scientific progress is determined not just by which problems get solved, but also by which ones get ignored. Isn't that your point, Imp?

▷ **Imp:** Exactly. In fact, let's codify this as Imp's First Principle: Most data are accumulated to confirm existing theories using existing techniques and are therefore unlikely to lead to a discovery or to the necessity of building a new theory. Or, stated as briefly as possible: Data cluster around knowledge. Implication (by contradiction): Scientific breakthroughs could best be stimulated by focusing research efforts away from what we know how to do into areas of ignorance.

▷ **Ariana:** So you might view science as a tree of knowledge, on which the growth is always from an existing branch. Older branches accumulate leaves more rapidly than new branches. But the tree shouldn't interest us—that's what's known. Research should focus on what we artists call the "negative space" in the picture—the gaps between the branches and leaves. Those are the places where surprises await.

▷ **Richter:** Perhaps. But I don't see what all the fuss is about. We live in the country that produces more Nobel Prize winners per year than any other. More scientific literature, in fact. What makes you think the present system isn't working? What are you after?

▷ **Hunter:** The same thing you are, Richter: fundamental, revolutionary breakthroughs. Publication explosion, yes; but information explosion? I think not. The number of papers goes up, but not the number of *important* papers.[33] Yes, we produce many Nobel laureates. But where were most of these scientists trained? Not in the U.S. In England, Germany, Austria, France.

▷ **Ariana:** Exactly. Ever notice how MIT, Caltech, Harvard, Stanford, and places like that tout how many Nobel laureates they have on their faculties? But when you look at where those people were trained, you have to start talking about Oberlin, Rutgers, Swarthmore, CCNY, Case Institute, Lafayette, Oregon State, and places like that. Why is that?

▷ **Hunter:** I've found exactly the same thing with the physical chemists I've been studying. They're trained at backwater institutions, but they end up in Berlin or at Harvard because those institutions have money and prestige. That doesn't mean those places do a good job of turning out mavericks.

And that's my point about the U.S. I bet if you divided the number of Nobel laureates per country by that country's population, the U.S. wouldn't be first. It would lag behind tiny countries like Denmark, Switzerland, Sweden, the Netherlands, and the U.K.

▷ **Constance:** It's true. By population, the U.S. ranked eighth in 1962; perhaps a bit higher today. We're even further down if you divide by GNP instead of population. The Soviet Union, which produces more Ph.D. scientists than any other nation, is down near the bottom of both lists, along with Japan. Same when you look at the impact of published papers, that is, the ratio of citations to publications. The U.S. is sixth after Denmark, Sweden, Switzerland, and so on.[34]

▷ **Hunter:** So the production of Nobel Prize winners, and I assume first-rate science in general, is not a direct function of the number of scientists a society produces. You've got to have the right conditions for scientists to flourish, and for all we know having a smaller number of more independent and better-funded scientists might be better than having an excess of unemployed ones.

▷ **Imp:** Money won't do it either. Ideas drive science. Listen to Hans Selye, who invented the concept of physiological stress:

> *As the years went by, I managed to acquire every available facility that modern science can offer in the way of the most up-to-date techniques of history, chemistry, and pharmacology. I have been given the means to construct one of the best equipped institutes of experimental medicine and surgery in the world and have acquired a staff of 53 trained assistants, technicians, and secretaries. Yet today, as I look back upon those early observations in 1936 I am ashamed to say that, despite this help, I have never again been able to add anything comparable in its significance to those first primitive experiments.*[35]

Albert Szent-Györgyi says something similar.[36]

▷ **Ariana:** And how many others might say the same if they were honest?

▷ **Jenny:** Okay. So money and manpower won't suffice to stimulate scientific breakthroughs. What will?

▷ **Richter:** Perhaps nothing. Go back to Ariana's tree analogy. How do we know that problems don't get smaller the further science progresses? Just as limbs become branches and branches, twigs. It would then take more time, energy, money, and people to solve ever smaller and more difficult—less interesting—problems.

▷ **Ariana:** Your own list of unsolved problems belies you.

▷ **Richter:** I did not say I believed what I said. But why should we believe that science is anything but a historical anomaly—like the Stone Age—that will come and go? Imp warns us that science may soon stop growing, that it will stagnate. Perhaps that is necessary. Perhaps we are nearing the end of what can be known. We will be reduced to merely increasing the accuracy of what we already know. Bentley Glass, for example, believes biology holds no more surprises. Macfarlane Burnet and Neils Jerne have declared that we are awaiting the end in immunology. All that remains is to reduce biology to chemistry and physics.[37]

▷ **Imp:** Richter, you don't really *believe* that?

▷ **Richter:** That is not the point. Burnet and Jerne could be right. Why, then, should we bother trying to increase the rate of discovery? Why should we train more scientists or give them any more money?

▷ **Hunter:** If Burnet and Jerne are right, there is no reason. But we've heard this same thing before. Just before Darwin, and just before Einstein and Planck and Schrödinger and Heisenberg. Several times in chemistry. And every time we think we've reached the limits of knowledge, we find that all we've really reached is the limits of a particular perception of nature. If history is any guide, the proclamations of Glass, Burnet, and Jerne may simply herald the next revolution.

▷ **Constance:** Here's an interesting historical aside: Did you know that from about 1890 to 1910, Albert Michelson had printed in the Register of the University of Chicago a passage by Lord Kelvin claiming that no new marvels could be expected in physics; all that was left was to carry out measurement to further decimal points.[38]

▷ **Hunter:** Need I point out that the period in which the Chicago register ran the Michelson-Kelvin paean to precision was precisely the most revolutionary period in modern physics? Yet Kelvin refused to accept the existence of radioactivity almost up to his death in 1907.[39] Utter nonsense.

▷ **Jenny:** But it can't be *utter* nonsense. I seem to recall a passage from C. P. Snow's novel *The Search* where he has some physicists around 1930 or '40 say something about the fundamental laws of physics and chemistry being laid down forever so that physics and chemistry are finished sciences. You don't happen to have that passage in your notes, do you, Constance?

▷ **Constance:** I don't read novels. I have too little free time as it is . . . But I do have another passage bearing a similar message.

Listen to Lagrange, the greatest of French mathematical physicists, who called Newton not only "the greatest genius that has ever existed, but also the most fortunate, for there is but one universe and it can happen to but one man in the world's history to be the interpreter of its laws."[40]

▷ **Richter:** Exactly.

▷ **Constance:** To which Robert Millikan, the Nobel laureate who measured the charge on the electron, replied, "What an assumption!" He even says that Michelson later upbraided himself for quoting Kelvin. Millikan perceived science as Hunter seems to. He goes on to say, "True, Newton himself was too great a man to make such a blunder, for he . . . described himself as a child picking up a seashell from the shore of the great ocean of knowledge."[41]

For Millikan, Newton, and lots of other scientists, every answer simply generates a whole set of new questions. There can be no end.

▷ **Hunter:** It's a matter of perspective. Look, as long as Imp's inventing principles, I'd like to add another: The more we know, the less we know. Or stated another way: Ignorance increases in direct proportion to knowledge.

▷ **Imp:** A paradox! I love it!

▷ **Hunter:** I like them, too—a taste that probably developed from my training in physics. Bohr once said there are two kinds of truths: shallow truths and deep truths. Shallow truths are those whose opposite is demonstrably false. Deep truths are those whose opposite is also true, so that the two truths are complementary.

▷ **Imp:** So much for my game of Contradictions!

▷ **Jenny:** The idea is a bit older than either you or Bohr. Vauvenargues said something like that a couple of centuries earlier. But here, I found the passage in Snow:

> *It is two hundred years since Newton talked of our being in the search for knowledge like children who picked up pebbles on the beach. This man who spoke of "finished sciences" was Newton's successor. As I heard his clipped, impersonal voice, saying what was to him an evident fact, I realized for the first time how far science had gone. We were* not *picking up pebbles from the beach any more; instead we knew how many pebbles there were, how many we had picked up, how many we should be able to pick up. They had found the boundary to our knowledge.*[42]

▷ **Hunter:** Which is precisely the kind of thinking that led Snow out of physics just as the great advances in atomic physics and crystallography were about to be made. He's no paragon of perception, either.

▷ **Ariana:** So it all comes down to personality, doesn't it. Some of us, as Stephen Vincent Benét suggests in "Schooner Fairchild's Class," are wigglers and some of us are sticks-in-the-mud.[43]

There are those who seek novelty, and those who resist it. Those who like a closed universe; those who envision an infinite one. And these polar types rarely get along.

▷ **Jenny:** But it's more than just two different sorts of personality, isn't it? People enter fields for different reasons, after all; some for the excitement of the work, some for the excitement of the career, and some for want of anything better to do.[44] So think about Imp's data, especially Price's exponential growth of science. Maybe all Price did was to document the professionalization of science. My impression is that there were no professional careers in science until the mid or late nineteenth century. All the great scientists prior to that were really dedicated amateurs—men who earned their living at something else and did science for the love of it. Now virtually everyone is a careerist. You have to be; tenure depends on grant dollars and numbers of publications, not quality of research or number of insights. At least if I believe Imp.

▷ **Constance:** Well, that's not absolutely true. Some scientists *were* paid for their scientific work before 1850, especially in France—even during the eighteenth century. Berthollet, for instance.

▷ **Jenny:** But most were like Lavoisier: independently wealthy aristocrats or their friends. The point is that they had unusual freedom to explore within the old systems of amateur research and patronage. Quite unlike what is possible with peer-reviewed grants and contracts these days.

▷ **Constance:** Well, there was certainly no recognized profession of chemistry before around 1870. Norman Lockyer, the first editor of *Nature*, was a civil servant, not a professional scientist. And then there were outright amateurs such as Darwin, Mendel, Rayleigh, and so on, right up to the 1930s.[45] Nobody could tell them what to think or fire them for doing unpopular or unfundable research. That's true.

▷ **Richter:** So? What is the point? What has this to do with discovering?

▷ **Ariana:** Lots, if we're interested in who discovers! Do the kinds of discoveries made by the decimal-point mongers differ from those made by the "every problem yields ten or more problems" types? Do you have to be a professional scientist to make a discovery? Is amateurism still possible, or even desirable? Can the lone scientist still compete with organized "Big Science"?

▷ **Imp:** And what about the implications for science policy? I mean, does training more scientists necessarily lead to more discoveries, or does it just create a larger pool of stick-in-the-mud careerists who close ranks against the ill-mannered wigglers? When you pay a person to produce results, doesn't that interfere with the sorts of results he or she produces? Should we all strive to be amateurs who can produce what we like as we choose? And what about peer review and granting procedures? What if the number of papers or the size of grants is no measure of the importance of a scientist's work—everyone at this table knows it's not—then what? Then how do you recognize the handful of seminal papers by a Mendel or an Einstein in all the sewage flowing out of word processors and computers these days?

▷ **Richter:** Enough, enough! You're getting totally carried away. Where in god's name did you pick up this romantic myth of the great scientist? Do you really believe that a Mendel, Darwin, or Einstein could exist if all those lesser minds, as you seem to think them, were cast out of science? I think not. The fertile spadework done by these less imaginative souls may be just as critical as the great theoretical leaps. You jump to far too many conclusions.

Look! You want to question the assumptions underlying our general notions of discovering. I approve. I am as interested in such problems as any of you. Otherwise I would be elsewhere. But what I want are problems that yield solutions, not problems that yield more problems.

▷ **Imp:** You may find that it is easier to count the angels that can dance on the head of a pin. But, you're right, my friend. We *are* generating too many problems. What we need is the right problem. So why don't you raise a few particular problems for us to hone our eager minds upon?

▷ **Hunter:** Or even better, suggest methods for defining the sort of solvable problems you'd like to address. After all, one of the goals of these sessions should be to get at general methodological questions as well as particular solutions.

▷ **Jenny:** Excellent idea! But first, why don't we break for a few minutes to pour some coffee and collect our thoughts? There's a limit to how much I, for one, can assimilate at one sitting.

▼ JENNY'S NOTEBOOK: Implications, Contradictions

At the first opportunity I turned to Ariana and said in a low voice, "Richter getting to you?"

"I can never tell whether he means what he says or says it to provoke," she replied. "He seems to disagree with everything on principle. Is he always like that?"

"Sometimes worse." I smiled. "Nonetheless I wonder if he doesn't have a point about this period of history being an anomaly. Not in the way he meant it, of course, but certainly in terms of social change. What I've heard so far this afternoon is that, for better or worse, the nature of science has been characterized by its explosive growth over the last few generations. That's bound to change the nature of the enterprise.

"Diderot talks about this sort of thing in *D'Alembert's Dream*. He says that the way a monastery, or any religious organization, keeps its character is by replacing its membership slowly. For every new monk, there are a hundred old ones to show him the ropes.[1] That's the way science used to be, from what I can gather. An apprenticeship system. But imagine what would happen if for every old monk there were suddenly twenty new ones to be trained! Novelty would overwhelm tradition. What would happen to the organization then? Wouldn't the old monks try to create a system that would keep them in control? And isn't that really what the grant system is?"

Ariana nodded. "And tenure. Ways to retain control of the hordes of young turks. The irony is that research is a young person's game. It's the old fogies who need to be excised from the system."

"Hunter says not necessarily," I pointed out. "He says some scientists retain their creativity."

"Perhaps," said Ariana, "probably like artists—Picasso comes to mind—because they keep moving from one thing to the next. Nothing worse than overworked art or science. But how do you convince a successful mogul to give up his pet project for one that might not work?"

"Sic Imp on 'em!"

Indeed, at the other end of the room Imp was baiting Richter (to the immense pleasure of Hunter) about the finality of science. "Don't tell me you actually subscribe to this Burnet-Jerne thing about awaiting the end of immunology!" Imp exclaimed.

Richter replied, "The Burnet clonal selection theory is an excellent example of what a good theory should be. It explains the known facts simply and coherently. It predicted a number of things that led Medawar to demonstrate the phenomenon of acquired immunological tolerance. That discovery led to organ transplants, among other things. And between Jerne's work on T and B cells, and his network theory for explaining why our immune system does not attack our own bodies, there is really little left to know."[2]

"You're crazy!" shouted Imp gleefully. "Sometimes the body *does* attack itself. Autoimmunity. Neither Burnet nor anyone else has an explanation—"

"Certainly he does," interrupted Richter. "He says that autoimmunity occurs when a self-cross-reactive clone is accidentally activated by a bacterial or viral antigen having a sequential homology to a self protein."

"Accidentally, my foot!" exclaimed Imp. "That's just a way of saying he doesn't know how autoimmunity occurs, but won't admit it. In fact, neither he nor anyone else has explained even the simplest facts of self-recognition. Like why pregnant women don't normally reject their fetuses, but Rh-sensitized mothers do. Antibodies aren't supposed to pass the placenta, right? Why do you recognize your own sperm cells as foreign though they're made up completely of your own DNA and proteins?"

"Imp, as usual, a little knowledge for you becomes a dangerous thing," growled Richter. "Jerne's network hypothesis clearly states that all possible antibody sequences are elaborated at birth. Self-reactive clones are then eliminated prior to the activation of the immune system. Transplants, too, are accepted because the buildup of T suppressor cells eliminates the T killers."

"But the whole T suppressor cell idea is crazy, too! Now you've got *two* immune systems! One to get rid of foreign antigens and one to get rid of self-reactive clones! How do suppressor T's know what to suppress, and when? Why don't their attempts to suppress killer cells just end up in a huge immunological conflagration? All you've done by invoking T suppressor cells is to push the question of how self-recognition occurs back a step. But you've still got the problem of how to determine which clones are to be eliminated. *And*, you've still got the question of how these suppressor T's are bypassed in autoimmunity!"

"No," replied Richter. "I really don't think you understand . . ."

I begin to realize that while both of them play "Implications" and "Contradictions" constantly, they have different styles. Imp is always skeptical of the current consensus. Richter is always skeptical of anything new. I'm not sure either of them actually believes a word he says. It's just a game for them. But what a game! They held forth like God and Jesus Christ for a good while longer before we got them both to sit down again.

◀ TRANSCRIPT: What's Worth Investigating?

▷ **Imp:** Well, that was refreshing! Gerald Weissman and Lewis Thomas would be proud of us. They've both suggested that we learn consciously to enter the "Age of Ignorance."[1] We must learn to perceive what we don't know before we can understand what we do.

▷ **Hunter:** Exactly. As J. C. Maxwell liked to say, the purpose of research is "to drive us out of the hypotheses in which we hitherto have taken refuge into that state of thoroughly conscious ignorance which is the prelude to every real advance in science."[2] A passage I have hanging over the door of my lab.

▷ **Jenny:** Well, I'd say we're succeeding! The little I thought I understood about science is now in doubt.

▷ **Richter:** Good. Then we must complete the process. You asked for problems. Begin at the beginning. You, Imp, and Hunter and Ariana have been throwing around terms like "discovery," "innovation," "creativity," "genius" as if you knew what they meant. I do not. I seriously doubt you do either. And, frankly, I doubt that these terms can be safely used in any discussion. What is a discovery? Who is creative? How do you know these things exist? I say these are terms like "god" or "love" or "vital force" that simply put a label to our ignorance. Let me explain.

Begin with creativity. Allow me to quote from the proceedings of a symposium entitled "The Creative Process in Science and Medicine," of which I happen to have a copy. Heinz Maier-Leibnitz, the German physicist, says:

> *It seems we are now near the question, what can we teach about the creative process. I think every one of us who has discovered something in his life is unhappy because his students do not discover things all the time. He does not understand why they do not, that is, he does not know how to teach them. So I would very much like to learn how I can teach my students to discover something, to be creative.*

Eh, Hunter? Imp? But Jacques Monod responds:

> *I did not think we were going to discuss teaching creativity, as I do not believe this is possible. Anybody who has experience of American universities knows that one of the basic principles of American education is that you can teach any subject, any subject can be taught. There is not a single college in the United States that does not give a course called creative writing. How do you do that, Leon?*

Leon Eisenberg, the sole American present, who was then chairman of the department of psychiatry at Harvard Medical School, responded,

"You will never find the answer."[3] That is what I say here and now: You will never find the answer. Creativity is unique to every individual and to every individual creative act. You cannot define it. You cannot say anything useful about it without trivializing it. You cannot teach it.

▷ **Ariana:** No. I'm sorry. You mistake the lack of an answer with the impossibility of answering. I agree there are problems with defining creativity, but that doesn't mean it can't be done. As for teaching it—I assure you, from my own personal experience and from the experiences of every great master who has studied with any other great master in any creative field, creativity can be taught. By example, if nothing else. And that, it seems to me, is Hunter's point about learning science by apprenticeship. The creative aspects of science, too, can be inculcated and nurtured in receptive minds.

▷ **Hunter:** Obviously I agree with Ariana. But let us accept your point that we must come to grips with what we mean by creativity and what elements of it may or may not be amenable to pedagogy. Please continue.

▷ **Richter:** Right. I have much the same problem with the term "discovery." A colleague of mine likes to tell the following story, which illustrates my concerns. It was supposedly told originally by Chaim Weizmann, who was a chemist in addition to being the leader of Israel. One of his graduate students ran into his office one day babbling about how he had just discovered the mechanism of a chemical reaction. Wonderful, remarked Weizmann. A week later the student reappeared having completed further tests. He sadly announced that his discovery of the week before was a mistake. Weizmann scratched his head. "I am puzzled," he said. "Are you announcing one discovery or two?"[4] Or, I might add, none? Was this student creative or stupid?

▷ **Imp:** Or typical?

▷ **Constance:** I have a similar story—it was told by Richard Willstätter, about his teacher, Adolf von Baeyer. Willstätter writes:

> *How delighted Baeyer could be at the success of a preliminary experiment! And how optimistic! He would rise, take a deep bow, and say, "The problem is solved." And he would take off the hat he always wore in the laboratory [German laboratories were often virtually unheated during the nineteenth century] and wave it . . . Of course, Baeyer sometimes came to my desk afterwards, even before his lecture, and said, "It was a false alarm again, I'm sorry."*

▷ **Jenny:** So if we look at this a bit irreverently, we might say that error is an unavoidable part of the process:

> *A scientist who worked in East Lansing*
> *Made a discovery that set her a-prancing.*
> *"Eureka!" she cried,*
> *But next day she sighed:*
> *An err'r made it much less entrancing.*

▷ **Constance:** But Willstätter goes on to say:

> *The mistakes of greater and cleverer men are a lesson and consolation to us. Ampère [the father of electrodynamics] had two cats for company, one larger than the other. But at times they disturbed the great physicist in his work because he had to open the door so that they could go in and out repeatedly. So he simply equipped the bottom of his door with two smaller openings—a big one for the large cat and a little one for the smaller cat. Let us be grateful to Ampère.*[5]

▷ **Imp:** Wonderful! So, the problem of discovery reduces to the question of how do you know when you have hit on the simplest, verifiable solution to a problem; and how can you differentiate between the *belief* that someone has hit upon such a solution and instances in which one has stupidly cut two holes in the door? On the broader plain, which Ariana would have us study, are you being creative if your creation is "wrong" or lacks insight?

▷ **Ariana:** Or, getting back to Jenny's point: Can one create without making errors?

▷ **Richter:** Just what I am trying to prevent. So before you two run wild like old Baeyer, allow me to finish stating the problem. Who discovers? No clearer than who is a genius. Your supposed discoverers are, in my opinion, nothing more than front men taking credit for events and circumstances beyond their control.

▷ **Imp:** Historical determinism again.

▷ **Jenny:** Tolstoy on Napoleon in *War and Peace*!

▷ **Richter:** Or recall Francis Crick's statement that the DNA double helix made him and Jim Watson, not they it.[6]

▷ **Imp:** Wait a minute! Surely there's a distinction to be made between the discovery of the properties of things that exist—such as the structure of DNA or the charge on an electron—and the invention of a novel theory or explanation, such as the concept of evolution by natural selection or of relativity of time and space. The former are observable, physical entities; the latter, concepts that can only be imagined, relationships. *Things* are sure to be discovered eventually, and so their discovery makes their discoverers, whereas *concepts* need never be invented, and so the invention is made by the inventor.

▷ **Richter:** I thought that was what we had gathered here to find out. You assert the very things you wish to know.

▷ **Imp:** Okay. You're right. Let me rephrase my problem, then. What I hope we can do today is to define a program of study by which to resolve these questions: Is the process of discovery comprehensible? If so, are there strategies and tactics that increase the probability of discovering? Who is most likely to be able to make best use of this knowledge? How must they be trained? What must they know? What conditions will promote their research, and what obstruct it?

▷ **Richter:** You assume too much again. How do we know that the questions you have posed are valid ones; if valid, answerable; and if answerable, useful?

▷ **Imp:** Oh, come on, Richter! Valid? They strike right at the heart of what we all do! Answerable? Who knows? Nobody does when they begin something new. That's the nature of research. Part of our research—part of *all* research—must be directed at finding out whether the questions are valid, and, if so, answering them in a useful way. Personally, I'll consider it useful to find out that there's absolutely nothing that I can do to train more creative scientists or improve the probability of discovering something important—if, in fact, that's the case (which I doubt). Then at least I can stop worrying about it. On the other hand, if better ways of doing these things do exist, I want to be in a position to benefit by *that* knowledge, too!

▷ **Richter:** Fine. But just how do you propose to investigate these questions?

▷ **Constance:** May I make a suggestion? In my opinion it's always best to begin with what one knows the most about so that you work from knowledge to ignorance.

▷ **Richter:** But I thought the point of this meeting was to expound upon our ignorance.

▷ **Ariana:** "To be conscious that you are ignorant of the facts is a great step to knowledge." Though I wouldn't expect you to have read Disraeli!

▷ **Constance:** Anyway, I've spent the last few months carefully compiling and culling my notes and running computer searches for all the information I could find concerning discoveries, scientific creativity and innovation, scientific frauds, funding, and so on. I could present it—

▷ **Richter:** Oh, god! That will take years!

▷ **Constance:** —and we could proceed from there. Is there something wrong with that approach?

▷ **Richter:** My dear, you are committing the Baconian fallacy—assuming that all you have to do is have all the facts and the answer will be obvious. What facts do you collect? How do you tell what is relevant and what not? No. Theories are not determined by facts. Rather the other way around. Induction yields possibilities only. Since you seem to have read almost everything, try looking up your notes on Hull, Hempel, Popper, Poincaré, Duhem, Carmichael.[7] They all say the same thing.

▷ **Ariana:** Philosophy aside, I'm afraid I must agree, too. As Paul Valéry says, "Collect all the facts that can be collected about the life of Racine and you will never learn from them the art of his verse."[8] Same here: Collect all the facts about discoveries and discoverers and you will still never learn how discoverers discover, or even what a discovery is. It isn't a listing of facts we need here, but an understanding of the process by which those facts came into being. It's the difference between reading about how to do medicine, and trying it yourself.

▷ **Hunter:** Or reading a mathematical proof and actually inventing one.

▷ **Ariana:** Think about your own position, Constance. You deal with patents day in and day out and probably know more about what's been invented than any of the inventors who come to you for advice. But I'll wager you've never invented anything.

▷ **Constance:** Well, of course not. That's not my job, is it?

▷ **Jenny:** Now wait a minute. You scientists may all agree that facts don't determine their explanation, but I can assure you that we historians and social scientists, and I understand psychologists, too, regularly employ Constance's methodology. Why, the historian's dream is to discover a complete and unutilized archival source—to have all the facts at one's fingertips! And I understand from my psychology friends that it's common to collect all the data they can using every test available and then, using statistical analysis or something, to determine afterward whether their data mean anything. The idea seems to be that truth will emerge from correlations. Completeness and quantifiability: that's our goal!

▷ **Richter:** But how does one know what has been measured without a theory to direct the measurement? How does one know what to correlate without expectations? How does one interpret the results without a conceptual framework?

▷ **Jenny:** On the other hand, how can you do anything new without knowing what's already been done?

▷ **Imp:** Wait, wait! Constance first. She's the one being attacked here, and we haven't yet given her a chance to respond.

▷ **Constance:** Well, I'm confused. Just what are you suggesting? If we ignore the accumulated facts, don't we risk wasting time reinventing the wheel, so to speak?[9]

▷ **Hunter:** No doubt. But we also risk seeing nothing new. Look, I've done a bit of research on the problem, and so far I haven't found a single instance of a major scientific figure advocating a thorough knowledge of the literature as a prerequisite to discovering or inventing. Most maintain the opposite position. Mach derogated the bibliographic mind as uncreative. Einstein rarely read, let alone cited the relevant literature. J. J. Thomson and his son G. P. both warned that knowing the literature precludes having original ideas and fostered the technique utilized by J. J.'s teacher, Osborne Reynolds, which was to work through an idea for oneself before reading the literature.[10]

▷ **Imp:** A common opinion among biologists as well: Charles Richet, Macfarlane Burnet, Peter Medawar.[11]

▷ **Jenny:** But *you're* both citing the literature. Aren't you doing just the opposite of what you're advocating?

▷ **Imp:** Jen, Jen, Jen. You know me better than that. When have I ever boned up on the literature before starting a project? I jump to some conclusion based on the two bits of information I do have, work out the implications, and then check to see what other people say on the subject. Since that's how I work, I assumed other people did too, and when I looked, I found confirmation. It's just a matter of where you start—with private knowledge or public. Either way, your conclusions must fit the available data.

▷ **Hunter:** The point is—and this is a message I spend a lot of time trying to get across to my students—you've got to begin with some point of view, even if you have to modify it later. Because if you don't, you won't know how to evaluate your data.

Look at it this way, Constance. I'll bet in all those notes of yours you could find experts advocating any pair of contradictory propositions we'd care to imagine.

▷ **Constance:** More or less.

▷ **Hunter:** Looking at scientific data is the same: from one perspective, they look this way; from another, that. Using one technique, one finds *x*; using another, *y*. Nothing in the data themselves will resolve the contradictions, so you've got to bring a point of view, a perspective, a theory *to* the data before you can evaluate or make sense of them.

▷ **Constance:** So, are you telling me I've wasted my time doing all this work?

▷ **Imp:** No, no, no. Nothing could be more valuable than all the information you've collected. But it's got to be used as a complement to imaginative hypothesizing. As a check upon speculative nonsense. Not as a starting point, that's all.

▷ **Jenny:** You still haven't addressed Constance's point about wasting time rediscovering things.

▷ **Imp:** Because it doesn't really need a response.

▷ **Jenny:** Imp!

▷ **Imp:** What I mean is, who cares if you do rediscover something already known? I've done it lots of times! Started as a teenager. Predicted that if *a* was true, then *b* had to be true, too. Looked it up. Sure enough, I was right! It's great! It gives you practice discovering. It gives you confidence that you can solve problems just as well as the next guy. And best of all, it means you don't just know the accepted answer, you understand how to invent it! You don't have to take the word of some authority, because you've convinced yourself first.

▷ **Richter:** Moreover, in my experience one rarely reaches the consensus response when one thinks for oneself.

▷ **Ariana:** Then how are we supposed to arrive at your precious universal truths? Doesn't dissent bother you?

▷ **Imp:** Why should it? Szent-Györgyi suggests there are two fruitful ways to begin a new line of research. The first is to invent a huge novel theory and then attempt to disprove it. The second is to begin with a very old and well-established observation and repeat it with a critical perspective, using your own hands and eyes. So Szent-Györgyi, far from being afraid of rediscovering known facts, suggests that research can start with the process of recreating the research that led to a known discovery. An investigator repeating an experiment or observation with an eye toward discovering what previous investigators have overlooked is likely to find something different because he's looking for something different.[12]

▷ **Jenny:** Of course, it *is* true that historians rewrite history every generation. Intellectual historians look at the French Revolution as a material exercise in the philosophical ideas of the Enlightenment; social and economic historians tend to look at it through the class-conscious eyes of Marx, and so on. The "New History" focuses on numerical analysis, the "New New History" on anthropological approaches to popular culture. Same material, different perspectives.

▷ **Imp:** Exactly my point. One of Szent-Györgyi's most important discoveries occurred when he reproduced a fifty-year-old experiment of extracting myosin from muscle. Myosin was thought to be the protein that allowed muscles to contract. By the time Szent-Györgyi repeated the experiment, muscle physiologists had determined the exact electrochemical conditions necessary for contraction of isolated muscle fibers. Szent-Györgyi tried these conditions on myosin, but they didn't work. So, he went back and redid the old extraction of myosin, looking for some component that was discarded in the process. The result: actin, another muscle protein. It's the combination of actin with myosin that forms a contractile muscle fiber. So the fifty-year-old fact that myosin was *the* contractile muscle protein turned out to be only partially true, and one had to go back to the very beginning—myosin isolation—to find out what had been overlooked.

▷ **Richter:** To put the problem another way: All data are valid, but they must nonetheless be evaluated. Data become facts only in relation to theories.[13]

▷ **Jenny:** Clear as mud. Can you give us a concrete example?

▷ **Richter:** Look, one of the things a theory does is to set boundary conditions and criteria for data evaluation. Boundary conditions delimit the circumstances under which data may be collected and interpreted. Think of Galileo's law of falling bodies. The law says that two bodies, regardless of their size, shape, or weight, will fall at the same rate in a vacuum. The phrase "in a vacuum" sets a boundary condition. Thus, if you were to drop a feather and a lead weight right here and now, your observation that the two fall at different rates would say nothing about Galileo's law. The data are valid. They are meaningless in this context. Unfortunately, not everyone is so explicit about boundary conditions, especially in the soft sciences. Confusion results as to what is a valid test of the theory.

▷ **Imp:** Absolutely. Claude Bernard gives the example of studies of the sensitivity of some nerves in the spine. In 1822 Bernard's teacher Magendie found the anterior spinal root to be insensitive; in 1839, sensitive. A certain Longet, who repeated these experiments in 1839, found that they were sensitive, but in 1841 found that they weren't. Longet then claimed that Magendie had misled everyone by reporting in 1839 that they were. Bernard himself then undertook to repeat the experiments, and he found that sometimes the nerves were sensitive, sometimes not. What were the facts?

Well, Bernard had a most ingenious answer: *All* the observations were facts under certain conditions; the problem was that neither he nor Magendie nor Longet understood the conditions that yielded one result or the other. Further experimentation finally allowed Bernard to describe the exact conditions necessary to yield both the positive and the negative results so that Longet and other physiologists could replicate the result at will.[14]

▷ **Ariana:** John Eccles had a similar experience. He dreamed the answer to a problem he'd puzzled over, did the experiments to verify the answer, and then published it. The result was accepted by other scientists. Four years later he disproved it—only to find subsequently that he had been right originally, but only for a special subset of cases.[15]

▷ **Hunter:** So the difficulty was not that he was right or wrong, but that he failed to recognize the boundary conditions of his theory.

But I have another question for you, Richter. What about artifacts? How do they fit into this scheme?

▷ **Imp:** Ah! A major—no, *the* major—stumbling block to the advancement of science.

▷ **Richter:** Artifacts—experimental findings that appear to mean one thing but in fact mean another—result from not controlling the conditions of observation, or from a failure to understand the boundary conditions of the theory the data are being collected to test. In other words, artifacts are data that are inapplicable to testing a given theory. For instance, the observation that a feather falls slower than a lead pellet of equal weight is an artifact of the experimental conditions with regard to testing Galileo's law. But it is a fact of great significance for a theory of aerodynamics.

▷ **Jenny:** So every datum is a fact for some theory and an artifact for another?

▷ **Richter:** With one very important exception: anomalies. Anomalies are data that contradict a theory even though they are collected under appropriate conditions and with due respect for the boundary conditions of the theory. Such anomalies reveal that the theory is an inaccurate or incomplete description of nature. All theories ultimately fail, much to the chagrin of us theoreticians. Though of course science would never progress otherwise.

▷ **Constance:** Okay, but now you've really confused me. You begin by telling me I can't induce ideas by just looking for patterns in data. Now what I'm hearing is that we can't start with theory and deduce consequences either. A pure theory is just as likely to be incomplete—or wrong—as an induction from pure data.

▷ **Ariana:** And it predisposes us to perceive the world one way, and not to be open to contradictory facts or ideas.

▷ **Jenny:** This doesn't seem to be getting us very far.

▷ **Richter:** On the contrary. We're just where we want to be: at a paradox. We've found that we can start neither with theory nor with data. Does this mean we cannot start? Not at all! Science always begins with problems defined by a mismatch between theory and data. The specific nature of the mismatch defines the problem and sets criteria for recognizing its solution. The more clearly defined the theory and the more specific the anomalous data, the more likely it can be solved.

▷ **Hunter:** Exactly. We all start doing science with previously acquired sets of data, theories, and techniques, not from scratch. Berthollet provides a good example.

▷ **Richter:** The important thing is that the problem be nontrivial, clearly defined, accessible.

▷ **Constance:** Oh, I see. So you'd agree with Werner Heisenberg that "asking the right question is frequently more than halfway to the solution of the problem."[16] Or with Albert Michelson, that "knowing what kind of problem it is worthwhile to attack is in general more important than the mere carrying out of the necessary steps."[17]

▷ **Richter:** That's right. We can all be taught to solve formalized problem statements. Few of us know how to invent them.

▷ **Jenny:** Fine. So how do you do that in practice? How is the question related to the answer?

▷ **Hunter:** Try an example.[18] I use this in my classes. Suppose you're in charge of a tennis match involving 1025 players. Each time a person loses, he or she is eliminated from the competition. How many matches must be played, assuming that you have no ties?

▷ **Imp:** Well, 1025 is 1 + 1024. 1024 divided in two is 512. So 512 players pair off. The winners are split in two. 512 by 2 is 256, et cetera, et cetera. So, 1 + 512 + 256 + 128 + 64 + 32 + 16 + 8 + 4 + 2 + 1. Whatever that is.

▷ **Richter:** Which you could have solved much quicker by recognizing that 1024 is part of a power series.

▷ **Hunter:** But either way, you'd still have to calculate the number. And if I put you in charge of another match having 637 players, neither of your solutions would help us solve our new problem. You'd have to calculate all over again for every tournament. No, no. There's a much more elegant way to solve the whole class of such problems—one that requires no calculation and which anyone can do in their head. But first you have to recast the problem into a different form. Anybody?

Okay, try this. Who wins the tournament?

▷ **Ariana:** The person who never loses a match.

▷ **Hunter:** Meaning what for the other contestants?

▷ **Ariana:** They all lose once.

▷ **Hunter:** So the answer is?

▷ **Ariana:** I see! The answer is that one must always play one fewer match than there are contestants! Everyone loses a match but the winner. So 1024 matches for 1025 players, 636 matches for 637 players.

▷ **Hunter:** Yes, but notice the form your answer takes—you've invented not only a specific numerical answer, but more importantly a rule governing the solution to the whole class of problems.

▷ **Richter:** An algorithm.

▷ **Hunter:** And the algorithm is good for any contest operating under the same rules of elimination. That's what we're after in science: questions that allow one to arrive more easily at a more useful, elegant answer.

▷ **Richter:** Indeed. One should always prefer algorithm-generating problem statements because they are the only ones that lead to deeper understanding. The other, ad hoc solutions, as you rightly point out, Hunter, tell us nothing but the solution to one problem. The hard work of calculating is no replacement for thinking—a lesson I wish those who use computers would learn. An overreliance on computing power and automated equipment makes too attractive the head-on attack so that no search is ever made for the elegant solution. No theory. No simplifying technique. Just number crunching and data generation. In consequence, no insights into nature.[19]

▷ **Constance:** Interesting. That's what Max Perutz was saying about James Watson's description of his work on DNA, in *The Double Helix*:

> *People sneer at Jim's book, because they say all he did in Cambridge was play tennis and chase girls. But there was a serious point to that. I sometimes envied Jim. My own problem took thousands of hours of hard work, measurements, calculations. I often thought that there* must *be some way to cut through it—that there must be, if only I could see it, an elegant solution. There wasn't any. For Jim's there was an elegant solution, which is what I admired. He found it partly because he never made the mistake of confusing hard work with hard thinking; he always refused to substitute one for the other. Of course, he had time for tennis and girls.*[20]

▷ **Richter:** Tennis and girls aside, your passage is to the point. As important as Perutz's crystallographic work has been, each protein must still be photographed and analyzed individually. We still have no algorithm for understanding protein structure.

▷ **Hunter:** Not entirely true. Perutz did invent some techniques—heavy-atom substitutions, for example—that are very widely applicable to protein studies. And Pauling and Corey's work went a long way toward providing general solutions to particular structures within proteins—alpha helixes, gamma helixes, beta sheets, and so forth.

▷ **Richter:** But no general solution, such as the Watson-Crick double helical structure for DNA. Those are the kinds of algorithmic solutions we want to find: an algorithm for protein folding, in this case.

▷ **Imp:** Which means learning to ask the right kind of questions. The problem of problems in a nutshell! How do we do it?

▷ **Richter:** If you will allow me to continue, that is a question I will address. I have, as a matter of fact, published an article on the matter of raising and defining problems.[21] Not definitive, but a badly needed start. I will not burden you with all the details here, but some general points may help us on our way.

In the first place, most people fail to realize that there are different classes of problems and that each class requires a different method for its solution. The ten most important classes of problems and their corresponding methods of solution are as follows—Imp, where is chalk?

Now, there may possibly be more (or fewer) classes of problems. That is, in itself, a definitional problem that requires taxonomic methods. All I insist on here is that the same method will not suffice to solve all ten classes. This should be obvious—

▷ **Ariana:** Nothing is obvious. First rule I was taught in med school.

▷ **Richter:** —but it is not. Experimentalists particularly seem to think their methods omnipotent. How often I hear them say, "There is no use speculating"—they always use the derogatory term for theorizing—"until all of the data are in; and then we won't need a theory anyway because the answer will be evident." Such nonsense they spout!

▷ **Imp:** Hey! Let's not get personal here!

▷ **Richter:** You I except. You at least appreciate theory, even if you do not always understand it.

▷ **Imp:** A left-handed compliment, if I ever heard one!

▷ **Richter:** The point is that no amount of observation or experiment will serve to solve anything but a data problem. Suppose there is a theory that predicts the existence of such-and-such an effect in the UV region of the spectrum of some chemical as it interacts with another, but the interactions are too fast to be measured using existing equipment. You can experiment and observe all you want, to no avail. The solution to the problem requires invention of a new technique of measurement.

▷ **Hunter:** Just as concepts such as energy and entropy and velocity are inventions for explaining observation, but are not necessitated by observation.

▷ **Richter:** To hammer the point home, consider some absurd cases. One cannot demonstrate that statistical mechanics is equivalent to thermodynamics for a given situation by inventing a new experimental technique. One must demonstrate that both theories lead to the same prediction. A matter of comparison. One cannot increase the sensitivity of a measuring technique by inducing a new theory. One cannot solve an artifactual problem by any method whatsoever, because such problems are incorrectly stated in the first place.

▷ **Constance:** I have a question. Are your artifactual problems the same as Max Planck's "phantom problems"?[22]

▷ **Richter:** Give me an example.

▷ **Constance:** Well, Planck defines a phantom problem as one that is "totally incapable of any solution at all—either because there exists no indisputable method to unravel it, or because considered in the cold light of reason, it turns out to be absolutely void of all meaning." He goes on to say, "There are many such phantom problems—in my opinion, far more than one would ordinarily suspect—even in the realm of science."

As an example, a very simple example, he says look at the walls beside you. From my perspective, that is the right-hand wall and that is the left. But for someone facing me, say, Ariana, it's exactly the opposite.

Problem Types

Type	Examples	Methods of Solution
Definition	What is energy? What species is this?	Invention of concept or taxonomy
Theory	How do we explain the distribution of species? What causes objects to fall?	Invention of theory
Data	What information is needed to test or build a theory?	Observation, experiment
Technique	How can we obtain data? How do we analyze it? How may the phenomenon best be displayed?	Invention of instruments and methods of analysis and display
Evaluation	How adequate is a definition, theory, observation, or technique? Is something a true anomaly or an artifact?	Invention of criteria for evaluation
Integration	Can two disparate theories or sets of data be integrated? Does Mendel contradict Darwin?	Reinterpretation and rethinking of existing concepts and theories
Extension	How many cases does a theory explain? What are the boundary conditions for applying a technique or theory?	Prediction and testing
Comparison	Which theory or data set is more useful?	Invention of criteria for comparison
Application	How can this observation, theory, or technique be used?	Knowledge of related unsolved problems
Artifact	Do these data disprove a theory? Is the technique for data collection appropriate?	Recognition that problem is insoluble as stated

Which of us is correct? Both of us, or neither of us, depending on whether you will accept two complementary answers or not. Or we could obviate the question by choosing to adopt a convention, such as everyone determining right and left with respect to the door rather than to ourselves. In one case there is no single answer to the problem. In the other there is no problem.

▷ **Richter:** Yes, Planck's phantom problem sounds very similar to my artifactual ones.

To continue: class is only one of two properties of problems. The second property is order. Every problem exists within what James Danielli called a "problem area" or "major or first-order problem."[23] Major or first-order problems are questions like "How does the immune system work?" or "What is cancer?"

▷ **Imp:** Or "How do scientists discover?"

▷ **Richter:** Exactly. Such general problems are never directly addressable. Note! They must be broken down into subproblems. Problems of a lower order. George Polya has written a beautiful book about how to do this in math, called *Mathematical Discovery*.[24] I recommend you read it. It is accessible even to high school students.

In our problem area of immunology, for example, there are definitional problems of what constitutes the immune system; technique problems of how to describe the system; theoretical problems of how to explain the function of the parts; integration problems of whether the mechanisms of immune reaction are compatible with the mechanisms proposed for other cellular functions; extension problems, such as whether immunological "memory" provides a basis for understanding mental memory; and so on. Obviously, one cannot theorize without data, integrate without a theory, or extend without a definition of memory. So certain problems must be addressed before others. The result: what I call an "ordered problem tree" of questions.

▷ **Imp:** Yes! Beautiful!

▷ **Richter:** Now, the position of a problem in a "problem tree" is extremely important. Danielli and Polya have pointed out that problems may be solved only when the technique exists to solve them. That makes everything a technique problem. Because I recognize different *classes* of subproblems, which they do not, I prefer a broader statement: Problems may be solved only when the appropriate techniques, data, theories, or concepts are invented to solve them. The trick of problem solving becomes an ability to construct a tree of logically connected problems so that one or more branches connect with something already known. The connections may be direct, as to a body of data, or indirect, as by analogy to a similar problem tree whose set of solutions is already known. Models, either real or mental, may provide theoretical or conceptual frameworks capable of grafting into the tree at a relatively high level of order. And if one has constructed a particularly well-ordered and strongly connected problem tree, the solution of any single subproblem at any level may create a "chain reaction" that solves an entire problem area.

But—not all logical problem trees are solvable! It is possible to construct logical trees that never connect with anything that is known or that can be known. They are functionally useless.

▷ **Constance:** Planck's "phantom problems" again. Or artifacts?

▷ **Richter:** Not necessarily. The problem may be valid, but no means for observing relevant data may exist. The rate-limiting step, to borrow a term from chemical kinetics, in solving the general problem is the technique subproblem. Until that subproblem is solved, work on all other

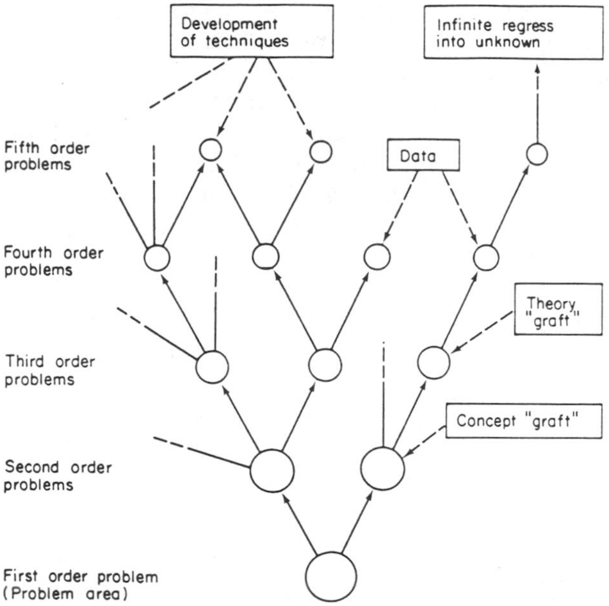

Schematic "problem tree." (Root-Bernstein, 1982c, adapted from Danielli, 1966)

parts of the tree is useless. So, I suggest a number of criteria for evaluating problems. First, avoid unsolvable problems—

▷ **Ariana:** Obviously.

▷ **Richter:** Nothing is obvious. Your words, I believe. Unsolvable problems are of two sorts: one, those that regress infinitely into the unknown; two, those that are so general that they cannot be reduced to solvable subproblems. Second, avoid trivial problems—

▷ **Hunter:** And how do you determine which are trivial? There've been a lot of supposedly useless discoveries that have turned out years later to be very important—electromagnetic induction, which led to electric motors and generators; relativity theory, which led, at least indirectly, to atomic power . . .

▷ **Richter:** You confuse lack of immediate utility with triviality. Triviality concerns where a problem is in a logical problem tree. Trivial problems are subproblems that have no connection with lower-order or more general problem areas. In other words, they are twigs or leaves that attach to no branch. They offer no hope of revealing anything larger than their own answers, the nonalgorithmic type of solutions we discussed before.

Electromagnetic induction and relativity theory, on the contrary, concerned problems of the most general nature. Electromagnetic induction served to link two previously disparate general problem areas: electricity and magnetism. Relativity theory served to transform the shape of the entire general problem areas of time and motion by questioning assumptions about them. It is, in fact, the very generality of these discoveries that prevented their immediate application. Their implications were so general that their utility lay many levels of subproblems away.

But no problem, no matter how theoretical or apparently useless, can be trivial if it is strongly connected to a general problem area or first-order problem. Since theories of time and motion were already being applied by engineers at the time of Einstein's work, one could be sure that his reformulations would one day find application. Physics is not trivial. Relativity theory is not trivial to physics. Thus it had to find applications. That is one of the few lessons of the history of science, in my opinion.

▷ **Jenny:** Might you say, then, that one way to identify important problems—at least of the sort involving revitalization of old knowledge—is to see what historically important scientists worked on?[25]

▷ **Richter:** Plausible. Certain traditions of important problems are worked over by succeeding generations of great men in mathematics and philosophy. That which has been fundamental tends to remain fundamental. Hunter?

▷ **Hunter:** I agree. One of the things that seems to characterize a great investigator is his (or her) "nose" for important problems. So if you haven't got the nose, or you're still developing it, it would make sense to rely on existing problem statements. One of the questions we should consider is whether, historically, great investigators actually have used Richter's set of criteria or not. I have a question on that account: if I understand you, the real question is whether there have ever been any trivial problems—in your sense—that have turned out to be inordinately useful?

▷ **Constance:** Hmmm. I've been looking through my notes. You say that trivial problems don't yield useful results—

▷ **Imp:** *Inordinately* useful results. After all, if you set out to design a new mousetrap and you succeed, then your trivial problem has yielded a useful, if unimportant, result.

▷ **Constance:** Yes, okay, but I *do* have some notes here about important discoveries that began with trivial observations—what one of my professors in college, Michael Mahoney, calls the "sublimity of the mundane." Things like: F. A. E. Crew, who was director of the Institute of Animal Genetics at Edinburgh University, observed that testes are almost always on the outside of a mammal's body, wondered why, and found that at body temperature spermatogenesis is prevented. Albert Szent-Györgyi, whom Imp's already mentioned, began studying oxidation reactions by asking, "Why does a banana turn brown when I bruise it?" He won a Nobel Prize when his studies yielded the fact that vitamin C prevents oxidation, and he elucidated the structure of the vitamin. C. T. R. Wilson, who won a Nobel Prize in 1927 for inventing the cloud chamber, which was so important to studies of subatomic particles, began his research when he observed coronas (the colored rings surrounding a clouded sun or hazy moon) and glories (the circular rainbow around a shadow cast into a cloud). He tried to reproduce these in J. J. Thomson's laboratory and observed that you could see ion trails in the cloud chamber as well.

And Richard Feynman's Nobel Prize winning work began when he observed the precession of the Cornell University insignia around a plate tossed in the air in the university cafeteria, and decided, as a game, to try to describe the rate of precession mathematically. Later he found that the same equations applied to the spin of electrons in atoms, a fundamental unsolved problem of physics. And I've found lots of other examples of sublimity of the mundane. I mean, if such everyday things can yield such important results, what does that say about whether something is trivial or not?[26]

▷ **Jenny:** I've got a related question. These men all seem to have discovered one thing while studying another. How does that fit into your scheme of problem trees?

▷ **Richter:** By the connectedness of the logical tree. Precisely because the tree is strongly connected, many apparently unrelated leaves will have logical connections at the branches or at the trunk. Thus, the phenomenon of the "sublimity of the mundane," as you call it, hardly contradicts my analysis. Nor does one thing unexpectedly lead to another. On the contrary, anything so fundamental to nature that it appears as a commonplace around us—that is, in many leaves of the tree—must operate upon basic principles unified at the trunk. To learn what those principles are is certain to yield knowledge of the greatest importance and utility, because the trunk divides in many directions, each of which is an application. These applications will be important because they are general. Useful because nature has already put them to use.

▷ **Hunter:** You subscribe, I take it, to the school of thought that maintains that nature operates on a very limited number of principles employed in a multitude of ingenious ways and combinations.

▷ **Richter:** I do. And therefore the more common the phenomenon observed, the simpler and more basic the principles employed are likely to be.

▷ **Imp:** And therefore the more accessible to analysis.

▷ **Ariana:** *If* you can perceive them in the first place! For isn't there another side to the "sublimity of the mundane"—that we overlook the most common things out of habit? A classic example is the invention of Ringer's solution, a standard saline solution with many physiological and medical applications. Sidney Ringer, like all his colleagues in cardiac physiology, performed his experiments using a solution made of sodium chloride in triply distilled water. This solution, when perfused in the hearts of the frogs he experimented on, kept them beating for perhaps half an hour. One day, Ringer's lab assistant got lazy and made up the solution using tap water. The perfused hearts continued beating for several hours. Ringer tracked down the cause—tap water contains traces of calcium, magnesium, potassium, and other ions essential to cellular processes—and then optimized the ingredients to create Ringer's solution.[27] I use this example in my classes to impress on students how dependent we all are on habit. Anyone perfusing hearts could have—indeed, *should* have—asked why one had to use a solution made exclusively with distilled water and sodium chloride. What would happen if . . . One can be certain that if the response is, "Well, that's what everyone else does," there's something important to be learned.

Surely part of discovering—be it discovering problems or solutions (pun intended)—is perceiving that which is always there but is overlooked through habit, lack of interest, or an untrained eye. Why should it take a Feynman to see an unsolved physics problem in a wobbling plate?

▷ **Constance:** Well, one answer to that might be what Alfred North Whitehead said: "Everything of importance has already been seen by someone who did not discover it."[28]

▷ **Ariana:** I like that. Discovery is not just perceiving something new, but perceiving what it *means.*

▷ **Imp:** Good. Then let's see what this means for us. We've got a general problem area—the process of discovering. What hidden assumptions are we overlooking through habit or by looking at the subject from a limited perspective? How do we get at them? When we do, where do we attack them? At the trunk, utilizing Szent-Györgyi's advice of inventing a huge, new theory to try to disprove? Or do we start at the leaves, by repeating an important observation that we know is strongly connected to a major theory, and hope we'll find something new? Implications or contradictions?

▷ **Hunter:** Why not both? Trunk and leaves simultaneously and work to the center. That way we'll avoid the Scylla of pure empiricism and the Charybdis of pure theory. I certainly have lots of historical cases of discoveries and inventions we could reexamine. The question is what we want to look for when we reexamine them. What's our big theory?

▷ **Constance:** May I make another suggestion? Hopefully better than my previous one?

▷ **Imp:** By all means.

▷ **Constance:** Okay. I assume—correct me if I'm wrong—I assume from what's been said so far that what you'd really like is a paradox or a contradiction.

▷ **Imp:** Nothing better!

▷ **Constance:** Then how about this: Some scientists think there's a logic to discovering and inventing, and others don't. Some scientists think that genius is comprehensible; others that it's just a matter of being in the right place at the right time. For example, Gerald Holton says the problem of scientific genius reduces to defining some special way in which the genial scientist works or thinks.[29] According to both Holton and Horace Judson, though, nobody's ever actually looked for systematic structures or algorithms in the thinking of scientists.[30]

▷ **Richter:** My dear, that is for the simple reason that geniuses have no special way of thinking. Your algorithms do not exist.

▷ **Ariana:** Don't exist? Or haven't been sought? Again you confuse what hasn't been done with what can't be done.

▷ **Richter:** While you assume that success will follow simply from trying.

▷ **Ariana:** Oh, come on—

▷ **Imp:** Now hold on, Richter. It's one thing to be skeptical, but don't be obstructionist. I think this has real possibilities. Why not propose as our huge, new theory that an algorithm for scientific discovery must exist—that there is a logic to discovery, and that we can uncover that logic by carefully examining how scientists actually work.

▷ **Hunter:** Still too broad.

▷ **Richter:** Right. So let me help you forward on your quest by contradicting you. It *is* impossible. Most discoveries occur by chance, not planning. Historical study will reveal nothing but stumbles and mistakes. So techniques, algorithms, strategies—whatever you call them— do not exist to be learned. Discovering cannot be taught. The best you can do is to be prepared for the accidents by knowing what the expectations and the outstanding problems of your field are. You piece the problems together. You order them. You identify the key elements, the rate-limiting steps, and then you wait for chance to hand you the missing piece. Or a lazy lab assistant.

▷ **Ariana:** Oh, come on! Ringer's invention could have been made with intention, and you know it. Why do you insist on misinterpreting everything? On making us all passive participants in an uncontrollable process? Sit and wait for the lucky break, you say? People are dying out there, Richter, *dying* while we sit and wait! I see them every day, and I have to act, through ignorance or not. So I can't wait.

And I can't accept your philosophy either. As the man who taught me surgery said, "Never, *never* accept anyone else's evaluation of the state of the art, including mine. I'm going to show you the best way I know to do surgery. Fifty years from now, my techniques will be considered barbarous. But if I have my way, you—the students I'm training in this room—will be the ones who will make my techniques look barbarous. Then at least my teaching, if not my practices, will justify themselves. So never assume that what I show you is the best way to do anything. Never assume that any idea you may have has already been tried. Surgery will only advance when and if you look at what I and my colleagues do, and say to yourself, 'That's not so hot. I bet I could do it better.'"

I don't accept your analysis. I'll look for myself, thank you!

▷ **Richter:** Please do. But where will you look?

▷ **Imp:** Perhaps Hunter can help us out. Following Richter's advice, I've been constructing a problem tree for us, and it yields the following as the desirable characteristics for our course of study. First, we need a definition of discovery or discovering. Second, we need to address the assumption that discovering is not merely a random or chance or accidental phenomenon. Third, it is preferable that whatever we study be something well known and important that we can look at with fresh eyes. Fourth, whatever material we begin studying should have definite and obvious connections to other material on discovering so that we have some hope of being able to go on to address the other subproblems we've posed, such as who discovers, under what conditions they do so, and what can be done to foster discovering. Hunter, what have you got in that manuscript of yours that fits those criteria?

▷ **Hunter:** Several things, actually. In the research I've done for my book, I've come across a couple of discoveries that are supposed to have been made by chance. There's one made by Louis Pasteur in 1858, for example, that meets all of your criteria. In 1858 Pasteur discovered the selective fermentation of asymmetric compounds by microorganisms. It's a classic example of a discovery supposedly made by chance; it's well known, and it turns out to be better documented than anyone had heretofore guessed: I've found a series of unpublished, uncited notebooks in the Pasteur archives in Paris that chronicle the development of the experiments leading up to selective fermentation. So, we can take a new look at old, well-known material in light of new research. Since the particular discovery of interest is part of a series of discoveries by Pasteur, we'll have ample material upon which to try out various definitions of discovering. Moreover, Pasteur's research starts in the Berthollet tradition—so we can develop the material we've already discussed—and it leads off in several directions, two of them quite important in the history of science: one, to van't Hoff's invention of the tetrahedral carbon atom and from there to the thermodynamics of chemical reactions; and the other, to studies of spontaneous generation and the germ theory of disease. So Pasteur's research is itself important.

▷ **Imp:** And I think we can all agree that Pasteur was a maverick if there ever was one—inventing the germ theory, vaccines, and so forth, in the face of resistance from virtually the whole medical community.

▷ **Richter:** Fine. But I have three worries. First, I've read a few books about discovering myself over the last few months, and as I. Bernard Cohen says, they made passing references to many individuals and episodes in the development of science, but nothing more.[31] Of course, Cohen is one to talk in light of his own panoramic study of *Revolution in Science*, but let that pass. If we are to do this, do it right. We must get down to the nitty-gritty.

▷ **Hunter:** Absolutely: Material from laboratory notebooks, private and professional correspondence, research papers, autobiographies, biographies, historical studies, and, in some cases, my own or other people's attempts to reproduce the conditions under which an observation or discovery was first made.

▷ **Richter:** Then that raises my second and third worries—also problems of technique, or methodology, if you like. Second: If you are going to present your case studies in detail, how can you present enough to make meaningful generalizations? Satisfy the law of statistical averages? And third: What makes you think nineteenth-century science can tell us anything about modern science or how modern science ought to be done? We already know that science is not the amateur, unorganized activity it used to be. What will studying "Little Science" tell us about doing "Big Science?"[32]

▷ **Hunter:** To answer your third question first, perhaps that's why the rate of discovering doesn't keep up with the rate of growth as determined by manpower, research dollars, or publications. Maybe science is *too* big. Too organized. Maybe a nineteenth-century perspective will illuminate that question.

▷ **Imp:** Besides, I want us to concentrate on "Little Science." I want to know what *I* can do better—not what NIH or NSF should do.

▷ **Richter:** As if the two were separate.

▷ **Hunter:** As to your second question, we'll all have to do our best to determine whether my case studies—and hopefully others the rest of you can add—are typical or not. That's where Constance's material can be so valuable. We may not want to start with the accumulated wisdom of the ages, but we certainly can't do without it.

▷ **Imp:** Or experience—personal experience. That should be the real test. After all, I didn't ask you all here to take a backward look at discovering but to learn how to discover more efficiently in the future. History, if we're going to use it, has to be an analytical tool for us.

▷ **Jenny:** If it can be.

▷ **Imp:** Why bother doing it, otherwise? The point is, the true test of the validity of our conclusions will never come through the numbers of discoveries we examine, but through the application of what we learn to our own, real-life laboratory problems. What we look for in history must be determined by our needs as scientists, much as that perspective may shock Jenny and Constance.

▷ **Jenny:** A purist I'm not. Don't worry about me.

▷ **Constance:** Me neither. Actually, I think it would be kind of neat if we could find a use for history of science. Then maybe I wouldn't feel that I'd wasted all those years in grad school.

▷ **Imp:** You didn't. I'm sure. But enough for today. We're all obviously getting tired and cranky. We've debated enough problems. Too many perhaps. So let's retire to the beach, or emulate Watson and hit the tennis courts and enjoy the rest of this day. Louis Pasteur and his accidental discovery of asymmetric fermentation can wait for next week. Till then, think good thoughts!

▶ IMP'S JOURNAL: Dogma Denied

> *Many of the experiments I performed in those days were provoked by [Hans] Bethe's conclusions. If he said a phenomenon would never be observable, I wanted to prove him wrong, which would make both of us happy. In several significant instances over the next four years I did.*
>
> *—Luis Alvarez, physicist, pianist, aviator (1987)*

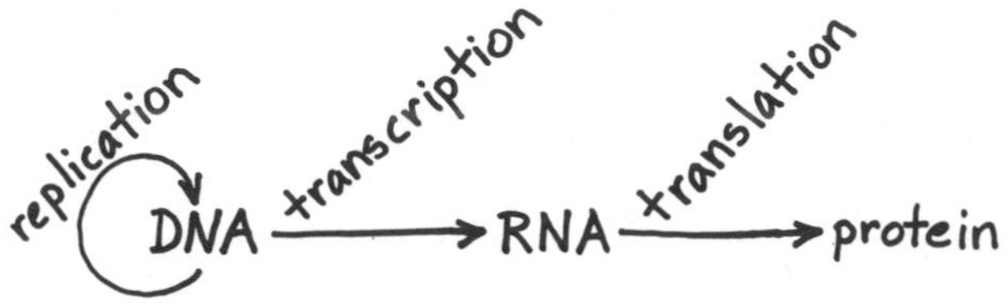

The Central Dogma (actually, the "sequence hypothesis") as illustrated in most texts.

So a "dumb" question defined the problem area: information transfers in biological systems. But how refined?

Research—stress the *re.* By going back over what I thought I understood and finding I didn't. By entertaining the impossible. By wanting to prove Francis Crick wrong about something important.

Confusion. That's what I found. Confusion concerning not only term "biological information" but term Central Dogma of molecular biology. Watson (*Molecular Biology of the Gene*) says Central Dogma is fact that DNA codes for proteins.[1] Lehninger, Temin, and Fraenkel-Conrat all say same.[2] But that's not dogma—that's experimentally verified, theoretically predictable fact!

And it's not what Crick says, either. He says DNA → RNA → protein is "sequence hypothesis," *not* Central Dogma.[3] And he should know—he formulated Dogma!

Crick: Central Dogma is "a negative *hypothesis*, so it's very very difficult to prove. It says certain transfers *can't* take place. That information never goes from protein to protein, protein to RNA, protein to DNA. It's not the same as the sequence hypothesis, which is much more explicit, and says that a certain transfer of information, the overall transfer from nucleic acid to protein, takes place in a certain way. The central dogma is much more powerful, and therefore in principle you might have to say it could never be proved . . . My mind was that a dogma was an idea for which there was *no reasonable evidence*. You see?!"[4]

Do I see? You can't even *test* a negative statement, let alone prove it! Makes no predictions—is based on no evidence—has no theoretical underpinning! The closest thing I know is the ban on perpetual motion in thermodynamics; but at least that ban not imposed until hundreds of people had failed to create perpetual motion machines. *And* ban is based on well-established theory: the laws of thermodynamics.

What was it Michael Foster said? "The man who constructs a hypothesis without supplying an adequate programme for its trial by experiment, is a burden to science and to the world; and he who puts forward hypotheses which by their very nature cannot be so tried is worse, for he is a purveyor of rubbish."[5] Rubbish! That's what this is! An untestable, unenlightening assumption I refuse to accept! And why should I? Science isn't religion! It should have no dogmas. It should have no untested—untestable!—assumptions (or, at least, as few as possible).

Yet, late 1970s, we find Crick himself stating: "*Nobody* tried to go from protein sequence back to nucleic acid because that wasn't on. You see. But I don't think it was ever *discussed*."[6]

Not discussed! Just assumed! This is science?

And listen to Crick's rationale: "In brief, it was most unlikely, for stereochemical reasons, that protein → protein transfer could be done in the simple way DNA → DNA transfer was envisaged. The transfer protein → RNA (and the analogous protein → DNA) would have required (back) translation, that is, the transfer from one alphabet to a structurally quite different one. It was realized that forward translation involved very complex machinery. Moreover, it seemed unlikely on general grounds that this machinery could easily work backwards. The only reasonable alternative was that the cell had evolved an entirely separate set of complicated machinery for back translation, and of this there was no trace, and no reason to believe that it might be needed."[7]

What kind of reasons are these? They come down to this: (1) "*I* don't see any way that a protein can act as a template for a protein." So? Maybe someone else can. (2) "Protein → RNA or DNA transfers would require an alphabet shift (translation)." Again, so? RNA → protein transfer requires translation, too. That, in itself, argues *for* the possibility of protein → RNA transfer, not against it. (3) "But the translation machinery is so *complicated* that it couldn't run backward." Yes, an electric motor is complicated, too—yet with small changes, it can be run backward as a generator. We know the molecular machinery—enzymes—are catalysts. They speed *rate* at which chemical equilibrium is reached, but do not change the equilibrium. Given that equilibrium is, as Berthollet suggested, a balance between a forward and a backward reaction, then under appropriate conditions, "back translation" should be possible. Unless Crick wishes to deny that molecular genetics obeys the laws of chemistry—which he wouldn't dare, being a strict reductionist![8]

(4) "We have not seen back-translation nor any machinery, special or otherwise, to run it." Neither you nor anyone else ever *looked* for it: "*Nobody* tried to go from protein sequence back to nucleic acid, because that wasn't on." How can you find something you won't look for? (5) "The reason it wasn't discussed is that there is no need for it." So *you* say! Since when do scientists tell Nature what she can and can't do, need or need not do? What temerity! What crap!

Besides, *I* can think of a need: need to decode protein sequence to determine whether it's antigenic ("not self") or "self" protein. We already know that lymphocytes have special, uncharacterized "machinery" for producing antibody diversity, which functions in ways not yet understood. Maybe that's the place to look.

Probably a wild goose chase, but I remember a certain satisfaction in finding others disgusted with this reliance on dogmatic faith instead of experiment. Barbara McClintock, for one, didn't believe dogma;[9] Erwin Chargaff never accepted it: "I remember vividly my first impression when I saw the two notes on DNA that appeared in *Nature* 21 years ago

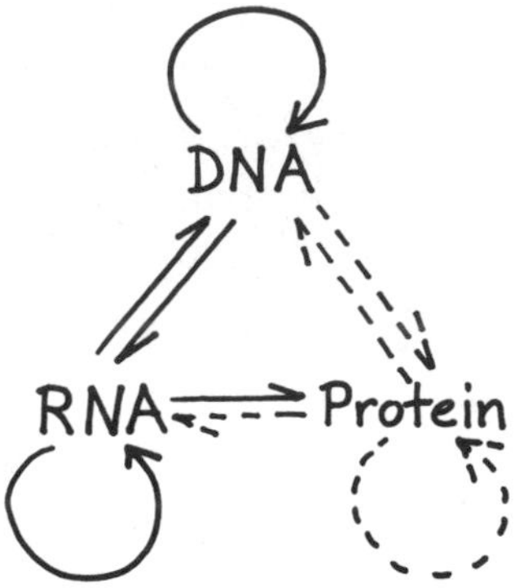

The Central Dogma denied.

[1953]. The tone was certainly unusual: somehow oracular and imperious, almost decalogous. Difficulties, such as the even now not well-understood manner of unwinding the huge bihelical structures under the conditions of the living cell, were brushed aside, in a Mr. Fix-it spirit that was later to become so evident in our scientific literature. It was the same spirit that soon brought us the 'Central Dogma' to which I believe I was the first to register my objection, never having been very fond of gurus with a Ph.D. I could see that this was the dawn of something new: a sort of normative biology that commanded nature to behave in accordance with the models."[10] Precisely! What I don't want to see doesn't exist. What I do see is all there is. Ha!

Only one thing to do: Turn it on its head. Assert possibility of protein → protein and protein → RNA transfers. Invent model predicting how such reactions would occur. At least such a model would be testable, a damn sight better than Crick's Dogma. And who knew; right or wrong, if I kept my eyes open, I could always hope for a surprise. And was I surprised!

Additional note: Now I understand why Watson and others so easily confused. Illustrations of Central Dogma are the problem. The Sequence Hypothesis is arrows drawn in (solid lines). Central Dogma is arrows left out (dotted lines).

As Ariana says, trick to recognizing the problem is perceiving what *isn't* there! Not obvious!

DAY TWO

Planning or Chance?

If the hypothesis is a much abused weapon, it is also an instrument of logic without which not even observation itself, by its very nature passive, can be realized . . . Even so-called accidental discoveries are commonly owed to some guiding idea which experience did not sanction but which had the virtue, nevertheless, of carrying us to little-explored or untouched territory.

—Santiago Ramon y Cajal, neuroanatomist, artist, and photographer (1893)

The illusions of the experimenter form a great part of his power. These are the preconceived ideas which serve to guide him. Many of these vanish along the path which he must travel, but one fine day he discovers and proves that some of them are adequate to the truth. Then he finds himself master of facts and new principles, the applications of which, sooner or later, bestow their benefits.

—Louis Pasteur, physicist, chemist, immunologist, artist (1880)

►

► IMP'S JOURNAL: Alternative Hypotheses

In science the primary duty of ideas is to be useful and interesting even more than to be true.

—Arthur Pardee, biochemist

In fact any starting hypothesis may be fruitful if it leads to original calculations or new experiments . . .

—René Taton, historian of science

So how did I invent amino acid pairing? By analogizing, modeling, and pattern forming.

Analogy to DNA structure and info transfer. Each amino acid in a protein should "pair" with an amino acid in the complementary protein. Only two reasonable alternatives for complementarity in proteins, either "parallel" or "antiparallel" interactions. [Editor's note: that is, both molecules are read in the same direction or in opposite directions.] Spent a week trying to invent such structures before checking structural chemistry texts. Actually came close to reinventing the Pauling-Corey beta structures but failed to put the correct bends in. At least I knew there weren't any other possibilities.

Both parallel and antiparallel beta ribbons exist and these align side chains in pairs. Wonderful! So, can amino acid side chains interact in pairs across such ribbons analogously to DNA base pairs?

Right off, antiparallel form has problems: some side chains in ribbon point at each other, others away. Any side chain that can interact when pointing at another can't when pointing away; side chains large enough to interact when pointing away from one another are too large to fit when pointing toward one another. Could flex the ribbon so not impossible, but strains structure. Not good.

What about parallel beta ribbon? Now that's pretty! (Didn't Crick and Watson insist that a good model be pretty?) Every side chain exactly same distance from its pair no matter where the pair is on ribbon. Good sign.

But which side chains pair? With twenty amino acid side chains there are four hundred possibilities. And that's ignoring that many can assume more than one conformation. Model building could take months and yield nothing. Has to be a key I'm overlooking.

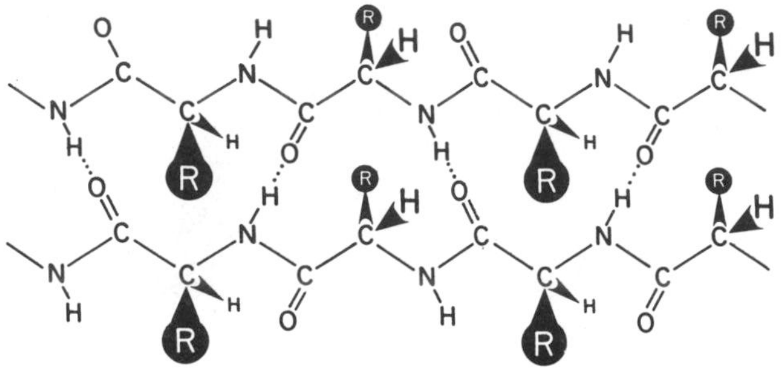

Parallel beta ribbon. The R groups represent amino acid side chains.

Ah! Try an assumption: if amino acid pairings exist, should be in genetic code.

Right. Then why hasn't anyone else found them?

Well, check genetic code. Maybe it'll become obvious.

Sure enough. Look at the way the code's displayed! Pattern's such you couldn't find amino acid pairs if you wanted to! It's arranged to conserve space and display the "wobble" in the third base. [Editor's note: the "wobble hypothesis" is an attempt to explain why the third base can vary without altering the amino acid encoded.] Pattern *assumes* that there's no reason to compare codon pairs or amino acids.

Got to rearrange code to reveal what I'm looking for. First, try reading codons as if paired in standard antiparallel manner. What amino acid side chain pairs result?

That's terrible. How can Pro interact with something as small as Gly and as large as Arg or Trp? And look at Val! It pairs with Asn, Asp, His, and Tyr? Not only are they different sizes, they're different charges! That's ludicrous! And Gly is supposed to pair with Ala, Pro, Thr, and Ser. Almost as bad. These "pairs" make no sense stereochemically or in terms of bonding. Try again.

So, rearrange code to read codons as if paired in parallel. A novel idea certainly, but what the heck. Means one strand is "backward" from the usual sense, but maybe that explains why one strand of DNA usually not read. Can't hurt to try.

Now this isn't bad! Pro now pairs only with Gly (structurally most intrusive, with the smallest side chain); Val with only His and Gln; and so forth. The only real oddities are the hydrophobic-hydrophilic pairs such as Leu with Glu and Asp. Still, mightn't an uncharged carboxyl nestle into a hydrophobic pocket? Or even form a C-O, H-C bond? Well, maybe there are more likely possibilities. Try some models.

Antiparallel beta ribbon still has problems. Not impossible, but ugly. Parallel beta ribbon lots better. Not bad at all! The Pro-Gly pair makes stereochemical sense, as do others. Can even make reasonable Leu-Glu pair without doing anything weird! Begin to think I am onto something![1]

The standard format for displaying the genetic code. Amino acid pairings are not evident

	U		C		A		G	
U	UUU	Phe	UCU	Ser	UAU	Tyr	UGU	Cys
	UUC	Phe	UCC	Ser	UAC	Tyr	UGC	Cys
	UUA	Leu	UCA	Ser	UAA	Term	UGA	Term
	UUG	Leu	UCG	Ser	UAG	Term	UGG	Trp
C	CUU	Leu	CCU	Pro	CAU	His	CGU	Arg
	CUC	Leu	CCC	Pro	CAC	His	CGC	Arg
	CUA	Leu	CCA	Pro	CAA	Gln	CGA	Arg
	CUG	Leu	CCG	Pro	CAG	Gln	CGG	Arg
A	AAU	Ile	ACU	Thr	AAU	Asn	AGU	Ser
	AUC	Ile	ACC	Thr	AAC	Asn	AGC	Ser
	AUA	Ile	ACA	Thr	AAA	Lys	AGA	Arg
	AUG	Met	ACG	Thr	AAG	Lys	AGG	Arg
G	GUU	Val	GCU	Ala	GAU	Asp	GGU	Gly
	GUC	Val	GCC	Ala	GAC	Asp	GGC	Gly
	GUA	Val	GCA	Ala	GAA	Glu	GGA	Gly
	GUG	Val	GCG	Ala	GAG	Glu	GGG	Gly

Amino acid pairings as represented in an antiparallel reading of paired codons

5′→3′			5′→3′
UUU	Phe	Lys	AAA
UUC	Phe	Glu	GAA
UUA	Leu	Term	UAA
UUG	Leu	Gln	CAA
CUU	Leu	Lys	AAG
CUC	Leu	Glu	GAG
CUA	Leu	Term	UAG
CUG	Leu	Gln	CAG
AUU	Ile	Asn	AAU
AUC	Ile	Asp	GAU
AUA	Ile	Tyr	UAU
AUG	Met	His	CAU
GUU	Val	Asn	AAC
GUC	Val	Asp	GAC
GUA	Val	His	CAC
GUG	Val	Tyr	UAC
UCU	Ser	Arg	AGA
UCC	Ser	Gly	GGA
UCA	Ser	Arg	CGA
UCG	Ser	Term	UGA
CCU	Pro	Arg	AGG
CCC	Pro	Gly	GGG
CCA	Pro	Arg	CGG
CCG	Pro	Trp	UGG
ACU	Thr	Ser	AGU
ACC	Thr	Gly	GGU
ACA	Thr	Arg	CGU
ACG	Thr	Cys	UGU
GCU	Ala	Ser	AGC
GCC	Ala	Gly	GGC
GCA	Ala	Arg	CGC
GCG	Ala	Cys	UGC

Biological implications: According to normal reading frame of DNA, complementary proteins appear to be encoded backward—which means more than one way to read code! Suggests more order in genetic code than people have suspected. Particularly Crick, who's claimed code arose by a "frozen accident." What kind of a theory is that anyway? Another of his "I can't figure it out so it has no explanation" theories? Makes more sense to think code arose from intersection of various sets of chemical interactions: set of possible base pairs (or codon-anticodon pairs); set of possible amino acid—nucleotide pairs (or side chain–anticodon interactions); and set of possible amino acid–amino acid pairs. That's a chemical basis. And it's testable.[2]

Methodological implications: detoured again. Started with immunological problem of info transfer and what am I discussing? Origin of genetic code! Well, why not? Immune system has to be based on code, right? But more important: Go back to question of why other people haven't suggested amino acid pairs before. Patterns. That's the stumbling block. Pictures, tables, graphs can be dangerous things. Revealing one

Amino acid pairings as represented in a parallel reading of paired codons. Parenthetical entries represent known "exceptions" to the genetic code

		5′3′	5′3′		
	Pro	CCU	GGA	Gly	
	Pro	CCC	GGG	Gly	
	Pro	CCA	GGU	Gly	
	Pro	CCG	GGC	Gly	
	Gln	CAA	GUU	Val	
	Gln	CAG	GUC	Val	
	(Gln)	UAA	AUU	Ile	
	(Gln)	UAG	AUC	Ile	
	His	CAU	GUA	Val	
	His	CAC	GUG	Val	
	Arg	CGU	GCA	Ala	
	Arg	CGC	GCG	Ala	
	Arg	CGG	GCC	Ala	
	Arg	CGA	GCU	Ala	
	Ser	UCU	AGA	Arg	(term)
	Ser	UCC	AGG	Arg	(term)
	Ser	UCA	AGU	Ser	
	Ser	UCG	AGC	Ser	
	Cys	UGU	ACA	Thr	
	Cys	UGC	ACG	Thr	
	(Trp)	UGA	ACU	Thr	
	Trp	UGG	ACC	Thr	
	Tyr	UAU	AUA	Ile	(Met)
	Tyr	UAC	AUG	Met	
	Phe	UUU	AAA	Lys	
	Phe	UUC	AAG	Lys	
	Leu	UUA	AAU	Asn	
	Leu	UUG	AAC	Asn	
	Leu	CUU	GAA	Glu	
	Leu	CUG	GAG	Glu	
(Thr)	Leu	CUA	GAU	Asp	
(Thr?)	Leu	CUG	GAC	Asp	

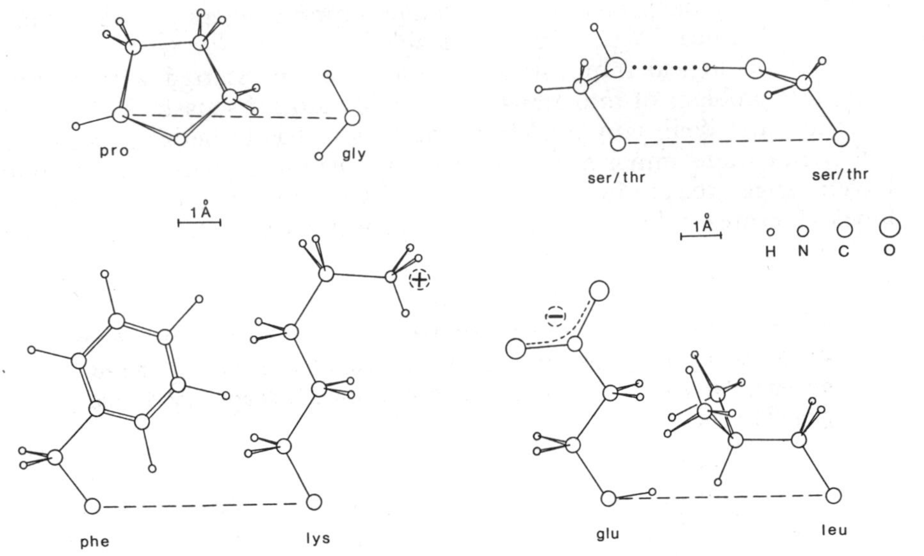

Amino acid pairs as they might interact in a parallel beta ribbon. (Root-Bernstein, 1982a)

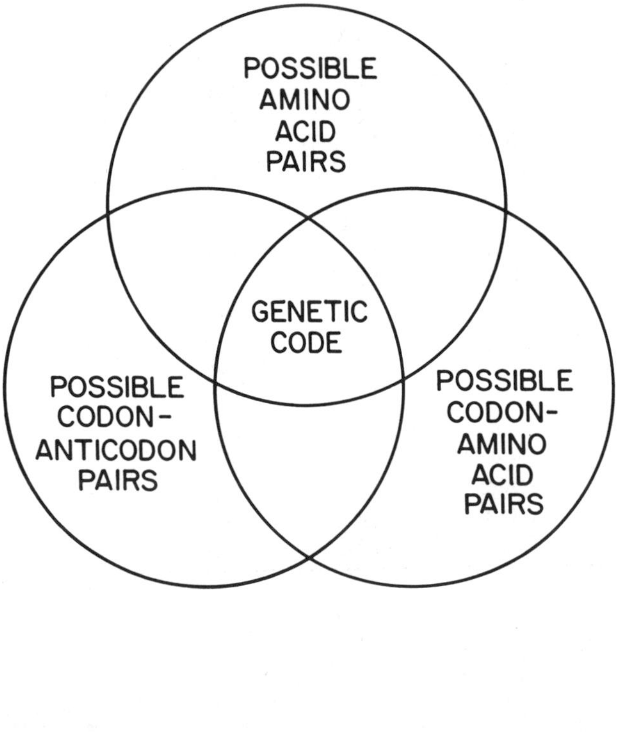

Possible origin of the genetic code. (Root-Bernstein, 1982b)

point, they hide assumptions, eliminate possibilities, prevent comparisons—silently, unobviously. Thus, a pattern makes sense of data but also limits what sense it can make. Rearranging can reveal new meanings without altering data. Implies possibility of making discoveries without new data and limits role of empiricism.

Lesson: Never trust a table, figure, or pattern until imagining other ways to arrange data. Search for new information by elaborating alternatives! Otherwise, you'll only see what the pattern maker *wants* you to see!

◀ TRANSCRIPT: Planning

▷ **Imp:** Welcome again, friends. Today, as promised, Hunter—who it appears has brought us some apparatus for a demonstration—will present his version of a supposedly accidental discovery made by Louis Pasteur in 1857. Bear in mind Richter's comments last week. He maintains that there's no algorithm, no logic of discovery; that discoveries are chance events whose workings cannot be rationally explained. If he's correct, then my hopes of developing a useful understanding of scientific discovery are dashed. Attempts by artificial-intelligence experts to program "discovering machines," and efforts of policy makers to plan rationally for scientific development, are equally unlikely to succeed. We shall simply have to rely on large numbers of investigators, lots of money, and the laws of probability to yield us our discoveries.

I'm not so pessimistic as Richter, however—

▷ **Richter:** Having dedicated yourself to this venture, you cannot afford to be.

▷ **Imp:** —for Hunter assures me that a credible case can be made for the rationality of the discovery process. If so, we must attempt to define that process as clearly as possible and draw out the parts that might be amenable to manipulation. In short, we must focus upon the questions of what constitutes discovering, and whether it is a rational process. Do I state the problem fairly, Richter?

▷ **Richter:** Yes. But I would like to clarify one point before we begin. My skepticism concerning a logic of discovery is not based on pure perversity. My opinion has been formed from well-established philosophical teaching. The standard books on philosophy of science almost all maintain that discovery is an illogical or even irrational process.[1]

▷ **Jenny:** Could you define our terms?

▷ **Richter:** Certainly. I limit logic to induction and deduction. Following Feyerabend, I define as irrational a belief in anything that has not been logically tested and verified.[2] All hypotheses are, by definition, irrational because they have not yet been tested. Hence, we must make a distinction, first proposed by Schiller, I believe, between the "context of discovery" and the "context of justification."[3] The "context of discovery" refers to the unique psychological, sociological, and historical context in which an investigator accidentally stumbles on an improbable idea in unusual circumstances. A paraphrase of Hanson.[4] The "context of justification," on the other hand, refers to the logical inductive and deductive tests used to determine the relative truth or reliability of a previously invented proposition. Conjecture and refutation. The point is that there can be no manual of discovering, no algorithm for inventing scientific ideas, because the "context of discovery" is irrational. Chance, inspiration, luck, accident—these are the scientist's only friends.

▷ **Constance:** Sure. Lots of scientists agree: The unconscious, chance event is the key to discovering.[5]

▷ **Jenny:** But wait a minute. If I heard Richter right, that means that psychology, sociology, and history are all the study of irrational events. That doesn't make sense. Why is an event in history more or less irrational than one in science? And how do you get from untestable irrational ideas to testable rational propositions?

▷ **Imp:** Ah, another lovely paradox! Science, the ultimate objective, rational undertaking, the epitome of rational thought, turns out to be a product of subjective, emotional, irrational flashes of chance insight and accidental observations![6]

▷ **Hunter:** Which in my opinion means we've made an incorrect assumption. For example, on the connection between hypothesizing, testing, and believing. I gather that the philosophers you cite perceive these to be separate steps.

▷ **Richter:** Correct.

▷ **Hunter:** I don't buy it. When I buld a hypothesis, I do it in relation to existing problems, data, and tested theories. The problem doesn't precede data and theories, as you seem to imply with your problem tree; it's a consequence of the juxtaposition. The hypothesis doesn't precede the testing; it's a consequence of testing—the failure of data and theory to match in a particular way. The particular mismatch then determines the range of hypotheses that may solve the problem. As you said yourself last week, Richter, a good problem statement limits the possible answers and provides criteria for evaluating them.

So what I'm saying is that I hold in my head simultaneously problems, hypotheses, and data, that I compare, test, and refine as I invent. It's a recursive process. It's also a rational one. When at last I have hypotheses, I have already determined that they're possible solutions to my problem statement according to the criteria laid out in the problem statement and according to my knowledge of existing data. In short, I have confidence in my hypotheses—I have reason for believing them, if you will—because they're tested in their creation. Thus, according to the definition of rationality suggested by Feyerabend and his colleagues, hypothesis generation must be a rational process.

▷ **Richter:** Perhaps we do not all invent as you do. But more to the point: Initial tests are limited. They do not provide grounds for rational belief. Furthermore, you suggest that you do not generate one hypothesis from your so-called test of theory against data, but multiple hypotheses. Some of these surely will contradict one another, and most or all will turn out to be incorrect. Therefore, you cannot possibly believe all of them.

▷ **Hunter:** And why not? Think of the wave-particle duality. It's much more useful to believe that light is both a wave *and* a particle than either one or the other. Only the "both and" formulation accounts for all the data at present. Complementarity again.

As for your point about initial tests being limited—sure, but all tests are limited. No matter how many white swans you've seen, there might be a black one around the next corner. So where are you going to draw the line for believing a hypothesis to be rational?

▷ **Imp:** Exactly. I mean, come on, Richter! Aristotle invented the rules of modern logic, yet most of his conclusions were empirically incorrect. Most subsequent science has turned out to be incorrect, as will most science in the future. You can't use correctness as a test for rationality or logicality.

▷ **Richter:** Then what criteria shall you use?

▷ **Hunter:** How about Polya's? You suggested last week that we read him, so I did. He defines as rational any argument based upon clear reason that can be articulated step-by-step. Thus, even a feeling or intuition for which arguments can be mustered would be rational.[7]

▷ **Richter:** You could not build a philosophy on that.

▷ **Hunter:** Maybe not. It's a problem of criteria. My criteria are not those of a philosopher. I'm a scientist. I'm not looking for truth; I'm looking for probabilities—ideas I'm willing to gamble on. Surely you'd accept statistics as a rational science even though it can't identify certainties?

▷ **Imp:** But isn't there more to it than that? Richter, you said a moment ago that there could be no handbook to ensure discovery. No algorithm. But let's not confuse things. There are algorithms—sets of rules—for playing chess. Computers can be programmed to play the game. Yet the algorithm, the knowledge of the rules alone, assures neither a human being nor a computer the ability to win. Even the masters lose. Why? Because one must know not only the rules of the game but strategies for employing the rules. When do you castle and when not? That's what I'm after: strategies. How do master scientists apply the rules, as Hunter says, to increase their probability of discovering?

▷ **Richter:** Then I hardly see what logic or philosophy have to offer.

▷ **Hunter:** Not much if they're going to insist on proof or disproof as criteria of rationality.

▷ **Ariana:** Frankly, all this discussion of philosophy drives me up the wall! No one ever knows everything about anything —nothing is ever proven in real life—so it seems to me that by the philosophers' definition we're all irrational about everything! But if science is as illogical and irrational as they claim, then what good is all the formal training we give to scientists? Does that training have nothing to do with whether we discover anything new? Could any layman do as well? Are all discoverers sleepwalkers who accidentally bump into great ideas in the dark?

▷ **Imp:** She's right! I mean, why should apprenticeship have any influence if science is irrational and illogical? But if strategies of discovering can be learned, then doesn't that argue for a rational, logical component?

▷ **Richter:** Yes, dammit, but only at the testing stage! You cannot just brush aside the distinction between the context of discovery and the context of justification.

▷ **Hunter:** Of course we can. It creates an unnecessary problem—an artifactual problem in your terminology. What makes that conceptual taxonomy inviolable?

▷ **Richter:** Experience. Even Poincaré maintained that "it is by logic we prove but by intuition we discover."[8]

▷ **Ariana:** So? Does that necessarily make intuition irrational?

▷ **Constance:** Actually, some scientists and philosophers disagree with Richter's position, you know. Claude Bernard said, "Nothing is accidental, and what seems to us accident is only an unknown fact whose explanation may furnish the occasion for a more or less important discovery."[9] And when one colleague congratulated Baeyer on his extraordinary luck, he replied, "It is not a matter of luck. I have no more luck than you have. The only difference between us is that I experiment more than you do."[10] Similarly, Irving Langmuir said that while you may not be able to predict discoveries precisely, you can certainly *plan* work that will lead to discoveries.[11]

Now, these comments suggest that the logic of discovery is in the setting up of the research, rather than in the testing of the hypothesis, and some philosophers are coming around to that position, too. For example, Nicholas Maxwell maintains there is a logic of discovery that makes possible a rational method for increasing the probability of discovery, but that to create such a logic involves recognition that science is an aim or goal-oriented activity focused upon the personal search of the scientist for beauty, harmony, simplicity, intelligibility, wonder, joy, and value. Until we understand these aims not as psychology of discovery but as method, Maxwell claims, we will never understand science.[12]

▷ **Richter:** O, god! Look—what I find beautiful you may not. Aesthetics and other personal values are no basis for *any* philosophy, let alone philosophy of science.

▷ **Hunter:** Michael Polanyi wouldn't agree. Consider his philosophy of personal knowledge as a basis for science.[13]

▷ **Ariana:** And your own search for truth asserts an aesthetic, doesn't it, Richter? Why do you value truth over, say, utility, or deep questions over trivial ones?

▷ **Imp:** In truth beauty, beauty truth!

▷ **Richter:** No platitudes. My position has nothing to do with beauty. I desire understanding and control of nature.

▷ **Ariana:** And if I studied the same problems as you do with a different aim? Such as an empathetic sense of the harmony and beauty of physiological systems?

▷ **Richter:** Then you would be an artist, not a scientist.

▷ **Ariana:** Are they so different?

▷ **Richter:** To my mind, absolutely different.

▷ **Ariana:** At least our positions are clear.

▷ **Imp:** But what about this idea that there may be other ways of perceiving the logic of discovery? Is Maxwell the sole pioneer, Constance?

▷ **Constance:** Well, no. Actually, there seems to be an increasing philosophical interest in the logic of discovering. But I won't go into it now if it's going to provoke an argument.

▷ **Imp:** On the contrary, an argument is precisely what we need. When people sit around coldly and rationally discussing things without getting excited or angry, not laughing or frowning, then they aren't discussing anything very important. You've got to feel it in your stomach, your heart, *and* your head, or it isn't worth doing. Challenge us.

▷ **Constance:** Well . . . The earliest reference I have to a logic of discovery is an article by N. R. Hanson in 1961. He simply pointed out that scientists reason their way to hypotheses somehow, and that "somehow" should be amenable to philosophical investigation. Similarly, in 1967, Peter Caws argued that scientific discovery is no less logical than deduction. He says that when philosophers utilize logic to prove or disprove an argument, they do the same thing scientists do (or musicians or any other human beings) when they reason from A to B about anything. What trips up the philosophers, according to Caws, is that they ask how a scientist performs his reasoning from A to B, but they never ask how they do it themselves. Where does position A come from? In consequence, they fail to recognize that their own deductions are just as fraught with aesthetic, intuitive, creative—and, Maxwell would argue, aim-oriented—behavior as anyone else's thinking.[14]

▷ **Imp:** Which is why philosophers can be wrong.

▷ **Constance:** The problem Caws sees is that his colleagues tend to view ideas as arising ex nihilo. Caws agrees with Hunter that a scientist begins with the accumulated problems, techniques, data, and theories of his or her predecessors. In this sense, the scientist is like everyone else in being part of a culture and in utilizing the same thought patterns and processes. In short, the logic of discovery or invention is, for Caws, the logic of everyday thought.

▷ **Richter:** Whatever *that* is. I thought this formalized-common-sense view of science had been discredited ages ago.

▷ **Constance:** As far as content goes, yes; but as to method, no. Errol Harris has made an argument similar to Caws's, claiming that the logic of discovery and the logic of proof (or justification, or disproof, depending on your philosophical school) coalesce if we simply consider science to be a process of question and answer like any other. Questions, answers, and tests alike, for Harris, are things that are not induced or deduced, but rather constructed or invented. Gary Gutting has taken the same position. Unfortunately, at least from my perspective as a historian of science, these philosophers and their colleagues assert the possibility of a logic of invention, but they don't demonstrate how it works. Very frustrating.[15]

▷ **Hunter:** Your point being that you don't have to be a revolutionary to suggest that neither the inductivist nor the hypotheticodeductive-justification model of science is adequate?

▷ **Constance:** I'm no expert. But look at Nickles's volumes on *Scientific Discovery*—you'll find several dozen philosophers maintaining the need for a new model.[16] There's certainly recognition of an outstanding problem.

▷ **Imp:** Then let me play "Implications" and suggest that it isn't a new model but a new kind of philosophy that's needed—that our failure to understand discovering, inventing, questioning and answering, or whatever we want to call it results from formal logic being too limited to address these processes. We need a new way of thinking that will do for logic what logic does for mathematics![17]

▷ **Richter:** Absurd. Induction, deduction, and inference are to be thrown out the window?

▷ **Imp:** No, of course not. But every theory and every technique has limitations. Boundary conditions. You said so yourself. Why not philosophies? Why not count the inability of logic to describe the creativity of the scientific process as a failure of philosophy, rather than casting aspersions on inventing by relegating it to the experts in the "irrational": the psychologists, historians, and sociologists?

▷ **Ariana:** I agree. Isn't it important to know what, if any, the limitations of logic are?

▷ **Richter:** Surely. But it is one thing to be skeptical in an attempt to test assumptions or elicit stronger evidence for a position. It is quite another to take skepticism itself beyond the bounds of reason, as Imp here is doing. If we abandon logic, we shall have no basis for discourse.

▷ **Imp:** I don't believe that's what I just suggested.

▷ **Richter:** Look, this is getting us nowhere. Let's get down to specifics. Let's hear what Hunter has to say about Pasteur and be done.

▷ **Hunter:** Or have begun. But I'll let you be the judge of that. I assure you that we'll shortly return to these issues with a vengeance. And to the question of whether history and the social sciences are merely fact-gathering enterprises! My Pasteur material is perfect for our discussion, if I do say so myself.

The discovery—or shall I say "observation and interpretation"—that I'd like to discuss today was made by Pasteur in 1857. It is, as I said last week, a classic example of a so-called chance discovery. Pasteur himself is famous for the statement that "in the field of observation chance favors only the prepared mind,"[18] so our problem is whether to put our emphasis upon the word "chance" or on the words "prepared mind." Most historians have emphasized chance. But it's worth recalling another one of Pasteur's epigrams: "Without a theory, practice is only routine driven by habit. Theory alone can cause the spirit of invention to appear and develop."[19] Possession of a theory implies use of intellect, and intellect implies rationality. Which will it be: rationality or chance?

First off, let me dispel any misimpressions you may have about Pasteur's being a doctor. Despite his invention of the germ theory of disease and of vaccines, he was trained as a physicist and chemist. Heir, in fact, through a series of master-apprentice relations, to Berthollet. This process of training conveyed not only techniques and styles of research, but problems as well. Pasteur's doctoral dissertations—he wrote two—concerned apparent exceptions to Berthollet's law of mass action.[20]

Now to understand what Pasteur did in 1857, I'm going to have to describe his first insight in 1848, and this in turn requires an understanding of the link to Berthollet. Since Pasteur's 1848 discovery has been

Louis Pasteur as a student at the Ecole Normale. (Musée Pasteur, Paris)

described by J. D. Bernal (who studied Pasteur's notebooks) as "entirely logical," whereas the 1857 discovery is universally attributed to chance, we can compare the two for ourselves.[21]

So, imagine, if you will, Pasteur as a young man of twenty-four sitting in the small laboratory of Professor Balard in the Ecole Normale in Paris.[22] The room is virtually devoid of equipment because Balard's philosophy was that simplicity was the highest scientific virtue and no one should utilize an instrument he or she couldn't make.

▷ **Ariana:** Which requires significant manual dexterity.

▷ **Hunter:** Absolutely. And a working knowledge of tools. Near Pasteur sits the goniometer he built with his own hands for measuring the angles of crystals, and the polarimeter he built for measuring the rotation of polarized light.

The nearsighted young man squints through a microscope at some tiny crystals of racemic acid. Racemic acid is a form of tartaric acid named for its source: grapes, or *racemus* in Latin. Both racemic acid and tartaric acid, as well as cream of tartar, which you use in baking, are byproducts of wine making. The acids are deposited on the inside of wine casks during fermentation. Their availability and the ease with which they crystallize made them common subjects for research by the 1840s, and a great deal was known about them.

▷ **Jenny:** Why was Pasteur looking at the crystals?

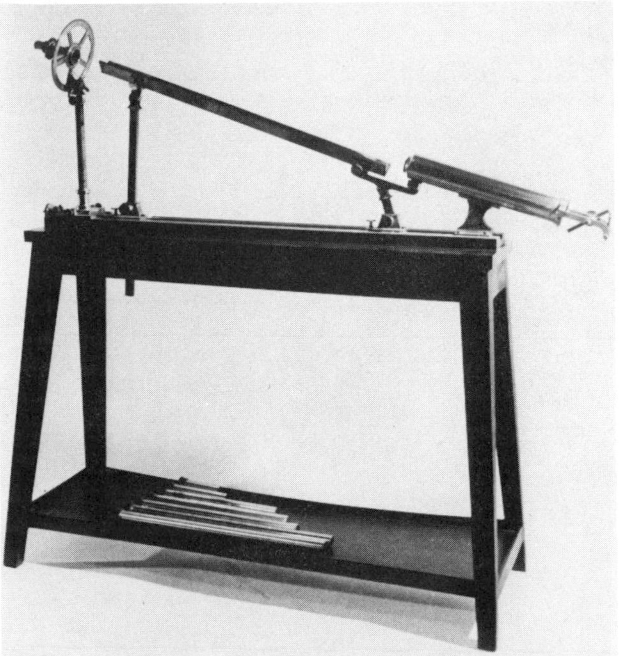

Polarimeter built by Pasteur for his studies of the connection between optical activity and crystal structure. (Musée Pasteur, Paris)

▷ **Hunter:** Because he was puzzled by them. Pasteur already knew quite a bit about crystals. When he was a student at the Ecole Normale in Paris between 1843 and 1846 he'd learned the techniques for studying crystals from Delafosse, one of Balard's colleagues. Delafosse had been a student of René Just Haüy, perhaps the greatest mineralogist of the time, and one of the first men to attempt to attribute macromolecular form to micromolecular organization. Delafosse in turn imbued Pasteur with both an understanding of the principles of crystallography and a love for the subject.

In 1844 an event of importance occurs. Jean Baptiste Biot (a member of Berthollet's crowd) communicates a memorandum from the chemist Eilhard Mitscherlich to the French Academy of Sciences concerning a comparison of racemic acid and tartaric acid, or the tartrate. Mitscherlich claims that both racemic acid and the tartrate are absolutely identical in all of their properties except one. They have the same molecular weight, the same constituent atoms, the same melting point, the same crystal structure. Everything is the same except that the tartrate rotates polarized light and the racemic acid does not. Pasteur finds intriguing and puzzling the idea that two chemicals can be completely identical except for this one property of rotating or not rotating polarized light.

▷ **Jenny:** Explanation, please.

▷ **Hunter:** Well, you've certainly heard of polarized light, perhaps in reference to Polaroid glasses. One of Berthollet's younger collaborators, Malus, had discovered how to polarize light, that is, how to align the light rays in a single direction by passing them through certain natural crystals or specially prepared lenses scored with a myriad of tiny parallel lines. Biot had found that polarized light was rotated to the right by some crystals of quartz; other crystals bent the light to the left; and still other quartz crystals didn't deviate polarized light at all. This observation suggested some difference in the internal structure of the crystals. Hauy had reported that while most quartz crystals are symmetrical in shape, some have extra facets which make them asymmetric. That is, a mirror image of one crystal can't be made to fit the form of the other crystal no matter how it's rotated or twisted. Your hands are asymmetric with respect to each other—you can't fit your right hand into a left glove.

The English astronomer John Herschel pulled all these observations together, demonstrating that right-handed crystals rotate polarized light to the right and left-handed crystals rotate polarized light to the left. Then, beginning in 1815, Biot demonstrated that many organic compounds, such as sugars and proteins, could also rotate polarized light. Since only the products of living organisms were optically active in solution, Biot concluded that only organic compounds—those produced by living organisms—could possess optical activity.[23] Clear so far?

▷ **Richter:** That suggests that tartaric acid was organic in origin and racemic acid artificial.

▷ **Hunter:** As a matter of fact, racemic acid was first isolated in 1820 by an industrialist named Kestner from a batch of tartaric acid—the natural organic compound—that had crystallized after industrial processing. So it could have been either. That's one of Pasteur's problems. Okay. Three events conspire to pique Pasteur's interest in the tartaric and racemic acids. The first is Mitscherlich's strange report of 1844 concerning the virtual identity of tartaric and racemic acid except for their action on polarized light. In light of Biot's correlation between organic compounds and optical activity, this raised the question of whether racemic acid was natural or industrial.

A second event occurs two years later. Pasteur meets another young chemist, named August Laurent, in Balard's lab.[24] Laurent is a fine chemist, although later a series of controversies virtually ruins his career and his health. Laurent influences Pasteur by confiding in him his theories of chemistry. In particular, he shows Pasteur some crystals of tungstates of soda, inorganic salts that have three crystalline forms. Laurent explains to Pasteur his own theory that the physical properties of compounds are always mirrored in differences in their crystalline forms, and proceeds to demonstrate these differences to Pasteur. In essence, Laurent is developing Hauy's earlier studies of crystal structure.

Pasteur quickly recognizes an incompatibility between Laurent's theory and Mitscherlich's report. If Laurent is right—and he can demonstrate the validity of his idea at least for the compounds he shows Pasteur—then Mitscherlich's racemic acid should not have the same crystalline structure as the tartaric acid. The one showing optical asymmetry should have asymmetric crystals, and the one showing no optical activity should have symmetric crystals. Physical properties should mirror crystalline form. Yet Mitscherlich had reported that racemic acid had the same crystal structure as tartaric acid. So, it turned out, had a M. de la Provostaye.

▷ **Imp:** Ah! A contradiction or paradox! The verities shall dance on a tightrope for us now!

▷ **Hunter:** Definitely. Here's our hypothesis-data-test wrapped up in one. External form mirrors internal structures; deviation of polarized light correlates to external form; yet the reported forms of tartaric and racemic acids are identical. So, in 1848, Pasteur, intrigued by this apparent contradiction, sits down to repeat Mitscherlich's and de la Provostaye's observations one by one. But, unlike them; he looks for *differences* between the racemic acids and the tartrates. And he finds them.

▷ **Ariana:** A case of repeating a well-established experiment looking for something new!

▷ **Hunter:** Precisely. But Pasteur doesn't find what he expects. He expects that the tartrates, which are optically active, will have asymmetric crystals which Mitscherlich and de la Provostaye had overlooked. And sure enough, the crystals do have tiny asymmetric facets. But when he looks at the racemic acids, which he expects to be symmetrical (because they don't rotate polarized light), their crystals, too, are asymmetrical!

▷ **Jenny:** In other words, neither Mitscherlich's nor de la Provostaye's "facts" were correct, nor was Pasteur's deduction from Laurent's theory.

▷ **Hunter:** That's right.

▷ **Jenny:** But, how could Mitscherlich and de la Provostaye have made such a mistake?

▷ **Hunter:** Quite easily. Few crystals are perfectly formed, and the asymmetrical facets on the tartrates are incredibly small. They could easily be interpreted as imperfections. Also, to recognize the asymmetry, you have to align the crystals and specifically search for planes of symmetry. That requires having a preconceived notion of what to look for. See for yourselves. I've brought tartaric acid crystals and copies of Pasteur's models.

▷ **Ariana:** *[looking through microscope]* So if you expect imperfections, you see them. If you expect asymmetry, it's there.

▷ **Hunter:** Right.

▷ **Jenny:** But what does that say for scientific objectivity?

▷ **Ariana:** That you see with your mind, not your eye!

▷ **Imp:** Which clearly makes the object subjective! You see what you expect to see. No more, no less.

▷ **Constance:** Speaking of which, Mitscherlich had his preconceptions, too, you know. He had invented the concept of isomorphism, literally "same form." Essentially, he believed that compounds having the same elemental composition—isomers such as tartaric and racemic acids—should have the same crystal structure. In many cases, this is true.[25] He expected it to be true of the racemates, too, and saw what he expected.

By the way, Jenny, you might be interested to know that Evan Melhado, who studied Mitscherlich's invention of isomorphism, maintains, as you do, that there's a match between the personal knowledge of the individual scientist and the problem he solves. Melhado thinks that only Mitscherlich had the necessary background to invent his theory of isomorphism.[26]

▷ **Jenny:** *[looking through microscope]* That's nice. But personally I don't see how Mitscherlich or Pasteur saw anything! How can you tell which of these crystals are asymmetric or not? Or which *way* they're asymmetrical?

▷ **Ariana:** You're having problems because you're not using what artists sometimes call "the eye of the mind." You have to imagine first what you expect to see.

▷ **Richter:** Oh, really!

▷ **Ariana:** Wait! If you're so skeptical, Richter, how about looking through the objective here—no pun intended—and answering Jenny's questions, if you can.

▷ **Richter:** It is enough for me to know that it can be done. I need not be able to do it myself.

▷ **Imp:** The essence of being a theoretician!

▷ **Ariana:** Oh, right. Yet knowing that a bicycle can be ridden is not the same as being able to ride the bicycle. The ignorant eye, like untrained balance, is untrustworthy if not downright dangerous.

▷ **Constance:** Duhem's philosophy exactly. Do we merely look through a microscope, he says, and report what we see? We should make the gravest blunders if we did.[27]

▷ **Ariana:** As anyone knows who's taught microscopy. First thing the students invariably focus on are the bubbles in the preparation or the dust on the cover slips.

▷ **Constance:** Sure. Ritterbush gives several historical examples. Theodor Boveri, an artistic cell biologist, was constantly correcting the observational errors of the theoretician Hans Driesch, who explicitly denigrated observational skill—

▷ **Ariana:** I hope you're paying attention, Richter.

▷ **Constance:** —and Matthias Schleiden, one of the founders of cell theory, once pillorized H. F. Lind for employing an objective artist—that is to say, a man totally unfamiliar with botany—to guarantee the correctness of his drawings. Schlieden maintained that the illustrations accompanying Lind's publications were inaccurate junk, full of artifacts and lacking essentials.[28]

▷ **Hunter:** Same problem in the physical sciences. First thing students see in X-ray diffraction photographs is the large black (or white) spot at the center, yet it's totally meaningless, since it's where the X-ray beam passes through the crystal without being diffracted. Everything they see is data, of course, but the trick is to figure out what part of it's *meaningful* data.[29]

▷ **Ariana:** Which requires the skills of an artist, such as abstracting: the elimination of unnecessary elements to highlight the important ones. Abstraction is characteristic of all great scientific drawings: Ramon y Cajal's, Boveri's, Harvey's—you name it.[30]

▷ **Imp:** That reminds me of something I read about Bernal, one of the founders of X-ray crystallography. He was friendly with many artists and used to do crazy things like take the artist Ben Nicholson over to C. D. Darlington's biology lab to show the artist how much a trained scientific eye could see that perhaps only a trained artistic eye could appreciate.[31] I have some vague recollection that Darlington himself took up drawing.

▷ **Ariana:** Then you might be interested to know that Pasteur himself was an artist.

▷ **Imp:** What? Paul Muni, how could you let me down like this!

▷ **Hunter:** More to the point, how could Vallery-Radot, Dubos, and the rest omit such a fact from Pasteur's biographies?

▷ **Ariana:** They didn't. *You* thought the fact was insignificant, so you overlooked it. There are clues in Vallery-Radot: a few sentences about a youthful pastel of his mother. Reproductions of portraits of his parents and a friend in Dubos. All easily passed by without comment. Unless, of course, you happen to like art. So I dug around a little and found a book called *Pasteur Inconnu* by Pasteur Vallery-Radot and a couple of

Portrait of his mother by Louis Pasteur, age thirteen. (Musée Pasteur, Paris)

articles by Denise Wrotnowska. It turns out that some twenty-five or thirty pastels and pencil drawings by Pasteur—all portraits—are known to exist. Some are at the Musée Pasteur at the Pasteur Institut in Paris. All done before he was eighteen. And, I should say, some quite excellent! There's also a lovely sundial he made as a teenager, demonstrating his mechanical skills.[32]

▷ **Hunter:** Now that you mention it, I seem to recall that Malus was a poet, and Laurent was also artistically inclined—a painter and musician, I think.[33]

▷ **Richter:** You're trying to tell me that Pasteur and Laurent had better perceptual abilities than the average scientist? Because they could draw or paint?

▷ **Ariana:** Yes! That's certainly part of having a prepared mind. Why not?

▷ **Richter:** Because artists don't train scientists—

▷ **Ariana:** Unless you happen to be Baron Cuvier, Ramon y Cajal, Edwin Goodrich—they were all trained by artists.[34]

▷ **Richter:** All anatomists, if I am not mistaken. Minor exceptions. Look, people with perfectly normal perceptual abilities—people without training in the arts, certainly—make important discoveries. So art is not necessary to scientific investigation.

▷ **Ariana:** I didn't say it was. I said a trained observational faculty is necessary, and art is one way of training it.

▷ **Richter:** Now you waffle. Try it this way. You claim the objective eye does not exist, correct? You can miss what might be seen by an error of omission? Yes?

▷ **Ariana:** Or a failure to distinguish important from unimportant features.

▷ **Imp:** Yep! Another argument against pure induction, if you'd care to adopt it.

▷ **Richter:** True enough. But answer me this: If the theory you hold is incorrect, do you see things that aren't there? Errors of commission?

▷ **Hunter:** It's happened, as you well know. Just think of N-rays or polywater! Great new discoveries that existed only in the minds of their discoverers!

▷ **Jenny:** Would someone please explain N-rays and polywater?

▷ **Constance:** Oh, sure. Shortly after X-rays were discovered, a French scientist named Blondlot claimed to have observed a second sort of emanation from X-ray sources, and called them N-rays. These N-rays supposedly affected things like the brightness of electric sparks. French scientists reported N-rays from all sorts of sources, but scientists in other countries couldn't reproduce the results. Finally a British scientist, R. W. Wood, visited Blondlot's laboratory and played a trick on Blondlot—pulled his power source or something—revealing that Blondlot perceived the rays whether his apparatus was working properly or not. They were all in his mind and in the minds of his friends.[35]

▷ **Hunter:** Polywater is the same sort of story. A Soviet scientist reported a polymer-forming type of water, sort of like "Ice-9" in Vonnegut's *Cat's Cradle.* Millions of dollars, pounds, and rubles were spent investigating its properties before it was finally demonstrated to be an artifact caused by impurities.[36] Irving Langmuir calls this sort of thing "pathological science" because the investigator inevitably employs a subjective technique to work at the very limits of perception and refuses to consider the possibility of artifacts or alternative interpretations.[37]

▷ **Richter:** So. Problem: How do we know that what Pasteur saw was any more real than what Mitscherlich and de la Provostaye saw?

▷ **Hunter:** Very simply because a number of people, including Pasteur's mentor Biot and Mitscherlich himself, took a skeptical look at the crystals again. Even though they hoped to find the young man out in some error, they now saw the facets for what they were: significant characteristics rather than accidental imperfections.

▷ **Imp:** Reproducibility, my good man! Reproducibility even by critics!

▷ **Hunter:** Exactly. But back to Pasteur. He has now located the expected asymmetric facets on the optically active tartrates but, unexpectedly, has also identified them on the optically inactive racemic acid. So, he is faced with another paradox. If the racemic acid crystals are asymmetrical, why doesn't a solution of these crystals display optical activity? Pasteur quickly realizes that his deduction—now proved wrong—that lack of optical activity correlates to crystalline symmetry is only the simplest of several possible hypotheses. It is also possible that, like Laurent's tungstate of soda crystals or quartz crystals, the racemic acid crystals have more than one asymmetrical crystalline form. Pasteur aligns the crystals with reference to a plane perpendicular to himself, and sorts the crystals into groups according to their asymmetry.

▷ **Jenny:** He did this all by hand? One crystal at a time?

▷ **Hunter:** That's right. It's the only way it could be done at the time.

▷ **Imp:** No worse than transcribing documents in the Carnavalet archives, my dear—or the Bibliothèque Nationale.

▷ **Hunter:** Certainly not. And sometimes the results are very rewarding, as we shall see. When Pasteur finishes, he has a pile of right-handed crystals and a pile of left-handed crystals. He finds that the right-handed crystals are identical in form to tartaric acid. Both rotate polarized light in the same direction. The mirror-image, or left-handed, crystals of racemic acid rotate light in the opposite direction.

This, then, is the answer, thinks Pasteur: Racemic acid does not deviate polarized light because the right-handed and left-handed molecules nullify each other's effect. "Pursuing my preconceived idea in the logic of its deductions," as he wrote later, he puts the idea to the test.[38] Taking equal amounts of right- and left-handed crystals, he puts them in solution together. The combined solution does not rotate polarized light! This time he is right.

I might add that Pasteur is so pleased with his observations that he runs out of his laboratory telling everyone of his great discovery, and whisks one of his colleagues off to a café, where he expounds on how this will change the course of crystallography.[39]

▷ **Imp:** Just like Crick and Watson running into the local pub to tell everyone they'd found the secret of life![40]

▷ **Hunter:** Or Crockroft running down the street proclaiming for everyone to hear: "We've split the atom! We've split the atom!" Very exciting!

Pasteur's models of the left-handed and right-handed crystals composing racemic acid. (Musée Pasteur, Paris)

▷ **Richter:** Getting back to the subject of chance or logic: I presume you quoted the line about "logic of preconceptions" specifically to provoke me. Fine. Then answer me this. If Pasteur was so logical, why didn't he find what he was looking for?

▷ **Hunter:** Ah, the crux of the logic question. But consider the alternative, if you will. If Pasteur was irrational, lucky, or what have you, how do you explain why he bothered to look, what he saw, and why he attached any significance to it? As Mitscherlich is supposed to have said to him: "I had studied with so much care and perseverance, in their smallest details, the two salts which formed the subject of my note to the Academy, that, if you have established what I was unable to discover, you must have been guided to your result by a preconceived idea."[41] Can you make any discovery without having a preconceived idea?

▷ **Imp:** Or overlook one without holding the wrong preconception?

▷ **Hunter:** Can you have ideas without thinking? Can you think without being rational or logical?

▷ **Richter:** Don't ask me. Ask Ariana. She's the subjective, irrational *artiste* here. Can one just *feel* what it's like to be a crystal, my dear?

▷ **Ariana:** Oh, go to hell, Richter. If you can't take any point of view but your own seriously, there's no use discussing it.

▷ **Jenny:** Well, perhaps this is a good time for a bite to eat and a few moments to reflect? Hunter?

▷ **Hunter:** Fine by me. When we move on to Pasteur's 1857 "chance" discovery, there'll be little opportunity to pause.

▼ JENNY'S NOTEBOOK: Personal Knowledge

> *It may be odd to claim the same personal engagement for the scientist [as for the artist]; yet in this the scientist stands to the technician much as the artist stands to the craftsman.*
>
> *—Jacob Bronowski, mathematician, humanist, and poet (1958)*

Ariana jumped up from the table and angrily left the room. I was just about to follow her when Hunter unexpectedly turned to Richter and said, "You ask whether it's possible to feel what it's like to be a crystal. From the tone of your voice, I presume you consider it impossible. Don't be so sure. Peter Debye wrote that he did indeed solve some of his problems by pretending to be a carbon atom and then asking what he would like to combine with or how he would react or rearrange himself under particular conditions."

I went to fetch Ariana back.

Hunter continued. "If I recall the passage correctly, Debye actually said something about having to use his "feelings—what did the carbon atom *want* to do?"[1] And he's not alone either. Another one of my colleagues—a quantum chemist—can actually *feel* what quantum equations are expressing. When he's at a meeting, you can watch him and tell from his body what he thinks of a speaker's ideas. If the molecular interactions being expressed are too loose, he sort of slumps or sprawls in his chair." Ariana and I walked in just in time to watch Hunter comically demonstrating his point. "And when the interactions are too closely packed, he becomes restless, and looks as if he's wearing a suit of clothes several sizes too small." Again Hunter mimed. I wish we could have seen Richter's face, but we were standing behind him.

Constance joined the act. "It's not limited to chemistry, you know. Charles 'Boss' Kettering, the guy who invented the electric starter for car engines, various types of internal combustion engines, and gasoline additives, and oversaw the work that led to freon refrigerants, and—well, enough—anyway, when his colleagues became too technical, as was all too frequent, he liked to respond, Yeah, but have you ever *been* a piston in a diesel engine? According to him, you didn't know anything until you could imagine being the thing you were trying to invent."[2]

Ariana entered the fray. "Exactly. Read Evelyn Keller's essays or her biography of Barbara McClintock, *A Feeling for the Organism*. McClintock's success was an ability to identify so completely with her material, whether it was *Neurospora* genes or individual corn plants, that she was no longer outside the system, but in it, part of it. The important thing, she admonished us, is to lose yourself in your material. You don't understand it until you can do that. Anna Brito says the same thing in June Goodfield's study, *An Imagined World*: If you're going to study tumors, you've got to *be* a tumor. It's this damned male invention of objectivism that screws everything up."[3]

"Hey!" Imp objected. "I won't deny a touch of male chauvinist piggishness, but let's not get sexist here. Hunter and Constance already gave us some examples of male scientists who feel their way into their work. And I've come across a few examples myself. Ramon y Cajal *felt* the urges of nerves striving to connect with one another. Jonas Salk *felt* what it was like to be a virus or a cancer cell and how it would act and what it would try to do.[4] Michael Polanyi wrote a whole book called *Personal Knowledge* about the necessity to extend one's body to include the object of study so that it's no longer external to us and we dwell within it.[5] But probably the most striking passage is something Joshua Lederberg wrote." Imp had been searching through his books as he spoke and now quoted from the one he was holding.

"Lederberg writes that the scientist 'needs the ability to strip to the essential attributes some actor in the process, the ability to imagine oneself *inside* a biological situation; I literally had to be able to think, for example, What would it be like if I were one of the chemical pieces of a bacterial chromosome? and to try to understand what my environment was, try to know *where* I was, try to know when I was supposed to function in a certain way, and so forth.'[6] Playacting, in essence. Learning a part in nature's cosmic drama so completely that one can act out a character other than one's own. That's the problem we need to address here, not whether women can do it better than men."

"Unless the scientists you've mentioned are psychologically—not physically, mind you, but psychologically—less masculine, or perhaps I should say more sensitive, than the majority of their colleagues," rejoined Ariana heatedly. "But let that pass. I shouldn't have raised the issue in the first place because it diverts us from the question that began this whole discussion: Are there ways of knowing—understanding—that transcend verbal and mathematical understanding?"

"Change 'transcend' to 'augment' and I'd definitely agree," said Hunter. "You must know the famous letter from Einstein to Hadamard explaining how he thinks. Not, he says, using words or symbols, but using visual images and kinesthetic feelings. What's it like to fall in an elevator at the speed of light or to ride a light wave and observe another beam of light beside you? That sort of thing. The mathematics and the words, said Einstein, are only the end products of the process and have to be sought for laboriously in a secondary stage after the initial insight is reached by these other means."[7]

Constance had been searching through her cards and piped up with, "Stan Ulam talks about solving mathematical problems by visual images and kinesthetic feelings, too: 'attempts to calculate, not by numbers and symbols, but by almost tactile feelings combined with reasoning, a very curious mental effort.'"[8]

"Very curious indeed," I commented. "So why doesn't anyone ever mention this tactile-kinesthetic-playacting sort of thing in math and science courses? I've certainly never heard of it before."

I should have kept my mouth shut. This was Richter's cue: "Because almost nobody thinks that way but a few oddball geniuses. And because no one would know how to teach it anyway. No language exists for communicating feelings, impressions, or images."

"Not true! Not true!" cried Ariana. "You're talking from ignorance again, Richter. I'm sure I've seen a study of eminent American scientists demonstrating that a third of them use three-dimensional images to do their problem solving; a third use verbal thoughts; and another third use imageless, nonverbal thought.[9] If I recall correctly, only a few used kinesthetic feelings, but still, I think there's only one conclusion that you can draw: most scientific thinking is not verbal or mathematical, and until we get that through our heads, we aren't going to understand how scientists *do* think or how to teach them science!"

"I'd like to see that study," said Richter, meaning the opposite, I'm sure. "But you still haven't addressed the problem of how to teach these skills. Or is that because you don't know how to?"

"Dammit, Richter, one step at a time," flared Ariana. "You yourself said that we have to define the problem first if we're going to achieve a reasonable answer. So, no, I don't have a pat answer at the moment. It's been hard enough just to get you to listen to the problem."

"Well, I'm listening now."

And thank heaven for that. I was afraid we'd never get things calmed down enough to return to the stated business of the day. Even so, it was a while before tempers were assuaged sufficiently for us to continue.

I think Richter is developing some respect for Ariana. When's Richter going to earn Ariana's respect?

◀ TRANSCRIPT: Chance

▷ **Imp:** All right! Enough small talk. Hunter?

▷ **Hunter:** Good! We've got a number of questions on the table. Is there a logic of discovery? If Pasteur's 1848 discovery is logical, why did he find something other than what he expected to find? How would the structure of a chance discovery differ? And, in a general sense, what do we mean when we contrast logic with chance? I've tried to organize my presentation of Pasteur's 1857 "chance" discovery of asymmetric fermentation to bring these questions out explicitly, so with your cooperation let's see what we can do to clear them up.

Now, there are two ideas I want you to carry forward from the 1848 observations we just discussed. The first is the existence of mirror-image organic compounds, or "racemates," as they came to be called after racemic acid. The second is August Laurent's hypothesis that physio-chemical properties mirror crystalline structure.

The so-called chance discovery made by Pasteur in 1857 concerned a method for separating the left-handed tartrates from the right-handed ones by fermentation. Traditional accounts of the discovery classify it along with two others made by Pasteur: the one in 1848 when he found he could manually separate right-handed crystals from left-handed ones; and another in 1853 when he found that by adding a different asymmetric compound to a racemic mixture, either the right- or the left-handed form of the racemate could be made to crytallize preferentially.[1]

▷ **Jenny:** I'm lost.

▷ **Hunter:** Don't worry about it. Essentially, Pasteur found a chemical means to separate racemates. It's still being used today.[2] Asymmetric fermentation, which he observed in 1857, provided yet a third means of separating racemates.

▷ **Jenny:** Why did Pasteur want to find new ways of separating racemic acid? Was it too laborious and difficult to do by hand? Or not very reproducible?

▷ **Hunter:** On the contrary, it was very reproducible. In fact Biot, who had examined the racemic acid himself without observing the asymmetry of the crystals, made Pasteur prepare racemic acid for him so that he could verify for himself that Pasteur's observations were correct. No. The problem didn't involve separating racemic acid at all. But I'm jumping ahead of myself.

I want to start with what might be called the legend of Pasteur's discovery of asymmetric fermentation. Dubos in his classic biography of Pasteur gives a typical account. Let me read it. He says that this method of separating out the constituents of racemic acid

> *was the result of one of those chance occurrences which are observed and seized upon only by the prepared mind. It had long been known that impure solutions of calcium tartrate occasionally became turbid and were fermented by a mold during warm weather, and Pasteur noticed one day that a tartrate solution of his had become thus affected. Under the circumstances, most chemists would have poured the liquid down the sink, considering the experiment as entirely spoiled. But an interesting problem at once suggested itself to his active mind: how would the two forms composing the paratartaric acid [that is, the racemic form] be affected under similar conditions? To his intense interest, he found that whereas the [right-handed] form of tartaric acid was readily destroyed by the fermentation process, the [left-handed] form remained unaltered.*[3]

▷ **Constance:** Yes, and I found the same sort of account in several other biographies of Pasteur.[4]

▷ **Richter:** Irrelevant. Historians can be wrong. What does Pasteur say he did?

▷ **Hunter:** That's the problem. Pasteur doesn't say. Neither in his published papers written during 1857, nor in his scientific correspondence to his patron Biot and J. B. A. Dumas in September 1857 when he made the discovery, nor later did Pasteur say why he did the experiments concerning asymmetric fermentation—or how he noticed the phenomenon.[5] He just says that if a tartaric acid solution is allowed to ferment . . .[6]

▷ **Richter:** Then Dubos and the others might be right.

▷ **Hunter:** Except for one thing: Pasteur's laboratory notebooks suggest a very different story.

▷ **Jenny:** You're not saying that the historians you just mentioned fabricated a story, are you?

▷ **Hunter:** Let's just say that they were all working with an incomplete set of data, and instead of stating forthrightly that no evidence seemed to exist as to the origin of Pasteur's observation of asymmetrical fermentation, they resorted to the concept of chance. In my opinion, when someone says that an event occurred by chance it's simply another way of saying "I don't know how it happened" without explicitly admitting it.

▷ **Imp:** Like "spontaneous mutations"? Dobzhansky maintained that the adjective "spontaneous" is simply a delicate way of saying you don't know anything about what causes mutation.[7] But give it a name and it no longer needs an explanation!

▷ **Ariana:** Molière's point about the doctor who attributes a potion's efficacy for producing sleep to its somnogenic influence. Medicine is riddled with such tautologies.

▷ **Richter:** Unrecognized problems. I agree on that point. But surely you cannot maintain a purely deterministic *Weltbild* in this day and age. Evolutionary theory is predicated on chance. Jacob and Monod built their philosophy of feedback inhibition systems around chance. Modern physics is founded on chance. Why not discovering?

▷ **Hunter:** Your characterizations lack subtlety, Richter.

▷ **Richter:** Never my strong point.

▷ **Hunter:** You are confusing chance with randomness. When and where a particular subatomic particle will decay is unpredictable because the events are randomly distributed in time. That the decay *will* occur is a matter of probability, or even certainty. Given enough decaying particles, we can even predict with great accuracy the rate of decay. So chance in this instance is not equivalent to lack of predictability.

▷ **Imp:** The same's true of mutations in biology, Richter. They occur randomly because their causes don't operate continuously or unvaryingly (this is true whether the cause is a chemical mutagen or X-ray), nor do they operate on a uniform set of objects (each living thing being unique). The randomness of the appearance of mutation is not, therefore, due to chance, but to the uneven distribution of causative agents acting on the uneven distribution of characteristics in a population.

▷ **Richter:** Caused, in turn, by the chance segregation of a particular set of genes during meiosis and the chance fusion of gametes during reproduction. You can't get off so easily.

▷ **Ariana:** Nor can you. Just because the result is unique doesn't mean the mechanism is. Every snowflake is supposed to be different, but there are general physical rules governing how they're built and limiting their possible forms, right, Hunter?

▷ **Jenny:** God doesn't play dice with the universe?

▷ **Hunter:** Unless you're Niels Bohr. Perhaps we're looking at this all wrong. Einstein and Bohr never did agree about the role of chance and probability in physics, yet both made tremendous contributions. Let's agree to disagree. Perhaps, at different levels of analysis, both positions will be useful, so let's allow Richter his chance events and see what he can develop from it. For my part, I have a preference for more classical explanations; I can't accept chance as an explanation for anything. Indeed, the reason I searched out the Pasteur notebooks was recognition that calling this discovery a chance event indicated that no one knew what Pasteur had actually done. My criteria for an acceptable historical explanation differed from those of the historians. So I had to ask, How might I find out what Pasteur *did* do? What other sources might exist? And how do I get access to them?

▷ **Jenny:** Exactly. Historical research involves more than just gathering facts: You've got to analyze data and question it as well. And, of course, hypothesize to know what data to look for.

▷ **Imp:** My goodness, Jen, you've been around me too long! Sounds almost scientific.

▷ **Richter:** So what's your version of the discovery, Hunter?

▷ **Hunter:** Well, one of the assumptions of all previous accounts is that Pasteur set out to discover ways of separating racemic acid into the constituent tartrates, and that when he accidentally observed asymmetric fermentation he was prepared to perceive its potential as a means of effecting the desired separation.

My account differs from the very outset. I don't think that Pasteur was interested in separating racemic acid at all in 1857. I think he discovered asymmetric fermentation while looking for something completely different—in the same way that he expected to find that crystals of racemic acid were symmetrical but found instead that they possessed two forms of asymmetry.

▷ **Richter:** Yet Pasteur was candid enough about finding something unexpected in his first discovery. Why was he not equally candid about how he discovered asymmetric fermentation?

▷ **Hunter:** That's a problem my description of the discovery accounts for. But your question cuts both ways. Had the experiment occurred by accidental contamination, he might also have said so. Yet he didn't. One of the advantages of my account is to explain why.

▷ **Imp:** Forward, then!

▷ **Hunter:** The problem that I think puzzled Pasteur was Biot's observation that the molecular constituents of living organisms are generally asymmetric, whereas the chemical products of the laboratory are invariably racemic or symmetric. How did asymmetric molecules come into existence? Biot, recall, had already ascertained that laboratory products never rotate polarized light. So clearly, simply asymmetric molecules did not originate by the chemistry Pasteur had learned as a student. Something was missing from chemistry. And being a physicist by training, Pasteur looked to physics for the missing pieces.

By 1852 Pasteur had hypothesized the existence of a cosmic asymmetric force that he supposed to be active during the elaboration of asymmetric molecules.[8] And as early as August 1851, he wrote that "in my recent research, I have shown that substances having an action on polarized light would have to be assimilated into those assemblages so frequently found in the vegetal and animal realms, which have asymmetry such that one can imagine other plants and animals whose forms are identical but not superimposable."[9] That is to say, plants and animals, like his racemic crystals, might exist in mirror-image forms. Those composed of right-handed compounds would have one asymmetry; those composed of left-handed compounds, another. Pasteur had extended Laurent's correlation of form with molecular constitution to the realm of living things!

▷ **Imp:** Incredible! You mean to tell me that he was thinking of the possible existence of plants and animals composed of, say, left-handed sugars and right-handed amino acids instead of the usual ones?

▷ **Hunter:** That's my interpretation of the passage.

▷ **Ariana:** Sounds like *Through the Looking-Glass!* Would such plants and animals be mirror images of the usual ones?

▷ **Hunter:** Presumably. Pasteur doesn't say explicitly. But he did try to find out. What physical forces, he pondered, are asymmetrical? Clearly, polarized light is asymmetrical since it acts differently on right- and left-handed compounds. Biot had recently shown that a powerful electromagnetic machine called a Ruhmkorff apparatus could intensify the ability of crystals to rotate polarized light. Perhaps electromagnetism was a source of molecular asymmetry, thought Pasteur. Certainly all plants and animals are daily subjected to terrestrial magnetism by the earth's magnetic field. So, perhaps chemical reactions or living reactions subjected to an intensified electromagnetic force would display new forms of asymmetry.

By the winter of 1851, he was privately planning electromagnetic experiments on plants to see if he could influence the formation of their asymmetric products so as to produce mirror-image forms. First the winter and then the cost of the required apparatus forced him to postpone the experiments. In the meantime, Pasteur asked Biot to help him procure more money for his racemic acid research.[10]

On the seventh of December, 1852, Biot wrote to Pasteur to say that the Académie des Sciences had granted him twenty-five hundred francs for his racemic acid research. Pasteur wrote back saying he wished to use the money to undertake his electromagnetic experiments on plants. Biot conferred with the crystallographer Senarmont and other members of the Académie, and then wrote back to Pasteur that the proposed research would represent an unacceptable use of the money. The idea was too crazy.

Pasteur replied at length. On December twenty-eighth, Biot, Senarmont, and another eminent patron of Pasteur's, the chemist J. B. A. Dumas, met at Biot's home to reconsider Pasteur's proposal. They decided that Pasteur should use the money to complete his crystallographic studies of racemic acid and how to separate its two asymmetric components. If any money and time were left, he could then proceed to such unlikely studies as inverting the asymmetry of plant constituents.[11]

Dumas, however, was sufficiently interested in the idea that terrestrial magnetism might account for the asymmetry of organic molecules to request some information from a beet-sugar manufacturer.

▷ **Jenny:** Why beet sugar?

▷ **Hunter:** Beet sugar was known to be asymmetric, and beets could be observed easily in their natural growing conditions. Biot passed along the information from Dumas to Pasteur in an unpublished letter that still survives in the Pasteur archives:

> *According to this manufacturer, beets, considered vertically . . . emit a series of rootlets which are associated vertically on opposing faces. A transverse section . . . represents them thus. For he maintains that*

> *he has noticed that these two series of rootlets appear to be naturally oriented north and south, one to the other. If this thing is true, one could suppose that it is an effect of terrestrial magnetism that one could reproduce artificially using [electromagnetic] currents.*

"If it is true," Biot continued, "it will go very well with your projected studies."[12] But he advised Pasteur not to think about it until he had completed his other research.

▷ **Ariana:** So what Pasteur did, and when he did it, were at least partially controlled by his patrons? A sort of informal peer review.

▷ **Hunter:** That's right. In fact, a couple of years later, Pasteur actually scrapped a book on asymmetry because Biot felt it was too speculative, and Pasteur couldn't afford (literally) to antagonize his patrons.[13] As we'll see, their opinions continued to exert an important influence on Pasteur for another ten years or so.

Pasteur began his proposed asymmetry studies in July 1853. He used both organic and inorganic material, and subjected the material to various "asymmetric influences," as Pasteur sometimes called them—things like polarized light; magnetism, with its north and south poles; electricity, with its positive and negative charges. He began to talk about possibly becoming a new Newton or Galileo if he could isolate his asymmetric force. It's not clear from existing records how much of this research his patrons knew about.[14]

Except for some crystallizations of inorganic formiate of strontium on a bar magnet that consistently yielded asymmetric crystals, none of these experiments was successful. He was unable to duplicate the formiate of strontium results with organic substances such as the tartrates or sugars. By December 1853, he had begun to speak privately of his experiments as "crazy."[15] In January, Pasteur had a breakdown.[16] His hopes were shattered. His patrons had been right: there was nothing in this line of research for him. He was not, apparently, destined to become the Newton or Galileo of asymmetric forces.

Pasteur took an extended vacation, and afterward was transferred from the University of Strasbourg, where he had been working, to the one in Lille. During 1854 he privately, and without telling anyone, resurrected his dream and set up new experiments. To test whether the earth's rotation affected the asymmetry of plant constituents, he contrived a piece of clockwork intended to keep a plant in continual rotation. To see whether the motion of the sun across the sky from east to west created molecules of one asymmetry in preference to another, he used a heliostat and a mirror to reverse the sun's rays as they struck the shoots of newly sprouting plants.[17]

▷ **Richter:** Ridiculous experiments. No one has ever created plants comprised of mirror-image constituents. For that matter, no one has ever produced nonracemic compounds in the laboratory without starting with an asymmetric substance. Basic chemistry.

▷ **Hunter:** You're right. But Pasteur didn't know that. And it's largely as a result of his research that we can say today that his experiments were misconceived. Pasteur at least had the intellectual courage to admit his ignorance and to attempt to replace it with knowledge. Biot, Dumas, Senarmont, de la Provostaye, and Mitscherlich—who were fully cognizant of the same mystery that plagued Pasteur—nonetheless made no attempts to solve it. In fact, Biot actively discouraged Pasteur. Biot firmly believed in what historians call "vitalism"—a philosophical perception that living things differ fundamentally from nonliving things due to intrinsic vital properties. One of these intrinsic vital properties, for Biot, was the asymmetry of organically synthesized molecules. He did not believe that Pasteur could ever create asymmetric molecules, and indeed had much to lose in terms of his own theoretical preconceptions if Pasteur ever succeeded.

▷ **Constance:** So several extrascientific factors, such as belief systems, patronage, and so on, affected Pasteur's work and its development.

▷ **Hunter:** Absolutely.

▷ **Richter:** Yet if Pasteur had listened to Biot, he wouldn't have wasted his time. He never did succeed. We still cannot synthesize asymmetric molecules de novo.

▷ **Hunter:** True. But had Pasteur not tried to do these impossible things, he would never have found some of the other things he did find. And that's one reason Pasteur's discovery of asymmetric fermentation is so interesting to consider in its historical perspective. For Pasteur *did* discover asymmetric fermentation—at least in my reconstruction of the discovery—during experiments designed to create his imaginary mirror-image plants. So trying impossible things can turn out to be useful in unexpected ways.

You see, despite one setback after another, Pasteur continued to cherish hopes of "synthesizing the diverse, natural principles of life," and he was even bold enough to tell his students at Lille in 1856 of his hopes.[18] Also, about this time, Senarmont wrote to him about some mold he saw growing on the icing of an old cake. It struck Senarmont that one could easily mix one's own growing medium so as to be selectively "favorable to their germination. What advantages could not accrue from this, if one could thus follow the developments of a vegetable seed or the egg of an animal into organizations that are infinitely variable and thus individually isolable."[19]

Pasteur was to use Senarmont's concept in various ways. The most immediate was his attempt to grow one group of beets on right-handed tartramide and another group on left-handed tartramide. In light of his Laurentian theory of form reflecting composition, he probably hoped to create beets having different forms—one composed of left-handed constituents, the other of right. In fact, what he obtained were normal beets from the right-handed compound and shriveled beets lacking sugar from the left-handed compound. He recorded these results in an experimental notebook but didn't publish them.[20]

▷ **Ariana:** This, then, is the discovery of asymmetric fermentation, right? Beets could metabolize the right-handed tartrate but not the left, and that's what's meant by asymmetric fermentation . . .

▷ **Hunter:** Well, that question is the crux of the matter. What do you mean by discovery? Pasteur himself seems not to have realized the implications of his results immediately. In light of his reason for performing the experiment—that is, to create beets having racemic components—the experiment was a total failure. So his initial response seems to have been to record the experiment and forget it.

▷ **Imp:** I've done that myself.

▷ **Ariana:** In other words, he had discovered asymmetric fermentation, but his preconceptions prevented him from perceiving the discovery.

▷ **Hunter:** Again I'm not sure I like the word "discovery" here. He had evidence that could be interpreted as demonstrating asymmetric fermentation. But he didn't yet have the *concept* of asymmetric fermentation and so couldn't interpret the data. To my mind, the experiment doesn't represent a discovery in and of itself, since the data, per se, were at the time uninterpreted and might have remained so.

▷ **Jenny:** That's somewhat confusing.

▷ **Hunter:** What I'm saying is that the same data can have very different meanings in different contexts. If you expect to find that beets will absorb both right- and left-handed tartrates equally and will reflect that absorption by showing differences in structure, then data showing no absorption of left-handed tartrates mean that your theory is probably incorrect. The data are contradictory facts. That's one meaning. If someone suggests an alternative hypothesis, such as that absorption of chemicals by living things is a selective process, then your data take on a very different meaning: that beets can select between two mirror-image forms of a compound. Then the data are confirmatory facts. This, as we shall see in a minute, is exactly what happened to Pasteur.

▷ **Constance:** You know, René Taton devotes a whole chapter of his book on *Reason and Chance in Scientific Discovery* to experiments made by famous scientists which were left uninterpreted, or which were interpreted in the light of one theory but had much more interest perceived through a different theoretical framework. For example, Ampère observed and even published an experiment which clearly (in retrospect) demonstrated induction of an electric current by a moving magnetic field. But he himself subsequently wrote to Faraday—who repeated the experiment and interpreted it as showing induction—that it never entered his mind to test for induction of a current because he'd designed the experiment with a very different question in mind.[21]

▷ **Imp:** Another good argument for repeating old observations in the light of new concepts!

▷ **Constance:** Sure. Here's another example: It appears likely that Galileo, while searching fruitlessly for the moons of Venus, saw Venus going through its phases but didn't attach any significance to his observations. Only when his friend Castelli wrote him a letter suggesting that, if the heliocentric theory of the solar system were correct, Venus should display the same phases as the moon, did Galileo understand or impart importance to what he had seen and then publish his observations.[22]

▷ **Jenny:** So what you're saying, in essence, is that you only see what you expect to see. The same point we were arguing about earlier.

▷ **Hunter:** Exactly. In fact, I always pay more attention to the results of control experiments than to positive results, because the controls often contain many more surprises than the data formally being interpreted. Now that I think about it, this must be because people report only positive results that support their preconceptions, and generally fail to analyze their controls at all. They don't expect them to have any significance.

▷ **Ariana:** So you maintain that discovery is not an event but a perceptual process involving the invention of interpretations of data.

▷ **Hunter:** That's right.

▷ **Ariana:** Then, at what point does Pasteur discover the concept of asymmetric fermentation?

▷ **Hunter:** I'm not sure that's even the right question to ask. But let's see what Pasteur did, and then discuss how best to make sense of it. As far as I can tell, an unrelated development brought his attention back to the beet data a short while later. Sometime after August 1957, Pasteur read a book review by the chemist M. E. Chevreul in the *Journal des savants.*

▷ **Ariana:** Chevreul being one of Berthollet's successors at Gobelins, right?

▷ **Hunter:** That's right. Now, how did you know that?

▷ **Ariana:** A visit to an art museum!

▷ **Hunter:** Of course . . . In any event, Pasteur made the following note concerning the review in a small booklet of ideas:

> *In his research on vegetation, Th. de Saussure has stated that when equal weights of different salts are dissolved in the same water, they are absorbed in different proportions.*
>
> *It seems to me from this to be very useful to find out if, by means of this absorption which is most probably something of a vital nature, one can decompose racemic sodium ammonium tartrate.*
>
> *If that works one could try the method on certain other compounds such as citric acid which do not have the same manner of construction as racemic acid.*
>
> N.B. *Does endosmosis [that is, selective absorption] by a membrane decompose the sodium-ammonium racemate?*[23]

Beside that entry he wrote "very useful" and underlined it. Shortly thereafter he appears to have gone back and reappraised the results of his beet experiments and made some additional notations. It was probably then that he realized the results' implications for fermentation. On December 21, 1857, he made the following rather tentative and obscure report to the Académie, apparently for the sole purpose of establishing priority to the discovery without revealing more than was necessary of the details: "I have discovered a mode of fermenting tartaric acid which is especially applicable to the ordinary right-handed acid and which is very bad or completely inapplicable to the left-handed acid."[24] He went on to conjecture that such fermentation might provide a way to separate racemic acid so as to produce purified left-handed acid.

▷ **Ariana:** So at *this* point he has discovered asymmetric fermentation! He uses the words "discover" and "fermenting."

▷ **Richter:** Yes, but what of the graduate student I mentioned last week whose discovery was no discovery?

▷ **Hunter:** Definitely a problem. In fact, Pasteur hadn't yet demonstrated that his observations were repeatable or that they were correct. He had not yet isolated pure left-handed tartrate by his method; he only suggested that he might be able to do so. Beets were not suitable for such experiments, however, since one could not isolate the left-handed acid very easily from the soil in which they grew. An organism that could be grown in solution, preferably a microorganism, was called for. Here Senarmont's suggestion of growing molds may have played a catalytic role in Pasteur's thought. Pasteur had seen such microorganisms growing in tartrate solutions during an 1852 tour of Europe in search of the origins of racemic acid—which, by the way, was of natural organic origin, as befitted a double asymmetric molecule.[25] Pasteur evidently recalled that observation shortly after making the note about Chevreul's book review. He wrote a few pages further on:

> *Study the microscopic vegetal productions [i.e., molds] from right and left tartrates. Are they identical? Nothing would be more interesting in this study than if they conveyed differences of form, structure, composition . . .*
>
> *Would it be possible to separate these productions, should they differ, so as to nourish even a small animal so that it would probably undergo a very different development according to whether it was nourished on one or the other type?*[26]

▷ **Richter:** Wait. That is no asymmetric fermentation experiment, but another misconceived mirror-image organism experiment!

▷ **Hunter:** In fact, it's both. But that's what I've been trying to warn you about. You, like the historians who've written about Pasteur, insist on believing that Pasteur was looking for asymmetric fermentation because that's what he found. My point is that Pasteur discovered asymmetric fermentation while looking for something else. Even having observed the selective fermentation of the right-handed tartramide by beets, and even having realized that he might therefore be able to purify the left-handed form, he still continued privately to hope that feeding microorganisms on the left-handed tartrates would yield microorganisms of novel structure.

▷ **Ariana:** You used the word "persistent" to describe Pasteur before, but I'm only beginning to appreciate what you meant!

▷ **Hunter:** And Pasteur's persistence paid off. But, again, in a way he didn't expect. For these new attempts to create new forms of life, Pasteur chose the yeast he'd been using to study fermentation. On March 28, 1858, he reported to the Académie that the yeast used right-hand tartaric acid in preference to the left, and that it was thus able to separate racemic acid into its constituent right- and left-handed forms. He didn't report the reason he'd undertaken the study or the fact that the experiments had been a failure from his private point of view. Laurent's theory of form reflecting constitution had not yielded new forms of life.

He did, however, conclude with an important new observation. He had demonstrated that asymmetric forces played an important role in the organic realm, and that asymmetry operated in the organic realm in the same way that it operated in the inorganic. Living organisms can chemically separate racemates just as Pasteur had found he could do manually in 1848 and again in 1854 using other asymmetric molecules such as quinine as chemical separators.[27]

▷ **Jenny:** And it's this report of Pasteur's that misled subsequent historians, who didn't know about his private ideas or his notebooks?[28] By linking the asymmetric fermentation results to his previous chemical separation results, he forged a logical but historically inaccurate link.

▷ **Hunter:** That's my contention.

▷ **Ariana:** The important question is how his own perceptions were changed. In reinterpreting his fermentation experiments for public consumption, did Pasteur also reinterpret his own ideas regarding asymmetry?

▷ **Hunter:** Yes and no. Yes, to the extent that he now knew more about the chemistry of living organisms. He now understood that their asymmetric constituents allowed them to select one asymmetric form of a chemical from another. But he still didn't know how an organism's original asymmetry was created. So no, he did not give up his attempts to create new forms of life from constituents of different asymmetries. His notebooks reveal that he continued to plan, and apparently to carry out, asymmetric force experiments through 1861.[29] In 1860, for example, he announced that penicillium too could separate racemic acid to yield a pure right- or left-handed form. Once again, his report failed to explain the rationale for undertaking the experiments.[30]

Only once, in 1883, at age sixty-one, did Pasteur actually admit publicly to carrying out some of these asymmetric force experiments.[31] And, in consequence, only someone who's been through Pasteur's private correspondence and his laboratory notebooks, as I have, is likely to be aware of the unproven and unpublished speculations that guided his research. Only with reference to such manuscript evidence does it become clear that Pasteur discovered asymmetric fermentation in the process of uniformly unsuccessful attempts to create new, asymmetrical forms of life. That's important in terms of Imp's "science of science": a research program doesn't have to succeed to be extremely fruitful.

▷ **Imp:** If you are willing to make detours.

▷ **Jenny:** Okay. But go back a minute. If you're right in linking Pasteur's discovery of asymmetric fermentation to his cosmic asymmetric force hypothesis, then the reason he didn't explain the origin of his experiments was to avoid divulging his private speculations. Is that right?

▷ **Hunter:** That's right. Remember, he'd already antagonized his patron Biot with the subject several times before. And he never did achieve any clear-cut evidence for a cosmic asymmetric force. So why speculate in public?

▷ **Jenny:** But you don't think that Pasteur discovered asymmetric fermentation by chance or accident?

▷ **Hunter:** Well, he certainly did *not* pick up a flask full of tartrates one day, notice that there was mold growing inside and then, on some inexplicable whim, decide to measure the optical activity of the remaining solution.

In fact, he couldn't reasonably have done so under any circumstances. He would have had to set up control solutions so that he could determine how optically active the tartrate was alone as well as in the presence of the mold; he would have had to demonstrate that the mold itself did not interfere with the polarimetry measurements; and he would have had to investigate whether the mold itself was optically active in such a way as to cancel out the optical activity of the tartrates. A simple measurement of an accidentally contaminated solution would have meant nothing.

No. His notebooks show that he very carefully planned the experiments that he performed, and was probably quite surprised by their outcome. And it's that fact of surprise that I think needs to be stressed as we discuss what discoveries are and whether they occur by chance or planning, for the problem of surprising results gets us back to Richter's question of whether Pasteur was acting logically if he discovered something other than the thing he was looking for.

▷ **Constance:** Well, didn't Bernal conclude that Pasteur's first discovery—the one involving racemic acid—was logical? Yet it, too, was a surprise.

▷ **Hunter:** Yes. And that's why I wanted to compare the two discoveries. It seems to me to be clear that in both cases Pasteur used logic in setting up his research program. But the key element in his discoveries was that they couldn't be (or at least weren't) predicted beforehand because his inductions and deductions were all based on inadequate or incorrect data. Remember, he was a physicist, not a biologist. He clearly knew very little about the basic facts of cell biology or biochemistry—but, then, neither did anyone else at the time. Any speculation is an improvement over ignorance if it results in controlled experimentation.

▷ **Ariana:** Okay. Then a possible working hypothesis might be that there is a logic of research, but that discoveries are surprises that are unpredicted by that logic due to the incompleteness of the information used to invent the hypothesis.

▷ **Imp:** Or the failure to elaborate and test all the possible hypotheses.

▷ **Ariana:** And, in consequence, discovery is the recognition that things don't fit some pattern you have in your head, so that you must invent a new pattern.

▷ **Imp:** I like it!

▷ **Hunter:** So do I. Think about some alternatives. Suppose Mitscherlich's and de la Provostaye's observations were correct. Then Pasteur would have verified them, and doubt would have been cast on Laurent's theory. Or, Laurent's theory could have been correct and racemic acid might simply have been symmetrical, as Pasteur predicted. In neither case would there have been a surprise. Both outcomes were predictable beforehand. And in verifying an extant theory there would have been no discovery, merely confirmation. There would then be no question of their logic. It would be obvious.

▷ **Imp:** Yes, yes! But there's another element to Pasteur's research that you seem to have overlooked. Isn't it true that in both his 1848 work on racemic acid and his 1857 work on asymmetric fermentation, the research was set up so that no matter what the outcome of the experiments, some new fact would have been forthcoming? I mean, whether Mitscherlich or de la Provostaye was correct, or Laurent, Pasteur's 1848 research was a valuable test of the two apparently contradictory sets of theories and data. And the 1857 research is an even more striking case. Whether Pasteur's beets and yeast and penicillin metabolized one asymmetric form in preference to the other, both equally well, or even grew into new asymmetric forms—any result would have been a novel addition to the corpus of scientific knowledge. Prior to Pasteur's work, nothing at all was known about the metabolism of racemic compounds, right? So, it seems to me that one of the key elements that made Pasteur a successful scientist was his ability to recognize new problems whose elucidation was guaranteed to yield novel results regardless of the outcome of the experiments. Right or wrong, Pasteur's hypotheses *had* to yield discoveries!

▷ **Jenny:** Well, one of Napoleon's maxims was "Always have two plans; leave something to chance." Or should I say the unexpected? Strategically sound science, too, it seems.

▷ **Ariana:** Indeed. A recent study showed that the most successful medical diagnosticians are those who invent a variety of hypotheses early on.[32] As the saying goes, "keep your options open."

▷ **Constance:** There's much written on the utility of multiple hypotheses.[33]

▷ **Richter:** Wonderful. But I still fail to see the logic in predicting things that turn out not to be true, or in designing experiments that could yield any result whatever.

▷ **Imp:** Stop being so perverse, man! You purposefully misunderstand. The logic is not in the conclusion but in the *way* the research proceeds, the protocol you write before beginning an experiment. The experiments are designed specifically to yield not just any result but one of, say, three possible results, any one of which will be novel. Besides, before you call Pasteur's ability illogical, I suggest that you try to design this sort of experiment yourself. I haven't seen a comparable experimental protocol in years. Some research programs test a hypothesis: yes or no. Only the yes results are generally considered to be significant. And most experiments are designed to describe, characterize, or measure—like Mitscherlich's and de la Provostaye's reports on the tartrates—not to yield surprises. Surely you know better than anyone here how few experiments are designed, let alone interpreted, in light of any existing theory. Why, it's one of your most cherished complaints!

▷ **Richter:** Which only goes to show that most scientists never use logic at all. But I still do not see that basing a research program on a daydream like Pasteur's nonexistent asymmetric force is logical. It is not logical to act on hunches or intuition or daydreams! Such things are fictions and must be treated as such.

▷ **Imp:** Let me ask you this, Richter. Do you believe that a geometrical point is dimensionless, that an infinity of these dimensionless points forms a line, and that two straight, parallel lines never meet?

▷ **Richter:** Of course. But what does this have to do with anything?

▷ **Imp:** A lot. You see, I've never believed in those things. I've used the postulates to solve problems, of course, but I don't think they're true. Just one of an infinite number of useful fictions, which includes the non-Euclidean geometries.

▷ **Richter:** Yes, well, we all know you are a bit different in the head, but so what? What are you getting at?

▷ **Imp:** I'm getting at believing, logic, and rationality. Surely you, of all people, know that modern mathematicians are generally agreed that logic and truth are totally independent. Pure mathematicians don't worry about whether the postulates they assume or the conclusions they deduce from them are true, but whether the alleged conclusions are the necessary logical consequences of the initial assumptions.[34] Surely by these criteria Pasteur was being logical. That his cosmic asymmetric force doesn't seem to exist, and that he couldn't prove his predictions *a priori*, seem to me to be irrelevant. The fact is that, given his assumptions, his subsequent research into the question *was* logical.

Case in point, going back to mathematics: It's been proven that it's impossible to prove that two parallel lines never meet. And we postulate non-Euclidean geometries in which parallel lines do meet. So if you believe that two parallel lines never meet, your belief is just as illogical—irrational—as Pasteur's belief in his cosmic asymmetric force. You believe in something that can't be proven. It must be assumed. The question of logic, it seems to me, is not what you believe, but how you act upon that belief. Anyone may dream of imagined worlds, but only people like Pasteur, whose dreams produce surprises, change the course of science.

▷ **Constance:** Sure. That's just what J. J. Thomson said: "Though a theory might be Bohemian it might be the parent of very respectable facts."[35] And T. H. Huxley once wrote: "A great lawyer-statesman and philosopher of a former age—I mean Francis Bacon—said that truth came out of error much more rapidly than it came out of confusion. There is a wonderful truth in that saying. Next to being right in this world, the best of all things is to be clearly and definitely wrong, because you will come out somewhere."[36]

▷ **Ariana:** Claude Bernard said much the same.

▷ **Hunter:** And so we come back to the difficulties caused by distinguishing the context of discovery from the context of justification. Listen to Sir Peter Medawar, himself a professed follower of Popper: "What shows a theory to be inadequate or mistaken is not, as a rule, the discovery of a mistake in the information that led us to propound it; more often it is the contradictory evidence of a new observation which we are led to make *because* we held that theory. Error or insufficiency is shown up by a critical process applied in retrospect."[37] The tests of one hypothesis *are* the context of discovery of another, and vice versa. Pasteur's discovery of asymmetric fermentation was, in this sense, a discovery only because, in retrospect, it required Pasteur (and his colleagues) to rethink and re-search relevant areas of science—to recognize, in light of new evidence found because he held a questionable theory, the inadequacy of both his knowledge and his theory. The discovery occurs precisely because the theory is wrong or inadequate, although the incorrectness and inadequacy only become apparent later—sometimes much later.

▷ **Ariana:** So there's a real danger in relying too much on logic. I'm thinking here of that passage from Hardy's *Far from the Madding Crowd* in which Oak's sheep dog, who's been trained to run after sheep, figures that the more he runs the sheep, the better. In the end he runs the sheep off a cliff, and then returns to his master asking for a reward for his extraordinary efforts. You have a copy, Jen? Hold on a sec while I find the passage. Here we go. The dog "had done his work so thoroughly that he was considered too good a workman to live, and was, in fact, taken and tragically shot at twelve o'clock that same day—another instance of the untoward fate which so often attends dogs and other philosophers who follow out a train of reasoning to its logical conclusion, and attempt perfectly consistent conduct in a world made up so largely of compromise."[38] Scientific research is a constant compromise between what we'd like to find and what we do find.

▷ **Constance:** Hmmm. Speaking of fatal logic reminds me of an absurd strategy used to keep diabetics alive before the isolation of insulin. Diabetics usually died of complications resulting from high blood sugar. A Dr. Allen therefore reasoned that diabetics could be kept alive if their blood sugar was kept normal. It turned out that to keep the blood sugar normal, the diabetics had to be kept on diets with so few calories that they eventually starved to death!

▷ **Richter:** But, my dear, at the risk of contradicting my own position, I must say that the Allen diet was not as irrational as you make it out. Diabetics could be kept alive in a semi-starved condition for several years longer than if they ate as they pleased. And while insulin was being transformed from a laboratory curiosity into a life-saving drug, the Allen diet kept many diabetics alive long enough to profit from it.[39] I will admit, however, that to treat diabetes successfully, the developers of insulin therapy did have to reject the Allen diet and initiate a very different path of research.

▷ **Imp:** So you're coming around, eh? Seeing the limitation of your own logic? It's like Lewis Thomas says: "Every now and then, something turns up in the course of exploration that's worth—as the guidebooks say about restaurants—a detour. I think that's when really important observations are made."[40]

But you see, don't you, Richter, that the concept of a detour presupposes a planned path. Had Pasteur, for example, stuck by his logic and avoided all detours, he would never have made his discoveries. He would have been run off a cliff by the logic of his deductions: Racemic acid doesn't deviate polarized light so it must be symmetrical—or something like that. Had he steadfastly maintained that course, he would never have realized that by turning just a bit left or a bit right a whole new world of asymmetry awaited his gaze. Fortunately, he abandoned his *a priori* logic in the light of relevant, contradictory observations. In my experience, few scientists are so wise.

▷ **Constance:** Well, failure to deviate certainly characterized a chemist named Theile and the immunologists Driesch and Wright. When faced with evidence contradicting their cherished theories, they inevitably suppressed it.[41]

▷ **Richter:** So—my original position is weak. Let us detour, then, dammit. Go back and choose a different turning. For instance, Imp said a few minutes ago that the logic of research follows given assumptions. Where do the assumptions come from? That is one problem we bypassed. Second, you all prattle on about planned research and logical thinking, but based on what? A single—no, let's be fair, a pair—of putative discoveries by Pasteur. Hunter conveniently provides us with *a posteriori* logic for these two discoveries. But what really went through Pasteur's head? No one knows. So, what have we learned? That Hunter is very clever. Nothing more.

▷ **Hunter:** Now, hold on a minute. Let's start taking these objections one at a time. Start with my reconstruction of Pasteur's actions. You seem to think I'm being somehow nonscientific in my approach by thinking myself into his head. Why is that any less scientific than learning to think about why an apple falls and not the moon, or how it feels to move at the speed of light or to act like a carbon atom? The same principles apply. Constance, do you have that passage from Claude Bernard you showed me, about the similarity between understanding other people's actions and understanding nature?

▷ **Constance:** Sure. Hold on. Here it is.

▷ **Hunter:** Thanks. Look, Bernard addresses the problem of understanding other people's actions explicitly:

> *When we reason about our own actions, we again have a sure guide, because we are conscious of what we are thinking and of what we are feeling. But if we wish to judge the actions of another man and to know the motives which make him act, then it is quite different. Doubtless we see before our eyes the man's movements and the acts which, we are sure, are expressions of his feeling and his will. What is more, we also admit that there is a necessary relation between actions and their cause; but what is this cause? We do not feel it ourselves, we are not aware of it as in our own case; we are therefore forced to interpret and imagine it from the movements we see and the words that we hear. So we must verify the man's acts, one by another; we consider how he behaves in such and such circumstances, and in short, we turn to the experimental method.*[42]

Bernard goes on to say that precisely the same process of imagining a series of causes whose consequences can be observed serves as the basis of the experimental method of physiology. So Bernard, at least, sees no difference in the methods one uses to investigate the actions of people and the actions of people's physiology.

▷ **Ariana:** Perhaps that's because Bernard was a playwright before he became a physiologist. He had learned not only how to playact, in the Lederberg-Polanyi sense, but also how to investigate psyche as well as matter.

▷ **Hunter:** Moreover, Constance once told me that Thomas Kuhn insisted that his students reconstruct a scientist's work by going through the records in the order they were produced, imagining at each step the process of thought and action that yields those records, experimenting with ways of understanding until they could actually predict what the next experiment would be about. Then you know that you understand the scientist. This is the same way we know when we understand nature.

▷ **Constance:** That's right. But I recall one more step. Kuhn insisted that any inconsistency or gap in our reconstruction of a scientist's thought indicated a failure on our part to correctly understand the scientist. Kuhn maintains that just as nature is consistent, so are people, and that the historian's job is to understand the person so well that no inconsistencies occur in the explanation of that scientist's actions.

▷ **Richter:** A debatable point. Most people are inconsistent.

▷ **Ariana:** Not on important things.

▷ **Hunter:** In any event, I find it very striking that Kuhn's and Bernard's criteria for investigating human actions are the same as those used to investigate scientific subjects. This suggests the possibility of actually constructing Imp's "science of science." We can apply scientific criteria to reconstructing how scientists discover. Given Dubos's and other historians' version of Pasteur's discovery of asymmetric fermentation by the chance observation of an accidentally contaminated flask of tartrates, and my rational reconstruction of his research program, which best accounts for the extant data in the simplest, most coherent manner? Which accounts for the greatest number of facts? Which is most consistent? Which leads to testable conclusions? Which is most useful as a guide to understanding science and how to do it? I think the rational explanation wins out.

▷ **Imp:** Hunter has you there, Richter. But even you, Hunter, haven't pushed the implications to their limits yet. If historical explanation is amenable to scientific evaluation, then philosophical explanation should be, too. Think about that for a minute. That means we could treat the pronouncements by Popper, Feyerabend, and colleagues concerning the irrationality and illogicality of discovering as a theory. As such, it should have the very characteristics that Popper, Feyerabend, et al. attribute to scientific theories, right?

Now, what do Popper and company say about theories? Baldly stated—perhaps too baldy—they say a theory should make predictions that are falsifiable, disprovable. So, is the theory that the context of discovering is irrational and illogical testable in this way? No! The king of falsification has himself proposed a nonfalsifiable theory! It makes no predictions and therefore can't be disproved. On the contrary, it acts only to inhibit research on relevant questions. (My objection to many of Crick's so-called theories, too, by the way—such as his ideas about the extraterrestrial origins of life and his new thing about the functions of dreams. All untestable.)

The point is that claiming that discovering is unknowable gets us nowhere. To "no" is not to know. So the falsificationist philosophers, and you, Richter, would be much more in keeping with your professed beliefs if you'd hypothesize that there *is* a logic of discovery and then try to disprove it.

▷ **Hunter:** Well. I agree with your intentions, Imp, but not with your conclusion. As I said before, I'd prefer to hypothesize that there's a logic of *research*—not of discovery—and that discoveries are, intrinsically, unpredicted surprises that occur as a result of the application of the logic of research to an inadequate or flawed set of principles and observations. Thus, all discoveries are unanticipated detours from *a priori* logic. What I want to know is to what extent a logic of research can increase the probability of making a discovery, and to what extent failure to use an appropriate logic of research can prevent discovery. In other words, is there a way of planning for the unexpected?

▷ **Imp:** Ah! Just like today! Had we accepted the pronouncements of Popper and company, we would never have searched for a logic of discovery and so never uncovered a logic of research.

▷ **Richter:** Presuming, of course, that is what you've uncovered! I still insist that we have learned nothing until your working hypothesis is checked. Two cases do not a sufficient data base make. There are dozens of other examples of putative chance discoveries besides Pasteur's. Surely, Hunter, to make your case a reasonable one you would have to demonstrate some logic or planning in every one of these chance discoveries—Minkowski, Richet, Fleming, Roentgen, Becquerel, dozens of others. Until then your hypothesis is hardly credible, let alone true.

▷ **Ariana:** Yet I say you must act as if an idea were true to test it.

▷ **Richter:** But search for cases in which it may not be true to test it properly.

▷ **Imp:** Isn't life full of pleasant contradictions!

▷ **Hunter:** Certainly. But Richter is right once again to insist that we not extrapolate from very limited evidence. We don't want to fall prey to thinking that all apples are red just because the first couple we pulled from the barrel are Romes, Red Delicious, or Macintosh. But it's getting too late today to do a proper job of questioning the wider literature on chance discoveries. I therefore suggest that we adjourn for today and reconvene next week to discuss precisely what we mean by a logic of research and the surprise of discovery, and whether these are adequate concepts to describe the phenomena we want to discuss.

▷ **Imp:** I propose we go further and try to describe in detail what this logic of research looks like and how it operates. Let's see how far we can push our ideas. Find the bounds.

So, till next week, remember the prepared mind, and be inventive!

▼ JENNY'S NOTEBOOK: The Eye of the Mind

> *How much sooner would the eye—accustomed to observe and estimate closely the differences of color, aspect, weight, and symmetry—learn to gauge their aberrations as the signs which make up the* facies *of the disease; how much better [would] the hand, trained to portray them accurately, be able to direct with precision and safety the course of the knife!*
>
> *—Sir Francis Seymour Haden, surgeon, anatomist, etcher, and painter (ca. 1890)*

Ariana dropped by tonight. "Do you have a few minutes?" she asked. "I've got a little something I want to try out on you two before I risk presenting it to the likes of Richter." We said we'd be delighted.

"As you know, I've got this notion that artistic creativity isn't so different from creativity in science. So I've been trying to see how far I can push the idea. I've just found the most intriguing article, and I want to see what you'll make of it."

With that, Ariana produced a biographical sketch of the nineteenth-century astronomer William Herschel. Herschel, she informed us, was a musician, conductor, and composer before he turned to astronomy and telescope building.

"You know how I keep harping—excuse the pun—on how the eye of the mind controls what the eyes in the forehead perceive. Case in point: these versions of Saturn. Well, Herschel takes the idea in a different direction. In response to criticisms that he perceived things other astronomers didn't, he remarked that 'seeing is in some respect an art, which must be learnt. To make a person see with such a power is nearly the same as if I were asked to make him play one of Handel's fugues. Many a night have I been practising to see, and it would be strange if one did not acquire a certain dexterity by constant practise.'[1]

"You see, this gets us back to Pasteur. Richter doesn't want to admit the role of personal perception in discovery; and Hunter, much as I usually agree with him, wants to limit what you perceive to theoretical conditioning. But I want to extend it to Herschel's position: that some people, especially artistically trained ones, such as Pasteur, actually perceive more because they have extensive practice at it. Sound crazy?"

"Let's hope so." Imp grinned. "But it makes sense. I've come across a couple of relevant things, too. First of all, there's something called 'personal equation' that used to be a big topic of discussion seventy-five years ago. Suppose you give ten different scientists the same set of peas from, say, a Mendelian cross of green-wrinkled seedcoat with smooth-yellow. Most of us would predict, I imagine, that, save for a pea or two, every scientist would find exactly the same number of green, yellow, smooth, and wrinkled peas. Not at all! Every investigator's personal

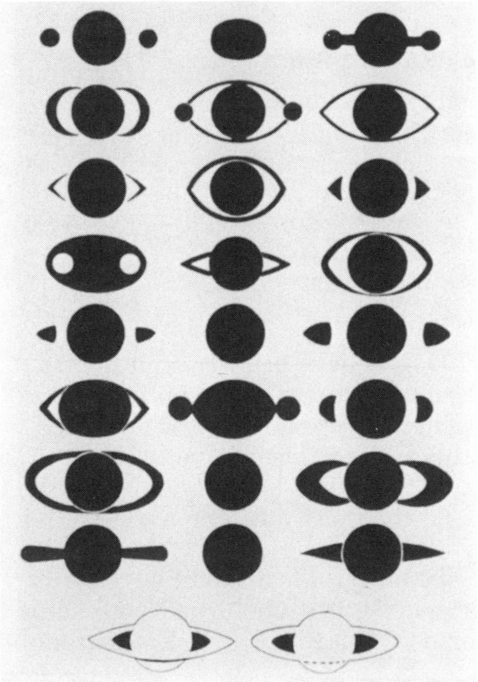

Various perceptions of Saturn, from Galileo to Huygens, 1610–1656. (Taton, 1959)

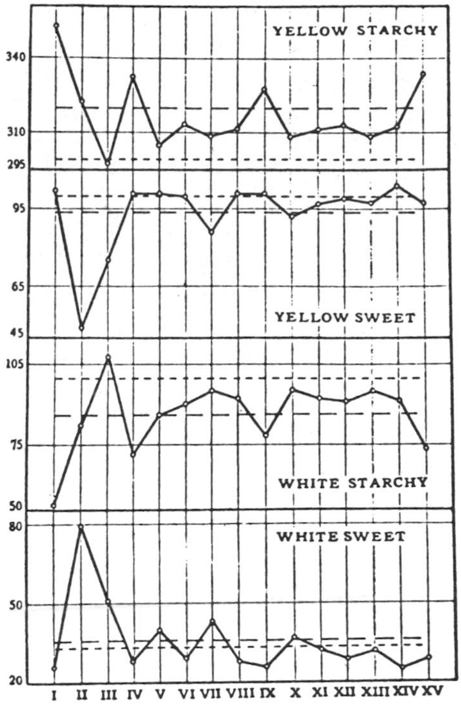

An example of a "personal equation." The diagram shows counts of kernel types in one ear of corn by fifteen observers. The dotted line is the Mendelian expectation and the dashed line the average of the counts of the observers. (Pearl, 1923)

count will be unique. Look, here're the results of a study by a statistician named Raymond Pearl, confirmed by several other leading statisticians of his day. And some guy recently showed that Mendel's results, which are statistically too good to be true, probably resulted from much the same phenomenon."[2]

"Wait a minute," I said, shaking my head. "You mean to tell me that no two scientists, even given the same theoretical background, will make identical observations?" Imp nodded. "Then how are observations verified?"

"By being in the same ballpark. By being compatible. That's why theories are so important. If one theory says the answer is over there and another says it's here, then the question is in which area do the data cluster. Because you'll never get observations made by different scientists at different times to agree exactly. Even using fancy measuring equipment, you get a distribution of results."[3]

"But getting back to that passage from Herschel," continued Imp, "it seems to me we have to ask how much of one's personal equation is due to simple uncertainties of measurement and categorization, and how much is due to training or lack thereof."

"Obviously Herschel would say that much of it is due to practice," prompted Ariana.

"And I'd agree," responded Imp. "When I was in high school we read an essay by a nineteenth-century zoologist named Nathaniel Shaler. In English class, mind you, not biology. Such is our educational system. Anyway, Shaler was trained at Harvard by Louis Agassiz, and in his autobiography he describes 'How Agassiz Taught Me to See.' Essentially, Agassiz sat him down in front of a series of specimens and made him report what he observed. Almost without fail, Agassiz sent Shaler back to look some more. He quickly learned the lesson that a single hour, a single day, even a week was insufficient to see all there was to see in any given specimen. The obvious characteristics are rarely the important ones. Not until Shaler had observed an anomaly in one of Agassiz's own classifications was he rewarded with the comment, 'My boy, there are now two of us that know that.'"[4]

Ariana nodded. "I read a similar story about one of T. H. Huxley's students, Patrick Geddes, who caught Huxley out in an error. Huxley congratulated Geddes and had his observations published as a correction to Huxley's own work.[5] Geddes went on to become quite famous himself, so the training apparently stood him in good stead."

"But the point," rejoined Imp, "is the number of students each had who did *not* observe what Shaler and Geddes observed. That's the phenomenon we need to explain."

I thought a moment. "Well, maybe it's the same sort of thing as learning to play an instrument. Most people who begin to learn the piano quit relatively quickly. Some go on to play as the mood strikes them. But the great ones, the rarities, are those who not only have talent but are willing to put in four or six hours a day practicing, day in and day out, year after year."

"But in this case the instrument is your own senses," said Ariana, "and the question is, how many people have the stick-to-it-iveness that Shaler or Geddes had, to learn how to really play their instruments! Which leads right into the second passage from Herschel that I wanted to bounce off you. Here's another analogy, but this time between playing a musical instrument and playing a scientific instrument. Herschel writes of his telescopes: 'These instruments have played me so many tricks that I have at last found them out in many of their humours and have made them confess to me what they would have concealed, if I had not with such perserverence and patience courted them. I have tortured them with power, flattered them with attendance to find out the critical moments when they would act.' All this, he says, '*to screw an instrument up to its utmost pitch* . . . pardon the *musical phrase*.'[6] So not only does he maintain that you have to train your eye to observe just as you must train your ear to be able to play music; you must also learn the techniques of making the instrument perform to its limits (and yours). You have to learn to *play* it!"

Ariana went on. "Peter Kapitza, the Soviet physicist, made a similar analogy. Scientific research 'is like the Stradivarius violin,' he said; 'this is the best violin in the world, but to play one you have to be a musician and know music. Otherwise, it will sound no better than any other violin.'[7] Give a Stradivarius to an average violinist, and he'll still sound like an average violinist. But give an average violin to a master violinist and you'll still know he's a master. It's not the instrument that makes the music; it's the violinist. And what I hear Herschel saying is that the same is true in science. It's not the quality of the machine that makes the breakthrough, but the skill of the scientist using it. Great experimentalists acquire a rapport with their instruments indistinguishable from that achieved by great musicians. For example, Fritz Houtermans, a Dutch physicist, used to bring his instruments roses on days he needed them to operate just so![8] Albert Michelson explains why: 'One comes to regard the machine as having a personality—I had almost said a feminine personality—requiring humouring, coaxing, cajoling, even threatening!'"

"Shall we forgive him his comment?" I interpolated.

"'But finally one realizes that the personality is that of an alert and skillful player in an intricate but fascinating game who will take immediate advantage of the mistakes of his opponent, who "springs" the most disconcerting surprises, who never leaves any result to chance, but who nevertheless plays fair, in strict accordance with the rules of the game. These rules he knows, and makes no allowance if you do not. When *you* learn them, and play accordingly, the game progresses as it should.'[9] Indeed, Helmholtz said he spent more time and energy getting his instruments to work than he did inventing them or the experiment he intended them for."[10]

Imp had been nodding his agreement. "You know, all this reminds me of when I was learning to operate an NMR machine. Like the one you use at the hospital to do brain imaging, but I was using it to generate 'fingerprints' of chemical compounds. You can even show what atom is bound to what. Very useful for my research on peptide-peptide and peptide-neurotransmitter interactions.

"In any case, what I noticed was that some operators were very good and others were lousy. For example, you have to 'tune' the magnetic field—or 'shim' it, as we call it in analogy to woodworking—to make it uniform around the sample. There are computer programs to do this, but no one who's any good uses them, as far as I can tell. Good operators get a *feel* for how well shimmed the machine is by looking at the shape of radio pulse that comes back from the sample and from the quality of the spectra. It's something you learn by experience. Some people don't seem to learn. They can't generate clean spectra.

"Then there's another step. With some NMR techniques, you get a spectrum that has a really wavy baseline that's difficult to analyze, so you have to correct the baseline to make it flat. You do this by adding a continuous mathematical function to the function describing the spectrum so that the two add up—the computer does this part—to a flat baseline. We used to have to do this correction using a series of sine functions controlled by knobs on the display panel. You added the functions by turning the knobs right or left, more or less. In essence, you had to have a visual model in your head for how sine functions add. It was clear that some folks can do this and others can't. I watched one guy struggle for half an hour with a correction that took me two minutes. I found out later that he usually gives up."

"In other words, he can't play the machine," I said.

"Not well," continued Imp. "Fortunately for him, there're now good computer programs to do baseline corrections, so there's little skill involved at that stage."

"That's always been the purpose of technology, hasn't it?" I asked. "To minimize the amount of personal skill, power, or resources the individual must muster to be able to do something."

Ariana looked pleased. "Yet you can never eliminate skill completely. So, assuming that one must learn how to see, and how to 'play' scientific instruments, is the opposite also true—that those never trained to see and those lacking training, skill, patience, and persistence can't play the instrument and can't see what they get if they do?"

"But isn't there something else to consider, too?" I said. "Might there be an equivalent to physical sight for theoreticians? Insight? I mean, are there some people who just see equations as a string of numbers, and others who see equations as pictures—or whatever—of the universe?"

Ariana agreed. "I think there probably is the ability to transform an equation into a picture, or a picture into words, or words into a model. You're bound to get better with practice."

"Yes, but think of the *implications* of what you're saying!" Imp got up and began striding around the room, talking loudly. "Think about it! How do we teach science? Or medicine for that matter? We give students some lectures, in the course of which they don't learn to use their senses at all. Then, if they're lucky, we show them an NMR machine or an HPLC, we plunk them down in front of a microscope or petri dish or jar of fruitflies and within some very short time—an hour or three—we ask them to do some simple task or make some simple observation. Now what can one learn about an instrument, what can one observe in an afternoon? Or even in a week? Nothing at all worth knowing! Yet I've made just such requests of hundreds of students—"

"We all have," said Ariana. "But remember what Jenny said about the role of desire on the student's part. You can't become a great pianist without trying. A lot of my students don't try. They want me to point it all out for them. 'Lay it on a platter for me, Prof. Show me all I need to know. Isn't that why I pay fifteen thousand a year for the privilege of attending medical school?' And, of course, I can't show them anything because they don't have the skills to perceive what I perceive.

"And then there are other doctors who are threatened by the idea. Nobody on the faculty will give me time to train students how to see.

'Too many facts to learn to worry about teaching basic skills, like how to see, for god's sake. What d'you think this is, an art school?' Nobody listens when I point out that virtually every great surgeon, anatomist, and diagnostician has been an artist. Most illustrated their own publications. I mean, it's a list of who's who in medicine: Gowers, Bright, Charcot, Richer, Cushing, His, Hodgkin, Bell, Lister, Banting . . . I could go on and on.[11] The list of musicians and poets and novelists is equally long. Things are certainly a lot different in scientific education today."

"It goes back to Hunter's point," I said, "that apprenticeship works great when you're apprenticed to a master, and it works lousy when the player you're copying can hardly play the instrument himself. In the latter case, you've got to train yourself."

We had a good time batting that idea around for the rest of the evening. We all agreed that we need to draw up a new list of "tools of thought," as Ariana calls them, to redefine what constitutes scientific literacy. Reading, writing, and 'rithmetic—even throw in computer competency—just won't suffice. Ariana's also going to look for more stuff like the Herschel passages among the writings of other scientists. She's really excited.

The only problem with these conversations is they leave me feeling less and less competent myself . . .

► IMP'S JOURNAL: Surprises

Wouldn't you know, first thing I find is something I wasn't looking for. Talked to several people about testing amino acid pairing. Mainly got bad advice. "Find a computer modeling system that can compute minimum energy conformations." Right. I've already modeled, thanks. I need *evidence*, not more models. "Try demonstrating protein synthesis off of a protein template using a cell-free enzyme preparation from lymphocytes." Too complicated! I'd like to know that chemical model's right before jumping to physiological complexity. Finally talked to Gustav. He said, "Look, simple physicochemical techniques ought to do it. Osmotic pressure. Electrophoresis. Start simple. Look for binding. If there's chemical complementarity as you suggest, the molecules will bind specifically to one another. That's how we tested complementarity of base pairing for DNA structure back in the fifties." Nice to hear your own advice in someone else's mouth!

Got to work. Tried some things. Found various peptides do bind specifically. A discovery in and of itself. Most of the interactions were ones I'd predicted, but others too. Nobody'd believe it though. Kept saying all we had were artifacts. God! The frustration! Eventually had to generate four sets of congruent data—osmotic pressure studies, electrophoresis, chromatography, and nuclear magnetic resonance spectroscopy—before anybody'd pay attention. As the saying goes: The more extraordinary the idea, the more convincing must be the proofs. And the more technically sophisticated tests, I find. An experiment that would have convinced a scientist in 1920 won't wash these days because it doesn't require thousands of dollars' worth of black boxes. As if fancy equipment yields more reliable results than simple apparatus! Experimental design and choice of controls determine reliability of results, not machines.

Anyway, solved different problem in meantime. Was bothered by interactions *not* predicted by amino acid pairing. Not that mine was only plausible mode of binding. Could have plain old attractions between oppositely charged peptides. Several cases of that. But found unexpected interactions between neutral peptides having aromatic side chains, too. Why? Gustav suggested intercalation. Perhaps the aromatic residues stacked. He was working on a molecule himself—myelin basic protein—that seemed to stick to itself. Largely a linear molecule. Might be pieces that interacted by intercalation of aromatic side chains, or maybe by amino acid pairing. How about giving it a shot?

No luck with either hypothesis. That's life. But in course of reading about myelin basic protein, came across reference to possible serotonin binding site on it. [Editor's note: serotonin is a flat, cyclic molecule that is a metabolic product of the amino acid tryptophan; it acts as an immunomodulator when released by platelets and as a neurotransmitter when released by nerves.] Hypothesis: Perhaps serotonin intercalated between aromatic residues on protein. Since we had peptide fragments of protein around from earlier work, figured we might as well find out. *That* worked. Quite strong, specific binding. Ran controls, other peptides with different sequences. Mostly negative. Tried a set with configurations of aromatic residues very similar to myelin basic protein. Got another surprise. Luteinizing hormone–releasing hormone [controls ovulation and spermatogenesis] and adrenocorticotropic hormone ["stress" hormone] both bound serotonin also. Both known to be modulated by presence of serotonin in body. Common sequence of all three peptides was a pair of aromatic amino acids separated by a hydrogen-bonding amino acid. Just right size for intercalating serotonin.

Invented the name "molecular sandwiches" for the feature (following advice of college embryology teacher who said best way to make a name for self in science is to invent new words!). Just like stuffing slice of meat (serotonin) between two slices of bread (aromatic residues on peptide), with addition of a few rules about what kind of meat could go between what kind of bread. Then identified variety of different "sandwiches" as common feature of the active sites of most hormones.[1] Probably act like keys to open locks of receptors. So, two years of work on amino acid pairing and what was I doing? Writing a paper on serotonin binding sites of peptides and possible mechanisms of neurotransmitter regulation of peptide activity. Now, is that typical, or is that typical?

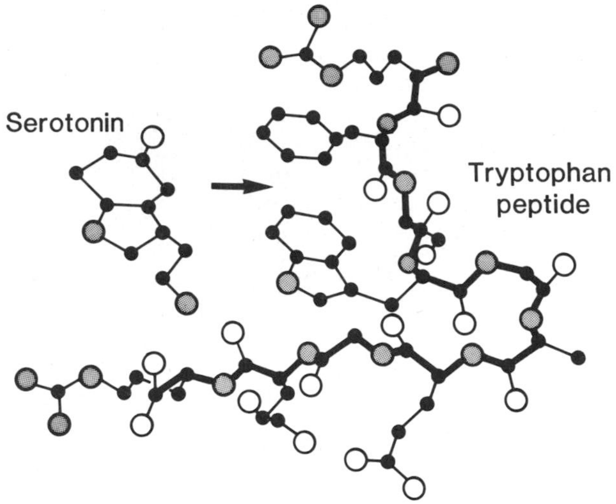

A "molecular sandwich": serotonin intercalating between the phenylalanine and tryptophan side chains of the tryptophan peptide of myelin basic protein. (Root-Bernstein and Westall, 1984c)

Well, it made tactical sense, anyway. Neurotransmitter interactions with peptides were recognized as a problem ready to be solved. Amino acid pairing? That was a problem I was inventing. If nothing else, it got me into the lab looking at things in new ways. Can't ask more of a hypothesis than that.

DAY THREE

Logic of Research, Surprise of Discovery

Formal logic, whose history as a rigorous system started with Frege and ended with Gödel, represents a refinement and specialization of the principles of everyday argument; the logic of scientific discovery, whose rigorous formulation is yet to be achieved (not that it holds out the hope of completeness once entertained by deductive logic), will similarly prove to be a refinement and specialization of the logic of everyday invention. The important thing to realize is that invention is, in its strictest sense, as familiar a process as argument, no more and no less mysterious. Once we get this into our heads, scientific creativity will have been won back from the mystery mongers.

—Peter Caws, philosopher of science (1969)

IMP'S JOURNAL: Competition

Keep several projects going at once and one will always be fruitful. Like my peptide-neurotransmitter research. Devoted most of '83–86 to that. (After all, got to publish!) But never dropped amino acid pairing. Finally got theory published. But not without trouble!

Christ! You'd think I'd attacked infallibility of the Pope! Several referees recommended rejection because amino acid pairing "contradicted the Central Dogma." Tell me something I don't know! Better: Read the first paragraph of my paper, dummy! I *say* it contradicts C.D.! At least my hypothesis is testable, for chrissakes! Another referee was more politic. Suggested rejection to "save author's reputation." Thanks, but I can look after myself. A third wrote: "If amino acid pairing existed, we would have seen it already." Really!? When did you look? A fourth stated (absolutely incorrectly) that if amino acid pairing could occur then Lamarckian inheritance possible. Since no Lamarckian inheritance, can't be amino acid pairing. Wow! Finally James Danielli chose a couple sympathetic reviewers (guys who'd never been in the molecular biology "tie club"), and paper appeared in *JTB*.[1]

Hoped it might spark some interest and did, initially. But no experimental tests. People loath to work on any but own ideas, at least in biology. So did what experiments I could. And a good thing, too! Crazy I may be, but at least I have company. As word got out about my heresy, word got back about competition. First, Mekler in U.S.S.R. Following same line of reasoning I did from problem of immune regulation to protein complementarity, he'd suggested possibility of protein-to-protein information transfers. N. D. Cook and Kaufmann followed. Far as I can tell, none of us knew about others. In 1980, two years before my first publication, Mekler published specific amino acid pairings based on complementarity of genetic code. Same set I'd rejected as implausible at molecular level. Then Blalock and Smith reinvented same pairings as Mekler.[2]

Was I wrong? Did I opt for the wrong alternative? Damn, that would hurt! But no—I'm the only one who's worried about physicochemical problems of interaction—only one who's built models and defined boundary conditions. My previous analysis of the alternatives holds. I *know* I'm right. Or at least right that if amino acid pairing exists, it's my pairings, not the Mekler–Blalock–Smith set. Besides, Grafstein reinvented my set of pairings from a different set of assumptions, didn't he?[3]

Oh, hell! Why bother even thinking this way? Only thing that counts are experiments. Run sufficient controls; test all alternatives; and nature will tell who's right—or produce another surprise. So get back to the lab!

▼ JENNY'S NOTEBOOK: The Body Is Part of the Mind

Thank goodness Ariana called tonight. I've been wanting to call her for several days, but I don't like to burden other people with my problems. She drew me out, though, like a true friend.

"The problem is that Imp's sick again—so ill, he's in bed. Won't see a doctor. Never will. Claims he knows too much to trust one."

Ariana laughed. "I don't blame him."

"Well, I wouldn't worry, except that this is the third time he's been this sick in four months. But it's not just that," I continued. "He gets in these foul moods, too. They seem to precede his illnesses. Thinks of nothing but work. Snaps my head off if I so much as dare to suggest relaxation, or a movie, or dinner with friends. Acts like I'm trying to sabotage his precious research. As if he doesn't spend enough time on it already. And the slightest thing can set him off.

"Like yesterday. He gets this questionnaire from the Institute for Scientific Information about some new atlas of something-or-other they're planning. It's supposed to keep investigators informed of the latest techniques, the emerging fields of interest, what areas are getting the most funding—and the least—what institutions are getting the most money, what journals are most cited, et cetera. Well! You'd have thought the world was ending! 'Christ!' he shouts. 'Here it is, all in one easy-to-order packet: How to become a professional money-grubbing crowd follower!' As if there was anything he could do about it. 'It's not what I can do about it,' he yells, 'it's what this kind of crap *does* to me! It keeps me from getting my work done because I refuse to stand in line to kiss ass for money and prestige. But those assholes at the universities! Oh, they're just wonderful—Can we seriously consider Joe Blow for tenure when he's never been published in *Science* or *Nature*? Is writing a textbook really a contribution to the field? How much overhead has he brought in over the last five years? That's all that interests them—money, prestige, doing the "in" thing. I may as well become a stockbroker as put up with this crap.' That's how he gets. And then he collapses into these illnesses. Sometimes I just can't stand it anymore."

Ariana just laughed. "Look, Jenny," she said, "how many years have I known you and Imp? Since you first met, right? Does he have a history of depression? No. Does he have a history of chronic illness? No. Is he under unusual stress and tension at the moment? Yes. So, like any normal human being, he's finding ways to cope. That's all. Very common. Very normal."

"But why take it out on me?" I asked. "And why collapse into these moody illnesses?"

"Why take it out on you?" echoed Ariana. "Because you're the only person he can trust enough to show his real feelings to. Even the blackest of them. That's actually a compliment, you know."

"You could have fooled me," I replied. "But what I want to know is whether there's something I need to worry about."

"Medically?" Ariana paused and chose her words carefully. "It comes down to this. The mind-body dichotomy simply doesn't hold water when you observe the human being from a doctor's perspective. You read D. H. Lawrence or Anaïs Nin or watch an X-rated movie and your body responds appropriately. Ideas are manifested as bodily states. Remember what Imp told Constance Saturday about engaging the material—feeling it in your head, your heart, your stomach?

"Now apply this to Imp himself. What's he trying to do? From what you tell me, something that most of his colleagues consider foolhardy or impossible—both in his research and in this Discovering Project. He's walking a tightrope, taking risks, questioning things, putting together a myriad of data into a simple, coherent pattern. It takes courage.[1] Sometimes the feeling is excitement. That's the side he shows us on Saturdays, for public consumption. He's daring the scientific community, or daring nature, the same way an athlete dares his competitors or dares a mountain he wishes to climb. It's fun. He gets an adrenalin rush. The flush of excitement. A heightened sensibility. It's wonderful!

"But sometimes you fall, the questions remain unanswered, the data won't coalesce; well, the body reflects that state of mind, too. It becomes increasingly ill at ease—with itself, with outside distractions. When I have a patient I can't diagnose, I feel at odds with the world, mentally *and* physically. Everything becomes a distraction under such circumstances, because the last thing the mind-body needs is yet another bit of data to add to an already unmanageable heap. Finally, the mind-body begins to wonder why it puts itself into such an imbalanced, distracted state, and begins to look for ways out. Maybe it blames the world. Maybe it blames itself.

"In Imp's case, I suspect he doubts the sanity of his undertaking, and so he doubts himself. But he's not the sort to admit that. So he takes out his anger at his own impotence by flailing at the world. That's his way of working out the mental-physical frustration he's experiencing. Unfortunately, stress is highly correlated with illness. Increased stress hormones, such as ACTH; they increase corticosteroids; corticosteroids decrease immune responses, and so forth. So he is frequently ill, too. Very normal. Better you save your worrying for when he just sits and stares listlessly into space for days on end. Then you know he's defeated himself."

"You're very comforting." I sighed. "But I hate it when he gets this way. Sometimes I feel angry, sometimes helpless. Then I begin to wonder whether I'm all right. Maybe if I were more successful in my career, Imp wouldn't feel so pressured. Maybe if I could cope better . . . I don't know. Sometimes I just feel so imperfect."

"Then obviously you should have married a geologist," responded Ariana.

I puzzled over this seeming non sequitur. "Why?"

"Because they love faults," quipped Ariana. "Look, Jenny, you must be pretty tired yourself. You're taking this much too seriously.

"Let me try a different tack. You know, you're in good company. I've been looking into who discovers, with Constance, and the life of the spouses is never easy. For example, one bit of information we've come across is the large number of eminent scientists who've had mental or physical breakdowns during their careers. Remember Hunter mentioned that Pasteur had one around 1856, when all his experiments on his cosmic asymmetric force failed? Had to take off about six months to recuperate. His friend Claude Bernard became so depressed with his research that after six years in Paris he seriously considered returning to the provinces to become a country doctor. Imagine physiology without Bernard! Darwin spent months unable to contemplate work, and his friends Huxley and Tyndall both suffered periodic breakdowns when overwork and lack of new ideas coincided. Mendel had repeated nervous breakdowns, and the mathematicians Weierstrass, Riemann, and Jacobi worked so hard they endangered their health. Michael Faraday claimed that the strain of remembering everything he needed to know for his researches was too much effort for his small brain; and sure enough, every five years or so, just as he'd completed some huge synthetic work, he'd collapse.[2] Being married to such a person has to be tough. In fact, when Albert Michelson had a mental breakdown during his initial investigations of the speed of light, his first wife tried to have him committed to an insane asylum! And you can imagine what it must be like to live with fellows like H. J. Muller or Albert Szent-Györgyi, who get so depressed about their work that they contemplate suicide. There are more of them around than we care to admit."[3]

"This is supposed to make me feel better?" I asked.

Ariana laughed again, her therapeutic laugh. "But you haven't let me finish. The good side to all this, if 'good' is the right word, is that these depressions and breakdowns seem to be related to a striving type of personality. It's not easy to be or to act differently from everyone else, but you can't create if you conform."

"People who don't climb mountains never have mountain-climbing accidents. Is that the idea?"

"Exactly. Most people avoid hurting themselves. Yet the brain fatigues, just as do other organs. Einstein talked about that, a lot—even wrote a little rhyme about it: 'In thought to be absorbed, to strain / and to overtax the brain, / Not everyone will undertake / if at all he can escape.'[4] Imp's not going to take the easy way, precisely because it *is* easy. He knows easy things aren't worth much in the long run. So, he's going to tackle each rough spot as it comes along. For the challenge. I can appreciate that, given my career. So sometimes you'll have to help him a bit—or put up with him, as the case may be. Just thank your lucky stars you've found someone worth putting up with!"

"Well, I could do with a bit of pampering myself at times." I felt guilty for saying it, but Ariana took the hint.

"Couldn't we all. So let me take you to lunch tomorrow. And next time I'm down, you can reciprocate."

Just as Imp no doubt will, as soon as he rejoins humanity. I complain, but he's really not so bad . . . I wonder if I can channel his anger into something more useful, like belying Richter's irrationalist-chance position. Now, where's that list of classic chance discoveries?

◀ TRANSCRIPT: The Probability of Discovering

▷ **Imp:** Okay, let's go! I hope you'll forgive my hoarse voice and sniffles. But let not sickness deter us from our appointed destiny! Richter, convince us that discovering is an irrational, unpredictable game of chance abetted by accident and played best in the land of serendipity.

▷ **Constance:** May I interrupt? It might be worth getting our definitions straight at the outset. I know this is pedantic, but it might save us from confusion later. Serendipity is not strictly equivalent to chance or accident. Chance, if you read Laplace on probability theory, refers to events controlled by factors about which we know absolutely nothing. That's as opposed to probability, which refers to events about which we know some but not all of the controlling factors. Accident is an unforeseen, unintended, or unexpected event. There's a negative connotation associated with accidents. But it's not clear to me whether accidents could be foreseen or expected, but aren't, or—

▷ **Richter:** All right, all right. Forget accidents. What about serendipity?

▷ **Constance:** Well, you have to understand accident to understand serendipity. The word "serendipity" was invented by Horace Walpole in 1754. He had been reading a fairy tale about "The Three Princes of Serendip"—the old name for Ceylon, now Sri Lanka—and noticed that "as their highnesses travelled, they were always making discoveries by *accident* or *sagacity*, of things they were not in quest of."[1] That's his definition of serendipity.

▷ **Richter:** Wonderful. Accident *or* sagacity! That doesn't get us far.

▷ **Constance:** I think the key phrase is "things they were not in quest of." That's apt. It certainly describes what we know of Berthollet and Pasteur.

▷ **Ariana:** So the question becomes whether serendipitous events occur by chance, probability, or planning? Right?

▷ **Constance:** Well, actually, James Austin has addressed that issue.[2] He identifies four types of chance: Chance I, or blind luck, in which the investigator has no input or preparation—

▷ **Imp:** Impossible. Theory-directed observation, remember?

▷ **Constance:** Chance II, or the "Kettering Principle," named after the great General Motors inventor who maintained that action creates surprises. Therefore sufficient active curiosity is sure to turn up something interesting eventually.

▷ **Ariana:** Lewis Thomas advocates the same principle. He says that if you get the research going—get away from hypothesizing—something's sure to occur.[3]

▷ **Constance:** Then there's Chance III, or the "Pasteur Principle": The prepared mind perceives in the sublimity of the mundane the unexpected. Sort of like Richter's idea of knowing the outstanding problems of your field. And finally Chance IV, or the "Disraeli Principle," which maintains that being or acting distinctively as a person will cause unexpected phenomena to emerge. This requires the person to act in an individually unique manner. Perhaps related to Jenny's idea that discoveries are made by unusually trained people.

▷ **Ariana:** Notice that in all but Chance I, the individual investigator has a crucial role in making the surprise appear. That seems to indicate, at the very least, the existence of tactics for increasing the probability of discovery. What do you say to that, Richter?

▷ **Richter:** That there is nonetheless an unknowable irrational element in the way all new ideas are generated. Irrational in the sense that there is no way to predict them or plan for them beforehand. Irrational in the sense that data underdetermine theory. Irrational in the sense that there are elements in every idea that are not derived solely from data or tested theory and hence have no logical or empirical basis. An *Erdenrest* or residuum that cannot be accounted for.[4]

▷ **Ariana:** Only if you limit your definition of rationality to strictly quantitative and semantic forms of logic.

▷ **Imp:** Besides, Richter, apply your arguments to scientific explanation itself. Your position reduces to saying that rational thought can only be applied to what's known. Whatever's unknown requires irrational thought. So if you believe—as I do—that every answer generates two new problems, then the only way to keep up with the increase in ignorance would be to become even more irrational!

▷ **Ariana:** No wonder quantum physicists are turning to Zen!

▷ **Hunter:** Look, Richter, there's always an unexplained *Erdenrest* in any system. That's not a criterion for determining logical consistency or explanatory power. Even thermodynamics, which is very nearly a complete logical theory, cannot describe in detail the thermodynamics of most natural systems such as, say, the garden we see out the window here. It's too complicated. There are too many factors to make an exact computation. What we're looking for is an increasing ability to account for as many factors as possible with as much accuracy as possible, not perfection.[5]

In this vein, I want to get back to what Ariana said a moment ago about limiting our notions of logic to quantitative and semantic forms of logic. I remember reading about an argument that took place between I. I. Rabi, Leo Szilard, and Enrico Fermi, about some aspect of nuclear physics. Szilard put a mathematical argument on the blackboard. Rabi disagreed and wrote a new set of equations. Finally Fermi said, "No, you're both wrong." And when they challenged him to prove it, he countered, "Intuition." Mitchell Wilson, a young postdoc who was present, said he expected Szilard and Rabi to laugh. They didn't.[6] I think we must seriously consider *why* they didn't. Why should scientists of the caliber of Rabi and Szilard accept as valid critique of mathematically demonstrable arguments a notion springing from intuition?

That brings me to Ariana's question last week: Must intuition be irrational? That's a good question. Have we slipped into nominalism by labeling thought processes we don't understand as "intuitional" and then, having satisfied our taxonomic urge, relegating all such intuitions to incomprehensibility? I don't think we have to take that route.

▷ **Constance:** There are alternatives. While reviewing these questions last week, I ran across a remark Linus Pauling made in a lecture. Pauling advocates what he calls a stochastic method for solving complex problems.[7] Unlike the modern mathematical use of the word, which connotes randomness, the ancient Greeks used the word in the sense of aiming skillfully guided guesses into the unknown.

▷ **Imp:** Which is something we never teach—guessing.

▷ **Richter:** Nor do I see how you could.

▷ **Hunter:** I'm not sure we know how to do it, either—yet. Though Polya certainly made a stab at it. And I don't think it would be impossible to push further. Think about Fermi. What was he doing? Saying: That's not how the world works. He's evaluating the equations of Szilard and Rabi not on their internal logic, but in relation to everything else he knows about physics. He's asking himself a super-question beyond the particular question: Is this how physical systems function? He responds no, not because anything specific is incorrect, but because in his accumulated experience, systems in general just don't do that. Equations of the form Rabi and Szilard wrote don't pan out.

Hence intuition might be defined as informal patterns of expectation born of experience. That's the crucial point. Rabi and Szilard listened to Fermi only because he had experience. He was an expert. And I use this term in the way Bohr did, to mean someone who's already made every possible mistake in the field. Intuition is also a way of saying, "I've tried that approach on this sort of problem before, and it doesn't work." It's very difficult to say why something doesn't work, but knowledge of what doesn't work can often be as valuable as knowing what does.

▷ **Imp:** Another thing we fail to incorporate into science teaching. The mistakes of predecessors that could save us time today. We don't, in that sense, encourage the development of intuition.

▷ **Richter:** Which we cannot possibly do, given the limited teaching time available to us. Look, this is quite interesting and it might even lead somewhere if we had time to pursue it. But keep to the issue: Do discoveries result from irrational, chance events or by logical planning? I understand your strategy. You want to play a semantic game that will shift the genesis of any idea whatever into the realm of rationality. And as long as we talk in generalities, you can define words as you like. I doubt that you will fare so well when it comes to specifics.

Once again, my position is well established by empirical studies of the subject. There are a dozen books and articles I could quote, but I prefer to discuss cases instead. Everyone seems to have his own private examples, but certain chance discoveries appear on almost every list. For example: the discovery of animal electricity by Galvani. Oersted's discovery of the connection between electricity and magnetism. Claude Bernard's discovery of the nervous control of blood flow. Charles Richet's discovery of anaphylaxis (an extreme allergic reaction that can cause death). Pasteur's discovery of asymmetric fermentation—

▷ **Hunter:** Ahem!

▷ **Richter:** —and of immunization using attenuated microorganisms. Minkowski's discovery of the role of the pancreas in diabetes. The discoveries of most of the vitamins. Goodyear's vulcanization of rubber. Alfred Nobel's invention of dynamite. Perkins's accidental synthesis of the first aniline dye. Alexander Fleming's discoveries of lysozyme and penicillin. Roentgen's discovery of X-rays. Becquerel's discovery of radioactivity. There are many more, but that will do.[8] Tell me that all these serendipitous events are figments of the historical imagination!

▷ **Hunter:** They are! Either that or the result of confusing the unexpected with the accidental. Constance and I spent the week looking into as many of these discoveries as we could, and very few occurred as the popular accounts maintain. But, Richter, why don't you present your descriptions of some putative chance discoveries, and Constance and I will see what kind of new material we've got that might change the picture?

▷ **Richter:** Fine. Start with Claude Bernard and the nervous control of blood flow. I'll quote Cannon's account:

> *In the biological sciences serendipity has been quite as consequential as in the physical sciences. Claude Bernard, for example, had the idea that the impulses which pass along nerve fibers set up chemical changes producing heat. In an experiment performed about the middle of the last century he measured the temperature of a rabbit's ear and then severed a nerve which delivers impulses to that structure expecting, in accordance with his theory, that the ear deprived of nerve impulses would be cooler than its mate on the other side. To his great surprise it was considerably warmer!*[9]

▷ **Hunter:** Surprising, yes. Accidental, hardly. Cannon himself states that Bernard set up the experiments to test the influence of the nerves on blood circulation. So there's a logic of research. Bernard thinks, "I expect this, so to make it appear I shall do that."

▷ **Constance:** Right. Bernard's biographers fully confirm the planned nature of the experiments.[10]

▷ **Richter:** All right. What about Richet? He claims in his book *Le Savant* that he discovered anaphylaxis by chance. He was testing the effect of sea anemone toxin on animals. He reinoculated some animals that had already survived one dose of toxin, expecting them to tolerate the doses even better. His model, I imagine, was vaccination. He was amazed instead to observe that even tiny fractions of the dose that the animals had already survived were now deadly. If any of you are allergic to bee stings, you will empathize. Richet was so surprised he did not at first believe the results. Certainly, he said, it was nothing he had foreseen.[11]

▷ **Hunter:** But again, Richter, unexpected, not illogical. Surely there's a rationale for reinoculating the animals to see if they've become more resistant to the toxin. Whether Richet was reasoning from immunology or toxicology, previous experience would have led him to believe that previously treated animals would become more resistant, not less.

▷ **Imp:** And anaphylaxis is still one of the great mysteries of medicine—as is, in fact, the whole area of allergy. So one should *expect* surprises, wouldn't you say, Richter? I mean, otherwise we'd already understand the phenomena. Logic permits of no surprises, right?

▷ **Richter:** You bait me.

▷ **Imp:** Perhaps; but my point stands.

▷ **Hunter:** In any case, we can dispose of Pasteur's discovery of vaccination using attenuated microorganisms in the same way. The standard story is that the discovery was an accident, as you say.[12] Pasteur was studying chicken cholera in 1879, went on a vacation leaving his germ cultures to sit, and returned to find the cultures avirulent—that is, incapable of giving chickens cholera. So he acquired new cultures from a natural outbreak of the disease and reinoculated the chickens. Again they failed to become ill. But chickens not previously inoculated did become ill. So Pasteur recognized that the old, avirulent cholera culture had "immunized" the chickens. Where's the chance?

▷ **Constance:** But—

▷ **Hunter:** Is the chance that Pasteur went on vacation? At some time Pasteur surely would have produced an avirulent culture for the simple reason that a culture exposed to air, as Pasteur kept his cultures, would eventually become avirulent simply with the passage of time. Pasteur later proved this himself. So whether Pasteur had gone on vacation or not, his culture would eventually have become avirulent. Anyone doing cholera research would probably have observed that.

Constance, you wanted to add something?

▷ **Constance:** Well, yes. The story you just told is completely wrong. I found an article only yesterday by Antonio Cadeddu, who's had a look at the relevant Pasteur notebooks.[13] He demonstrates that the experiments were definitely planned, and they did not occur as the myth suggests. First off, the crucial flask of cholera that became attenuated was not even sown until October 28, 1879—after Pasteur had returned from his vacation. And the flasks of old cholera cultures were not left to stand over the summer but were maintained by Pasteur's assistant, Emile Roux.

▷ **Ariana:** So what *did* happen?

▷ **Constance:** Well, something like this, as near as I can make out: Cultures of cholera were made during July 1879, just before Pasteur left for vacation. Roux was left in charge of these cultures. Some became acidic, however, and the cholera in these cultures evidently was not growing. Roux inoculated two chickens with these acidic cultures to determine conclusively whether the cultures were virulent or not. The chickens did not become ill. Eight days later, Roux inoculated them with a more recent, and virulent, culture.

▷ **Ariana:** And the chickens still didn't become ill.

▷ **Constance:** No! Not at all. That's just the point. Both chickens died within four days. And the blood from one of these twice-inoculated animals was used to inseminate a new, virulent culture in "Flask X" on October 28th. So the experiments not only took place long after the summer, but they were carried out by Roux—not Pasteur—and did not yield the result suggested by the mythical version. You see, Pasteur was *trying* to "enfeeble"—that's his word—the cholera germ. The October experiments showed that this enfeeblement was possible, but it didn't result in a vaccine. In that sense it was a partial success. So, between December 1879 and March 1880, he and Roux tried all sorts of new things. They froze the germs at minus thirty-eight degrees Celsius. They put them in a strong acid culture medium. In the meantime they kept the old cultures, and again some became acidic, including "Flask X." Some of the acidic cultures killed the chickens, some didn't. Until February, the ones that didn't kill the chickens still didn't protect them against subsequent inoculations with virulent cultures.

Oh, yes, I should also add that Pasteur was trying his experiments not only on chickens but on rabbits and guinea pigs, too. He did things like pass the germs from one animal to another, as well as from flask to flask and flask to animal and back again. Even between species. Guinea pigs, for example, could transmit the disease without ever becoming ill themselves. None of which helped to produce a vaccine, but certainly demonstrates the direction his research was taking.

▷ **Hunter:** And later, with rabies, the same sort of experiments did pay off, for he found that the rabies virus attenuated as it passed from animal to animal.

▷ **Constance:** Yes. In any event, it's clear that only the acid treatment (which Pasteur, evidently holding to Lavoisier's theory of acidity, equated with the presence of oxygen) had any enfeebling properties for the cholera germs. So lots of acid flasks were set up to study the effects of the amount of acidity and length of time in acid, and various ways to create acidity (such as bubbling the culture media with oxygen). By March he had two series of acidic flasks, one artificially created (the "P" series) and the other created naturally by exposing the culture to air for several months (the "X" series), which both had properties of vaccines. The key seems to have been placing the microbes in a weakly acidic medium for a long period of time, but not too long, because then they became completely avirulent.

▷ **Jenny:** But why did Pasteur embark on the research in the first place? What was he looking for?

▷ **Constance:** That's what's missing from Cadeddu's article. From what I can find out, Pasteur had been reading about Jenner's method of vaccinating against smallpox with cowpox, and about attempts to prevent syphilis by a similar method, long before he began his own investigations. He had already noted what everyone knew, which is that people who survive a disease never get the disease again. His earliest cholera notes indicate an intense interest in why some chickens survived the cholera and others didn't, and he's clearly speculating on ways to make all the animals survive. I think it's clear that Pasteur knew what he had to accomplish to produce a vaccine even before he began: He had to produce a microbe powerful enough to cause illness and to protect against subsequent illness, but not so virulent that it could kill. As Richter says, the problem defined the answer. After that, Pasteur tried whatever he and Roux could imagine that might hurt the microbes, and kept their eyes open.

▷ **Jenny:** So what you're saying is that, if there's any chance involved, Pasteur created it himself. Or, using Pauling's stochastic principle, he knew which direction to guess in.

▷ **Ariana:** The "Boss" Kettering–Lewis Thomas principle.

▷ **Constance:** Pasteur said as much himself: "The illusions of the experimenter form the greater part of his power. These are the preconceived ideas which serve to guide him."[14]

▷ **Richter:** Pasteur, I presume, never informed us what his preconceived idea was. The records are once again incomplete?

▷ **Constance:** As far as I know.

▷ **Hunter:** Which explains the origin of the myth. It probably arose precisely because Pasteur himself never breathed a word of how the research developed. We have only the notebooks to guide us. Previous historians haven't even used them.

▷ **Constance:** Sure. You see, Cadeddu argues that Pasteur himself was not sure how the attenuation process worked, so although he announced the invention of the vaccine early in 1880, he uniformly refused to discuss the process, both in private and in public. Personally, I believe that Pasteur was waiting to patent the process. You can't patent a process that's in the public domain, so Pasteur often patented his processes, then published them, and usually turned the patent over to the public afterward. But the point is, he left historians nothing but chance to resort to as an explanation of his vaccine. Until Cadeddu's work, that is.

▷ **Richter:** Damn. I assume you have the same sort of explanation for Perkin, Oersted, and so on?

▷ **Hunter:** Planned experiments, unexpected outcomes. Applies to almost all the so-called chance experiments listed by people like Cannon, Beveridge, and, more recently, Austin.

▷ **Imp:** But you're not giving up already, are you, Richter? We've got so many more examples to examine.

▷ **Ariana:** Like Minkowski.

▷ **Richter:** Yes. Why not? You must have your little amusements. *My* understanding of the Minkowski story is that in 1889 Minkowski removed the pancreas from a dog to study the role played by that organ in fat digestion. Subsequently, the dog urinated on the laboratory floor. Flies were attracted to the urine, which in turn attracted the attention of a sharp-eyed lab assistant, who wondered why. Minkowski found the urine to be loaded with sugar. Thus, by a lab assistant's chance observation of flies attracted to urine, the connection was made between the pancreas and sugar metabolism. Minkowski had stumbled accidentally onto an experimental model for diabetes. We now know that the pancreas contains the islets of Langerhans, which secrete insulin, which in turn controls sugar metabolism.[15]

▷ **Hunter:** Okay, but isn't it odd that no one has ever named this perspicacious lab assistant, nor explained why Minkowski should have cared that flies were attracted to the urine?

▷ **Ariana:** Even more to the point, the discovery was hardly a thoughtless one even as you've stated it, Richter. This unknown lab assistant was at least bright enough to spot the anomalous behavior of the flies when he saw it. And Minkowski had to have some reason to think the anomaly worth investigating. How do you account for that? I mean, producing urine attractive to flies is not a classic symptom of any disease, let alone diabetes! There's a gap here you haven't explained—how you get from this "accidental" observation to the discovery.

▷ **Imp:** Ah! To make the point even more strongly—in order to discover the connection between diabetes and the pancreas, Minkowski had to find the sugar in the urine. But how did he know to search for sugar? Flies are attracted to a lot of other things besides sugar, as anyone who's visited a farm must know! So what's the thinking that makes the combination of flies and urine a significant indication of the presence of sugar rather than some other aromatic or proteinaceous substance?

▷ **Richter:** I have no explanation. I cannot tell what went through the mind of a man long dead.

▷ **Ariana:** You don't have to. Minkowski told us himself in an account of his discovery that only came to light many years later. First you have to know that Minkowski was trained as an M.D. and was an excellent surgeon. He says that he kept the depancreatized dog himself, since it was housebroken. But no matter how often he took it out, the dog still urinated in the house. In medical terms it had polyuria—unusually frequent urination. Polyuria is a typical symptom of, among other things, diabetes, and Minkowski says that his professor at medical school had taught him always to test for sugar in the urine in cases of polyuria. If there was none, then one was facing a case of, say, bladder infection. If there was sugar, one could be almost certain one had a case of diabetes. So Minkowski, following the standard procedure of medical diagnosis, treated the dog as if it were a human patient, tested its urine for sugar, and found the test positive. Subsequently, he was able to show that the dog's symptoms were, in all particulars, indistinguishable from the natural human form of the disease.[16] No chance at all. Straightforward medical logic. The only surprise was that depancreatization affected not just metabolism of fats but sugars, too.

▷ **Richter:** But, dammit, where do these myths come from?

▷ **Ariana:** Can't say for sure, but would you accept a guess? For one thing, Minkowski's announcement of an animal model for diabetes was initially met with tremendous skepticism. The pancreas is a meandering, diffuse organ, not easily located in its entirety. Claude Bernard, the most authoritative contemporary physiologist, had declared its complete removal an impossible operation. Minkowski, typically, responded that there was no such thing as an impossible operation—

▷ **Imp:** Ah! A man after my own heart!

▷ **Ariana:** —and he proceeded to perform it. His self-assuredness and lack of respect for authority made him somewhat unpopular among his colleagues. Indeed, despite what in retrospect looks like a highly inventive research career, Minkowski managed to survive in academia only with great difficulty.[17] So perhaps one of his better placed or more traditionally successful colleagues invented the fly story as a way to cast aspersions on Minkowski's personality. I'm sure we can all think of colleagues about whom it's claimed—usually in private—that they couldn't really have carried out the research that bears their name in its published form. Surely, it was some talented graduate student or postdoc . . .

▷ **Imp:** And sometimes it's true.

▷ **Jenny:** But isn't there something else going on, too? Isn't there a sense in which it's comforting to those who don't discover things to believe that those who do are simply lucky?[18] I mean, human nature is such that it's difficult for any of us to admit that we just aren't as smart or as perceptive, or whatever, as Susie Scientist down the hall. Right? So when Susie Scientist makes some great breakthrough, the natural reaction is jealousy.[19] It takes a very secure person to say, "Wow, that's great," and not follow it up with "What a lucky break!" It seems to me that "luck" is just the palliative of the mediocre.

▷ **Richter:** Nothing personal intended, I assume.

▷ **Jenny:** Of course not. I was just speaking in the abstract.

▷ **Richter:** I see . . . Then, speaking abstractly, of course, what about Fleming's discoveries of lysozyme and penicillin? He found lysozyme when a drip from his nose accidentally landed on a plate of bacteria and dissolved them. Penicillin was found due to an accidental contamination. You'll have problems telling me I'm wrong about these two examples, because I checked the latest accounts. Nobody has a rational explanation for either discovery.[20]

▷ **Imp:** I do!

▷ **Richter:** Right, Imp. *You* have an explanation for everything.

▷ **Imp:** It's my nature!

Look, would anyone mind if we went into the lysozyme and penicillin cases in detail? Richter's right: The standard accounts don't suffice, so Jenny and I put in a lot of effort to sort things out—my sickbed amusement this week.

Jenny, why don't you begin? I'll save my voice.

▷ **Jenny:** Certainly. First, what's lysozyme? Most of you know, but I didn't, so I'll begin by answering my own question. Lysozyme is an enzyme—that is, a protein that carries out a chemical reaction—that has antibacterial activity. It's found in virtually all tissues and body secretions, including tears, saliva, nasal mucus, and blood, and in eggs and lots of other things. In short, it's one of the body's major lines of defense against infection, along with the skin, lymphocytes, and antibodies.

Okay. Now the history. The accidental-drip-on-a-contaminated-plate hypothesis is based upon two accounts, plus some apparently unsolvable problems in reconstructing Fleming's research. The first account is by V. D.—

▷ **Imp:** "Venereal Disease"?

▷ **Jenny:** —Allison. Forgive Imp's sophomoric humor: He has a point. Two, actually. Fleming made his reputation treating syphilis with Ehrlich's "magic bullet," Salvarsan, so the initials "V. D." would surely have conjured up more than just his assistant. And second, just this sort of word play is essential to our reconstruction of Fleming's discovery.

But Allison's account first, since it purports to establish contamination of the plate that subsequently displayed the lysozyme activity. Allison was Fleming's laboratory assistant at the time he discovered lysozyme in 1922. Allison recounts that Fleming

> *was busy one evening cleaning up several Petri dishes which had been lying on the bench for perhaps ten days or a fortnight. As he took up one of the dishes in his hand, he looked at it for a long time, showed it to me, and said: "This is interesting." I had a good look at it. It was covered with large yellow colonies which appeared to me to be obvious contaminants. But the remarkable fact was that there was a wide area in which there were no organisms; another further on, in which the organisms had become translucent and glassy. Beyond that, again, were organisms which were in a transitional stage of degradation, between the very glassy ones and those which were fully developed with their normal pigment.*
>
> *Fleming explained that this particular dish was one to which he had added a little of his own nasal mucus, when he had happened to have a cold. This mucus was in the middle of the zone containing no*

Alexander Fleming in his laboratory, 1909. (St. Mary's Hospital Medical School, London)

colony. The idea at once occurred to him that there must be something in the mucus which dissolved or killed the microbes . . . "Now, that really is interesting," he said again. "We must go into it more thoroughly." His first care was to pick off the organism and stain it by gram. He found it was a large gram-positive coccus, not a pathogen, and not one of the known saprophytic [living on dead or decaying matter] organisms commonly met with, but obviously a contaminating organism which was more likely to have been in the atmosphere of the laboratory and may, of course, have blown in through the window.[21]

▷ **Constance:** The Pasteur story all over again.

▷ **Ariana:** But which Pasteur story? The myth? Or Hunter's version?

▷ **Jenny:** Let's see. W. H. Hughes, one of Fleming's later assistants (who was not present during the period of the lysozyme discovery) provides the evidence for the accidental drip of nasal mucus. We assume he was heir to an in-house story. He says that Fleming often had colds.

▷ **Ariana:** Or allergies?

▷ **Imp:** Perhaps. You work with fungi and such long enough . . .

▷ **Jenny:** Hughes claims that "one winter day when [Fleming] was examining some bacterial colonies growing on a culture plate his nose dripped onto some of them. These colonies had not been planted there by Fleming but had come by accident from the air. They were *contaminants*. The colonies splashed by the mucus melted away; there were others like them on the plate and he was able to *sub-culture* these onto another plate [to perform further experiments]."[22] So, combining these accounts, we have quadruple luck: First, an unusual microorganism accidentally contaminates a culture plate. Second, Fleming gets a cold at the same time. Third, of all the plates he works on, he accidentally drips on the plate with the contaminant particularly sensitive to lysozyme. Fourth, the putative contaminant is one of a very few microorganisms extremely sensitive to lysozyme. Is that a reasonable espousal of the standard account, Richter?

▷ **Richter:** It will do.

▷ **Imp:** And we're actually supposed to believe that? Fleming accidentally contaminates a plate not once but twice, and still he keeps it as a valid experiment. Would you do that, Richter?

▷ **Richter:** Not me. But since genius is said to verge on insanity . . .

▷ **Ariana:** Oh, come on, Richter! Stop playing games. Allison wouldn't have treated these events as a valid experiment, nor would you, but Fleming did. Why? What's the difference between your view of the world, your habits of thought, and his? What did he perceive in this apparent series of accidents that you can't?

▷ **Richter:** That's not my problem. You supply the answer if you think it exists.

▷ **Hunter:** No, it *is* your problem, Richter, if you insist on defending the distinction between context of discovery and context of justification. What has Fleming discovered, in the sense of what does it *mean*? Lacking any theoretical expectation, does it mean anything? How does he know that his nasal mucus won't dissolve any and all bacteria? How does he know that it isn't a chemical effect of pH or of osmotic pressure? How does he know that this particular bacterium, which he's never seen before, isn't water soluble? It seems to me the best we can say of the chance account is that it explains not how Fleming discovered lysozyme, but how he discovered the problem that leads to lysozyme. This is the point at which the irrationalist paradigm fails, where the separation between context of discovery and context of justification creates an insuperable barrier to the understanding of scientific research. Because, in this philosophical context, you can't get from the observation to the meaning.

▷ **Richter:** I confess I see no way out of this quandary.

▷ **Imp:** Which means we have to rethink the whole matter. Get into Fleming's head. And having tried to do that, I say Fleming *planned* the experiment. The evidence is in his notebooks and his first publication concerning lysozyme.

▷ **Jenny:** But that wasn't where we intended to start, was it?
▷ **Imp:** No, you're right. I'm being too impulsive, as usual. Go ahead.
▷ **Jenny:** Thank you. The most striking thing about Fleming as a person was that he always played games. He was raised in a family that played everything from poker and bridge to ping-pong and quiz games. He'd pitch pennies with his secretary when there weren't any patients to see; play croquet, bowls, bridge, or snooker at his club; shoot competitively and play water polo as a longtime member of the London Scottish Rifle Volunteers (it's said that although he was a very small man, he could outmarch most of his outfit carrying a forty-pound pack). He took up golf, but rarely played a straight game—instead he'd putt using the club like a snooker cue, or invent dozens of wacky rules to make the game more interesting. He even "invented"—if that's the right word—various forms of miniature golf for play indoors. Needless to say, he was popular with children.
▷ **Hunter:** Sounds rather like Richard Feynman's description of himself in his autobiography—always playing.[23]
▷ **Jenny:** And then there were Fleming's painting exploits.
▷ **Hunter:** Even more like Feynman.
▷ **Jenny:** He was the only nonartist member of the Chelsea Art Club and actually took up painting so as to sell a picture to qualify for membership.[24] He wasn't very good, in my opinion. But the point is, he'd try anything, apparently, and never take it too seriously.
▷ **Imp:** Don't forget his germ paintings. Talk about playing with microbes! Get this: Fleming would take microorganisms that developed colors and

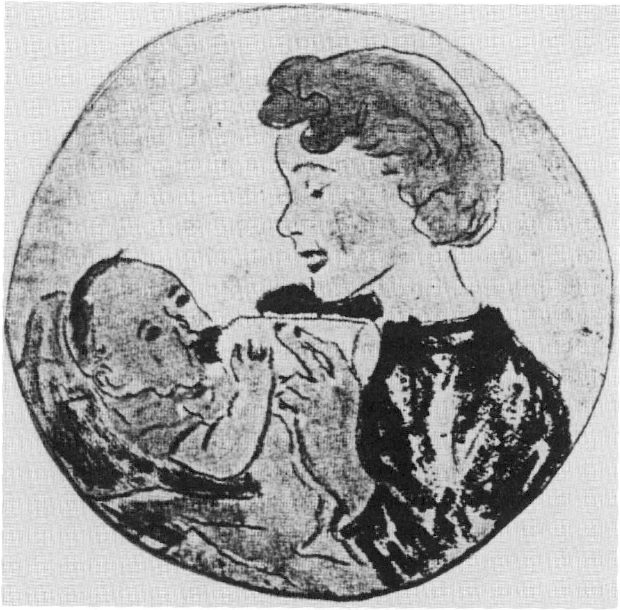

One of Fleming's "germ paintings." (Maurois, 1959)

"paint" them on a petri dish and then incubate it. After a day or two a "painting" of a ballerina, a Union Jack, his house, a mother feeding a baby, or the fleur-de-lys of St. Mary's Hospital—that's where he worked in London—would appear. Look at this! And, Richter, before you ask "But what good is it?" or "So what?"—the Duchess of York, now the Queen Mother, beat you to it. She asked him in 1933. Apparently her question amused Fleming. Personally, I think his germ paintings are very important. Think for a minute about how much bacteriology you'd have to know to do one. You have to know how fast the bacteria grow, whether they'll grow on the same medium, whether they'll interfere with each other's growth, and under what conditions they develop their colors. You have to maintain cultures to play with, and constantly search for new cultures to expand the range of your palette.

▷ **Ariana:** And don't forget hand-eye coordination. You get only one shot with a germ painting, I would think. One slip and the picture's worthless.

▷ **Imp:** Not to mention foresight and planning. The original "drawing" is done with nearly colorless wipes of a sterile loop on agar. You have to imagine the result that you want to achieve.

▷ **Richter:** Fine. So he's a skilled bacteriologist. I would have thought we could take that for granted.

▷ **Imp:** But isn't that the point of research, not to take anything for granted? One of my objections to most accounts of discoveries is that they imply that anybody could have made the discovery in the same context. We assume that all observers observe the same thing.

▷ **Ariana:** A corollary of viewing science as an objective enterprise.

▷ **Imp:** Yet we know for a fact that other scientists in Fleming's lab (Almroth Wright, Allison, Hughes) and in other labs (Pasteur's, Lister's, John Tyndall's, and so on) saw many of the phenomena that indicated the existence of lysozyme and penicillin, and didn't make the discoveries. Why is that?

▷ **Ariana:** Because everyone has a different personality—a different set of ideas, preconceptions, skills, desires, commitments, experiences.

▷ **Imp:** Exactly. As you are so fond of saying, what you know determines what you perceive.

▷ **Richter:** Observation is theory-directed. I agree.

▷ **Ariana:** Then push a step further. If we want to know who discovers and who doesn't, and if we admit that every individual has a different set of theories and problems, then every individual will observe different aspects of nature and interpret them differently.

▷ **Constance:** Making discovery unexplainable, or totally dependent on psychology?

▷ **Imp:** On the contrary, meaning that we have to know what individual observers know and how they habitually act to understand what they observe and to explain why other people don't observe the same thing.

▷ **Hunter:** Just like last week with Pasteur, Biot, and Mitscherlich. Mitscherlich's commitment to isomorphism and Biot's commitment to vitalism prevented them from perceiving the asymmetry of the tartrates, whereas Pasteur's commitment to Laurent's theory of form as a mirror of physical properties led to his observations. It's not irrelevant that Laurent was a personal friend of Pasteur's. Sometimes *who* you know determines *what* you know.

▷ **Richter:** So rather than making personality and experience a psychological haven for the inexplicable in science, you want to make it a rational contributor or controller. A clever inversion of the usual position.

▷ **Imp:** Without which Jenny and I don't believe you can understand how Fleming found lysozyme or penicillin.

▷ **Jenny:** Yes, because Fleming's penchant for play spilled over into his work. His most recent biographer, Gwyn Macfarlane, writes that "Fleming's natural level was indeed play, and if he ever ascended into the stratosphere of higher thought there are no signs of it. 'I play with microbes'—his often repeated description of his work—was literally true. Most of his research was a game to him and indeed most of his enjoyment came from games of all kinds."[26]

Fleming himself told another biographer that "I was just playing" when he discovered penicillin.[27] And in 1945, shortly after receiving his Nobel Prize, Fleming said again, "I play with microbes. There are, of course, many rules to this play . . . but when you have acquired knowledge and experience it is very pleasant to break the rules and to be able to find something nobody had thought of."[28] Just as he often did in his games.

Now, it's our contention that Fleming's need to play provides the essential key to understanding the "logic"—or, as Ariana calls them, the "tools of thought"—underlying his research program. You won't appreciate Fleming if you don't appreciate fun.

▷ **Constance:** Oh, I'm so glad somebody brought this up! You know, science is always portrayed as a serious activity, but I've got dozens of notes here on scientists who loved play and fun. Like Albert Michelson, who when asked why he performed his experiments measuring the speed of light, began by answering with platitudes about the "Value of Science, about Contributions to Knowledge," and so forth; abruptly he interrupted himself and retracted his previous statements. Then he laughed and said, "But the real reason is because it is so much fun!"[29]

Similarly, the entire Copenhagen school of physics, which included Niels Bohr, Werner Heisenberg, and their colleagues, has been described as "famous for a joy in the contemplation of nature that could lead at times to flippancy." One visitor became so annoyed by this lack of respect, he said to Bohr: In your institute nobody takes anything seriously. Bohr replied: That's quite true, and even applies to what you just said.[30]

A lot of scientists would agree with his sentiment.[31] There have even been joke issues of journals, such as the July 1911 *Zeitschrift für Elektrochemie.*

▷ **Imp:** Still are—*Journal of Irreproducible Results.* Even *Perspectives in Biology and Medicine* encourages humorous articles as essential to "not taking ourselves too seriously."

▷ **Richter:** And why not? One certainly questions nature at one's peril.

▷ **Ariana:** But it's not necessary to be solemn to be serious, Richter.[32] No play, no creativity. How can anyone accomplish anything worthwhile otherwise—as a scientist, musician, artist, historian, or doctor?

▷ **Richter:** Some do it with hard work.

▷ **Imp:** That's no contradiction! You work harder when you're having fun!

But back to Fleming before we get totally lost.

▷ **Jenny:** As I was going to say, not only did Fleming's habit of playing with microbes provide him with an unusual sense of the expected, he also developed various strategies for cultivating the unexpected. Allison reports that when he first arrived in Fleming's lab in 1922, Fleming "started to pull my leg about my excessive tidiness. Each evening I put my 'bench' in order and threw away anything I had no further use for. Fleming told me I was a great deal too careful. He, for his part, kept his cultures sometimes for two or three weeks and before getting rid of them, looked very carefully to see whether by chance any unexpected or interesting phenomena had appeared. The sequel was to prove how right he was."[33]

▷ **Richter:** The key word being "chance."

▷ **Imp:** The key being the methodological *cultivation* of chance. Fleming was bright enough to understand how relatively little bacteriology he or anyone else knew, and so he took steps to let nature teach him. Allison wouldn't let himself be taught.

Look, it's just what Konrad Lorenz said once about himself: "Contrary to your Shakespeare, there is madness in my method!"[34] Truly! If you're always meticulously careful and everything is planned, you can only see what you expect to see. And if you do see what you expect to see, it isn't worth doing the experiment. So you've got to inject some chaos into the system to create the conditions under which the unexpected becomes probable. That's what Fleming and Lorenz did. Like Fleming, Lorenz kept all sorts of birds, fish, mammals, and reptiles as pets, got to know the habits of each individually, and then watched for unusual behaviors. He never planned experiments. He never measured. He never once published a graph. Yet he received a Nobel Prize! All of which prompted Desmond Morris (himself a far from typical investigator) to comment: "If a group of typical cautious, orthodox scientists were to make a careful examination of Lorenzian 'scientific method,' it is unlikely they would ever recover from the experience."[35]

▷ **Constance:** How so?

▷ **Imp:** Ah, a chance to tell one of my favorite stories![36] One of Lorenz's research foci was the concept of "intention movements"—that is to say, motions such as the head bobbing that is an alarm signal preceding flight in birds. His interest in the subject resulted from an unintentional experiment. He had trained a free-flying raven to eat raw meat from his hand, and he had been feeding it off and on for several hours one day as he walked around his farm visiting his animals. He would reach into his pocket and take out the meat, and the raven would swoop down to grab it with his beak. A form of Pavlovian training, I suppose. Lunch came, and having drunk copiously, Lorenz decided to relieve himself near a hedge. The raven saw him put his hand into what it took to be his pocket and pull out another morsel of meat, and swooped down, hungrily grasping this new mouthful in his beak. Lorenz howled with pain. But, as he said afterward, the event left a deep impression on him. About the importance of intention movements, that is.

▷ **Hunter:** A graphic demonstration of the concept.

▷ **Jenny:** Which, believe it or not, leads to the next point in our argument. Another facet of Fleming's research style was to invent simple ways to display complex phenomena—ways to make something swoop down and bite you, as it were. Something he learned from his mentor and boss at St. Mary's, Almroth Wright.

▷ **Richter:** May I interrupt? This is very amusing in its way. But is all this preparation really necessary for your explanation of Fleming's discoveries?

▷ **Jenny:** You'll never believe it otherwise!

Almroth Wright during World War I. (St. Mary's Hospital Medical School, London)

▷ **Richter:** I may not anyway.

▷ **Jenny:** A distinct possibility. But back to Fleming's tutelage under Wright. Wright was a master of technique, by all accounts.

▷ **Imp:** Absolutely. And since discovery always involves making the invisible visible, it's worth looking at some of Wright's technical tricks—like how to isolate lymphocytes from blood with just a pipette. Sound impossible? Not at all! Let the blood clot in the pipette. The lymphocytes migrate to the glass wall and stick there. Then the blood clot can be blown out and the appropriate medium for lymphocyte growth sucked back into the pipette. Ingenious! Or take his studies of gas gangrene during World War I.

▷ **Jenny:** With which Fleming helped.

▷ **Imp:** The problem was to determine whether individual patients were doing better or worse, and to test the effects of as many antiseptic agents as possible in as short a time as possible. The usual way to do this was to take counts of the bacteria from blood or serum samples, or to grow cultures and then count. Both were difficult and time-consuming. Wright figured out a faster quantitative measure. In gas gangrene the anaerobic microorganism produces gas. Wright took advantage of this characteristic. He'd put a serum sample in medium in a small test tube and plug the tube with vaseline, pushing the vaseline down into the tube till it touched the medium. Then he simply incubated it for a given time at a given temperature with several control tubes as references. The distance the vaseline plug was pushed up was a measure of the amount of gas produced, which was directly proportional to the number of bacteria. Brilliantly simple![37]

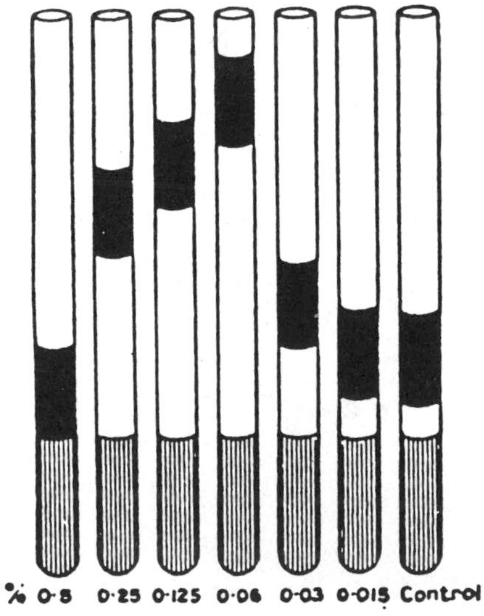

Wright's technique for measuring number of gas-producing bacteria. (Colebrook, 1954)

▷ **Jenny:** Fleming was no slouch at this sort of technical invention, either. For example, one of the things he and Wright noticed during the war was that while antiseptics seemed to help cut wounds, they were virtually useless against the sort of tear wounds caused by shrapnel. Why? Apparently Fleming imagined himself inside a tear wound and realized that there would be these sharp points of the tear ripping into the tissue—the sort of imagining we discussed last week.

▷ **Imp:** A cut wound, on the other hand, would be smooth.

▷ **Jenny:** Who's telling this, anyhow?

So Fleming took a test tube—Oh! I forgot to tell you that another of his little games was making things out of glass. Little glass mice, dogs, cats, unicorns—that sort of thing.[38] He also made his own glassware in the lab. So, he took a test tube and pulled long, sharp points out of it. Then he put bacteria in an agar culture in regular test tubes and in his specially prepared ones. Antiseptics were added in various concentrations to both sets of tubes, and the contents poured out after a time. New nutrient broth was then added to see if any bacteria had survived. Usually the bacteria in the plain test tubes didn't survive. The ones in the spiky tubes, however, always survived, no matter how concentrated the antiseptic.[39]

▷ **Imp:** Somehow the bacteria sporulated in the spikes or something. Anyway, it explains why topical antiseptics were useless for treating some kinds of wounds. Wasn't the type of infection, but the actual shape of the wound that was responsible.

▷ **Jenny:** And apparently such experiences led Wright and Fleming to other important conclusions, too. Fleming later told Ridley, another member of the St. Mary's team, "when I was a young doctor in the '14–'18 war, the Old Man [Wright] was very much concerned with the power of blood to kill bacteria by means of its own leucocytes and serums. But I realized that every living thing must, *in all its parts*, have an effective defense-mechanism; otherwise, no living organism could continue to exist. The bacteria would invade and destroy it."[40] Ridley believed that this was the basic idea which guided all of Fleming's subsequent research.

▷ **Ariana:** In other words, Fleming discovered lysozyme because he had a niche in his mind waiting for it.

▷ **Jenny:** Well, yes and no.

▷ **Imp:** Yes, Fleming had the right niche so he could accept what he found, but no, he wasn't looking for this hypothetical protective substance when he found it.

▷ **Constance:** So, serendipity. He starts out looking for one thing and discovers another, just like Berthollet, Pasteur, Minkowski, and the rest.

▷ **Imp:** And, like each of those men, against the prevailing dogma of his field. Elie Metchnikoff, probably the most famous immunologist of the early twentieth century, had stated that no internal protective agent such as Fleming presupposed actually existed. The skin acted as a mechanical

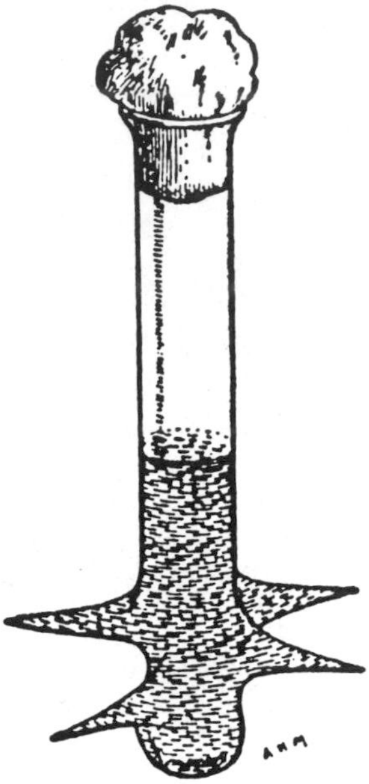

Fleming's "spikey test-tube" for testing the effectiveness of antiseptics in the treatment of tear wounds. (Colebrook, 1954)

barrier, and the lymphocytes killed anything that got inside. Nothing else was needed in his view.

▷ **Ariana:** Except that lymphocytes can't very well protect the eyes, nose, mouth, ears, and other orifices, can they?

▷ **Imp:** Apparently Metchnikoff didn't worry about that. Fleming and Wright evidently did, or were forced to by the war.

▷ **Jenny:** Imp, quiet! You're getting hoarse, and I'm perfectly capable of telling this.

So, once again to the history of lysozyme. Let's consider what Fleming did, given this new context of play, technical ability to display complex phenomena simply, and unsolved problems left by the war.

The first thing to consider is what Fleming said he did himself. Fleming's public account of his work is entitled "On a Remarkable Bacteriologic Element Found in Tissue and Secretions." It was published by the Royal Society in 1922. Fleming writes:

> *The lysozyme was first noticed during some investigations made on a patient suffering from acute coryza [a cold]. The nasal secretion of this patient was cultivated daily on blood agar plates, and for the first three days of the infection there was no growth, with the exception of an occasional staphylococcus colony. The culture made from the nasal mucus on the fourth day showed in 24 hours a large number of small colonies which, on examination, proved to be large gram-positive cocci . . .*[41]

▷ **Imp:** So who is this patient, and why cultivate his or her nasal secretions?

▷ **Jenny:** Imp!

▷ **Imp:** Don't stop me now! I'm having fun! If I can't talk tomorrow, so what? There's no doubt as to who the patient is: It's Fleming himself. On November 21, 1921, Fleming draws in his lab notebook a picture of a petri plate with a colony marked "Staphyloid coccus from A. F.'s nose."[42] So, Fleming did have a cold—or allergy—as Allison reported. That's important.

Note several other things. First, Fleming says he isolated on purpose the bacterium which Allison and Hughes claim was a contaminant. Since neither Allison nor Hughes have any way of knowing Fleming's intentions, I opt for believing Fleming. Besides, as we'll see, his story makes sense. Second, Fleming isolated other bacteria as well—not just the one contaminant. The rest were all staphylococci. Those bacteria are not affected by what turns out to be lysozyme. Fleming displays no interest in those at the time, but is fascinated by this unknown bacterium that's sensitive to lysozyme. Why? We'll get to that in a minute.

The important point here is that Fleming had some criteria for determining which bacteria were of interest to him. So we have to determine what those criteria were. And third, the crucial bacterium is not isolated until the fourth day of his very bad cold—that's an important point, too. He didn't just do his isolation once. He did it repeatedly. That indicates planning. He was looking for something.

The next important point is that Fleming often had colds. He had had his nose broken in an accident as a boy, and it led to chronic bouts of head colds.[43]

▷ **Jenny:** I already told them that.

▷ **Imp:** Sorry. So he was undoubtedly preoccupied with the question of what causes colds. Me too! My own nose was broken when I was twelve, and every winter is a continuous series of sneezes and sniffles, as I'm sure Jenny will attest.

▷ **Richter:** Is this really relevant?

▷ **Imp:** You bet! Every experiment has to have a motivation, and the personal involvement, intellectual or otherwise, of the investigator. Besides, it both poses the problem and provides the means of experiment.

Now comes the fun part! All of Fleming's notebook entries concerning what turns out to be lysozyme, from November 1921 through the beginning of 1922, are labeled as studies of "bacteriophage." That's "virus" for you modernists.

▷ **Constance:** Okay. So what?

▷ **Imp:** So what! Don't you see? Hunter? Richter? Ariana? Wow! Time to back up. Anybody remember how viruses, or bacteriophage, were discovered?

▷ **Ariana:** Felix d'Herelle. Around 1915, while studying a diarrhea of locusts.

▷ **Jenny:** Imp, you didn't tell me! All you said was diarrhea. But diarrhea of *locusts*. Who would think that such inconsequential work could yield such an important result?

▷ **Ariana:** He found an invisible agent, smaller than any known bacterium, that ate the gut bacteria. About the same time, a man named Frederick Twort also reported an agent that lysed bacteria. Then, a few years later, d'Herelle isolated from the stools of patients suffering from dysentery another invisible bacteria-eating agent that would pass through the finest filters known. He called it "bacteriophage"—a thing that eats bacteria.[44]

▷ **Imp:** Now can anybody see what I'm driving at? It hit *me* instantaneously!

▷ **Ariana:** Well, I've got an idea, but it's so crazy I'm embarrassed to suggest it.

▷ **Imp:** Then you've got it!

▷ **Ariana:** It's sort of a pun on two colloquialisms: Diarrhea is often called "the runs" and colds give us "runny" noses, so what if both have the same cause?

▷ **Imp:** Yes! Runny bottom, runny nose.

▷ **Richter:** You *must* be joking.

▷ **Imp:** Never more serious in my life! Why not? It's a perfectly respectable hypothesis: If bacteriophage invade the bacteria of the gut to cause diarrhea, why not bacteriophage invading bacteria of the nose to cause colds? Simple extrapolation. I mean, Fleming was almost right, you know! Colds *are* caused by viruses. It's just that they directly invade the host tissue instead of the bacteria. But nobody at the time knew that was possible. So Fleming used what he knew to form a pregnant analogy.

▷ **Jenny:** As Diderot once commented, "Ideas awaken each other . . . because they have always been related." What better way to describe Fleming's thinking?

▷ **Richter:** This is crazy! Even less likely than Hunter's account of Pasteur's "cosmic asymmetric force." I swear, you do a better job of demonstrating the irrationality of scientific investigators than I!

▷ **Hunter:** Why? What's so irrational about investigating an analogy? Given the same facts, both Imp and Ariana came to the same conclusion. And they can communicate their insight so we understand it. Sounds rational to me.

▷ **Richter:** No. Imp's analogy, as you call it, is—well, to be blunt, absurd, laughable.

▷ **Imp:** Then laugh! I did when I thought of it! Nobody said that the genesis of scientific ideas is intrinsically any more serious than the genesis of any other ideas. Why can't a hypothesis have the same genesis as a joke?

▷ **Constance:** Well, Arthur Koestler did suggest they have analogous structures. In fact, lots of new hypotheses have been deemed so crazy that they were perceived as jokes. For example, when H. J. Muller suggested in 1922 that viruses were little more than naked genes, Henry Fairfield Osborn thought Muller was joking and congratulated him on his sense of humor. And when J. J. Thomson announced the discovery of the first subatomic particles, electrons, a distinguished physicist later told him he thought the lecture was a gag.[45]

▷ **Hunter:** Yes. But we know who the joke was on!

▷ **Constance:** And Lewis Thomas has made the generalization that laughter is a sure sign of interesting and important developments, of surprise: "Whenever you can hear laughter and somebody saying, 'But that's *preposterous'*—you can tell that things are going well and that something worth looking at has begun to happen in the lab."[46]

▷ **Ariana:** Besides, who's ever met a creative scientist who didn't make puns, tease, or play practical jokes? Laughter's part of our lives! The point is that play with words—and the ideas and concepts they represent—is the stock-in-trade of the scientist as much as of the poet. And we expect the results of this play to be both rigorous and surprising; to avoid equally chaos and cliché.

▷ **Richter:** But, dammit, Imp's analogy is so weak! Runny bottom, runny nose. This is not a model of clear reasoning.

▷ **Hunter:** It's a model of inventive reasoning! Who cares whether the analogy is weak or strong if it helps us to perceive the problem in a new way? You yourself claim that it's not the hypothesis but the testing of the hypothesis that counts. So?

▷ **Ariana:** Ah! I have it! Look at it this way, Richter:

> *Assumptions, hasty, crude, and vain*
> *Full oft to use will Science deign;*
> *The corks the novice plies today*
> *The swimmer soon shall cast away.*[47]

That's A. H. Clough, 1840. You may not like the analogy, but it's a means to an end.

▷ **Constance:** Aha! Wait, wait! Yes, here it is. Sir Humphry Davy said in the very same year, 1840, that "in the early stages of discovery the imagination is often dazzled by the brilliance of new facts and trusts to weak or remote analogies."[48] Do you think there's a connection of some sort between Davy and Clough?

▷ **Ariana:** Why not? I doubt that Davy got his ideas of how to do science from Clough, but perhaps Clough read or heard Davy lecture.

▷ **Richter:** And perhaps this is all beside the point.

▷ **Ariana:** Why? Because two Englishmen in 1840 paraphrase one another and the similarities lead Constance and me to make a possible connection? It's precisely the sort of reasoning—recognition of a common pattern—that Imp proposes as the rationale for Fleming's work. Or is it the fact that—no, don't interrupt me—that these connections might turn out to be wrong? They might. So what?

Ever read about the invention of the stethoscope? Laënnec encounters a young woman in 1816 whose heart condition is impossible to diagnose using chest percussion because she's so fat. He recalls that sound travels through solid objects such as wood, so that if one end of a stick is scratched, the scratching can be heard when the other end is close to one's ear. He takes a sheet of paper and constructs an analogue to a stick by rolling it up. He places one end over the woman's heart, and the other end near his ear. He hears the heartbeat louder than ever before. Yet he never did realize that his analogy was incorrect.[49]

The point is that it made him try something he wouldn't have tried otherwise, and it's the *trying* that's important, because that's what yields surprises. Stop being so damned stuffy, Richter.

▷ **Constance:** Ariana's got a point. I understand Edison's inventions were all based on incorrect electrical theories, but the inventions still work.

▷ **Imp:** Exactly. It's not the correctness of the logic that counts, but the fact that there are reasons for trying something you have never tried before. And so we're back to Huxley's statement that next to being right, it's best to strive to be demonstrably wrong, for out of error will come enlightenment. But don't judge my analogy before you know the evidence upon which it's based and what kind of argument can be spun from it.

The diarrhea virus–cold virus analogy is not as ridiculous as it seems. First, consider appearances. The action of a drip of mucus on a plate of susceptible bacteria is to create a hole in the bacterial "lawn." Same with virus.

▷ **Richter:** But a time difference exists. Lysozyme works within minutes. Viruses may take a half an hour to several hours, depending on temperature.

▷ **Imp:** Absolutely. But that hadn't been established in 1921. Besides, how else can we explain Fleming's actions both before and after he adds mucus to his mysterious coccus? We know he has criteria for evaluating his experimental results as interesting or uninteresting, because he ignored the bacteria he isolated previously. The bacteriophage hypothesis provides the criteria, explains why he labels his notebook entries "bacteriophage," and suggests why he adds nasal mucus to the bacteria. D'Herelle's and Twort's procedures tell him how to look for his hypothetical virus, and how to recognize when he's found it.

Then again, look at what Fleming does a day or two after he observes the dissolving activity of his mucus, again noted under the heading "bacteriophage." The classic experiment in virology was and still is to grow your gut bacteria in a tube of broth, then add your filtered material containing viruses and incubate. Within a few minutes, and all in an instant (because the virus particles replicate as if by clockwork), the viruses rupture the bacteria, and the broth goes from cloudy to clear, just like that!

Now, what's the first experiment Fleming runs the day after he observes that his mucus lyses the bacteria he isolated from his nose? He grows the bacteria in a tube of broth. A day later, he filters a new mucus sample to get out the bacteria, adds the filtered material to the bacterial tube, and incubates. What happens? All of a sudden, almost instantaneously in fact, the tube goes from cloudy to clear![50] The same effect was observed with another nonpathogenic bacterium, and with cholera bacilli. Are these the actions of a man who knew what he was looking for and proceeded upon a planned course when his expectations were fulfilled, or those of a man who accidentally stumbled on an unexpected and puzzling observation and didn't know what to do next? I say Fleming knew what he was after: He labels the experiments "bacteriophage" and he does the classic bacteriophage experiment!

▷ **Richter:** A pretty story. But you've left out a key event. You claim Fleming isolated this strange lysozyme-sensitive micrococcus from his nose. That's impossible. Nasal mucus is a vast repository of lysozyme. The micrococcus couldn't have survived. It must have been an accidental contaminant. Chance strikes again.

▷ **Imp:** No! Think, Richter! Whether the micrococcus was isolated from Fleming's nose or arrived as an accidental contaminant is irrelevant. The outcome would be the same. The micrococcus couldn't survive in his nose if there were lysozyme there. It couldn't survive on the petri dish because Fleming had coated it previously with his nasal mucus.

▷ **Constance:** But—but then the whole experiment is impossible! It couldn't have occurred by chance and it couldn't have occurred the way Fleming recorded it in his notebooks!

▷ **Imp:** Let me give you a hint. Another of Imp's principles: At the root of every paradox is a false assumption. Does that help anybody?

▷ **Constance:** Well, are you hinting that perhaps Fleming added his mucus to the wrong plate or something like that? Such things have happened before.

▷ **Richter:** So chance or accident again.

▷ **Imp:** *Au contraire!* There is a possibility you've overlooked. Throw out the impossible and whatever's left, no matter how improbable, must be the answer. So: Perhaps there was no lysozome in Fleming's mucus when he had his cold.

▷ **Hunter:** Even I have problems buying that, Imp. You told us Fleming used his own nasal mucus to destroy the micrococcus.

▷ **Imp:** You aren't paying close attention. I said perhaps there was no lysozyme in Fleming's mucus *when he had a cold*!

▷ **Hunter:** Then what's the difference between mucus secreted during a cold and mucus secreted at other times?

▷ **Imp:** Exactly what I asked myself this week. Fortunately, I had a bad cold which provided both motivation and a source of experimental material.

▷ **Jenny:** You should have seen this guy! Sneezing and dripping into little vials for days on end. And running off to the lab yesterday, all excited.

▷ **Imp:** Hey! Why do you think I call this the "Phlegming" project—p-h-l-e-g-m? Anyway, it's certainly no worse than the early studies of digestion. At least I wasn't working with vomit or bowel movements.

▷ **Jenny:** Heaven be praised.

▷ **Hunter:** And?

▷ **Imp:** Well, what would you predict, Richter? Or Ariana? What happens to a secreting organ or nerve when you consistently stimulate it for a long period of time?

▷ **Ariana:** Cannon's classic work.[51] After a period of continuous stimulation,

Protein Content of Nasal Mucus during Severe Cold

Sample	Mg. protein/ml. mucus[a]
Normal mucus (avg. 3 samples)	87.7
Day 1 of cold (1:00 P.M.)	5.3
Day 2 of cold (10:00 A.M.)	7.6
Day 3 of cold (10:00 A.M.)	0.8
Day 3 of cold (9:00 P.M.)	0.6
Day 4 of cold (10:00 A.M.)	1.2
Day 6, normal (10:00 A.M.)	90.0

a. Determined using Biorad protein assay with serum albumin standard.

the organ stops secreting. There's a refractory period, and secretion resumes at a greater level.

▷ **Richter:** Same basic phenomenon in nerves.

▷ **Imp:** And the same thing with the secretion of lysozyme. Or, at least, protein. I didn't actually have a lysozyme assay to hand. By the third day of my cold—and I have to say, I was dripping almost constantly for four days—the protein concentration, and hence lysozyme, in my mucus was virtually nil! Presumably the same thing could have happened to Fleming. What do you think, Richter? Still implausible?

▷ **Richter:** No antihistamines? Other drugs?

▷ **Imp:** You impugn my scientific integrity!

▷ **Richter:** Then I can't object to your mechanism of action. But you've hardly got sufficient data to prove your point even in your own case—and no evidence whatsoever as to its applicability in Fleming's case.

▷ **Imp:** No, but I have enough data to make my hypothesis plausible, and the chance account implausible. I can explain the entire unfolding of the discovery without once having recourse to inexplicable events.

▷ **Jenny:** And if you keep this up, you won't be able to explain anything. You're beginning to sound like a frog. May I?

You see, unless the whole thing was a series of chance events, then the chance explanation doesn't make sense. As Imp shouted—or more accurately, croaked—at me, "One planned action screws it all up! If you plate the mucus, you have to have a reason for plating it. If you drop your mucus on a bacterial culture—whether it arrived as a contaminant or not—you have to do it on purpose!" And, when the mucus lyses your bacteria, you don't get excited for no reason. So either the discovery was totally planned or totally chance.

▷ **Richter:** Fine. I'll accept that conclusion. But I'll also point out again that any body of data can be given a rational explanation—

▷ **Ariana:** Or an irrational one, evidently!

▷ **Richter:** —but its rationality doesn't ensure its correctness.

▷ **Ariana:** Nor does its irrationality!

▷ **Hunter:** A matter of criteria of choice. Einstein once said that theories are like detective novels. We should prefer the ones that are internally consistent, follow a rational pattern, and don't depend on accidental or inexplicable events for their solution.[52]

▷ **Ariana:** Exactly. Aesthetics.

▷ **Imp:** It's more than that, even. It's a matter of recognizing that the whole debate over context of discovery and context of justification is based on the false premise that the two are clearly distinguishable. Look at what Fleming does: He looks for a virus, and in testing for virus activity discovers instead lysozyme. He never disproves his previous hypothesis that viruses cause colds—he doesn't falsify or verify it—he simply abandons it unresolved to explore a more interesting possibility. The tests *are* the discovery in this case. The tests both define the new problem and solve it. So there's no way to say that up to this point, the discovery is due to irrational elements of chance, history, psychology (or whatever) and then suddenly—boom—logic takes over; to this point we're dealing with the inexplicable context of discovery, and after this point, with the logic of validation. You just can't make the distinction. The two are inseparable parts of a continuous process of question and answer and question again.

▷ **Jenny:** A process I shall dare to interrupt before you lose your voice altogether. As hostess, I insist on a break.

▼ JENNY'S NOTEBOOK: Patterns in Mind Space

"That was very good," Ariana whispered to me. "I didn't know you knew so much history of science."

"I don't," I said. "Imp took care of the science; I worried about the history. Between us, it all worked out. He just needed distraction. Help me get the coffee."

As we went into the kitchen, I heard Constance ask Imp how Fleming determined that lysozyme wasn't a virus. He explained to her, as he'd explained to me, that the difference is fundamentally simple. A virus can replicate—it can copy itself. So if you put a few viruses into an appropriate bacteria culture, you'll soon have millions or billions more. Enzymes, such as lysozyme, can't replicate. Thus, you have no more lysozyme after it has destroyed the bacteria than before. In practice, this difference translates into a simple experiment. You take what are called "serial dilutions" of your bacterial cultures after the virus or lysozyme has been added. That is, you put in one milliliter of virus or lysozyme solution into ninety-nine milliliters of bacterial culture. You allow the virus or lysozyme to act. Then you transfer one milliliter of this solution to a new batch of bacterial culture. If the material is a virus, your one-milliliter solution will be at least as active in the new culture as in the old because the few viruses you added have replicated.

But if the active material is an enzyme or chemical, its activity will be one hundred times slower, because you've diluted it a hundredfold.

Ariana took in the cups. "Imp!" she burst out. "There's something I must ask you. I know Jenny won't like me taxing your voice, but: How did you invent your runny nose–runny bottom analogy?"

"Now that's an interesting question," I heard Imp say. "Let me think a few minutes."

When we were all settled again, Imp began. "Let's see. It went something like this—though not sequentially. What I experienced was an associative sort of thinking. First, I had this problem that the information concerning Fleming's discovery didn't fit together. Well, that's not precisely accurate. It fit together in various ways—the sorts of things we essayed today—but there was always some inconsistency, or a piece missing.

"Have you ever taken apart a clock, or something like that? And you put it all together again, but there's this piece left over. Well, that's what kept happening to me. And the pieces that kept getting left over were the ones having to do with Fleming's notebooks. Fleming says he was culturing his nasal mucus, apparently on purpose. Why? And if the lysozyme is on the plate, how does the micrococcus grow in the first place? These were problems that seemed to have no solutions. And then there was the mystery of the 'bacteriophage' headings to the experiments. Fleming discovers lysozyme, not a cold virus! What in heaven's name does bacteriophage have to do with anything? So I had these pieces of unintegrated information. But unlike a clock, which doesn't work at all if you leave out a key piece, I had no assurance that the notebooks were in fact key pieces in the structure of Fleming's discovery. I didn't know how many pieces there were supposed to be in my puzzle!"

Hunter nodded vigorously. "That's an excellent point which we must discuss in detail at some time. How, in real life, do you know which pieces are part of your intellectual puzzle and which are not? What we scientists are really faced with isn't a clock we've taken apart, but lots of clock pieces mixed up with pieces from all sorts of other machines. We have all the facts *and* all the artifacts of nature. We have to sort these out before we can put our puzzle together. Science isn't just a matter of putting a puzzle together, as Kuhn and some others imply; it's more a matter of determining what pieces belong in any particular puzzle. After that there's only a limited number of ways to put the puzzle together. Before that, the possibilities are almost infinite!"

"Yes," responded Imp. "Either infinitely large, because you can make virtually any combination if you're willing to discard a few pieces; or infinitely small, because there's no way to form a working pattern that includes extraneous elements."

"Right!" interrupted Ariana. "The bane of readers of detective stories. As Sherlock Holmes says, 'It is of the highest importance in the art of detection to be able to recognize, out of a number of facts, which are incidental and which vital. Otherwise your energy and attention must be dissipated instead of concentrated.'"[1]

"Exactly," continued Imp. "In this case I was quite certain that the bacteriophage piece fit into the puzzle because Fleming put it there himself. By writing it over and over again in his notebook, he effectively said, 'This is important. Don't ignore it.' I didn't." Imp coughed. "Excuse me."

"Okay, now, let's see. The next thing that happened was—no, wait. Really the first thing I did was to pretend I was Fleming. Like we talked about last week. I imagined I was sitting at his lab bench doing his work, thinking his thoughts. So I put myself in his position and kept running through scenarios that might lead me to write down the sort of jottings he left us in his notebook.

"Anyway, one thing I kept in mind was Hunter's statement that discoveries result when you plan experiments to look for one thing and see something else instead. So I backtracked mentally, and asked myself what I—Fleming—was looking for that led me—him—to find lysozyme. And that's when I realized the significance of the 'bacteriophage' headings in the notebooks. I—Fleming—was looking for bacteriophage! Obvious, once you see it. I had held the answer all along—I just lacked the right question!"

"A vindication of Richter's thesis about the importance of asking the right question," I added diplomatically. Richter grunted.

"Exactly!" continued Imp. "The next question was, Why look for bacteriophage? And that's when everything suddenly coalesced, all at once! I had all this information about Fleming's colds—I had a cold (that helped)—his culturing his mucus for bacteria, his labeling his experiments 'bacteriophage,' and then all this old knowledge about viruses emerged: that when *E. coli* in the gut are invaded by viruses, one gets lower-intestinal disturbances—if I may be crude, the 'runs'—and then it all fit together! Boom! The link was made—runs, runny nose, a viral cause for both. What could be simpler?"

"See, I told you you should read Wilson's *A Meeting at a Far Meridian*," I teased Imp. "The same sort of verbal pun provides a clue to solving the main character's problem. Something to do with gates or walls and electronics."[2]

Imp shrugged. "You don't need to turn to fiction—Ralph Lewis records having invented a hypothesis by punning. He was working on the dissemination of ergot fungus to rye by insects, but couldn't get the insects to pick up the spores in his experiments. Thinking about differences between the natural state of the spores and his experimental conditions led him to realize that whereas his spores were desiccated, the spores produced by the fungus were embedded in a liquid excretion known as honeydew. Lewis's immediate thought was: 'Would honey do?'

In fact, the fungal honeydew is a 2.33 molar sugar solution, and honey *will* do! The honeydew-spore combination attracts the insects, causes the spores to stick to them, and prevents their desiccation long enough for fungal development. Somehow Lewis managed to get his story into his dissertation, but it was not, of course, deemed appropriate for the published version.[3] I'll bet there are dozens of similar stories that never get beyond the lab—but that doesn't mean they're false or useless."

Hunter nodded. "Word similarities were important to Arrhenius, too. People in a variety of different chemical, electrochemical, and physical fields during the late nineteenth century talked about 'active' and 'inactive' molecules. Arrhenius was the first to investigate whether the verbal similarities might have a unified physical basis, too. Received a Nobel Prize for his trouble."[4]

But Ariana wasn't satisfied. "Your description is quite clear, Imp," she said, "but I want to know what was actually going on inside your head. What did it feel like when everything suddenly fit together? What *happened*?"

"I'll make you a deal," replied Imp. "I'll tell you what it felt like to me if you can describe what an orgasm feels like to you!"

"Oh, Imp!" I burst out, seeing Constance blush. But Ariana interrupted before I could scold him. "No, Imp has a good point. I suppose I am asking the impossible! I'll tell you what, I'll settle for less. When a novelist describes an orgasm, he or she doesn't describe what actually goes on in the mind, but rather provides a set of partial descriptions or analogies that bring to mind the desired feelings in any reader who has actually experienced an orgasm. Let's assume that discovering, or having an insight like yours, is a fairly common event, but limited in most people's lives to rather mundane problems like how to defrost a chicken in half an hour so I can make dinner tonight. Isn't there some common experience you can draw on to bring to our minds what you experienced?"

"Are you sure you're not a psychiatrist?" quipped Imp. "Look. I didn't use orgasm simply to shock you or to change the subject. The feeling you get when you've been working on an important problem for a long time and suddenly the answer comes to you in a flash is a very personal, very deep emotional and even physical experience. It verges on the sublime."

"It's what many of us live for," put in Hunter quietly. "Shuddering before the beautiful. I believe that's how Chandrasekhar expressed it.[5] You may experience it only once or twice in a lifetime, but I can tell you that it's a far more powerful addictor than any drug."

"That's it!" exclaimed Imp. "I *knew* I'd had that feeling before!"

"You've experimented with drugs, then," said Richter acidly. "It would explain a great deal."

"No, no," cried Imp, smiling gleefully, "it's the same sort of feeling, greatly magnified, that I get when I look at a stereoscopic view of a complicated molecule such as a drug! Maybe this analogy won't help some of you, but Hunter and Richter should empathize. Well, Hunter anyway," Imp teased.

"I've looked at stereoscopic drawings of chemical models, too," interrupted Constance, just as Ariana leaned over to me and whispered, "I doubt that Richter is capable of empathizing with anything!"

"Well, I haven't," I said, trying not to laugh. "Can't you dig out one of your books for me so I can empathize too?"

Imp searched for an appropriate example. "It's probably important to say that I get this big 'aha!' feeling only when I look at these drawings without a stereoscopic viewer. With the viewer everything appears three-dimensional almost instantaneously. But when I was a poor student, I used to look at these pictures by placing a notecard between the two images and putting my nose down to the card so that each eye saw only one image. It's much harder to get the images to merge that way, but more consonant with the feeling I'm trying to evoke. If that doesn't work, a couple other techniques are worth a try. Forget the note card. Focus on a distant object and then lower your gaze to the pictures, refocusing on the overlapping middle image. Or you can try crossing your eyes first and then refocusing on the middle image—that sometimes works for me.[6] Here, try this, Jen."

While I struggled to get the images in my right and left eyes to merge, Imp continued talking. "When you first look at one of these pictures you see two independent images, one in each eye. You *know*

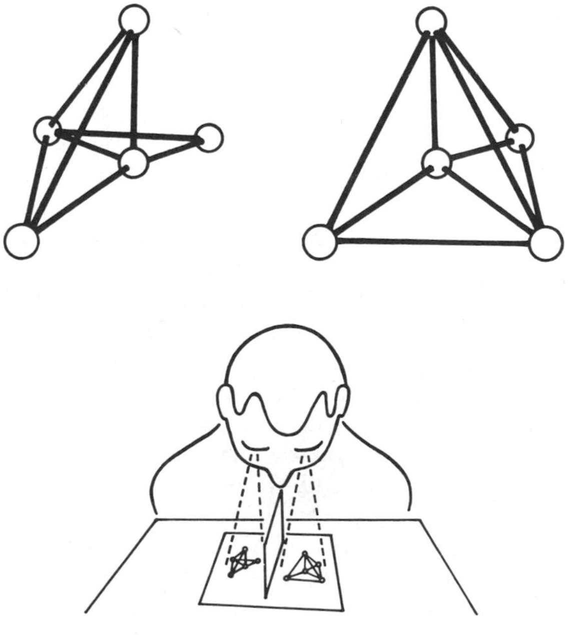

Top, *stereoscopic drawing;*
bottom, *directions for viewing.*

intellectually that the two images should overlap, but they don't. So there's a strain, a mental effort involved in making the images coalesce. With a big molecule, you can get a real headache!" I'll say!

Hunter nodded appreciatively. "Sometimes I even feel it in my stomach. Tension, I suppose."

"Right," continued Imp. "And then the two large images start to come together but, dammit, they don't quite merge. That's the frustrating part. Too many points. Sometimes you're not quite parallel to the images. Finally you hit the right way of focusing to make the first link, and presto! your two-dimensional pattern becomes three-dimensional. *That's* when you feel a tremendous wave of relief—a feeling right here in your gut—that this is the way it's supposed to be!"

"I *see* it!" I exclaimed. Talk about the sublimity of the mundane! Every day our eyes make a three-dimensional world out of two flat images, yet it takes a simple perceptual exercise like this to remind us of the process. "You've got to see this, Ariana!"

"Neat, huh?" Imp grinned. "Now just imagine the same sort of mental-physical resolution occurring not after a few minutes of frustration over some lines on a piece of paper, but after weeks, months, or even years of effort to solve some problem that you are convinced is fundamental. The elation, the physical release, is tremendous!"

Hunter nodded. "Absolutely. Remember Pasteur's reaction to his discovery of the handedness of the tartrates, or Crick and Watson proclaiming that they'd revealed the secret of life! Or the feeling of our own lesser successes. That's the feeling you yearn to have if you've never had it, and yearn to relive once you have!" Richter looked almost melancholy. Then Imp returned to business.

"Of course the stereochemical view of a molecule isn't a perfect analogy to what happened when I thought of the runny nose–runny bottom idea. In a stereoscopic image, both two-dimensional drawings show the same object, from slightly different orientations. With ideas, you're comparing two different images and searching for a very small region of overlap, which may or may not be there. Worse, your two images may not be aligned in the same perspective—as if your right eye sees one view of a molecule and your left eye sees the image turned one hundred and eighty degrees and twisted out of plane. You can't match them up that way. So you have to rotate the idea clusters around in your head, try various combinations of points to see if they overlap or not." I began to get a headache just trying to imagine idea clusters twisting in mind-space. "That's where clues, like knowing that bacteriophage is a common element, become crucial to aligning the data clusters. Such clues indicate probable regions of overlap and thereby minimize the number of possible orientations.

"There's another major limitation of the analogy, too. Stereoscopic vision can integrate only two data sets. What I was really doing in my head was comparing lots of data sets—data on bacteriophage, data on Fleming's notebooks, data on lysozyme's properties, data concerning other explanations of the discovery, data concerning unsolved problems, and so on. So the stereoscopic drawing is only a simplified analogy of what must be going on in the brain in n dimensions, where n is the number of discrete data sets that must overlap simultaneously to yield an n-plus-one dimensional picture."

"So maybe an n-dimensional Venn diagram would be a better model," suggested Ariana, looking up from the stereoscopic images.

"I don't know about better," responded Imp. "Different, certainly. The problem with a Venn diagram is that it doesn't yield the sudden 'aha!' feeling you get with a stereoscopic drawing when everything fits. And it doesn't give you the sense of added dimension in the final picture—the sense that the whole is more than the sum of the parts. I mean, we all had the parts of the Fleming puzzle before we had the whole picture, yet the whole picture gives us information that the separate parts don't.

"No, I don't like Venn diagrams in this case. But maybe there *is* an analogy to be made between the mental realm and the physical, along the lines of the second law of thermodynamics: It takes an excess of energy to bring about ordering; part of this energy is incorporated into the bonds linking the ordered parts; and this order, or 'negative entropy,' is a measure of the amount of information contained in the new construct. Maybe this sense of an added dimensionality is in some way a qualitative measure of the negative entropy of the new perceptual system."

Richter awoke from his reverie at this last statement of Imp's, stubbed out his cigarette, and scowled. But before he could say anything, Hunter added, "You might be interested in having a look at Brillouin's book *Scientific Uncertainty and Information.*[7] He's one of the founders of information theory, and he suggests that information can be measured by the reduction of entropy—that is, the disorder—in a local situation. Theories therefore create information by ordering facts so that they're less random than before."

"Which would explain why theories take so much energy to invent," nodded Imp. "You have to compensate for the increased order by an even greater dissipation of energy, right? Actually, I think Szent-Györgyi and Paul Weiss said much the same thing.[8] But then they did a fair amount of work on thermodynamics and theories of hierarchical organization, too."

"None of which has gotten us very far, in my opinion," countered Richter. "Look, you almost had me convinced that you were on to something. Then you ruined it with that last bit of nonsense."

"Oh, well." Imp grinned. "I always was one to push an idea too far!"

Richter shook his head. "Ignoring this thermodynamics nonsense, you've still overlooked a few things."

"Oh, good. I was so worried I'd having nothing to think about tomorrow!"

Richter sighed. "I see what you want to do: add pattern forming and pattern recognition to the pantheon of logical ways of thinking. Commendable in principle, damned difficult in practice—particularly if you insist on dealing with aesthetics. I'll leave you to it." He grimaced. "But look, you have two major problems. First, just because you make that jump from a two-dimensional to a three-dimensional image does not mean the link you make is right. You are, I assume, aware that the kind of impossible two-dimensional images Escher excelled at have stereoscopic counterparts as 3-D illusions.[9] I have several times had the feeling you speak of, only to find that I was wrong. Any two sets of randomly generated data are sure to have *some* overlap. Second, even if you do get a valid conjunction of ideas, how do you know it's the best possible conjunction? Do you stop with the first 'Eureka!' or do you look for more complete intersections? What are your rules—logical rules—for evaluating what you accomplish?"

"What can I say, Richter? You're absolutely right, of course. You can't stop with the 'aha!' feeling. You have to check it. You have to make sure all the essential points from your original patterns are incorporated into the new pattern. Lots of times they aren't, so you have to start again.

"As to whether you've found the ultimate overlap—well, isn't that what the history of science is all about? Each generation demonstrating a larger set of overlaps than the generation before? Personally, I'll be satisfied to come up with a new, reproducible image that either combines currently disparate elements or unifies some existing images. If I wait for the ultimate synthesis, ha! I'll be dead!"

"That's where we differ, you and I," replied Richter. "You have a capacity to tolerate endless ambiguity and incompleteness. I find them abhorrent." It struck me that Richter had uttered a profound truth about himself.

What could one say to that? The conversation lagged, and Ariana and I put away the stereoscopic images.

I forced Imp to be quiet and drink two more cups of tea before we continued.

◀ TRANSCRIPT: The Fun of Discovering

▷ **Jenny:** Imp, are you sure you want to go on? You're looking pale.

▷ **Imp:** Never too tired to have fun! Let's polish off this chance-versus-planning thing once and for all! I'd like to present my explanation of Fleming's discovery of penicillin, and then we can consider whatever remaining cases Richter and Hunter care to hash out between them. However, since my voice is going, I will make one concession to Jenny—I'll let her talk for me.

▷ **Jenny:** Until you can't bear your own silence, anyway. Well, to begin, the key passage upon which the chance discovery of penicillin is based is by Fleming himself, and appears in his first publication on the topic:

> *While working with staphylococcus variants a number of culture plates were set aside on the laboratory bench and examined from time to time. In the examinations these plates were necessarily exposed to the air and they became contaminated with various microorganisms. It was noticed that around a large colony of a contaminating mold the staphylococcus colonies became transparent and were obviously undergoing lysis.*[1]

The mold was *Penicillium notatum*, of course (not that I knew that a week ago!), and its product, penicillin, has saved millions of lives.

I must point out, however, that there are no contemporary sources, not even notebook entries, to corroborate this passage; and the next reference to the discovery in Fleming's corpus of writing is not for sixteen years—that is, not until 1944, when Fleming had already become a public figure as a result of the clinical success of penicillin.

▷ **Constance:** Yes, I dedicated some of my own research this week to Fleming's case, too, and from what I can gather it turns out that Fleming couldn't have done what he says he did. Hare writes that it isn't possible for the *Penicillium* mold to kill off bacteria when placed on a previously staphylococcus-sown plate, because penicillin kills only cells in the process of mitosis.[2] Once the colonies of cells on a culture plate are visible as colonies, it's too late. Often, *Penicillium* won't even grow on a plate on which pathogenic bacteria are already present. Macfarlane reports that several investigators discovered these things while trying to repeat Fleming's work during the 1940s.[3]

▷ **Ariana:** Is it possible that the contamination occurred at the same time the staphylococci were plated?

▷ **Imp:** Probably not. The work of Hare, D. B. Colquhoun, and Fleming himself shows that the *Penicillium* must actually be fully established several days before the pathogenic bacteria are introduced, if it's to be an effective antibiotic. And in that case, the mold would have been readily apparent to Fleming before he added the bacteria.

▷ **Hunter:** In which case, Fleming would have deliberately added the bacteria to a contaminated plate.

▷ **Constance:** Exactly what Colquhoun suggests.[4] But on the other hand, Hare claims that Fleming *did* do what he said he did.

▷ **Ariana:** How can that be?

▷ **Constance:** Well, like you, Hare suggests that the plate was contaminated at the same time the bacteria were plated. The mold grows better at room temperature, the bacteria at body temperature. Hare hypothesized that Fleming did not incubate his culture plate. Apparently one way to distinguish among staph strains is to grow them at room temperature; it brings out colors in them that don't appear when they're incubated at body temperature.[5]

▷ **Ariana:** A trick he learned from his bacterial paintings, no doubt!

▷ **Constance:** Anyway, if the air temperature had been particularly cold, mold growth would have been favored. If it then turned warm, the bacteria would grow. Such a scenario would permit the chance observation. As it turns out, Hare found that from July 27 to August 6 it *was* cold, about fifty-six to sixty-four degrees Fahrenheit. Then it got warm, with temperatures reaching seventy-five.

▷ **Ariana:** And that's a big enough difference to make Fleming's account plausible?

▷ **Constance:** Apparently. Hare set up the experiment and produced a plate that looks just like Fleming's; and Hughes got the same results when he reproduced the conditions, too.[6]

▷ **Richter:** Then Fleming's discovery was not only chance, but triple chance. Chance that the *Penicillium* spore landed on his culture plate. Chance that Fleming failed to incubate that particular plate. Chance that the weather was perfect to bring about the crucial result.

▷ **Constance:** Oh, but there's more! You see, Fleming supposedly failed to see that the mold had killed the bacteria. After he first examined the plate, he discarded it into a Lysol bath, where it somehow failed to be immersed. Then he invited a colleague, Merlin Pryce, to see some of his results and rescued the culture from the Lysol bath, whereupon he noticed that the bacteria were translucent, or lysed, in the region around the mold.[7]

▷ **Richter:** Answer that, Imp!

▷ **Imp:** With pleasure. Hare may be a good bacteriologist, but I question his historical acumen. Dates—you've got to pay attention to dates, Jenny always tells me! Assumptions, I shout back! Assumptions—excuse me; voice won't take this excitement—assumptions are the key! So Jenny and I have had a good look at dates *and* assumptions, and there's a lot that still doesn't hang together, right, Jen?

▷ **Jenny:** The first point is, as I started to say before, that there are literally no contemporary accounts of what happened in Fleming's lab. As far as anyone can determine, there are no notebook entries corresponding to the sort of experiment or accident reported by Fleming. There are no relevant letters. There are no journals or things of that nature. Basically, nobody—including Fleming, apparently—perceived the importance of his experiments. Thus, when Pryce and other visitors to the lab recalled what had taken place, it was after 1944, when Florey and Chain and their coworkers had transformed Fleming's toy into a miracle drug. So we're talking here about recollections made at least sixteen years after an event which no one took seriously at the time. In fact, no one was subsequently able to pinpoint when Fleming first showed him a penicillin plate, beyond the fact that it was in the autumn of 1928. So there's no way, either by personal recollection or by lab notebooks, to verify Hare's hypothesis.

And there's an important fact about Fleming's notebooks that you should know: They were looseleaf notebooks, not bound. And the pages weren't consecutively numbered until after acquisition by the archivists. So the fact that there are no entries for the origins of the penicillin work could be due to one of two factors: Either Fleming never made any entries because there was nothing to record; or he destroyed or lost the entries he had made.

▷ **Constance:** But why?

▷ **Jenny:** Well, consider for a moment. It's 1945 or so. You've suddenly become world famous. You've just received the Nobel Prize. In 1929 you wrote an account of your discovery that makes it appear to have been chance. But your notebooks show that you're playing around with something kooky, like Pasteur's cosmic asymmetric force idea. Since the pages aren't numbered, you "mislay" the crucial ones.

▷ **Imp:** Or you just take them out to examine them for some reason, like writing a speech or a review or something, forget to replace them, and they get lost, or misclassified in the archives, or who knows what.

▷ **Constance:** Or, in light of previous discussions, Fleming may have recorded what he did but in such a way that it looks like some other type of experiment.

▷ **Jenny:** Interesting thought. Bear it in mind while we consider more dates. For example, Hare believes that the contaminated staphylococcus plate was sown at the end of July 1928. Fleming was on vacation all of August, followed by the warm spell in September. So Hare suggests that Fleming probably saw his plate the first week of September. Problem one: The first notebook entry concerning any mold or its effects on bacteria is dated October 30. What was Fleming doing for two months? Problem two: The experiment that is recorded is not his observation of an accidental contaminant on a plate of staphylococci, but clearly a planned experiment in which the mold was carefully plated and its effect on eight different bacteria tested.[8] Why is the original observation missing, what was it, and when did it occur?

▷ **Ariana:** Maybe Fleming didn't realize the significance of his observation at first, so he didn't record it. Just kept the petri dish around and realized weeks later the significance. We have a precedent for that with Pasteur. Sorry, but I still don't see why we can't just accept Fleming's published version, as fleshed out by Hare. We accepted Fleming's published version concerning lysozyme . . .

▷ **Imp:** True. But consider: Even in the lysozyme story, Fleming covered up some of the facts. He never mentioned bacteriophage publicly. What he did publish, however, agreed with his notebook version, and the plausibility of both versions can be confirmed readily by experiment. In the penicillin case, Fleming's published account corresponds to nothing written in his notebooks and is experimentally so unlikely as to be questionable. Besides which, it seems to me that Hare's reconstruction represents a case of the highly improbable possible. Following the writers of detective fiction, I'd prefer a more probable impossible explanation!

▷ **Richter:** I fail to see why Hare's experimental explanation is "so unlikely as to be questionable." Do explain.

▷ **Imp:** The correlation of unlikely events, for one thing. I've nothing against a person perceiving a common, mundane event in a new light because

Reconstruction of Fleming's laboratory as it was in 1928. (St. Mary's Hospital Medical School, London)

he suddenly asks a new question. But really, a man who states he's looking for an antibiotic substance just happens to go on vacation at precisely the time of year and during the unusual weather conditions necessary to allow a plate of staphylococci to display sensitivity to an almost unknown form of *Penicillium*? Come on!

▷ **Jenny:** Besides, there are further problems with Hare's chronology. If we accept Hare's account, then two months pass before Fleming writes down an experiment in his notebook. You, Ariana, suggest that he didn't comprehend what he'd seen. Plausible, except that Pryce says, "What struck me was that [Fleming] didn't confine himself to observing, but took action at once. Lots of people observe a phenomenon, feeling that it may be important, but they don't get beyond being surprised—after which, they forget. That was never the case with Fleming."[9] According to Pryce, Fleming immediately took a piece of the mold and subcultured it in a tube of broth. So if the early September date is correct, then Fleming actually began experimenting with *Penicillium* almost two months prior to entering any results in his notebook. Curious, to say the least! Furthermore, his October thirtieth experiment assumes previous work. So what has he been doing?

▷ **Richter:** Figuring out how to reproduce the accidental contamination, perhaps. It might take that long to figure out that the *Penicillium* had to establish itself first.

▷ **Imp:** Damn! I hadn't thought of that!

▷ **Jenny:** It may be irrelevant anyway. There's reason to believe that Hare's chronology is incorrect in another respect. The text accompanying a drawing of the original contaminated plate, if that's what it is, reads: "On a plate planted with staphylococci a colony of mould appeared. After about two weeks it was seen that the colonies of staphylococci near the mould colony were degenerate."[10] "After about two weeks," says Fleming—but Hare claims Fleming was on vacation for about five. The time frames don't match. So we have to either ignore Fleming's account again, or question Hare's.

▷ **Hunter:** But can we believe this statement of Fleming's? If the *Penicillium* must be established before the staph, and Fleming says it occurred the other way around . . .

▷ **Ariana:** Well, what about the petri dish Fleming identifies as the original?

▷ **Imp:** Ah, yes, what about that petri dish? Jenny, why don't you give our friends a rundown on what we found while I set up the slides.

▷ **Jenny:** Okay. We've located four different versions of the original plate. That is to say, the petri dish illustrated in each is the same, but a slightly different text accompanies each. One version is a drawing; three are photographs.

Ready, Imp? All right. The first version is the photograph reproduced in Fleming's 1929 paper. It's labeled, and I quote: "Photograph of a culture-plate showing the dissolution of staphylococcal colonies in the neighborhood of a penicillium colony."[11] No mention that this is the original, contaminated plate. The only other reference to the plate is in the text, in which Fleming says that some plates "became contaminated with various micro-organisms. It was noticed that around a large colony of mould the staphylococcus colonies became transparent and were obviously undergoing lysis."[12] This passage can be interpreted in two ways: first, that the figure is an example of the sort of phenomenon Fleming observed, specially prepared for illustrative purposes; or, second, that this *is* the plate he observed. The point is that no direct reference is made in Fleming's paper to the plate in the photograph as having provided the original observation.

▷ **Ariana:** Wait! So your point is that everyone has assumed the plate illustrated by Fleming is the original, but it may not be?

▷ **Constance:** But I thought Fleming said somewhere it *was* the original!

▷ **Jenny:** Well, he does, but only many years after the fact. A drawing reproduced in Maurois is labeled "Anti-bacterial action of a mould (Pen-

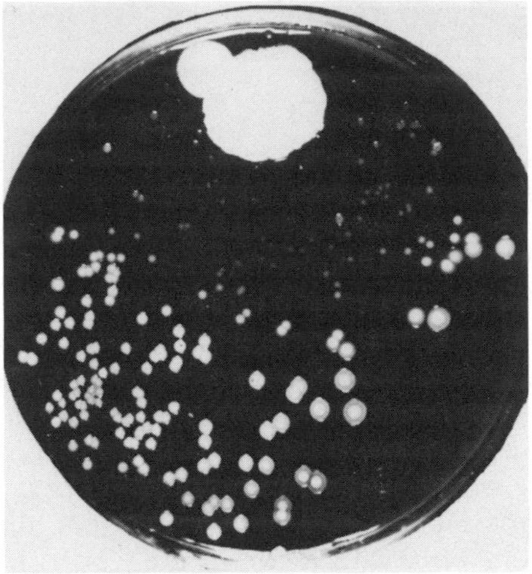

Photograph of "original" penicillin plate. (St. Mary's Hospital Medical School, London)

icillium notatum)."[13] The *P. notatum* label is the giveaway. Fleming thought his mold was *P. rubrum* until mid-1930, when Charles Thom identified it properly. So the drawing postdates 1930.[14] Other evidence (drawings in the same style) suggests it was prepared for a lecture after 1944, when the subject became "hot." A photograph reproduced in Hughes is labeled "Print of the culture plate which started the work on Penicillin" and states that the plate is twenty-five years old.[15] Another photo reproduced in Taton is of the identifying sheet for the plate when it was given to the British Museum sometime after 1945. It too says, "Original culture plate on which penicillin was observed."[16]

Finally, there's a reproduction of the 1929 photo in a 1944 paper by Fleming about how he discovered penicillin. Here, although Fleming states that "after a lapse of fifteen years it is very difficult to say just what processes of thought were involved" in the discovery, he nonetheless unequivocally describes the 1929 photo as the "original culture plate with the mould colony inducing dissolution of the staphylococcal colonies." However, he also states in the same paper, "What had originally been a well-grown staphylococcal colony was now a faint shadow of its former self."[17] So we're stuck. There is no unequivocal evidence from before 1944 that the 1929 photograph represents the original contamination. If we accept Fleming's further statement that the staph colonies were well established *prior* to the contamination and that he observed the mold after two weeks, then we must throw out Hughes's reconstruction. The discovery is literally impossible as Fleming presented it.

▷ **Richter:** So what are you suggesting?

▷ **Imp:** A case for Macfarlane's law. The same Macfarlane who wrote Fleming's most recent biography is a well-known pathologist and Fellow of the Royal Society. He once suggested that "when a number of conflicting theories co-exist, any point on which they all agree is the one most likely to be wrong."[18]

▷ **Ariana:** Exactly. In my experience, controversy occurs when neither of the two theories corresponds to the truth.[19]

▷ **Constance:** So on what point do the conflicting theories agree?

▷ **Jenny:** All accounts agree that the plate illustrated in Fleming's 1929 paper was the plate that was accidentally contaminated. Everyone who has attempted to explain Fleming's discovery—Bustinza, Colquhoun, Hare, Hughes, Macfarlane—have concentrated on explaining that "original" plate. Yet we have no direct evidence that the plate *was* the original, or what it was the original of: contamination? penicillin activity? By 1944, Fleming himself may not have known, if he kept no records. Furthermore, the plate we've just seen is so clear an example of bacterial antagonism by a fungus that nobody, let alone an expert like Fleming, could have overlooked it and discarded it into a Lysol bath! We need a different explanation. Something vague rather than striking. Something easily overlooked.

▷ **Constance:** I don't know. Sounds to me like you're discarding not only the plate but Fleming's later descriptions of it as well.

▷ **Imp:** No, those we reinterpret. Assumptions can be jettisoned at will, but heaven forbid we break Richter's "law" that all data are valid and must be explained in terms of fact, artifact, or anomaly. What I'd like to do is

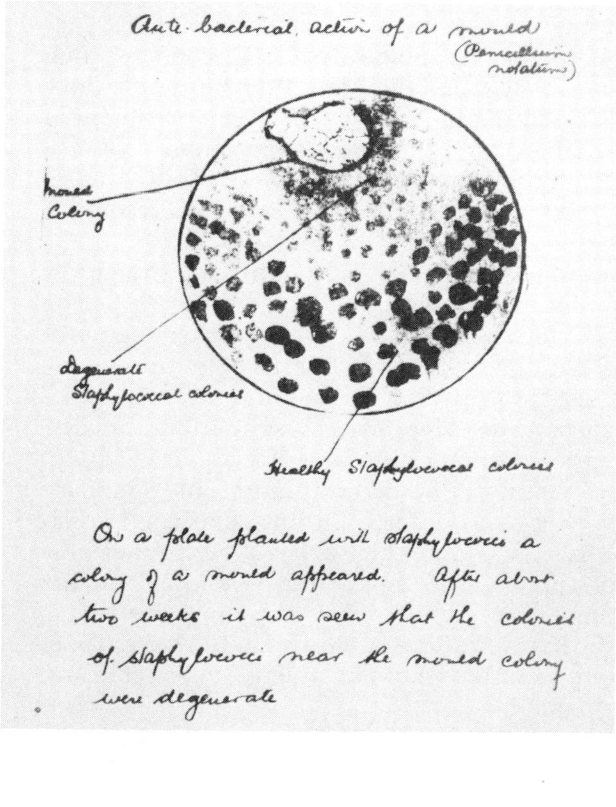

Fleming's drawing of the "original" penicillin plate, ca. 1944. (St. Mary's Hospital Medical School, London)

recategorize some facts as artifacts, and bring to the fore some facts that have been largely ignored in previous accounts.

For example, there's another striking feature of Merlyn Pryce's visit to Fleming's lab. According to Maurois, Pryce said that during his own bacteriological research, he himself "had often seen old microbial colonies which for various reasons had dissolved. He thought that probably the mold [on Fleming's plate] was producing acids which were harmful to the staphylococci—no unusual occurrence."[20] I stress the "no unusual occurrence." Fleming himself commented in 1946 that "the fact that bacterial antagonisms were so common and well known, hindered rather than helped the study of antibiotics as we know it today."[21] And Pappacostas and Gaté had published an entire book reviewing examples of such antagonisms in the very year of Fleming's research on *Penicillium*![22] So why should a fairly common occurrence, easily dismissed by Pryce, be so important to Fleming? Eh, Richter? The interpretation problem again.

▷ **Richter:** Fine. So what was this great expectation Fleming was dying to gratify?[23] Antiseptics?

▷ **Imp:** Lysozyme! I believe Fleming was looking for new sources of lysozyme. Fact: Fleming's colleagues were almost all under the impression that the mold Fleming showed them in the autumn of 1928 was producing lysozyme—"another of Flem's little tricks," they commented—and Macfarlane reports that "Fleming himself at first thought this was the case."[24] But Fleming would have known from his previous research that lysozyme had no effect on staphylococci. So how could he consider such a possibility now? Either he knew from the start that he didn't have lysozyme, because the mold lysed staph; or he hadn't observed the mold activity on a staph plate at all, but did the staph experiment *after* observing some previous intriguing activity displayed by the mold. I therefore suggest that Fleming's original observation resulted from a lysozyme experiment.

▷ **Constance:** But why? Why would Fleming stil be looking for sources of lysozyme when he'd discovered it seven years before?

▷ **Imp:** Ah! You persist in perceiving discoveries as events rather than processes. Look, you don't just sit down one day, observe something new, report it, and immediately begin to search for some other novelty. Discovering is a process that spans time. It requires confirmation of observations, reinterpretation of prior results, rethinking of hypotheses, making of predictions, further testing, the determination of boundary conditions, extrapolation to new cases, integration into existing knowledge, a change of normal activity in Kuhn's sense, the definition of new problems, the invention of applications, and so on. Fleming and his coworkers were still reporting new research on lysozyme into the 1930s. And they were obviously having fun doing it. Between 1922 and 1928, for example, Fleming's group examined human mucus, tears, sputum, many tissues and organs, blood, leucocytes, plasma, and serum. A fishing trip led him to try pikes' eggs, and then the eggs of dozens of other fish. He also examined the eggs of every bird he could lay his hands on. Tears were collected from horses, cows, hens, ducks, geese, and fifty other species kept by the London Zoo. Earthworms and snails had their slime investigated. Flowers whose parts were tested included tulips, sunflowers, carnations, lupine, buttercup, elder, dock, nettle, poppy, candytuft, and syringa. Among others. Large numbers of vegetables, plants, and their roots and tubers were tested.[25] Is it too much to suggest that he also examined any fungus that happened to come his way? Particularly since Papacostas and Gaté had just published a book stating that microbes have chemical defenses against one another?

▷ **Ariana:** All of these things yielded lysozyme?

▷ **Imp:** Yes.

▷ **Ariana:** I wonder. You know, of course, that just in the last decade three new classes of peptide antibiotics have been isolated: Boman found cecropins in moths; Lehrer, here at UCLA, found defensins in mammalian

Acquiring tears for lysozyme research. Drawing by J. H. Dowd, 1922. The tears were actually produced by squirting lemon juice in subjects' eyes. Egg white soon replaced tears as the major source of lysozyme. (St. Mary's Hospital Medical School, London)

neutrophils; and, most recently, Zasloff isolated magainins from frog skin.[26] I can't help wondering whether Fleming came across some of these, or others we don't know about, but mistook them for lysozyme.

▷ **Imp:** It's possible. He wasn't doing much more than characterizing his material as proteinaceous and antibacterial. Perhaps the time is ripe to take a second look at his results. Could be very exciting!

▷ **Richter:** Sounds plausible. But let's get back to penicillin. You do not deny that the *Penicillium* arrived by chance?

▷ **Imp:** No, why should I? Hell, Fleming clearly experimented with anything he could lay his hands on, wherever he found it. He tested lysozyme on any bacterium he found, at random, whether he knew what it was or not. That was part of his research style. Playing, remember?

▷ **Ariana:** Carefully planned disorder—that's what I call it.

▷ **Imp:** Absolutely. Look, if Fleming were investigating molds for lysozome activity, it explains virtually everything. Okay, yes: Some of Fleming's cultures *do* get contaminated. Okay, yes: Fleming thinks that's interesting and cultures the mold. But now—a blank. None of the records tells us what he did next. Fleming's account is impossible. It contradicts Hare's reconstruction. We have a two-month hiatus even if we do accept Hare's account. So try this instead. Fleming says, and I'll quote the passage again:

Fleming in his laboratory, 1944. (Bettmann Archive)

> *While working with staphylococcus variants a number of culture plates were set aside on the laboratory bench and examined from time to time. In the examinations these plates were necessarily exposed to the air and they became contaminated with various microorganisms. It was noticed that around a large colony of a contaminating mold the staphylococcus colonies became transparent and were obviously undergoing lysis.*[27]

Suppose all of this is true, but Fleming merely left out a few steps in the account. Yes, he observed some contaminants, including at least one he'd never seen before: *Penicillium notatum*. Was it on a staph plate? Maybe yes, maybe no. He says only that it was found *at the time* he was working with staph variants.

What next? Well, I think that we can now fill in what may have happened between mid-September 1928, when he probably first spots the mold, and 30 October, when he records his first *Penicillium* experiment. I think we can accept Hare's idea that the mold grew during a cold snap. That makes sense. Fleming finds it on one of his cultures. He transfers some of the mold to a new petri dish using ordinary nutrient agar and, once it gets established, adds some *Micrococcus lysodeikticus*—his standard test for lysozyme activity. The results are weakly positive, because penicillin does kill the micrococcus, but not very efficiently.[28] Fleming at first overlooks the dissolution (or more likely, the interference in the growth) of the micrococcus and almost discards the plate into the Lysol bath, as reported by Pryce. Upon reexamination, he observes the weak activity. This is the experiment he shows his colleagues. Not very novel. Not very impressive. Not even worth recording. Just another lysozyme experiment, with a mold this time. A couple of weeks gone.

Next, having observed the mold's activity, Fleming (as recounted by Pryce) takes a bit of the mold and cultures it in Sabouraud's medium. Indeed, Fleming tells us as much in his 1944 reconstruction of the discovery, and this makes sense. It takes a couple of weeks to get a well-developed culture. He now sets up controls to verify that the mold is secreting lysozyme by testing its activity on a series of bacteria whose sensitivity to lysozyme is already known. Standard procedure by this time. He sets up the test around the twentieth of October, growing the mold for five days before adding various *coli*, hay bacillus, and some of the staphylococci he's got lying about. Note: He doesn't use *lysodeikticus* because (I infer) he's already tried that, as we know from his 1929 paper on *Penicillium*. He records the results of this experiment in his notebook on 30 October because he gets a big surprise. The mold affects a whole host of pathogens, including staph, that aren't fazed in the least by lysozyme! He makes a special point to write beside his drawing of the plate, "Therefore mould culture contains a bacteriolytic substance for staphylococci."[29]

Now, I ask you: If you had first observed this mold clearly dissolving staph cultures in early September, as Hare maintains—if you already had in your hands the beautiful plate that everyone thinks is the original contamination—would you bother recording such a conclusion some six weeks later in the context of another experiment? I doubt it. I think the October 30 experiment is exactly what it seems to be within the context of the notebooks: the first penicillin experiment—the first recognition by Fleming that he's dealing with something unexpected and exciting!

▷ **Ariana:** And the so-called original plate?

▷ **Imp:** A carefully prepared demonstration plate, later misidentified either to simplify the logic of presentation or, after a fifteen-year lapse, out of forgetfulness. You see, we keep forgetting two things. First, that Fleming's an expert at making phenomena stand up and shout, "Look at me!" Can we really believe that this accidental contamination was not only the first but the *best* illustration of the phenomenon that Fleming observed during his six months of experimentation prior to publishing? I doubt it. I say the plate was carefully designed. And, second, you've got to remember that the so-called original plate still exists in the British Museum. That means Fleming had to fix it with formaldehyde within a relatively short period of time or the mold would have overgrown the plate. If he knew the plate was that important, why did he fail to record it in his notebooks, and why the conclusion concerning *Penicillium*'s antibacterial activity in the 30 October entry? Also, having rarely, if ever, worked with fungi before, how did he know that the plate could be preserved with formaldehyde?

▷ **Ariana:** Perhaps from his bacterial paintings, again.

▷ **Constance:** You know, Fleming did write that despite the role of chance in his observations, "All the same, the spores didn't just stand up on the agar plate and say 'I produce an antibiotic,' you know."[30] Which brings us back to Pryce's statement that he thought the mold was just producing some acids. In fact, in 1947, a guy named W. M. Scott reported making the same basic observations as Fleming, but he hadn't followed them up.[31] You can't just see a phenomenon and know what's happening, can you? Even according to Hare's account.

But I guess what I don't understand is why Fleming didn't just say he'd found penicillin while looking for lysozyme, if that's what really happened. Why the story about the "original plate"?

▷ **Imp:** Well, in the first place, if you read his 1929 paper carefully, Fleming didn't say anything about the so-called original plate. He says a culture got contaminated. He cites the photograph as an illustration of the phenomenon of interest. Second, as I'm sure you know, Constance, the logic of presentation rarely corresponds to the logic of discovery. Nobody actually writes out for peer-reviewed publication how they actually got their results, because, as Alan Hodgkin says, a lot originate in "perfectly dotty" ideas, which would be laughed at. So scientists invent simplified, logical descriptions for presenting their data in the least confusing and most convincing manner.[32] Better to leave out what really happened than raise doubts about one's intellect!

I mean, just imagine for a moment trying to write into a research paper the account I've just given: "While looking essentially randomly for organisms producing lysozyme, a common but unidentified mold was isolated from the air of the laboratory. Initial experiments showed that the mold appeared to have lysozyme activity, so controls were set up, including staphylococci, which I just happened to have been working on at the time. Much to my surprise, the mold had unexpected properties, so I was now forced to further characterize and identify the mold, et cetera. This subsequent research conclusively demonstrated that the product of the mold was not lysozyme, but rather a new substance having the following characteristics . . ." Too complicated and indirect. Much more to the point to start with the mold lysing the pathogen, because that's the important and novel observation.

Personally I'd have liked a version that read, "I like to play with microbes. One day, just to see what would happen, I . . ." But of course, that wouldn't do either. Scientists aren't supposed to be playing. So Fleming kept his playing private. As most of us do—but not all, I must say. Szent-Györgyi once discovered a new carbohydrate derivative, and, having no idea what its constitution was, named it "ignose"—"ose" being the suffix used by chemists for sugars and carbohydrates, and "igno" for "ignosco," or "I don't know!" The editor of the *Biochemical Journal* returned his paper, saying there is to be no joking about science. Szent-Györgyi responded by suggesting an alternative: "godnose." I'll let you guess the editor's response.[33]

Szent-Györgyi's substance, by the way, turned out to be ascorbic acid, or vitamin C, a discovery which helped win him his Nobel Prize. It's perhaps worth thinking about the sort of personality that can joke about something as important as that.

▷ **Constance:** Ah, that's all very well, but is your account of Fleming's discovery true?

▷ **Imp:** Godnose! Or, in my case, runny nose knows! That's not the point. I'm trying to demonstrate to you that any given set of data is compatible with a range of alternative explanations, each determined by the assumptions and the aesthetic criteria you employ in choosing, weighting, and organizing the data. After today, I hope no one will raise the awful chimera of "facts" to confuse the issue, because there are no such things except as defined by a hypothesis or theory. In short, the facts are context dependent, as much as it may hurt to hear it!

▷ **Hunter:** Which suggests to me that there may be yet another way to look at Fleming's work that relies on neither chance nor logic, but on probabilities. Try this. Forget the specific experiment we've been perseverating about. Think instead about its context and the probability that it would occur. Fleming clearly had a problem in mind. His research assistant at the time of the penicillin research, Stuart Craddock, recalled that "If [Fleming] told me once, he told me a hundred times, that the only usable antiseptic would be one which would arrest the growth of microbes without destroying tissues. On the day when such a substance should be discovered, he said, the whole treatment of the infections would be transformed."[34] So Fleming knew what he was after. Pasteur's prepared mind. In Richter's terms, he had defined a problem area and specific criteria by which to recognize a solution. That is not chance. So already Fleming has a higher probability of discovering penicillin than bacteriologists who don't have this problem in mind.

Then there's Fleming's mode of working. We've seen that he liked to play with microbes, by which he meant sometimes breaking the rules of standard practice. He often gave advice like "Never neglect any appearance or any happening which seems to be out of the ordinary: more often than not it is a false alarm, but it *may* be an important truth."[35] And he created opportunities for odd things to happen by keeping his plates around long after he needed them, by paying attention to contaminations, creating germ paintings, and so forth. Again, in comparison with Allison or Scott, Fleming's methods increased his probability of being surprised.

And then there's the fact that bacterial antagonisms were common. Constance, didn't you tell me that other doctors had tried to use such antagonisms clinically prior to Fleming?

▷ **Constance:** Sure. Pasteur and Joubert, Tyndall, Lister, Burdon-Sanderson—there are quite a few.[36] But none of their experiments worked out terribly well. Or they didn't persist if they did.

▷ **Hunter:** So Fleming is only one of probably hundreds of bacteriologists to observe bacterial antagonisms, and only one of probably dozens to attempt to utilize one or more of these antagonisms clinically. The question we must ask ourselves is: How many times before had the problem been investigated by other scientists? How may times had Fleming himself unsuccessfully investigated similar phenomena—perhaps without recording the experiments? And how many fungi-bacteria combinations would have yielded similar results? Remember, *Penicillium* acts on bacteria other than staphylococci, and other fungi produce other antibiotics.

You see, in this systematic context, the discovery appears to be almost inevitable. The thing that's confusing us is the particular historical instant in which Fleming did the work. So if we accept Imp's point about discovering being a process, then it's not the event we need to explain but Fleming's particular role in the process. Type IV chance: the Disraeli principle of acting distinctively. What characteristics of Fleming's made the event probable?

▷ **Richter:** Your point?

▷ **Hunter:** Try an analogy. Think of the second law of thermodynamics. People often read the second law as saying that systems move toward disorder. Naive observers look at the evolution of living things, which tends toward greater order or complexity, and conclude that evolution contradicts the second law. This is nonsense, of course. As long as the overall system tends to disorder, particular parts can tend toward order. In fact, David Hawkins and Victor Weisskopf have suggested a "fourth law of thermodynamics" which states that any warm material in contact with its surroundings will tend toward greater order, the order in the

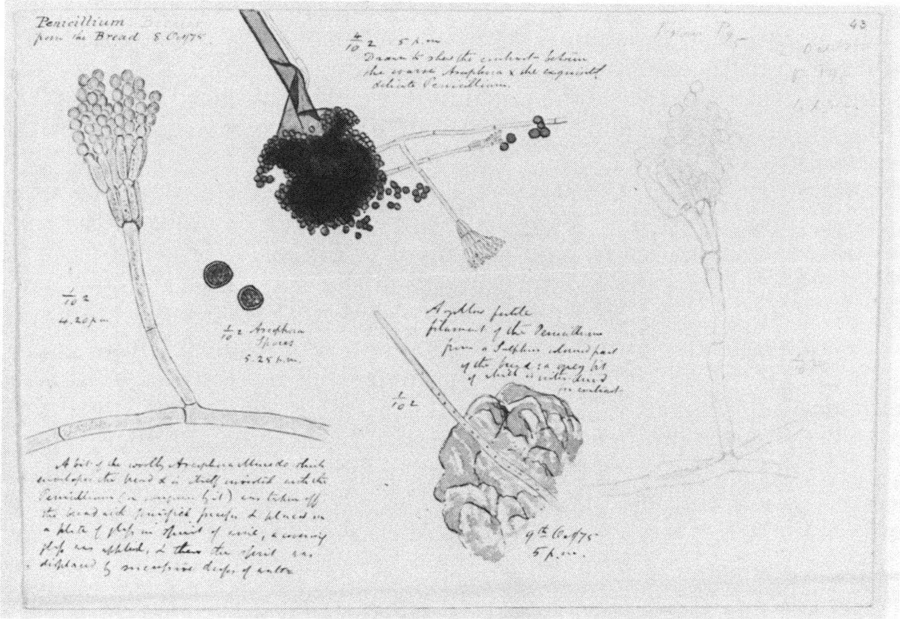

Joseph Lister's drawings of penicillium, October 1875. (Royal College of Surgeons of England)

material being overcompensated by the loss of order in the surroundings as the surroundings absorb the heat.[37] This "law" applies to everything from gases forming stars to molecules forming cells.

You see the analogy. We've been looking at a single discovery by a single observer and asking whether the event is "orderly" or "disorderly." We may conclude that the event is too disorderly to be rational. Yet in the context of the system of research involving the set of all bacteriologists, the set of all contaminations, the set of all problems being addressed, the set of methods adopted by Fleming, and so forth, the particular event becomes not only plausible but highly likely. Almost as if there were a "fourth law" for research, which might read something like "In any problem area in which a sufficient diversity of subproblems are being explored, techniques exploited, and observations made, a discovery is certain to occur."

▷ **Imp:** The key word, I take it, is "diversity"? It's not just that lots of research is being done, but that as many people as possible are looking at the same area in as many different ways as possible. Otherwise you just get the same thing over and over again. Clearly Fleming invented ways of introducing as much diversity into his own work as possible.

▷ **Ariana:** So if Pasteur's ability lay in imagining experiments that yielded surprises no matter what the outcome, we might say that Fleming's flair was to foster the conditions under which the unexpected, or even the unlikely, could flourish.

▷ **Hunter:** Right. In fact, Constance and I would like to posit two classes of chance discoveries that really have a high degree of probability when examined systematically. One class involves the sort of problem-oriented research conducted by Fleming, in which the investigator knows what to look for but not how to cause it to happen. Diversity of action, the "struggle against routine," eventually produces a result.[38] Think of Edison's research. He knew exactly what properties the thing he wanted to produce would have. "I want a fiber or filament that glows when electricity is passed through it, lasts several months, etc." He then had his workers experiment on the properties of literally tens of thousands of fibers, wires, threads, and so forth. Drug companies often operate in much the same way. They develop an assay, or test, for some activity they want the drug to have, and then they test tens of thousands of chemicals for this activity. Other criteria, having to do with the economics of production, product safety, of the ability to deliver the drug or to control the length of time the drug is active, supply additional tests for any putative drug candidate. The discoveries of ways to fix photographs and to cure rubber also fall into this category.

The limitations of this research method are that it's highly labor intensive and material dependent. You've got to have lots of time, people, and test material to use up, and an absolutely specific set of *a priori* criteria for recognizing success. As far as I can see, that limits the utility of the technique to the isolation or production of new objects like light bulbs or antibiotics. In short, it's of use only to what we might call "technologists," but not to those who investigate the causes of things (such as cancer or heart attacks) or who wish to discover new classes of phenomena, like magnetism or electricity.

Of somewhat more interest to basic scientists are a second sort of high-probability discoveries—so-called serendipitous (in the sense of unexpected) observations that occur to so many investigators so frequently that sooner or later their significance is bound to be recognized. Constance has been concentrating on those.

▷ **Constance:** Right. The key test for this type of discovery—though Hunter and I have somewhat modified the criteria—is given in René Taton's classic *Reason and Chance in Scientific Discovery*. Taton concludes that it's necessary to distinguish between what he calls "psychological chance" and "external chance." By "psychological chance" Taton means how, when, and why a set of ideas suddenly coalesce in an individual's mind. He believes that this psychological aspect of discovery is a constant in every discovery.

▷ **Imp:** Which, as we've seen with Pasteur and Fleming, is hardly chance at all.

▷ **Richter:** In your opinion. Go on, my dear.

▷ **Constance:** By "external chance" Taton means unplanned events that are so unusual or so rare as to preclude repeated observation. There are, he says, almost no examples of external chance operating in science.

▷ **Richter:** Perkin's synthesis of aniline? Roentgen's discovery of X-rays?

▷ **Hunter:** But Richter, Perkin had a perfectly logical research program! He was trying to synthesize quinine using aniline sulphate. He produced an aniline dye instead. Heaven only knows how many other chemists had done something similar and thrown it down the drain—it's a very simple and likely reaction. It's not making an aniline dye that made Perkin's reputation, but imagining what he could do with it.

▷ **Constance:** The discovery of X-rays was also a high-probability event. The essential facts are that Roentgen, while reproducing some experiments by Hertz and Lenard, noticed that when he turned on a Crookes cathode ray tube (the antecedent to the television tube), a fluorescence appeared on a barium platinocyanide screen nearby. Roentgen made a connection between the two events, investigated it further, and showed that whatever was being emitted by the cathode ray tube could pass through many solid objects, such as paper and the flesh of his hand, but not through the bones of his hand or various metals. The cathode ray tube emissions, which he named X-rays, could also fog photographic plates.

In retrospect, it appears that Crookes, the inventor of the tube, had also observed the fluorescence but attached no importance to it. Both Frederick Smith and Lord Kelvin had observed the fogging of photographic plates, but had attributed it to poor manufacturing by the photographic supply company. Kelvin had even sent the plates back with a complaint to the manufacturer! That's four scientists already who recorded seeing the same thing within months of each other.[39] So, anyone having a cathode ray tube and either a fluorescence screen or photographic plates would probably have observed the effect of X-rays. Surely you can see that Kelvin or Smith would eventually have been convinced by his photographic suppliers that the problem was his, not theirs, and would then have tracked down the source of the fogging. Crookes, Hertz, Lenard, and who knows how many other scientists would also eventually have been puzzled by similar problems, because they were using the same equipment.

▷ **Hunter:** The crucial point is that Perkin and Roentgen were the first to do these four things with their unexpected observations: one, to perceive them as interesting—remember Fleming's advice to keep your eyes open for such things; two, to divert their research to their investigation—think of last week's discussion; three, to persist in tracking down and characterizing the source of the phenomena; and four, to make something of the phenomena—that is, to interpret and use them.

▷ **Constance:** That's right. And there are lots of similar high-probability examples that fit this pattern. Classics such as the case of the floppy-eared rabbits.[40] Or my favorite, the story about Jocelyn Bell Burnell, one of the discoverers of pulsars—

▷ **Jenny:** Hold on! What's a pulsar, please?

▷ **Contance:** A stellar object that emits regular pulses of energy at radio frequencies. Dr. Burnell is a radio astronomer, someone who uses an instrument designed to focus electromagnetic radiation in the radio band. In the course of normal observations, she and her colleagues observed regular little blips of radio signals that shouldn't have been there. The blips were so regular that at first the astronomers thought they were noise or interference, but those possibilities were soon ruled out. And after systematically ruling out other known causes of such blips as well, the group started finding these periodic blips, or pulses, from other parts of the sky, too. Finally, other radio astronomers verified their existence and eventually the pulses were shown to display the characteristics predicted for hitherto hypothetical neutron stars. (Neutron stars are objects that have collapsed to such a dense state that normal atoms no longer exist—just neutrons.) But here's the key passage of Dr. Burnell's account, verbatim:

> *The discovery was almost totally unexpected. We learned later of a radio astronomer at another observatory—I won't say who or where—who several years earlier was observing a portion of the sky to the right of Orion, northward, where we now know there to be a pulsar. And he saw his pen [on the recording device] begin to jiggle. And he was about to go home for the day, and thought his equipment was misbehaving. And he kicked the table and the pen stopped jiggling.*[41]

▷ **Imp:** Afterward I bet he kicked himself!

▷ **Ariana:** This is very interesting. It means it's human nature not to want to perceive anything that disturbs your prearranged mental model of the world.

▷ **Hunter:** Sure. I have a friend who does research on computer software. He has a colleague who, every time a program does something odd, blames it on the hardware. My friend, on the other hand, always approaches the problem by asking, first, whether it is reproducible, and second, under what circumstances. Then he tries to figure out why the software acts that way rather than the way he expected. The result, of course, is that the computer program continues to function as before, but my friend's mental model of how the program works changes. The problem with people like his colleague is that they prefer their model of the world to the world itself, and so they invent ways of ignoring the disjunctions between the two. They mentally, if not physically, "kick the machine."

▷ **Imp:** Which means they miss making discoveries!

▷ **Jenny:** But not, as we saw last week, by refusing to make a detour in their research. In these cases, it's by refusing to perceive the phenomenon in the first place.

▷ **Constance:** Right. Macfarlane Burnet records that after reading d'Herelle's book *Le Bacteriophage*, he cultured a patient's urine and found it heavily infected with *E. coli*. There were two large clear areas among the colonies that he immediately recognized as being caused by bacteriophage. He goes on to say:

> *what impressed me was the fact that in all probability bacteriophage plaques must have been seen occasionally on urine cultures almost from the time Koch first developed solid jelly-like media for the growth of bacteria [some thirty years earlier]. No one, however, had apparently recognized that those clearings were significant and presented a phenomenon calling urgently for investigation. Even for an experienced scientist it is quite extraordinarily difficult to grasp the significance of an unexpected appearance.*[42]

Burnet also says he observed but didn't recognize the significance of hemagglutination, which Hare and McClelland developed into a standard assay for measuring viruses and antibodies.[43] So even the best—Nobel laureates like Burnet—can overlook what others (sometimes as in the case of Twort and d'Herelle, several others) find significant. Taton and Hadamard have extended discussions of the problem in their books.[44]

▷ **Imp:** Aha! That's it! I knew some little maggot was nibbling away at my brain. Turn it on its head! Not only are there missed discoveries, there's a whole class of multiple discoveries as well, isn't there?

▷ **Constance:** Yes, of course. Robert Merton's classic study found that the majority of all discoveries are multiple discoveries—which argues very strongly for discovery's being a high-probability process.[45] I thought everyone knew that. For example, perhaps a dozen people suggested the first law of thermodynamics (the conservation of energy) within a few years of one another. And at least three laboratories first isolated peptide hormones from the brain within a few months of one another. There are literally thousands of such examples.

Merton's interest was in priority disputes—who discovered what first, who was going to win some prize or secure some endowed chair somewhere. The sort of competition theme described in Watson's *Double Helix* or Nicholas Wade's *The Nobel Duel*.[46]

▷ **Imp:** And if many scientists are all arriving at the same conclusion or finding the same objects at the same time, then there must be a logic to discovering!

▷ **Jenny:** I don't follow you.

▷ **Imp:** Okay. Consider the alternative. Suppose most discoveries do have a chance or accidental component. If three or six or twelve scientists (or groups of scientists) all arrive at the same conclusion to their research at roughly the same time, then whatever the chance or accidental component may be, it must operate simultaneously in three or six or twelve minds or laboratories. Surely that's a farfetched idea! On the other hand, if we assume that most scientists maintaining the same theoretical presuppositions and observing the same general phenomena will logically reach the same conclusion, then the cause of simultaneous discoveries is no longer mysterious.

▷ **Richter:** Except that earlier today you argued that *not* everyone faced with the same phenomena perceives the same thing. Wright and Allison could have discovered lysozyme but did not. Pasteur, Tyndall, Scott, and those others could have found penicillin but did not. You cannot have it both ways.

▷ **Jenny:** Richter's right, Imp. Would anyone in Napoleon's situation have acted like Napoleon? You don't really want to argue that kind of historical determinism, do you?

▷ **Imp:** No, no, no! I must not be making myself clear. In the first place, Richter, I'm not talking about shared observations, but shared problems and preconceptions. Simultaneous discoverers must share the same theoretical and methodological commitments. And second, Jenny, anyone trained like Napoleon would have acted very like Napoleon. The question is, how unique was Napoleon's training and experience?

▷ **Ariana:** Or how unique is a scientist's set of tricks for eliciting novelties from nature?

▷ **Jenny:** So the major argument against chance discoveries—other than the actual historical documents, of course—is that you can't just see something; you have to interpret it, too. But you can't interpret an observation in the absence of either a well-formulated problem or an explicit theory that predicts what you expect to see.[47] And if lots of people maintain the same theoretical outlook and address common problems, then they'll very likely observe the same thing at the same time. Is that right?

▷ **Imp:** Bingo!

▷ **Ariana:** What you see, or don't see, depends on what's inside your head, not what's in front of your face! As T. H. Huxley wrote to his grandson Julian, "There are some people who see a great deal and some who see very little in the same things."[48]

▷ **Constance:** So we're back to where we were two weeks ago—the sublimity of the mundane and all that. Szent-Györgyi saying that "discovery consists of seeing what everybody has seen and thinking what nobody has thought."[49] Or Whitehead saying that "everything of importance has already been seen by someone who did not discover it."[50]

▷ **Imp:** No, Constance, we *are* making progress! I just realized something—something that we've overlooked because it isn't there to be seen. Consider Richter's list of chance discoveries. Or the probabilistic examples discussed by you and Hunter. Or any other so-called chance discoveries you can think of. In that entire group there's not a single theoretical discovery. No Copernicuses or Galileos or Newtons or Laplaces or Einsteins "accidentally" stumbling onto a new concept of the universe. Right? Am I right?

▷ **Richter:** No. You're wrong. You left out Darwin reading Malthus on population. No Malthus, no mechanism for evolution. No mechanism for evolution, no *Origin of Species*.

▷ **Constance:** Well, that's not quite accurate, Richter. Natural selection—Darwin's mechanism—is another multiple discovery. A. R. Wallace also thought of it after reading Malthus, who was, after all, hardly unknown in England. Besides, several other scientists have at least a partial claim to the idea of natural selection.[51] But Darwin's a poor example for another reason. By 1838 he was looking for a mechanism for evolution, as his notebooks and correspondence clearly show. So, saying that he found a clue in an unexpected place seems to me to beg the question. Darwin found what he was looking for.

▷ **Imp:** Think literally. This may be simplistic, but we don't have stories about Darwin hearing two people talking about pigeon breeding and one says, "But that's not *natural,*" and the other responds, "*Selection* is the prerogative of the breeder," and, voilà, the idea of "natural selection" is born! Now that would seem to me to be a chance confluence of events! Certainly Darwin would have had no control over the words of the people he overheard, nor are words "natural" and "selection" at all likely to follow one another on a purely random basis.

Most of the other examples you cited, Richter, are just as absurd. Fleming's discoveries controlled by unwanted contamination, unexpected weather, and just the right complementary microorganisms to display the needed effects. Minkowski's discovery controlled by the unwanted presence of flies attracted to urine. Unforeseen observations concerning unimagined problems. No, I can't buy it! Besides, can you think of any other theories discovered by so-called chance?

▷ **Constance:** Actually, I can.

▷ **Imp:** Goodness! You're as bad as I am! Between us, we could argue anything.

▷ **Constance:** Well, not quite. Anyway, the example I'm thinking of is John Dalton's invention of the atomic theory of chemical combination. Arnold Thackray has argued that the theory was developed "accidentally"—that's his word—and not by either induction or deduction. Essentially Thackray's argument is that Dalton set out to solve a meteorological problem and only recognized the application of its solution to chemistry as an afterthought. In other words, the invention of atomic theory was unintentional or accidental.[52]

▷ **Hunter:** No, I can't agree, Constance. It *was* intentional. Dalton did have a research program in mind. The fact that it had to do with meteorology rather than chemistry is irrelevant, because Dalton himself connects the two. He detours, just like everyone else we've studied.

▷ **Constance:** But even Holton calls the invention "irrational."[53] He claims that Dalton misread Newton; accepted the caloric theory; assumed without evidence that the particles of different gases were different sizes; concluded from these things that various aspects of meteorology were now explained which in fact were not; and based everything on inadequate experimentation.

▷ **Ariana:** Which gets us right back to Laënnec's stethoscope—

▷ **Imp:** The question of what is a "fact"—

▷ **Hunter:** Look: What bothers me most are statements such as "misread Newton" or "accepted the caloric theory." So what? Who can say what is the right way to read Newton in any particular context? What was Dalton supposed to do, invent a workable kinetic theory sixty years before Clausius, or forgo any attempt to understand the behavior of gases at all? Is anyone who accepted the caloric theory "irrational" for believing a theory we don't believe today? Where does this normative evaluation come from?

▷ **Richter:** Your point?

▷ **Imp:** This: If you can't invent a theory by accident, then you can't make *any* discovery by accident. Because to make sense of any observation, you must interpret it within a theory. And the only significant discoveries, as we've said before, are those that cause us to rethink and research what we thought we already understood.

▷ **Ariana:** Which means that we've been trying to subsume too much of the scientific process under the rubric of discovering, right? Perhaps we should seriously consider the possibility that a logic of discovery is actually a logic of invention. Think about how important the invention of new techniques was to Pasteur or Fleming—or to any emerging science. Or the invention of problems, or the invention of theories to solve them. Look back at Richter's problem types: most require invention of some kind, not observation or accident. So everything in science really is planned—is intentional—except that sometimes these investigations yield surprises instead of expected results. Discoveries are the subset of inventions that are surprises.[54]

▷ **Hunter:** Nice. You've just answered the question I was about to ask. There are inventions that are commonly called discoveries that are *not* surprises. Mendeleev's periodic table of elements, for example, predicted many missing elements and specified precisely the properties these elements should have, yet historians write about the subsequent "discovery" of these new elements. It seems to me that it would be more accurate to speak of their prediction and isolation. But their prediction and isolation depend upon the *a priori* invention of their categories by Mendeleev. Scientists then began actively seeking objects having the appropriate characteristics. In short, these so-called discoveries are predicted.

▷ **Richter:** Although they, too, yield surprises: the various isotopes, for example.

▷ **Jenny:** Okay. I've been trying to jot down where we stand, as I see it. It seems to me that we've been able to demonstrate, if not definitely, at least plausibly, the following things: One, many of the most commonly cited examples of putative chance discoveries were, in fact, based upon a rational research program that yielded unexpected results, or surprises, and required the investigator to make a detour from the preplanned path. Two, their apparently unlikely character is usually due to lack of information concerning the genesis of the observations rather than actual chance. Three, most, if not all, other so-called chance discoveries either are high-probability events awaiting a properly trained interpreter, or result from a nearly random search for objects having a very specific set of characteristics defined by a specific *a priori* problem. Four, further arguments against chance playing a role in the majority of discoveries are that most discoveries are made simultaneously by several scientists or groups of scientists; and that no observation can represent a discovery unless it's interpreted in light of a theory (theory representing a rational structure). Five, add to that Hunter's point that some discoveries are actually the result of verifying theoretically correct predictions. Then, six, we come to Ariana's position: that it's not the discovery, the surprise, that we must explain, but the process of invention that underlies the logic of research that leads to the discoveries. Seven, and finally, in practice we can't distinguish between the logic of discovery (or invention) and the logic of testing (or validation). The one is an integral part of the other.

▷ **Imp:** Sounds pretty good to me. What about you, Richter? No barbed comments to sting our complacency?

▷ **Richter:** Obviously we could go on discussing particular examples forever, but I don't see the point. I *am* surprised, however, that no one here has raised the most cogent objection to my position: You could have challenged me to prove that all discoveries and inventions occur solely by chance. That I obviously could not do. Therefore, I must concede that some discoveries have a logic, and we should certainly try to learn how to encourage these. In short, let us just say that I am no longer as fond of the idea of chance as the *deus ex machina* of discovery as I once was.[55] I am willing to accept a logic of research as long as it leaves room for the surprise of the unexpected.

▷ **Imp:** Then leave it we shall! But since Richter is vacating his chair as universal critic, I'll step in. We've had some tantalizing material these last weeks, and we've come up with a lot of ways to enhance the probability of discovery: identifying key anomalies; Berthollet's transfer of his expertise to a new context; Pasteur's dual-edged hypotheses that yield new insights whether correct or not; Fleming and Lorenz's controlled creation of chaos; the use of analogy by Fleming and Laënnec, which, whether you like the particular examples or not, illustrates Richter's point that one way to solve a problem is to graft from a solved problem tree to an unsolved one. We've seen that correctness in theory is virtually irrelevant to discovering as long as it's coupled with experiment.

Is this sufficient? Definitely not! Lots of ideas, but we haven't yet pulled them all together into a coherent whole. We still don't understand the process of discovering or invention. Where do we turn next? Any ideas? Hunter?

▷ **Hunter:** I think there's a lot more to be learned along the same lines. I've been going through the work of four Nobel laureates—van't Hoff, Arrhenius, Ostwald, and Planck—trying to figure out how they managed to place themselves at the emerging forefront of a new field like physical chemistry or quantum theory. How are they trained? How are they different from their colleagues and similar to one another? What are their "rules of thumb"? The material follows directly from Pasteur's asymmetry work—at least, van't Hoff 's work does—or from Berthollet's mass action theory. So we can build on concepts we're already familiar with to try to flesh out a general outline of the discovery/invention process and the extent to which it can be manipulated by individuals.

▷ **Ariana:** And the rest of us ought to find people who describe the process of doing science. Read good autobiographies, talk with colleagues . . .

▷ **Constance:** And may I suggest reading Peter Caws, Errol Harris, Nicholas Maxwell, William George—well, why don't I just send you each a brief bibliography?

▷ **Imp:** Sounds perfect. Anything else? Right, then. Off to bed with me. See the rest of you next week.

▶ IMP'S JOURNAL: Unexpected Connections

Kuhn says somewhere that discoveries must change how science is done. Importance of discovery measured by extent of change in normal practice. Fleming a case in point. Prior to W.W.II, penicillin used only as additive to agar to kill off pathogens that interfered with growth of other bacteria. Aid to isolation of bacteria otherwise difficult to identify. Didn't persevere on possibility of human use. So part of discovering is linking up solution to right problems. Doesn't always happen immediately. Solution and problem can both exist prior to their meeting.

My own work another case in point. Invented amino acid pairing as mechanism for protein-to-protein information transfers to solve a problem most immunologists don't accept as valid. Then, along comes Stanley Prusiner with his "prions"[1]—subviral particles causing diseases such as scrapies in sheep, and possibly kuru and Creuzfeld-Jakob disease in humans. Composed solely of protein. Yet they replicate! How? Nobody knows. But only two possibilities, if Prusiner's right: either direct protein-to-protein replication, or protein-to-RNA and back again. Either way, I'm sitting pretty and the Central Dogma ain't!

Of course, if Prusiner is wrong . . .[2]

No matter. Looks as if pairing hypothesis may apply to peptide hormones with complementary functions. Hormones regulating each other's activities based on stereochemically determined binding. Worth looking into. Didn't Ariana say something about two reproductive hormones that control each other? Have to ask.

Principle of inventing solution and then looking for problems holds in my other work, too. Same thing happened with "molecular sandwiching." Invented idea to solve serotonin binding site on a single protein. Unexpectedly found model works for some hormones. Then, in telling results to David Felten, find another possible application. "You know about the problem of cotransmission, don't you?" asks David. "No," I say. So he explains to me that neuroanatomists and neurophysiologists have been finding odd things about nerves. Every nerve has a neurotransmitter. At first thought to be monoamines like serotonin. Simple molecules. One type—serotonin, norepinephrine, dopamine—to a nerve. Then they found that peptides could be neurotransmitters—substance P and bradykinin [Editor's note: pain transmitters], enkephalins and endorphins [natural opiates]. Worse, they found *pairs* of these neurotransmitters in single nerves—costored, coreleased, often modulating each other's physiological effects. Why pairs? And why only *particular* pairs: substance P–serotonin, enkephalin–dopamine, and so on.

"That's the problem of cotransmission," David concluded. "Looks to me as if you've got a theory that could explain what these pairs of monoamines and peptides are doing together." And so began my attempts to extrapolate "molecular sandwiching" to cotransmitters and other drug-peptide interactions.[3]

It's clear: A solution is only a beginning. Answers raise questions. That's what you've got to look for. New questions for existing answers. Finding broadest range of applications for one's novelty. That in turn, requires broadest meandering in fields of nature, and knowledge of as many puzzles as possible.

So why the hell don't we all spend more time talking about what puzzles us instead of our precious and all too limited insights? Good knowledge of unsolved problems of science, and how we know that they're problems, would probably be better education for students than all the "facts" we could cram into their heads. Exploration—that's what we need to cultivate!

DAY FOUR

Creating Unity from Diversity

He would often begin with an idea which, after he had worked at it for some time, turned out to be wrong; he would start off on some other idea which had occurred to him while working on the previous one, and if this turned out to be wrong he would start another, and so on until he found one which satisfied him, and this was pretty sure to be right. He often started out in the wrong direction but he got to the goal in the end.

—J. J. Thomson, physicist, writing about his mentor, Osborne Reynolds (1937)

For the moment, I am blundering without precise method. I repeat old experiments in this field and demonstrate others which pass through my head . . . I hope that, among the hundred remarkable phenomena which I come across, some light will shine from one or another.

—Heinrich Hertz, physicist (in Taton, 1957)

Wassermann's basic assumptions were untenable, and his initial experiments irreproducible, yet both were of enormous heuristic value. This is the case with all really valuable experiments.

—Ludwig Fleck, bacteriologist and philosopher of science (1979)

►

► IMP'S JOURNAL: Thematum

So you stumble about, surveying whatever vistas reveal themselves. Look for applications, connections, patterns. That's it—patterns! One thing that makes me different is this desire—no, inner *need*!—to perceive how everything fits together. Principles, algorithms much more powerful than individual facts or unique solutions.

When did I realize this? Certainly as teenager. Even then, knew I didn't want to be a doctor. Didn't want to treat individuals. Rather be a Darwin, perceiving relationships between things. Emulate Pasteur and create general cures the doctors administer to individuals. Why? Who knows. Genetic curiosity? Probably biographies, too. Hagiography. Mythologies we create around Pasteurs and Darwins. A desire to emulate the best. (Is such mythology necessary to recruiting—even to forming—their equals?)

But what seeds the solution, what crystallizes global vision into specific action? Emulation again, I suppose. Exemplars. Having the double helix touted as the epitome of science. And from that, the distillation of simplicity, beauty, functionality, modeling, complementarity.

There. Complementarity. That's my theme. Complementarity of structure producing complementarity of function. Physical complementarity mirrored in functional complementarity; functional complementarity the reflection of philosophical complementarity. (When did I first read Bohr? No, it wasn't Bohr. It was Haldane senior. *Philosophy of a Biologist.*)[1] Recognition that function must be relative to something—that the function of a molecule is determined by its interaction with others. Already resistant to "master molecule" theories before learning them. At twenty, realizing that the statement "Measles virus causes measles" is false: Measles virus causes measles only in human beings. So it's not the virus that causes the disease, but the combination of virus and host. Or did I borrow that from René Dubos?[2] No matter. The principle holds. Extrapolate. Generalize. "DNA is the secret to life." Nonsense! DNA without all the proteins and cell structures is functionless. Complementarity again. Very nearly the end of my career as a biochemist! "You have no business in this department if you can't even understand the most basic concepts of molecular biology . . ." (*You* have no business being a teacher if you can't appreciate another point of view than your own!)

Implementing such a theme. Do I want to isolate yet another new molecule? NO! No. Aim into the void. An arrow to follow. A guess, shot from the bow of my preconceptions: Molecules with complementary functions will display structural complementarity. The yin-yang of biology. The test, breathtakingly simple: Molecules that affect the same system in opposite ways will be structurally complementary and so bind to one another. Lock and key. Antibody binds antigen. Each hormone has its structurally complementary hormone; each neurotransmitter, its structurally complementary transmitter. The whole, a self-regulatory system based upon the simplest precepts of chemistry. Elaborated, specialized, modified by evolution into byzantine and baroque receptor systems, specialized cells and intricate pathways. But still, at heart, in its historical roots, simple.

Guiding principles are always simple, even simplistic. Must be. Complexity just the elaboration of every possible combination of simple processes. Sebastien Truchet's half-color tiles arranged in an infinity of orientations.[3] Life evolving. One pattern into another. Simple chemical systems elaborated. But always directed by the same principles.

That's what I'm after. Something universal, basic, simple. So simple it's got to be wrong. So simple no one else has thought of it. And that's why the whole field's mine—and anything I find along the way!

◀ TRANSCRIPT: Modeling the Process

▷ **Imp:** So! Today we discuss the process of invention. Allow me to propose that the only possible model will be a stochastic one with heuristic and iterative elements. In consequence, there is no direct route to any important conclusion.

▷ **Jenny:** Really, Imp—such jargon! What are you trying to say, that you aim into the unknown using some preconception as a guide, and then you successively refine your ideas by learning from your mistakes?

▷ **Imp:** I'm simply trying to be provocative.

▷ **Richter:** Which is no excuse for being unintelligible. Still, you are essentially correct, except for your use of the term "iterative." If you are utilizing a mathematical analogy, the correct term would be "recursive." Iterative processes simply repeat the same operation on each term. Recursive processes utilize the previous term to determine what operation to perform on the next. But we already know this.

▷ **Imp:** *You* may know that the process is recursive and full of detours; *I* may know it. But where in the way we teach science, in the way we fund science, or in the way we model science do we account for recursive detours? Have you ever taught your students the errors and miscalculations that even a single scientist went through to reach his or her results? Of course not! Here's the standardized derivation for the result, and the evidence (carefully selected) to support it. Ambiguities? Mistakes? Contradictions? Hell, most textbooks don't even whisper about such things.

And peer review! Do you think you'd get a grant funded by writing, "I've got a crazy idea that is not at present justified by anything more than a hunch, but I'm certain that if I try enough things and make enough mistakes I'll find something important, even if it isn't what I expect to find"? No way!

So, sure, we all know how science really works, but do we *act* on that knowledge? On the contrary! We've acquiesced to a system that implicitly assumes, one, that surprises are an unimportant aspect of research; and two, that there is actually a high probability of your solving the problem you set out to solve. Historically, neither assumption is accurate. So whatever model we come up with is going to have very important policy implications. And if our insights turn out not to be new, then it's time the goddamn administrators start listening to old, well-established facts!

▷ **Hunter:** Absolutely. The object of planning for research must be to optimize the ability of the investigator to recognize and solve *some* problem, not to predetermine who will reach particular conclusions by specified methods. What we need is a model that incorporates strategies for cultivating the unexpected, methods for testing a hypothesis that will generate the surprising observations and questions of unanticipated discovery.[1]

▷ **Constance:** Well, this may be sort of presumptuous on my part, but I think I've got something that might fit the bill. May I? You see, I started out with a slightly different set of questions—things we'd left over from earlier discussions: What's a discovery? How does it differ from invention? Is it an event or a process? Are there any great discoverers, or are some people just lucky? That sort of thing. Anyway, I've come to the conclusion that discovery is a process that always requires invention, and I've invented a model to describe the process.

▷ **Imp:** Is it patentable?

▷ **Constance:** You're joking, of course, but that *is* a serious question. Why should an algorithm for processing information be protected only by copyright when a technique for processing, say, chemicals can be protected by patent? Why should a computer chip be patentable but not the scientific theory or the computer program that makes the production of such chips possible?

▷ **Jenny:** Isn't this a bit off the subject?

▷ **Constance:** Actually, no. You see, it's that sort of thinking that led me to my model. As we discussed last week, every discovery has an element of invention in it, as well as an element of taxonomy, an element of theory, and so forth. All of these are part of a continuous process. Why then do we say that only one element in this process is patentable? Why do we identify one individual who performed one step as the discoverer or inventor? If properly formulating a problem gets you more than halfway to its solution, why do problem inventors get no credit in our system? When two or more researchers who have been doing the same sort of research both apply for a patent at about the same time, and both have utilized the other's results in arriving at their own—as was recently the case with both the development of the AIDS antibody test and the production of the first high-temperature superconductors—who deserves the patent? When one person invents a theory that allows another to build a new machine, who is the inventor? I deal with this sort of problem all the time. The legal assumption is that it is always possible to demonstrate who actually produced any particular invention and when they produced it. In actuality, I've dealt with a lot of cases where it's not clear at all. The who, what, and when are not unambiguously identifiable.

Consider a classic historical example. Most people who write about the discovery of Uranus say that William Herschel discovered it in 1781.[2] However, when you look carefully at the history of the discovery, not only is Herschel's claim questionable, but it also isn't clear what was discovered or when it was discovered. Let me quote Thomas Kuhn:

> *On the night of 13 March 1781, the astronomer William Herschel made the following entry in his journal: "In the quartile near Zeta Tauri . . . is a curious either nebulous star or perhaps a comet." That entry is generally said to record the discovery of the planet Uranus, but it cannot quite have done that. Between 1690 and Herschel's observation in 1781 the same object had been seen and recorded at least seventeen times by men who took it to be a star. Herschel differed from them only in supposing that because in his telescope it appeared especially large, it might actually be a* comet*! Two additional observations on 17 and 19 March confirmed the suspicion by showing that the object he had observed moved among the stars. As a result, astronomers throughout Europe were informed of the discovery, and the mathematicians among them began to compute the comet's orbit. Only several months later, after all those attempts had repeatedly failed to square with observation, did the astronomer Lexell suggest that the object observed by Herschel might be a planet. And only when additional computations, using a planet's rather than a comet's orbit, proved reconcilable with observation was that suggestion generally accepted. At what point during 1781 do we want to say that the planet Uranus was discovered? And are we entirely and unequivocally clear that it was Herschel rather than Lexell who discovered it?*[3]

Now, in my experience most discoveries and inventions are of this nature. Many individuals participate in different parts of the process. The discovery of oxygen, the invention of the first law of thermodynamics, the DNA double helix—now that's a good one. Watson and Crick got the Nobel Prize, but their work was entirely dependent on the development of X-ray crystallographic techniques by the Braggs, Rosalind Franklin's application of that technique to DNA, Avery's demonstration that DNA rather than protein contains the genes, Chargaff's determination of the base-pair ratios, Jerry Donohue's knowledge of the correct base structures, Wilkin's crystallographic "proofs" of the structure, Meselsohn and Stahl's demonstration of the mode of DNA replication, and so forth. In this sense, all discoveries are to some extent communal efforts.

▷ **Richter:** So you're coming around to my view that there are no great scientists. We all build on what our predecessors have left us.[4]

▷ **Constance:** Well, I suppose that's one way of looking at it. But what I'm trying to say is that Watson and Crick played only a very small role in the overall process by which DNA structure was elucidated. They didn't set up the problem. They didn't invent the techniques. They didn't acquire the necessary data. They didn't provide any of the tests of their hypothesis. And their correct hypothesis appeared only after a number of incorrect attempts by them and Linus Pauling.

▷ **Imp:** If it *is* the correct hypothesis! There've been at least three subsequent models of DNA structure, you know. Each differs significantly

Linus Pauling's triple helical model of DNA (1952). (Science Museum, London)

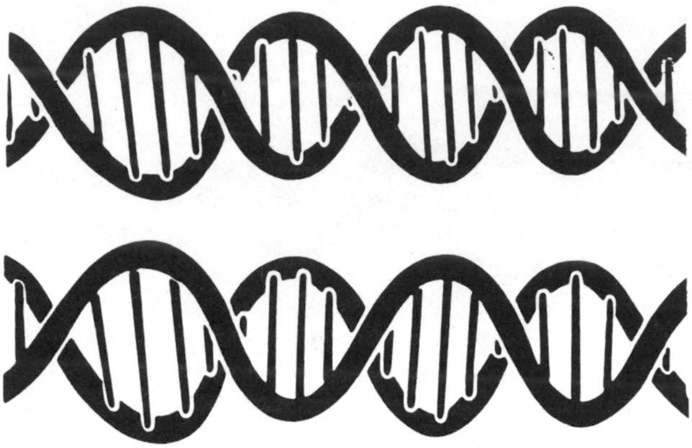

Simplified Watson-Crick double helical model of DNA (1953) compared with simplified "warped zipper" model (1976). (Rodley and Bates, 1976)

from the double helix.[5] The double helix is very pretty, but it has one major problem: How does it unwind so that replication can take place? That's never been solved.

▷ **Richter:** I thought there was an unwinding enzyme, topoisomerase or some such.

▷ **Imp:** Sure. And in my view, it's insufficient—doesn't explain the physics of the process at all. How does a molecule a mile long untwist in the space of a few microns? Think of the molecular chaos that would result! You can't unwind a double-helical, circular bacterial chromosome at all unless you nick it first, and there's no evidence I know of that that's what happens. Furthermore, if the preferred conformation of the molecule is helical, then where does the energy come from to keep it unwound until replication has been achieved? Besides, why should we believe that the packed crystal structure of DNA molecules in aggregate has anything to do with the actual structure of a single DNA molecule in solution as it exists in a cell?

Lots of problems. Watson and Crick even discuss some of these in their initial papers, and report having tried to invent other models to obviate the difficulties.[6] But they never solved them. So some guys in Germany invented a model that didn't hold up under scrutiny, and various people in Australia and India have proposed very similar models that do fit the accumulated data—so called "side-by-side" models that look sort of like a double helix, but are in fact complex "zippers" that can come apart without unwinding. Needless to say, Crick and Watson aren't fans of these models.[7]

But that's not the point. The point is that the process of discovering doesn't end just because you make a novel observation or reach a plausible solution to some outstanding problem. The first observation, the first plausible solution is just the beginning. Research starts with the answer. What does your model make of this sort of thing?

▷ **Constance:** Well, I think quite a lot. It's one reason I invented the model.

I looked at as many models of the processes of inventing and discovering as I could find—historical, philosophical, psychological, even computer models—but they all have major problems.[8] They tend to be directional—that is, one goes from a problem to a solution, but the solution doesn't raise any additional problems. Most are based on the Graham Wallas approach: preparation, incubation, illumination, verification.[9] These models make it appear that research is always successful and always ends with a discovery, which as Imp points out isn't the case at all. Most research is unsuccessful and solutions generally create more problems than they solve.

▷ **Ariana:** But we're a solution-oriented society.

▷ **Constance:** There *are* a number of people, such as Garrett Hardin and Fritz Hartmann, who have proposed cyclical models that yield new problems as the result of new solutions.[10] But these models aren't sufficient either: There are no *people* in them, just disembodied thought processes. I didn't think Ariana would accept that. And none of the models deals with the questions Ludwig Fleck, Tom Kuhn, Barry Barnes, and others raise, of how novelty becomes integrated into the larger community of science. Just reaching a solution in private isn't sufficient to make a contribution to science, because science is, at least at some level, public knowledge.

So that's why I invented this model. Here, I've brought a copy for each of you. There are four major inputs into the model: the cultural context in which the research is performed; codified science; science in

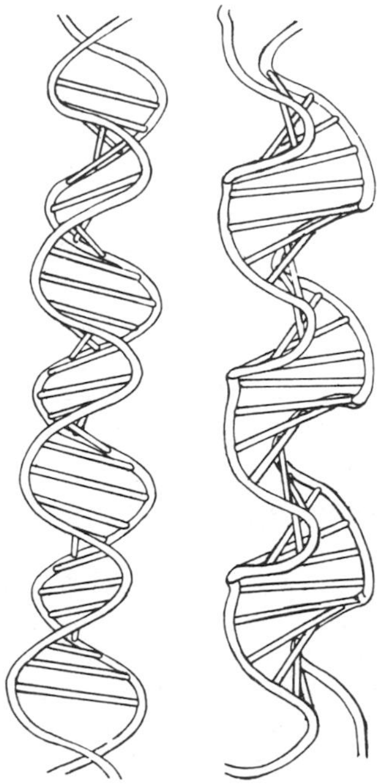

Details of the "warped zipper" model. (Rodley and Bates, 1976)

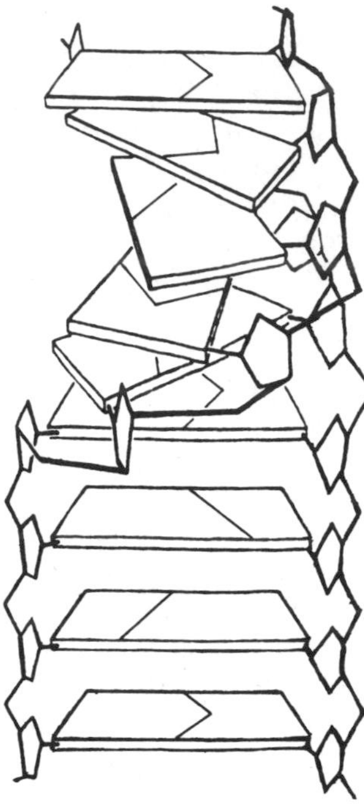

Cis-ladder conformation of DNA. (Cyriax and Gäth, 1977)

the making; and the individual scientist, who is a nexus for the other three. Every individual will represent a unique mix of hereditary proclivities and environmental experiences.

Now, I haven't drawn an explicit input for the cultural context of science. I've just sort of assumed that the process is embedded in a particular context, which includes everything from the economics and politics of science to educational policies, religion, and social organization. For example, language may provide possibilities and limitations. Think about last week and Imp's runny bottom, runny nose pun. Could a non-English-speaking scientist invent an idea based on "the runs"—I mean, is there an equivalent in German, French, or Chinese? I could be wrong, but I don't think so. Or consider politics. Rudolph Virchow, who was an ardent republican during the 1848 revolution in Europe, likened the development of an organism, from a single unspecialized cell to an adult form with a diversity of tissues and special organs, to the development of a republic of individual persons, each with a specialized function, yet cooperating to serve the commonwealth. Most historians agree that his political commitments and the available models of governmental organization influenced his biological theorizing.

▷ **Ariana:** Just as Stephen Jay Gould's dialectic materialism is inseparable from his biological theorizing. You can see it in his choice of topics.

▷ **Constance:** Next is codified science. That's like Kuhn's "normal science."

The textbook accounts, the standard techniques, the accepted definitions and experimental values, derivations, and synthetic theories. What we already know.

▷ **Imp:** What we *think* we already know!

▷ **Constance:** Okay, sure. The third input I call "science in the making." There's not much discussion of this kind of science in what I've read, but it's the part that seems to drive research. It's all the things we'd like to know or like to do, but haven't yet figured out—outstanding problems of all types, from definitional and technique to integrating diverse theories and codifying them.

Constance's model of the process of scientific invention.

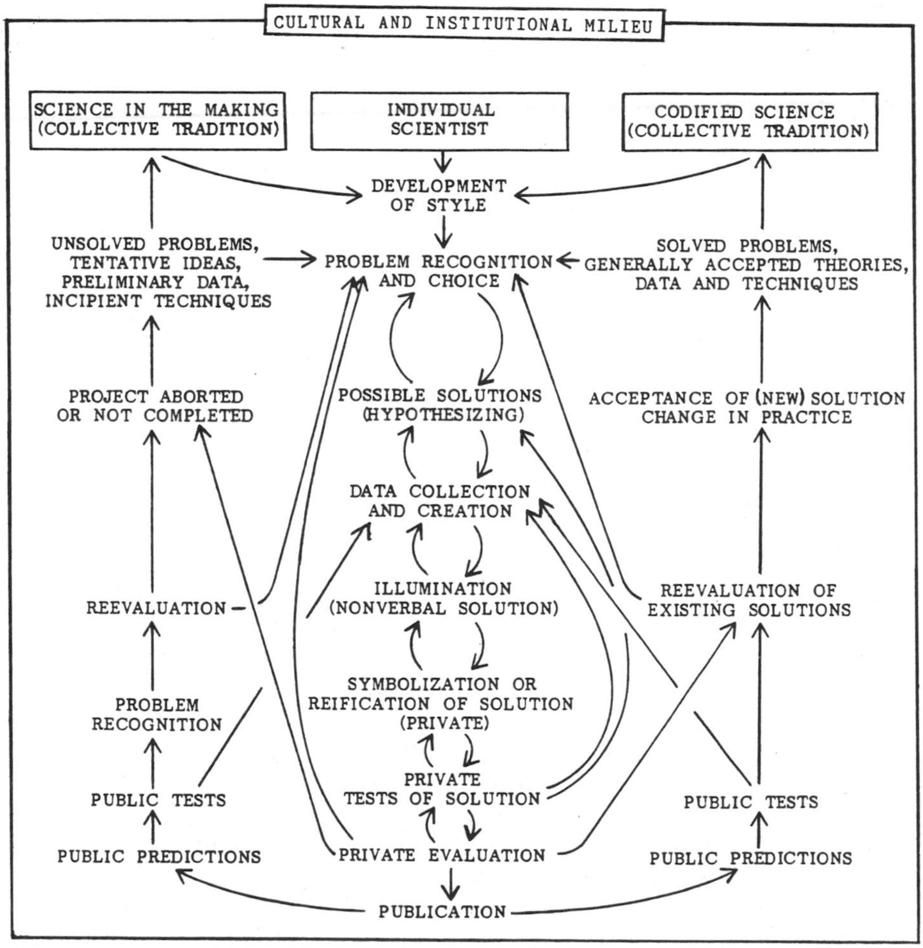

Finally, and most important, is the individual as the nexus for the other three inputs. As we've seen with Berthollet, Pasteur, and Fleming, each individual is a unique blend of hereditary potential, acquired skills, personal predilections, codified knowledge, science in the making, and cultural biases. These give the individual more or less potential to recognize and solve any particular set of problems.

Now, somewhere in this mix of cultural context, codified science, science in the making, and personality, the scientist finds disjunctions or gaps—problems, in short—that intrigue her (or him). Out of all of the problems the scientist encounters, she or he picks one or two to follow up. That's where other criteria come in. Will society value the results? Is the problem personally interesting? Are there time, money, energy, resources, techniques available for addressing it?

▷ **Richter:** Does it appear to be solvable? Is it strongly connected to important problem areas?

▷ **Imp:** Will it earn me tenure? Is it a hot topic with lots of funding? Is it someone else's territory? How much competition can I expect?

▷ **Constance:** Sometimes these criteria conflict, obviously. One balances various considerations, refines the problem into subproblems, and so forth. What data, techniques, theories already exist? What subproblems can be addressed immediately, and of these which should be addressed first? All of this occurs, as Imp says, through a recursive process of feedback. You start one place and head in one direction, only to find that way impassable; or you find a more inviting detour; or an incorrect hypothesis leads to fascinating results. The kind of heuristic thinking Polya recommends. You may go through the cycle a hundred times before a solution emerges, often in a sudden flash of illumination or insight.

▷ **Ariana:** And all this goes on inside a single individual's mind?

▷ **Constance:** Yes. And I know what you're about to ask. How do I describe the Herschel-Lexell or Watson-Crick et alia kind of discovery? Hold on a moment, and I'll get to that.

To quickly run through the rest of the process: The scientist puts his or her results through a series of private tests before deciding either to abort the research as a failure or to make the private efforts public through publication, a talk, letters, or some other means. At this point, the problem-data-resolution package is reviewed by other scientists, who, as a result of access to other data and theories or because they utilize different evaluation criteria, may accept, reject, modify, or further test it.

▷ **Hunter:** Or interpret and amplify it. It's incorrect to think that the inventor of a new theory perceives all or even most of its implications. Planck invented a solution to the equations describing black-body radiation but it was Einstein and Ehrenfest who recognized that his solution described the quantum.[11] Einstein, in turn, wrote the equation $E = mc^2$, but denied for decades the practical possibility of fission and fusion.[12] A discovery is a discovery only when it becomes part of scientific practice. Practice, in turn, modifies the initial discovery by creating new connections to other patterns.

▷ **Richter:** Your answer to my initial question of why a mistaken observation is not a discovery—that it is not reproduced or taken up by the scientific community. Plausible. But excuse us, my dear. You were going to answer Ariana's question.

▷ **Constance:** Yes. Actually, what you and Hunter just said leads right into what I was about to say. You see, each individual contributes some new elements to codified knowledge and science in the making by going through the individual process I've just outlined. If we look at a major discovery or invention involving many scientists, then the same model can be used for the process whole groups of researchers go through. As Claude Bernard pointed out, an individual scientist usually specializes in parts of the cycle.[13] One may recognize or invent an anomaly, gather data to demonstrate its significance, and convince his or her colleagues that its solution could be significant. Another may solve some subproblem such as the invention of a new technique needed to gather relevant data. Yet others—Pauling, Watson, and Crick in the DNA case—may invent a possible solution. Someone else invents tests of the solution. And then there are those such as Lexell or Einstein or Darwin who reformulate the contributions of their predecessors, recognizing in these results new problems and new potentials. So each scientist goes through the whole process of invention as an individual, but her or his public contribution may be only a step in the larger process, or simply the linking of some of the individual processes. It all depends on the type of problem—and here I'm referring to Richter's problem types—that the individual investigates.

▷ **Richter:** Very clever. You have, however, created some problems for yourself that you may have overlooked. For example, your feedback arrows are very ingenious, but what prevents one from entering an endless loop, without ever getting as far as hypothesizing or going public with one's results? Doesn't make much sense.

▷ **Imp:** Au contraire! It makes perfect sense. Haven't you ever had a student do that, Richter—suddenly freeze up after doing all the research and refuse to write it up because it doesn't make sense, or for fear that it might be wrong, or because there's no use hypothesizing until all the data are in? Or known a guy who solves a problem to *his* satisfaction, but can't be bothered to publish because he loses interest as soon as he knows the answer?

▷ **Hunter:** Ettore Majorana's story. According to Fermi, he worked out Heisenberg's theory of the atom before Heisenberg but, having satisfied himself, never wrote it up for publication.[14] And then there are the ideas that are so outrageous that you yourself question the propriety of making them public. Supposedly an Italian priest developed many of the theorems of non-Euclidean geometry a hundred years before Bolyai, Lobachevsky, and Gauss, but couldn't countenance his own results. Gauss himself was too chicken to publish anything so antidogmatic until his colleagues did, and even they had their problems: Bolyai's father, himself a mathematician, pleaded with his son to drop the subject before it destroyed his health, happiness, and career.[15] Just the sort of advice that prevented a number of other discoveries from coming to light: Wolfgang Pauli refused to let his student Kramer publish the idea of electron spin,[16] and Kurt Mendelssohn's professor refused to allow him to announce the discovery of gas degeneracy because the data were inconsistent with Sommerfeld's theory of the atom.[17] In consequence, these discoveries are credited to other people.

▷ **Constance:** Kuhn's point. Science is a social phenomenon as well as an intellectual one. You've got to change what scientists do and how they do it to have any impact.

▷ **Jenny:** Which implies that the popular conception of the genius working in isolation—the Sinclair Lewis *Arrowsmith* ideal—is a myth?

▷ **Hunter:** Not entirely. The inventor must be sufficiently outside the mainstream to perceive the world differently, yet sufficiently socialized—and sufficiently determined—to influence his peers.[18]

▷ **Ariana:** Jerome Bruner's definition of creativity, "effective surprise," fits our use of the term "discovering" exactly.[19]

▷ **Hunter:** Certainly. Think of Einstein. Definitely a peripheral member of the physics community, yet adept at manipulating its social and political aspects to achieve consideration of his ideas. Not all great innovators command the social or communication skills to be so successful. Consider Mendel, Herapath, Waterston, J. Willard Gibbs—even Darwin to some extent.

▷ **Jenny:** Mendel and Darwin I know a bit about. Who are the others?

▷ **Hunter:** Herapath and Waterston were two nineteenth-century British scientists who separately invented forms of the kinetic theory of heat well in advance of Clausius. Both had their papers summarily rejected by the Royal Society as nonsensical. After various difficulties, both bitterly turned their backs on organized science and, in turn, were conveniently forgotten by those whose ideas they'd challenged.[20] Gibbs was an American scientist, many think the greatest of the nineteenth century, who was almost as neglected as Mendel. He communicated his most important results in mathematics so complex that later physicists, including Einstein, complained of the difficulty of his papers. Moreover, he published these papers in the *Transactions of the Connecticut Academy of Sciences*, which was virtually unknown in Europe. Had not J. C. Maxwell and Wilhelm Ostwald championed his work, it would almost undoubtedly have been neglected and then rediscovered. As it was, both van't Hoff and Max Planck reinvented many of his results in more accessible forms before Ostwald drew their attention to Gibbs's papers.[21]

That's why I mentioned Darwin. Imagine Darwin without Wallace to spur him on to publication and Huxley to hawk his ideas to anyone who'd listen. It's not enough that you invent an important idea; you've also got to convince the scientific community to pay attention to it.[22] That's easy if you address an outstanding problem that everyone recognizes, such as the structure of the gene. It can be damn hard if you're a Mendel or an Einstein and you've invented the problem as well as the solution. Then you have to convince your colleagues to accept a shocking solution to a problem most didn't know they had!

▷ **Constance:** That reminds me of statements by both Newton and Darwin that they didn't realize when they brought their brain-children into the world they'd be committed to raising and protecting them forever after.

▷ **Ariana:** And like real parents, perhaps some scientists aren't fit to be good guardians to their ideas. Which means we may be expecting too much of individuals: that they be not only technical or theoretical geniuses, but social geniuses and ace communicators as well.

▷ **Jenny:** But doesn't this imply the possibility of abuse? If discovering is partially a social function, couldn't a very adept manipulator of public (or even scientific) opinion create the illusion of having made an important discovery when in fact he's found nothing at all?

▷ **Richter:** N-rays and polywater again. There are some unsavory stories circulating about several recent Nobel laureates as well: proclaiming discoveries before they had evidence in order to manipulate funding, taking credit for the research of underlings or competitors, hiring public relations firms to publicize their results, suggesting farfetched practical applications of their results in order to justify research that has no utility in the foreseeable future.

▷ **Ariana:** I've heard that the media people are even beginning to complain about being used. Now there's a joke! But getting back to my point, it seems to me that we need to understand more about the role personality plays in a scientific career. We have two choices: either we need to train scientists more effectively to deal with the entire process of discovering—not just the fund raising and problem solving, but communicating, collaborating, and all the rest; or we have to make it feasible to specialize in various aspects—allow people like Mendel or Herapath or Waterston to think their heretical thoughts and team them up with a Huxley or Maxwell or Ostwald who'll make sure that they're heard.

▷ **Imp:** Excellent idea, except that most of the people we've been discussing here aren't team players. They're mavericks, explorers, leaders in their own right. They may not want help. What might be more useful is to have a sense of how to live through Constance's model, to experience being a maverick, so that the objections, rejections, and dejections don't come as a surprise and won't be taken as personal affronts.

One of the most valuable things I've ever gotten out of reading history of science is the knowledge that my approach to research, as idiosyncratic as it may seem, is typical of famous scientists of the past, and that the difficulties I face in getting colleagues to pay attention to my work are hardly unusual. Hence my search for strategies by which the successful scientists carried off their quixotic quests.

▷ **Hunter:** And thus the importance—now, I don't want to sound like I'm tooting my own horn—of taking a look at the three men who founded physical chemistry during the 1880s: J. H. van't Hoff, Svante Arrhenius, and Wilhelm Ostwald. Max Planck also made important contributions. All four received Nobel Prizes, and I think all would qualify as your exploratory mavericks. To my mind, they represent the great men whose existence Richter doubts—I'd say for political rather than rational reasons.

▷ **Richter:** Meaning what, precisely?

▷ **Hunter:** Meaning that you wince every time we hold up anyone as a model of scientific virtue. It's as if you want to denigrate the idea that a Pasteur, a Darwin, an Einstein is in any meaningful sense qualitatively different in ability than anyone else. Clearly you believe we're too sophisticated to believe in geniuses or paragons worthy of emulation. After all, this is the age of democracy, right? Talent is only a reflection of opportunity. We could, any of us, make such important contributions to science if only we had the chance. Isn't that your line—glorification of the ordinary scientist as a replacement for the mythology of the frontiersman fighting the prejudices and dogmas of his time?[23] Well, I don't buy it. The scientists we honor by giving their names to phenomena or by placing their names in the headings of our textbooks *were* different. There *is* an unequal distribution of ability and potential. We aren't all pregnant with ideas.[24]

▷ **Richter:** But you cannot deny that were we to go back in time and eliminate any individual scientist, no matter how famous, the course of science would be altered not at all. As the phenomenon of simultaneous discovery demonstrates, scientific advance is inevitable.[25]

▷ **Hunter:** No, I'm sorry, I can't accept that. Look carefully at the history of science, Richter. What are we talking about here? If not Herapath, Waterston, and Clausius, then who? If not Darwin and Wallace, then who? If not Pasteur himself, or Fleming himself, then who? You're overlooking the multiple missed opportunities that preceded their breakthroughs. Few—very few—scientists have or have ever had the breadth of knowledge, the experiences, the intellectual skills to produce the great generalizations these men did. No, I'm sorry, I don't see thousands of scientists waiting impatiently in the ranks to step into these men's shoes. One or two; perhaps a handful. But how tenuous that line is! Eliminate that handful of scientists who work, act, and think differently, and the whole endeavor collapses.[26]

▷ **Richter:** And yet you sit here discussing how to cultivate such scientists. Make up your mind. Can they be cultivated or not? Are these unusual men, if we accept them as such, a result of heredity or environment? I am willing to allow that there are strategies for discovering that are more likely to succeed than others. But they aren't worth a fig if they can be used only by geniuses!

▷ **Jenny:** "Thou say'st not only skill is gained / But genius too may be obtained, / By studious imitation." Sir Joshua Reynolds, quoted—I believe accurately—from the notebooks of Thomas Young, in 1792. You're always trying to force things into either/or regimens, Richter. Why not both/and? Why can't some unusual scientists be unusual by birth and still be an example to those who wish to attain genius? Why not the Samuel Smiles self-improvement philosophy that produced Michael Faraday, George Stephenson, and so many other makers of the Industrial Revolution? Why should emulation of the training and working methods of those we call geniuses not confer genius?

▷ **Richter:** I don't know. You tell me.

▷ **Hunter:** Come on, Richter, let's not play games. Perhaps it's a sign of arrogance, but a lot of the physical chemists I've been studying decided quite young that they were going to get to the top of their profession and did so by forthrightly emulating the best scientists they could find, sometimes by taking advantage of the apprenticeship system, sometimes—Einstein and van't Hoff are examples—by the kind of educational self-help Jenny's espousing. One of the things van't Hoff did as a teenager was to read as many biographies of eminent scientists as he could find. He'd read some two hundred by the time he was twenty-five. He claimed there were mental traits and ways of working typical of those who leave their names in the chapter headings of textbooks that differentiate them from those who appear only in the footnotes.[27] Ostwald wrote a whole book, *Grosse Männer*, or "Great Men," on the same subject.[28] And really, isn't that the whole point of our discussion of apprenticeship?

▷ **Ariana:** Enough justification, Hunter. Let's get to the point. We—I mean Hunter, Constance, and I—spent the last week trying to get at the connection between personality, education, and apprenticeship in the work of van't Hoff, Arrhenius, and Ostwald. We looked for common themes in their training, their ways of thinking, the skills they brought to their work. In essence, ways in which these men maneuvered themselves through the process Constance has modeled for us to arrive at the most important problems emerging in nineteenth-century science. One case we looked at was how van't Hoff and a fellow named Joseph Achille Le Bel arrived simultaneously at the concept of a tetrahedral carbon atom. Nothing, to my mind, demonstrates more clearly the way in which the interactions between personality, training, and experience influence the form a theory takes.

▷ **Hunter:** Or the nonequivalence of even the most closely allied simultaneous discoveries. Richter, you say that we could eliminate any individual from science and not change its course. Let's see.

▷ **Imp:** Do it!

▷ **Hunter:** First, the essential biographical background of van't Hoff.[29] Born 1852, Rotterdam, Holland. Excelled at music, mathematics, and the physical sciences in his primary and secondary schooling. Won a host of prizes for his musical accomplishments, but a series of exciting high school lectures on chemistry attracted him to that profession. Since so few chemistry professorships existed in Holland, his parents convinced him to matriculate at the Delft Polytechnic School, a technical college,

Jacobus Henricus van't Hoff in his early twenties. (Bischoff, 1894)

where he studied applied chemistry, or what we call chemical engineering. Some practical experience in a local sugar factory convinced the young van't Hoff that this was not for him. Nonetheless, his two years at Delft were formative. He read William Whewell's *History of the Inductive Sciences* and *Philosophy of the Inductive Sciences*; Hippolyte Taine's *On Intelligence*; and August Comte's *Course of Positive Philosophy*. These men inspired him to begin reading the hundreds of biographies of scientific men that I mentioned a moment ago.[30]

Comte, in particular, had a lasting effect on van't Hoff. He would write to Ostwald in 1901 that "insofar as I approximate a philosopher, I remain attached to Comte."[31] As I'm sure Jenny could tell you, Comte's philosophy, positivism, eschewed all hypotheses and concepts that could not be observed directly, disavowed metaphysics, and maintained that the only true science is mathematics. Let's not argue about that now—I'm just telling you what Comte says, not what I believe. Essentially, Comte conceived of knowledge as a ladder, with mathematics at the top and the social sciences and history at the bottom. Each discipline could move up the ladder by acquiring the scientific accoutrements of the discipline above it. The original reductionist philosophy. What's wrong with chemistry, Comte asks? His answer: Berthollet was trained as a physician rather than as a mathematician. If only, he lamented, Berthollet's *Essay on Statical Chemistry*—"an almost incomparable monument to the power of the human spirit for the systematization of chemical ideas"—had been properly mathematized. Then, then! we should have a truly scientific chemistry![32]

The teenaged van't Hoff took Comte's words to heart. He completed the three-year curriculum at Delft in two years and moved to the University of Leiden to study mathematics. At the age of nineteen he knew what he must do: rework Berthollet's *Essay on Statical Chemistry* from the mathematical perspective. Unfortunately, he found that mathematical training grated on his personality. He was asked to learn a great deal by rote and was fettered by a standardized curriculum that was largely irrelevant to his needs. "I am not treated so much as a man," he complained to his mother, "but as a mere organ for the acquisition of knowledge. Everything is 'objective,' and if I had not recently been captivated by the [poetry of the] highly emotional [Lord] Byron, I would soon shrivel into a desiccated scientific accretion."[33] He began to realize how difficult a task he had set himself.

▷ **Ariana:** Not the least difficulty being that he wasn't training himself within standard disciplinary lines. But I want to stress something very important here: Van't Hoff not only reads poetry, he also *writes* it. One of the common themes running through the education of all creative people is an early experience with the creative process itself. It doesn't matter whether it's in poetry, music, painting, crafts, inventing, or science itself: The act of creating something for yourself is an assertion that you are not afraid to test yourself against tradition, or to express your personal view of the world, despite the perils of any such endeavor. As William Carlos Williams wrote of his two mistresses, medicine and poetry: "the one nourishes the other."[34]

▷ **Hunter:** Indeed. Roald Hoffmann, a recent Nobel laureate in chemistry, says much the same thing about the connections between his poetry and research.[35]

▷ **Constance:** Well, you have to remember that, as a result of the tradition of chemist-poets such as Humphry Davy, chemists were once defined as poets who had taken a wrong turning.[36]

▷ **Hunter:** In any event, poetry was not enough to sustain van't Hoff. He left Leiden after a year to study with August Kekulé in Bonn. You may recall that the biggest stumbling block in Laplace's application of gravitational attraction theory to chemical mass action was that the shapes of atoms were unknown. At the very small distances between atoms, these shapes could alter the gravitational field significantly. Comte, Cuvier, Whewell, Kekulé, and Adolphe Wurtz in Paris had all pinpointed the shape problem as one of the most significant facing chemists. Kekulé was one of the world's experts on molecular structure, so what better place to begin?

▷ **Constance:** Yes, but don't forget that a lot's changed in chemistry between Berthollet and van't Hoff. For one thing, the whole tradition of mass action studies virtually disappeared.[37] I mean, had van't Hoff studied in Paris or Berlin, for example, he'd probably have learned to scoff at Berthollet's work. He certainly wouldn't have learned the gravitational-attraction analogy for affinity. Chemistry had moved on to Berzelius's dualistic electrochemical attraction theory, and that in turn had come under attack after 1860 as a result of renewed support for Avogadro's hypothesis. Avogadro had postulated that equal volumes of gas at equal temperature and pressure contain the same number of molecules.

▷ **Jenny:** You mean equal volumes of the *same* gas? Or any gas whatever?

▷ **Hunter:** Any gas. That's the beauty of Avogadro's insight: there's nothing obvious about it at all!

▷ **Constance:** As decades of controversy made clear. One issue of contention was that Avogadro believed that some gases, such as hydrogen and oxygen, had to be diatomic molecules; others, such as nitrogen, monatomic. But if, as Berzelius maintained, chemical attraction was electrical in nature, then how could two electrochemically negative atoms such as oxygen or two electronegative ones such as hydrogen stick together? Like charges repel.

On the other hand, you couldn't go back to gravitational attraction because the phenomenon of valency had been established. Oxygen had a valency of two because it likes to make two bonds, often with two other atoms (H_2O); carbon's valency is four because it likes to make four bonds, often to four other atoms (as in methane, CH_4).

A number of chemists, such as Wurtz, who supported Avogadro's hypothesis, pointed out that there was no way to explain valency on the basis of the planetary movement of atoms around one another presumed by Berthollet and Laplace. Why two or four "planets" rather than one or ten? How did atoms combine to form molecules if they were whirling around all the time? Some chemists—Kekulé for example—went so far as to assert that atoms were simply useful heuristic devices without reality, and therefore that all these apparent contradictions were purely imaginary.

The point is that van't Hoff was really trained in quite an odd way. Not to jump the gun too much, but it appears that this is one of the themes common to many innovators—certainly the physical chemists Hunter's been studying: They combine old, even outdated traditions with new techniques or renew old approaches to solve new problems. Their mix of personality, codified science, and science in the making is very unusual.

▷ **Hunter:** As we'll now see. Van't Hoff left Utrecht to study with Kekulé. Initially he was enchanted by Bonn. "In Leiden all was prose; the surroundings, the city, the people," he wrote to his parents. "In Bonn, all is poetry."[38] The lab was exciting, as Kekulé was involved in a prolonged controversy over the structure of benzene, a cyclical compound composed of six carbon atoms and six hydrogens. The problem was that carbon was supposed to have four valencies—that is, it could combine

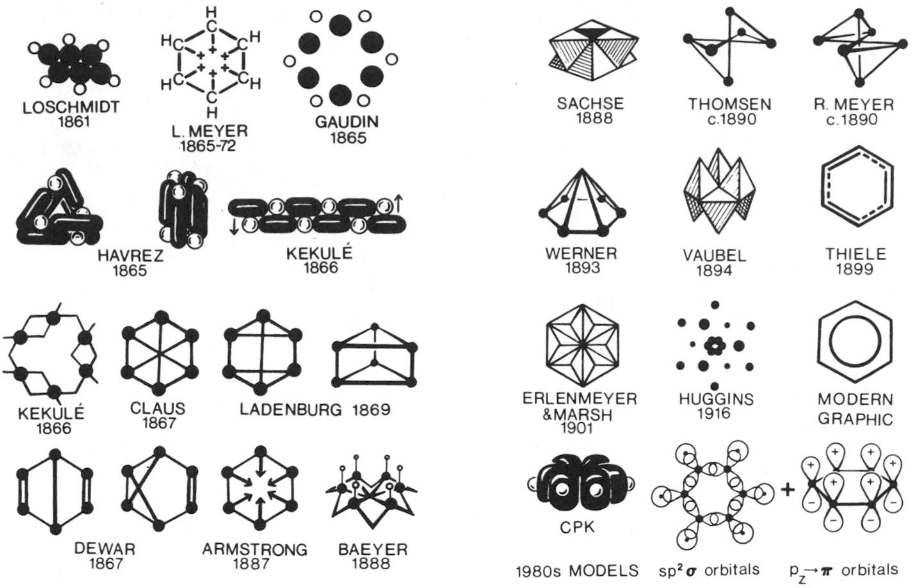

Models of benzene (C_6H_6), 1861–1980. (Based on Cooper et al., 1986; Koeppel, 1975; Levere, 1975)

with four other atoms—whereas hydrogen had a valency of one. What structural model best explained benzene and all its possible derivatives? Dozens of models were suggested (and, in fact, continue to be). Constance helped me put together illustrations of the various models, which I've got here.

▷ **Richter:** Most of these are invalid models, I take it.

▷ **Constance:** Oh, no—everything from 1865 on was considered to be plausible by some group of scientists or another. And I know that Kekulé's resonance model was still being used until at least the 1930s.

▷ **Richter:** Odd. Why proliferate models if you have one that works?

▷ **Hunter:** Hard to say. Each represents a slightly different view of bonding and suggests different explanations for valency and chemical reactivity. Partly it's a matter of scientists asking themselves why they should accept so-and-so's visual convention when they can think of one just as good.

These models also differed somewhat in their ability to explain another problem: isomerism. Isomers are compounds made up of the same atoms arranged in different ways. Pasteur's tartaric and racemic acids are one type of isomer. Kekulé had been studying others: the various

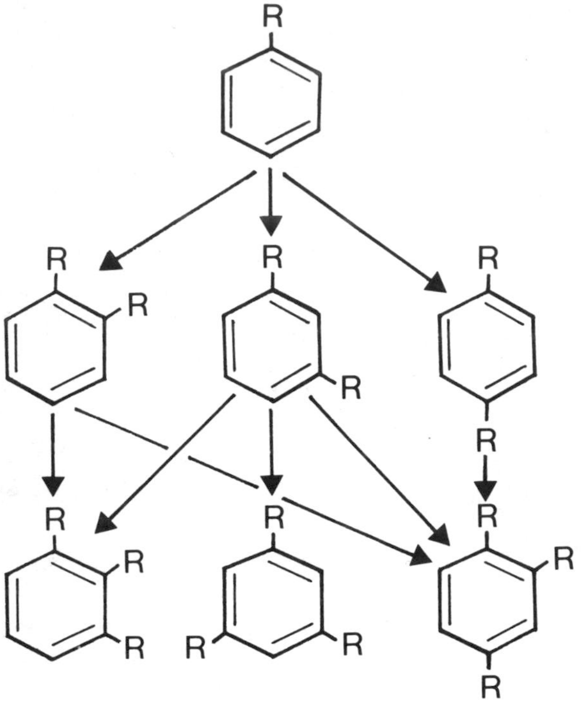

Benzene substitution, showing how such drawings can be used as "reaction diagrams" of the sort invented by Kekulé.

substitutions of benzene, for example. Depending on where a new atom is added to a benzene ring, you get different forms with different properties—para, meta, and ortho. You could determine which form you were dealing with by how many new forms you got when you added yet another atom. One from a para form; two from an ortho form; three from a meta form. Such studies allowed Kekulé to investigate the connection between composition and reactivity, and he was one of the first chemists to draw two-dimensional reaction formulas to describe his results.

▷ **Ariana:** It's worth mentioning that Kekulé was an architect before he became a chemist, so he was in the habit of thinking graphically.

▷ **Hunter:** Yes—it was an exciting time in organic chemistry in which a diversity of skills could be useful. Yet once again unhappiness dogged van't Hoff in the laboratory and classroom. Kekulé, who was rather cocksure, informed van't Hoff that chemistry had reached a dead end and was "without visible prospect for new advance."[39] Nothing but details remained to be clarified.

▷ **Imp:** The old "closed field" fallacy, eh, Richter?

▷ **Ariana:** Which van't Hoff was smart enough to ignore. He practiced what Walter Kaufmann called "the faith of a heretic": I cannot believe until I've seen for myself.[40]

▷ **Hunter:** Max Planck's philosophy as well.[41] But not one designed to make one popular, especially in an autocratic and hierarchical German university. Like Oscar Minkowski, Albert Einstein, Svante Arrhenius and all the others who decided they could do what their professors thought impossible, van't Hoff quickly found himself in trouble. Just before leaving Kekulé's lab a short year after he'd arrived, he wrote his parents another impassioned letter:

> *A thousand memories and two semesters have stormed over me like a simoom [a desert wind] and have transformed the whole into a desiccated agglomeration. I think of what they say behind my back. I was the youngest in the laboratory and the only one that opposed investigating camphor; I wanted to work on my own idea although I was educated at an unknown university. Ach, they have made me out to be an imaginative fool, a scientific Don Quixote, a misconceived eccentric. General enmity and jealousy make them ingenious; and then there are their number, size, and age.*[42]

▷ **Ariana:** We get an even deeper insight into his difficulties and his inner strength from another event that occurred at about the same time. A young woman who acted as Kekulé's lab assistant—he apparently liked to have pretty women around him—died when the apparatus she was operating broke, spraying her with potassium cyanide. Van't Hoff, deeply shaken, wrote a Byronic elegy in English, which was clearly as much about himself as his subject. "Thyne enemies were many and were strong," he wrote, joining her in their shared fight to break the fetters of

Kekulé von Stradonitz. (Bugge, 1930)

myopic prejudice: she, a revolutionary by the simple fact of her gender; he, because he would not join the fraternity. "Through thy brain such wild ideas flew, that thou couldst not resist." Both young mavericks, excluded from the herd for doing and thinking differently. Neither satisfied with the order of things. Both striving to accomplish something beyond the expectations of their colleagues: "To live and to be great or else to die."[43]

▷ **Hunter:** And yet the prejudice van't Hoff encountered strengthened his resolve and made him very explicit about how his approach to science differed from that of his colleagues. For them, facts were the object of research. They accepted that their job was to fill in the details of Kekulé's schema. For van't Hoff though, observations were just the beginning: "The Fact is the basis, the foundation," he wrote at this time; "Imagination, the building material; the Hypothesis, the ground plan to be tested; Truth or Reality, the building."[44]

▷ **Imp:** Amen!

▷ **Hunter:** Ideas fascinated van't Hoff, not experiments. As his research assistant Charles van Deventer said of him a few years later, "Soundness and solidity he certainly values, but he is in love with the idea in its general form, and his proofs are directed more towards establishing his idea in the world as a great rough block that cannot be overthrown, than to modelling and rounding it off—that he willingly leaves to others."[45]

▷ **Ariana:** Which suggests that there are two kinds of scientists, just as there are two kinds of sculptors: those who create forms by accretion, as with pieces of clay added to an armature, and those who create by subtraction, perceiving the hidden form within the block of stone and chiseling away the unwanted material to reveal it. This is a very useful way to look at the difference between van't Hoff and Le Bel. Van't Hoff was a subtractor, discarding the excess to reveal the beauty of the form lying within; Le Bel, as we'll see, was an accretor, attempting to put every fact in its proper place and tremendously frustrated if all the pieces didn't fit.

You see, of course, what we're doing here: setting up the connection between personality and style of invention. Who one is determines what one does and how one goes about doing it.

▷ **Hunter:** And thus the importance of contrasting van't Hoff and Le Bel. They met when van't Hoff moved from Kekulé's lab to Wurtz's lab in Paris. Wurtz, as I mentioned earlier, was another chemist who had pointed to the problem of the shape of atoms as the most important one facing chemistry. Coming from the Berthollet-Pasteur school of chemists, he was puzzled by the inability of chemical theory to account for most isomers. Sure, Kekulé could describe the various substituted benzenes, but two-dimensional reaction diagrams could not explain the differences between Pasteur's tartaric and racemic acids.

▷ **Ariana:** And Wurtz, who was artistically inclined, had enough visual imagination to realize the problem.

▷ **Hunter:** A new, fundamental idea was needed, he told his students.[46] By the time van't Hoff arrived in Paris, he'd solved the problem. So had Le Bel, already working in Wurtz's lab.

▷ **Ariana:** Now this is a very interesting discovery—invention would be a better word—because the answer actually preceded formulation of the problems and collection of the data, but no one prior to van't Hoff and Le Bel perceived the connections.

▷ **Jenny:** Run that by me again. If the solution was already in hand, why didn't everyone else—

▷ **Hunter:** Quite simply because nobody had all the information in their heads at the same time. We have this odd misconception that because there was less scientific literature a hundred years ago, scientists actually read and understood it all. Not true. They read what interested them, just as we do today.

▷ **Constance:** Besides, as Poincaré said, "To invent is to choose." What does the individual choose to bring together?

▷ **Imp:** Meaning that personality and experience contribute equally to invention: nature and nurture, in balance!

▷ **Hunter:** Absolutely. Compare van't Hoff with Pasteur. I've read just about everything Pasteur published on asymmetric compounds and quite a bit of the unpublished work as well. As far as I can tell, his interest was in two very specific questions: how many symmetric and asymmetric forms of a single molecule exist, and how the asymmetric forms are produced. Although he plays with ideas for explaining molecular asymmetry in a published lecture in 1860—tetrahedral shapes, spirals, helices, and so forth—he never goes beyond suggesting that any number of general models might explain the facts of asymmetry.[47] By limiting himself to the asymmetry puzzle, he effectively overlooks or ignores isomery, valency, and other aspects of molecular constitution that puzzle van't Hoff.

▷ **Constance:** You see, historically the invention of a viable theory of three-dimensional molecular shapes required three elements: one, belief in the existence of three-dimensional, real atoms; two, integration of experimental and conceptual models of chemical composition; and three, an explanation of valency.[48] Pasteur never really addressed the latter two, so he couldn't possibly solve the related problems.

▷ **Hunter:** Actually, I would add two more requirements for a full theory of stereochemistry: an explanation of isomery and an explanation of optical activity. Van't Hoff is the only person during the 1870s to perceive all five of these problems as facets of a single metaproblem. For reasons that I'll elaborate a bit later, he believes that everything has to fit together.

▷ **Jenny:** What about Le Bel? He doesn't?

▷ **Hunter:** No. And so his theory turns out not to be as general as van't Hoff's. But van't Hoff first.

As near as I can make out, what happened was this. Van't Hoff left Bonn with an intimate knowledge of Kekulé's two-dimensional reaction models of molecules, as well as of unsolved problems concerning valency and molecular shape. He probably had the optical isomer problem in his

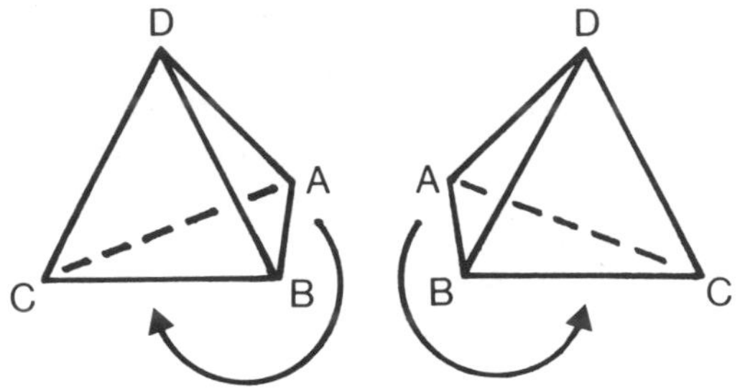

Mirror-image tetrahedrons showing right-handed and left-handed (mirror-image) asymmetry.

head, too, since he planned to go on to Paris. But first he returns to Utrecht, where he reads some papers by Johannes Wislicenus characterizing three forms of lactic acid.[49] One form rotates polarized light to the left; one to the right; and one is optically inactive and cannot be separated into right- and left-handed forms. In other words, it's not a racemic mixture, like Pasteur's racemic acid. In Pasteur's terms, there were right-handed, left-handed, and symmetrical forms. The difficulty is that one can't draw two-dimensional arrangements like Kekulé's that will yield three forms of lactic acid. Just doesn't work. So Wislicenus suggests that the atoms need to be arranged in a three-dimensional pattern "in space."[50] Van't Hoff goes for a long walk. It flashes upon him that the problem can be solved if you assume that the carbon atoms have four valences placed at the corners of a tetrahedron.[51]

▷ **Ariana:** An illumination!

▷ **Hunter:** Yes. But I want to stress what I said before: There was nothing original in this suggestion. A tetrahedral arrangement had been suggested by William Hyde Wollaston in 1808, Pasteur in 1860, Alexander Butlerow in 1862, Marc Gaudin in 1865, Kekulé in 1867, and Paterno in 1869.[52] Van't Hoff was undoubtedly aware of some of these. Each of these earlier suggestions was fraught with difficulties, however. Wollaston's was too early to address isomery or optical activity. Pasteur's was just one of several suggestions, and he failed to address the issues of valency or reaction products. Gaudin's suggestion was made purely for the aesthetic beauty of the symmetry involved and addressed none of the relevant chemical questions.[53] Kekulé's suggestion, like Pasteur's, was one of several, made as heuristic devices. He did not believe in the existence of real, three-dimensional atoms.[54] I'm afraid I don't know enough about Butlerow or Paterno to tell you the limitations of their theories, but I'd guess they were of a similar nature. It is clear, however, that none of them presented their suggestions in a manner designed to convince their peers of the utility of their ideas.

The point is that van't Hoff is the only person to suggest that a tetrahedral carbon atom would solve all five subproblems simultaneously, and to commit himself to the actual existence of such an atom.

▷ **Constance:** It seems to me that this is where his diverse background stands him in such good stead. Because he was educated in a backwater institution in Holland, he doesn't know that the Berthollet tradition is virtually dead elsewhere in Europe; he still believes in real atoms whose shape determines their chemical properties. His *Wanderjahren*—travel years—then provide him with direct experience of each of the major themes he is to pull together. Therefore, when he proposes that the carbon atom is tetrahedral in shape with valences at each of the four corners, he's able to demonstrate that his hypothesis explains virtually all the existing observations concerning isomery, optical activity, reaction diagrams, and the rest.

▷ **Ariana:** Coherently, in a single framework. The only new things about his idea are the pattern he invents—the ordering of the pieces—and the predictions he draws from that pattern.[55]

▷ **Richter:** The essence of being a theoretician. But he'd not be published today. No new data, no publication—even in many theoretical journals.

▷ **Imp:** Tuzo Wilson talks about that. I was reading a talk he gave recently on the joy of research. Wasn't all joy, though. He had one of his most important papers rejected precisely because it had no mathematics in it, no new data, and it contradicted accepted dogma. Therefore it had to be worthless speculation, right?[56]

▷ **Richter:** Tell me about it.

▷ **Hunter:** Well, whether van't Hoff couldn't get published or didn't try I can't tell, but his theory was privately published and circulated at his own expense to scientists he hoped would appreciate it.[57]

▷ **Ariana:** At least people read it. Would you take seriously a privately printed scientific paper today?

▷ **Hunter:** The Ramanujan riddle. How many kooks are there for every un-

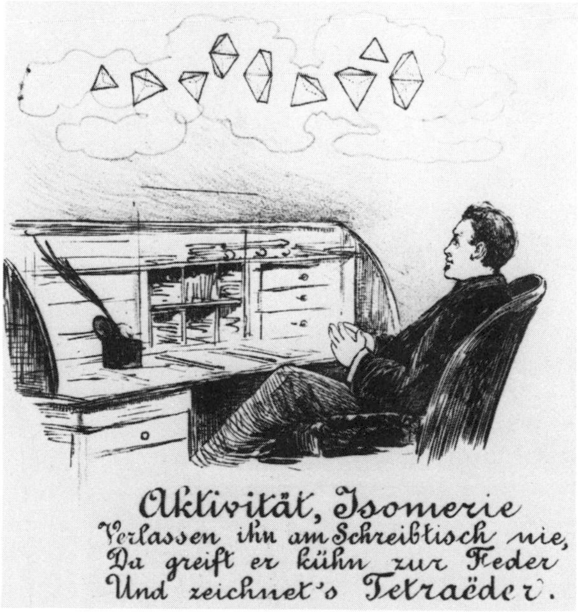

Van't Hoff inventing the tetrahedral model of the carbon atom. Drawing by his friend and fellow chemist Ernst Cohen. The rhyming caption reads: "Activity, isomery / never leaving your desk / you grasp the quill boldly / and tetrahedra sketch." (Museum Boerhaave, Leiden)

known mathematical genius sitting in a little town in India? How much effort is it worth to evaluate the nonsense that crosses your desk in the hope that something valuable might one day appear?

▷ **Jenny:** Okay. But what about Le Bel?

▷ **Hunter:** Le Bel. I wish I could tell you as much about him as I have about van't Hoff, but I can't. He's one of the hazy, unknown figures of the history of chemistry. A rich amateur, apparently trained completely within the Pasteurian tradition of optical activity studies. His research on the form of the carbon atom begins there, makes no use of structural formulas, and never addresses the problem of valency.[58] In consequence, although Le Bel invents the same tetrahedral model as van't Hoff, he demonstrates its application to only a fraction of the problems. For example, Le Bel does not place the four valencies of carbon at the corners of a tetrahedron. Rather, he talks about the atoms that bind to carbon *arranging themselves* symmetrically, much as Gaudin had suggested a few years earlier—a result of a balance of forces, not internal structure.

▷ **Ariana:** In other words, the differences in his training are reflected in the form of his theory. It also appears that Le Bel lacked van't Hoff's intellectual courage, a personality trait reflected in the nature of his theory. He couldn't commit himself completely to the idea of the tetrahedral carbon atom.

Joseph Achille Le Bel in his twenties. (Bischoff, 1894)

▷ **Hunter:** That was due to his training, also. Following Pasteur and Laurent, he believed that the shape of a crystal was determined by the shape of the molecules comprising it. Thus, tetrahedral molecules should crystallize in a cubic form. Le Bel, however, was aware of two carbon compounds, carbon tetrabromide and carbon tetraiodide, that crystallized in biaxial forms—exceptions. He could not imagine a way to build biaxial crystals from tetrahedra, and therefore rejected the notion that the carbon atom is intrinsically tetrahedral in shape. The danger of being an accretor. Instead he suggested that the atoms could also arrange themselves as a tetragonal pyramid, with the carbon atom at the apex. Unfortunately, the pyramidal model made a number of predictions that turned out to be incorrect. For example, a pyramidal model allows for asymmetric, optically active substances of the form CH_2R1R2 (where R1 and R2 are two different atoms other than hydrogen). No such optically active isomers are possible with a tetrahedral model. Le Bel spent fruitless years attempting to isolate such substances.

▷ **Jenny:** But doesn't that help to prove van't Hoff right? I mean, if we view science as a process of trying many things and eliminating those that don't work, then isn't Le Bel's contribution just as important as van't Hoff's? Doesn't he rule out other possibilities?

▷ **Richter:** Most interesting. You are arguing that theory validation is not a matter of internal testing, but of comparison to other available theories. I rebel against such relativism; yet this notion of creating a diverse set of hypotheses to be tested against one another intrigues me. It is not how we generally train scientists to proceed, yet it keeps reappearing in our discussions.

▷ **Ariana:** It is how we train doctors—at least at some hospitals. The best diagnosticians are those who conceive of the largest number of specific hypotheses during the early phases of diagnosis, and then systematically eliminate these possibilities one by one.[59]

▷ **Constance:** Recently there's also been discussion of the "method of multiple hypotheses" among certain historians and philosophers of science, too.[60]

▷ **Richter:** Fine. But what I was saying is that we scientists are trained to be "right." We—certainly I—have an aversion to being wrong. Yet comparison of a diversity of hypotheses—many untenable—may be necessary to data evaluation. One hypothesis may act as the control for testing another. If so, what are the consequences?

▷ **Ariana:** Or causes? Diversity of hypotheses; diversity of training. Again, how one thinks and what one thinks about are largely determined by one's background. Le Bel demonstrates the drawbacks of relying on a single tradition of research. Too close adherence to any school of thought can cripple you.

▷ **Constance:** But getting back to the pheneomon that started this discussion—simultaneous discovery—I think it's important to say that what we've just seen in the case of van't Hoff and Le Bel is typical. Kuhn has written an analysis of the simultaneous discovery of the conservation of energy, attributed to Helmholtz, Mayer, Liebig, Mohr, and as many as a dozen investigators during the 1840s. He writes of these so-called codiscoverers:

> *In the ideal case of simultaneous discovery two or more men would announce the same thing at the same time and in complete ignorance of each other's work, but nothing remotely like that happened during the development of the conservation of energy . . . no two of our men even said the same thing. Until the close of the period of discovery, few of their papers have more than fragmentary resemblances retrievable in isolated sentences and paragraphs. Skillful excerpting is, for example, required to make Mohr's defense of the dynamical theory of heat resemble Liebig's discussion of the intrinsic limits of the electrical motor. A diagram of the overlapping passages in the papers by the pioneers of energy conservation would resemble an unfinished crossword puzzle. Fortunately no diagram is necessary to grasp the most essential differences.*[61]

▷ **Jenny:** So simultaneous discovery is a myth?

▷ **Constance:** Well, I'm not sure you want to go that far. But certainly, even in instances of simultaneous discovery, one investigator could not simply be substituted for another without consequentially affecting the course of science. I mean, Wallace's theory isn't Darwin's when you come right down to it, Le Bel's isn't van't Hoff 's, and Lothar Meyer's periodic table of elements isn't Mendeleev's. In fact, that reminds me of something Sementsov said about why we remember van't Hoff and Mendeleev instead of Le Bel and Meyer.[62] Van't Hoff and Mendeleev had total faith in their theories, to the extent that if individual facts didn't fit they, as Ariana might say, chipped them off the block. Those facts will turn out to be artifacts, they asserted—correctly. Mendeleev left gaps in his table for elements he was sure had to exist. Le Bel and Meyer, though, created less appealing and ultimately incorrect patterns because they insisted on the sanctity of the data. Le Bel forced every datum into his model whether it fit or not. Meyer refused to leave gaps because there wasn't any evidence to indicate the elements existed.

▷ **Ariana:** One must trust in the purity of one's ideas, then.

▷ **Imp:** And let them be general! As we've said before: Go for the algorithmic solution. Notice that van't Hoff doesn't just solve Wislicenus's lactic acid problem. He asserts that the solution he invents for that one case must hold for all possible cases. That's a really important lesson. I guess I'm particularly sensitive to it because I studied with Art Pardee for several years, and one of his greatest frustrations was that he failed to win a Nobel Prize for inventing the idea of allostery as a mechanism for feedback inhibition in enzymes. This sort of thing happened to him several times, and there's no doubt that in each case he was onto an important idea. But as one of his colleagues (who shall remain nameless) said, he lacked the knack of generalization. He failed to say, "What I see in this particular instance must be true of all similar systems." The people who did make that point, and then went on to demonstrate it—Monod, Jacob, and Changeux in the feedback inhibition case—were the ones credited with the discovery.[63]

▷ **Jenny:** Well, I still think the most interesting lesson of the van't Hoff–Le Bel story is that you can invent something new utilizing nothing but existing data and concepts, that novelty can be generated by the organizing principle. There are a lot of historians and social scientists who should hear that message. We're swamped with data, but rather than search for unifying patterns in it, what do we do? We go out and find some more untouched documents! "Cherchez les manuscrits" instead of "Pensez profondément."

▷ **Richter:** So? Become a theoretician.

▷ **Imp:** Or simply set aside a day a week to ponder the Big Problems as I do.

▷ **Jenny:** Will fifteen minutes do? I need some coffee.

▼ JENNY'S NOTEBOOK: Falsification

Ariana followed me out to the kitchen and asked in a low voice, "What's the matter with Richter today? He isn't his usual negative self."

I nodded. "I know. I was talking to Imp about him last night. To my mind, the arguments he was provoking were getting out of hand. Actually Imp thinks that's great. No heat, no light, he says. Anyway, he thinks Richter agreed to join the Discovering Project from a personal need. He looked up Richter's publication record. Apparently he had a series of fairly important papers right out of grad school—still regularly cited and all that—and since then he's just been recirculating the same old ideas over and over again. He can't be much over forty, if that, but Imp thinks he's afraid he's burned out."

"That would explain his obvious discontent."

"So, Richter's got a conflict," I continued. "On the one hand, he doesn't want there to be any way to increase the probability of discovery, because that means he's wasted the last ten or fifteen years of his life. On the other hand, he can't face another twenty years of intellectual barrenness either, so he hopes we *do* come up with something useful. Imp thinks his combativeness results from this inner turmoil."

"Perhaps," Ariana conceded, "but it doesn't excuse his pigheadedness."

"No, but maybe his relative calm today means we're making some headway." We carried the coffee in.

That's when Richter popped his question, eyes half-closed like a cat, but alert. He reminded me of a professor I'd had, who used to fall asleep at seminars—we thought. Then, boom! the eyes would fly open and a cogent, precisely timed query would stun the unsuspecting speaker. "I'm sure I've missed something," he'd begin—and you knew he hadn't. Somehow he, and I suspect Richter as well, had mastered the art of daydreaming, allowing the droning words of the speaker to spin the web in which the speaker himself would be caught.

"Tell me, Hunter," Richter began, "if the gravitational attraction theory with its planetary motion for atoms was killed by mid-century by the notion of fixed spatial valencies, how did Rutherford and Bohr get away with it?" We all paused.

"Ah," replied Hunter, smiling, "so you've hit another philosophical snag, eh? Well, my answer may not satisfy you because it's not philosophical, but here it is: Rutherford and Bohr were either unaware of the chemical problem—after all, they were physicists—or they ignored the problem."

"Hold on a minute, you two," I interrupted. "If you're going to discuss this in front of the rest of us, you'd better fill us in."

"Certainly," Hunter said. "Essentially Richter's raising a point about the philosophy called falsificationism, right?"

Richter nodded. "The main proponent is Karl Popper.[1] He recognized back in the thirties the impossibility of proving any scientific conjecture. He argued that we can never have access to all possible data. The next observation or experiment might yield data disproving the hypothesis. Therefore, there is no point in attempting to prove anything in science. Example: One can never prove by observation that all swans are white, because the next one might be black. But one could disprove or falsify the statement that all swans are white by finding a black one. Thus Popper argued that the object of science, rather than cultivating inductionism, should be to disprove or falsify hypotheses as quickly as possible so that new and better ones may be constructed. Science therefore advances in a series of what Popper calls conjectures and refutations—a natural selection of ideas, if you will. The more severe our tests, the more quickly we approach scientific truth.

"Now, I have never liked this philosophy much. I fail to see that one can disprove any better than one can prove. As Duhem and Poincaré argued long before Popper, the basic ideas that direct science are definitional, not empirical, and in consequence axiomatic.[2] Moreover —"

"Hold on," I interrupted. "Sorry, but this is all news to me. What's axiomatic about the basic ideas of science?"

"Consider the very famous case of Hans Kolbe versus van't Hoff," Hunter volunteered. "Kolbe was a German chemist who refused to accept the existence of atoms because there was no direct observational proof or 'positive' evidence. After van't Hoff published his tetrahedral carbon atom theory, Kolbe wrote one of the most scathing attacks ever published in the history of science. He suggested that van't Hoff, who had found a position at the Utrecht Veterinary College by then, must have borrowed Pegasus from the college stables and taken a flight of fancy that would bring ruin to chemistry if taken seriously![3] A perfect case of conflicting axioms: Van't Hoff's chemistry presupposed the existence of atoms;

Kolbe's did not. Much like the difference between Euclidean and non-Euclidean geometries, and leading to similarly divergent results."

I was puzzled. "Okay, but if you reject the existence of atoms, how do you do chemistry?"

"No problem," replied Hunter. "You use the concept of 'equivalents' instead. Say you have a gram of carbon. The equivalent of oxygen would be the number of grams necessary to combine completely with all of the carbon—masses instead of particles. Or you could adopt the sort of thermodynamic approach to chemistry called 'energeticism,' suggested at the end of the century by Ostwald and Duhem. No atoms; just equations describing heats of reaction, entropy, and so forth. Quite interesting results come out of these alternative sets of axioms despite their limitations."

"The point is," continued Richter, "that one cannot prove or disprove an axiom. One judges its utility by the results one derives from it. The problem is how?"

Imp had to put in his two cents' worth. "There's another problem as well. Falsificationism is based upon the distinction between context of discovery and context of validation: conjecture, then refutation. But if the test is the discovery, as in so many of the cases we've discussed here, then the idea of falsification becomes irrelevant. Pasteur didn't falsify his cosmic asymmetric force hypothesis; he used it to generate experiments that led him to observe asymmetric fermentation. Fleming didn't falsify his conjecture about fungi producing lysozyme; he used it as a heuristic strategy to search for antibiotics. Proof and disproof have nothing to do with exploring nature. Hypotheses are concubines that are retained when fruitful and abandoned when they become barren. What they produce is irrelevant as long as they produce."

Richter didn't think so. "As usual, you overstate the case. Propositions and their logical tests certainly have a major role in producing codified science. However, codified science is rarely productive of new insights. But the point is this: What constitutes falsification? Lakatos, Harvey, and Kuhn point out that some empirical elements in all theories are born falsified.[4] Evidence is always available to contradict any theory. So naive falsificationism is untenable."

"So, because every theory will turn out to be wrong at some time during its testing, the test isn't whether it's right or wrong but how it compares with other competing theories?" I asked.

Richter nodded curtly. "My point in raising this issue with Hunter. One must know when to ignore the contradictory data and when not to. My question is: Why are the contradictions ignored in the case of the Rutherford-Bohr atom?"

"Why in any case?" Hunter pondered. "It seems to me it's a matter of balancing the benefits of the solution versus the problems it creates—cost-benefit analysis. Look at it from Bohr's perspective. The bare-bones story is this.[5] In 1897, J. J. Thomson discovers the electron. Thus, the atom is composed of positive and negative matter. The working model is like a raisin pudding—electrons (the raisins) are mixed into the positive matter (dough). Then, about 1912, Rutherford demonstrates that the positive matter is concentrated in a very dense nucleus surrounded by a very diffuse cloud of negatively charged electrons. He suggests that the electrons are like the planets orbiting the sun. Bohr goes on to demonstrate that such a model is stable only if the orbits are quantized, and that the energy levels in these quantized orbits correspond with the Balmer, Lyman, Paschen, and Brackett series of spectral lines produced by heating hydrogen to high temperatures. In short, the Bohr atom explains very nicely the physical properties of atoms circa 1912."

"But not the *chemical* properties," insisted Richter. "You can't explain the tetrahedral valency of the carbon atom using planetary orbits, quantized or not."

"No," conceded Hunter, "not very easily. You can, however, explain the general problem of valency: The number of open electron orbits predicts the valency number exactly. It's just the problem of their localization in space that must be overlooked. But again, I have to stress that Bohr doesn't do organic chemistry—he talks to physicists, he publishes in physics journals, he addresses physical problems. Until Schrödinger invents his quantum mechanical model of the atom, it's the best model around—meaning that it explains more evidence coherently with the fewest assumptions. No one claimed it was perfect. And it's just plain tactically stupid to throw out a theory that solves ninety-nine problems because it won't solve the hundredth."

"Unless you have a theory that solves the hundredth problem, too," Richter said.

"But even van't Hoff 's theory doesn't do that. He can't explain the spectral lines that Bohr does, right? Implication: Having a beautiful and useful theory is more important than having a theory that is in total agreement with all the facts," Imp asserted.

Hunter nodded. "Sometimes stated as Maier's law: If the facts don't fit your theory, ignore the facts.[6] At least until the facts can be reevaluated or demonstrated to be artifacts—another reason falsificationism doesn't work in day-to-day science. For example, when Feynman and Gell-Mann put forward their theory of beta decay in 1958, it was in stark contradiction to widely accepted data. Yet all of these data were later found to be incorrect or were recalculated and found to be in agreement with the new theory.[7]

"Or consider van't Hoff 's tetrahedral carbon atom. Not only was it 'falsified' (at least for Le Bel) by the existence of biaxial crystal forms, but Marcellin Berthelot had published data in 1866—seven years before van't Hoff 's idea—claiming that styrene is optically active. According to van't Hoff 's theory, styrene can't be asymmetric, and can't therefore be optically active.[8] Either Berthelot's observation was incorrect, or van't

Hoff's theory was. Another case of theory versus observation, precisely like the one that led Pasteur to his observation of the racemates. In this instance, however, there was no surprise, at least for van't Hoff. He prepared styrene and demonstrated that the optical activity was due to a contaminant called styrocamphor.[9] Berthelot, outraged, repeated his experiments without altering his method or reproducing van't Hoff's extractions, and reiterated his initial result."[10]

"'I'll see it when I believe it.'" Ariana chuckled.

"Exactly," agreed Hunter. "Raising the issue of how one knows whether a reproducible result is artifactual without a theory to guide one's evaluation. Until the theory is accepted as possible, who will bother doing the necessary work to test it and to reevaluate prior data? So, as Richter said before, you've got to treat your hypotheses as probable truths and try to validate them before you go about demonstrating their limitations."

Richter pursed his lips and grunted. "And even then, one may not know whether one black swan falsifies one's conjecture or merely places a boundary condition on it. Every statement is, as I've said before, bounded.[11] Popper and his colleagues seem to overlook the taxonomic possibility that the scientist may take care of the apparent anomaly simply by defining swans as white and placing black swans in a new category of their own. A black swan, the scientist might argue, is not a swan properly so-called and is thus irrelevant to testing the taxon 'swan.'"

Imp nodded agreement. "Such difficulties probably explain why a recent survey revealed that very few scientists attempt to falsify their own results."

Ariana was intrigued. "In other words, all this philosophical stuff by Peter Medawar about falsification as the method of biomedical research is nonsense, right?[12] Just as there can be no unambiguous proof of a theory, there can be no unambiguous disproof, either. The position that one counterexample disproves the hypothesis is untenable in practice, as Claude Bernard made clear with his example of the experiments that yielded one result one day and another the next."

"Absolutely." Imp grinned mischievously. "You ever noticed Medawar trying to disprove his own results or those of his colleagues such as Burnet or Jerne? You don't get Nobel Prizes for disproving ideas—you get them for establishing them."

"Sure," Hunter agreed. "I have some vague recollection that Bernd Matthias, the guy who discovered superconductivity, said that the first twelve experiments designed to demonstrate the phenomenon failed. The next one succeeded."[13]

"I've got an even more interesting example," put in Constance. "There was a physicist named Dayton C. Miller who repeated the Michelson-Morley experiment measuring the speed of light. If you remember, their experiment was designed to measure the effect on the speed of light of the 'ether' that was supposed to fill space. Instead, it demonstrated that there was no ether. The speed of light is constant. Well, in 1925 Miller reported that the speed of light was *not* constant. Why did this positive result not falsify the negative one?"[14]

Richter had the answer. "Obviously because it is irrelevant whether a test of theory is positive or negative. Every negative result is a positive result for another theory, every positive result a null point from another perspective. Besides, no theoretical statement or experimental test can be evaluated individually. We evaluate systems of propositions in light of systems of results. Miller's results cannot be taken seriously until there is some system of reasons available to convince physicists that the entire corpus of relativity theory is shaky. No single experiment, no single technique can do this."[15]

"But, looking at this from a rather naive point of view," I said, "you seem to be saying there aren't any experiments that definitively prove, disprove, or otherwise differentiate between theories. Is that right? I mean, what about all those 'experiments that changed the world' I learned about in high school?"

"Now, wait," interjected Hunter. "You've confused too many things here. There are certainly ways of differentiating between theories: sets of predictions that don't coincide, for example. An experimental result that accords with the predictions of one theory but not another may lead us to favor the former. On the other hand, you're right about 'crucial experiments.' They don't exist.[16] Indeed, many so-called crucial experiments turn out to be inconclusive in retrospect. Hertz, for example, convinced many scientists of the existence of radio waves by an experiment purportedly demonstrating standing waves just like the ones you get in optics or acoustics. Some years later, when it no longer mattered, his result was shown to be an artifact produced by the instrument used as a detector.[17] Ostwald and Walther Nernst performed a 'crucial experiment' to demonstrate the existence of free salt ions in solution that likewise suffered on later examination.[18] It just isn't that easy."

"Imagine if it were!" exclaimed Imp. "Think how nice it would be to design one experiment that said yes or no, and leave it at that! But experiments rarely yield a clear yes or no. And even if they do, you always have to worry about the technique. A Skinner box allows rats to respond in only one way, mazes in another. Therefore, the limits of the technique itself must be tested, and the limitations of these tests of the technique determined, ad infinitum. Makes you wonder why anyone bothers, doesn't it!"

I shook my head skeptically. "Particularly if it's all as subjective as you make out. Yet surely science advances—how?"

Constance had been trying to get a word in edgewise and my question finally gave her a chance. "Perhaps we're limiting ourselves too much by discussing only logical criteria again. I mean, this may sound a little crazy, but when I was in grad school I put together a list of criteria

Criteria for Evaluation of a Scientific Innovation

Logical
- Offers unifying idea that postulates nothing unnecessary
- Is logically consistent internally
- Is conceivably refutable
- Is explicitly bounded in application
- Is consistent with previously established theories and laws

Empirical
- Makes observationally confirmable predictions or retrodictions
- Confirming observations are reproducible by skeptics
- Provides criteria for interpretation of observations as fact, artifact, or anomaly

Sociological
- Resolves problems, paradoxes, or anomalies recognized by community
- Posits problem solving model (paradigm) for related problems
- Poses new problems
- Changes how scientists think about and perform their work
- Changes textbooks and training

Historical
- Demonstrates novelty in history of the field
- Meets or surpasses criteria set by predecessors or demonstrates they are artifactual
- Incorporates history of testing of previous theories or observations

Aesthetic
- Displays beauty and harmony
- Demonstrates technical skill in experimentation and communication of results
- Requires interpretation

for theory choice suggested by various historians, philosophers, and sociologists of science, and what I came up with is a bit messy, but I think accurate. As far as I can tell there are at least four categories of criteria scientists employ in evaluating new results and theories.[19] First are logical criteria, such as Occam's Razor, internal consistency, logical testability, and boundedness. We all know about those. Second are empirical criteria, such as experimental testability, consistency with data, reproducibility, and the ability to account for all data in terms of facts or artifacts. Again, nothing surprising, I think. Third are sociological criteria. Here some of you might disagree with me, but I think there's ample evidence to back me up. Sociological criteria are things such as whether a new result or a novel theory solves recognized problems, poses new sets of problems to work on, or provides new tools with which to address existing problems; provides a problem solving model or paradigm—perhaps you'd all prefer the term 'algorithmic solution' after our discussions; and provides new definitions, concepts, or techniques beneficial to the problem solving ability of other scientists. In short, the sociological criteria are essentially Kuhn's: Does the new science change how science is done by creating a new pattern of 'normal' work? If something doesn't offer the scientist new ways of doing science, I don't think it's accepted. That's why most scientists prefer an obviously incorrect theory to no theory at all.

"Finally there are historical criteria. These get at Jenny's question of how science progresses. Foremost is the requirement that a new theory meet or surpass all the criteria set by its predecessors or demonstrate these to be artifactual. In other words, novelty is relative to the past and is tested against it. So new science has got to do everything old science has done, but better. It's got to explain all the data gathered under previous relevant theories—or in philosophical terms, accrue the epistemological status acquired by the testing of previous theories. And it's got to be consistent with all preexisting valid scientific theories; for example, an evolutionary theory can't contradict thermodynamics at the juncture between the two. Does all that make sense?"

Ariana nodded vigorously. "I'd say it sums up what Hunter's been trying to tell us about the Bohr atom quite well. Particularly if you think about balancing these four sets of criteria. You aren't going to discard a physical theory of the atom because it doesn't explain a set of chemical problems if the corresponding chemical theory fails to account for the physics, and if the physical model allows you to be productive in the sense of raising and solving new problems.

"On the other hand, I do have a quibble. I do wish you'd add a category to your list: aesthetics. That makes five *quint*essential criteria. It seems to me that the only way to account for people's differing axiomatic assumptions is to say that people have different tastes. Some like a positivistic nature without atoms, others a mechanical universe full of little machines; some believe that God plays dice with the universe, others not; some want nature to be harmonious and symmetrical, others are willing to entertain a chaotic vision. So when it comes to understanding why scientists choose (or acquire) axioms, we've got to look at what they consider beautiful, harmonious, consistent, elegant, and so forth—and you simply can't do that without a sense of aesthetics."

Richter shook his head doubtfully. "You have an amazing ability to introduce your primary concern into every discussion. Well, perhaps you are simply illustrating your own point. But you must realize what such an eclectic, or shall I call it catholic, view of science means in practice. Not only do we not acquire scientific truth; its acquisition becomes impossible. We must discard both proof and disproof for—what shall I say—aesthetics? history? communal consensus?—"

"And dissension!" interrupted Imp. "A major oversight on the part of your philosophers, I might point out. As your own example of the Bohr atom so clearly illustrates, any theory of science has got to explain not only why theories are accepted or rejected but also why they are only accepted or rejected provisionally. As Bohr reportedly said himself, every statement must be construed simultaneously as a question."

"So, essentially you're an advocate of a conventionalist point of view?" said Constance tentatively. "Like Harré, for example. He maintains that both inductivism and falsificationism assume incorrectly that statements about nature are either true or false. He proposes instead that these statements be treated as conventions that are neither true nor false, but more or less economical, fruitful of new knowledge, and useful for solving outstanding problems. Then no theory need be perfect to be the best, and the role of experiment isn't to prove or disprove any particular theory, but rather to illustrate its potential or limitations."[20]

Hunter agreed. "The larger the number of experiments and concepts a theory integrates successfully, the more economical, illuminating, and useful it is. A mismatch between theory and experiment doesn't disprove the theory so much as indicate that another is possible."

"Or that the experiment itself was faulty or misinterpreted," Imp added. "I seem to recall some article claiming that the measured charge of the electron kept changing every time the value predicted by theory changed.[21] People had probably observed the other values before, but wouldn't report them because they didn't fit expectation. Observation is theory-laden and theories observation-dependent. The only way to look at it is as a symbiosis with the two evolving hand-in-hand.

"I must add, however, that I can't go along with Harré's statement that the purpose of experiment is illustrative. Some experiments may be illustrative—like Fleming's penicillin plate—but others are exploratory, like his search for bacteriophage in nasal mucus. Bacon's distinction between experiments of 'fruit' and 'light.' Only the latter promise surprises."

Richter wasn't pleased by the direction the discussion was taking. Our conclusions were very messy, he kept repeating. Where's our assurance that we're moving toward truth? What prevents several alternative theories from existing simultaneously, each having different adherents? Isn't that what we've just been talking about? Imp pointed out. Multiple models of benzene. Nonequivalent simultaneous discoveries. Physicists inventing theories that no organic chemist could possibly accept. Berthelot refusing to consider van't Hoff's results because they don't fit his preconceptions.

The conversation continued back and forth for another ten minutes. The only thing we all agreed on was that the current explanations of how scientific results and theories are evaluated were incomplete, and that the most important missing element was an explanation of how both consent and dissent were possible simultaneously. I'd say our little group might make a good test case.

◀ TRANSCRIPT: Global Thinking

▷ **Imp:** Back to work! Where do we go from here?

▷ **Jenny:** I seem to recall that Hunter and Ariana were about to present us with a prosopographical study of the founders of physical chemistry when we sidetracked them into the issue of whether simultaneous discoveries are identical or not. Why not let them finish?

▷ **Imp:** Retaliation for all the scientific jargon, eh? What, pray tell, is a prosopographical study?

▷ **Jenny:** Biographical study of a group of people aimed at determining similarities and differences.

▷ **Ariana:** That's certainly what we have been doing. Basically what we found is that who you are, what you know how to do, and the ways in which you think are as important to making discoveries as formal training. You resemble the thought you conceive. Moreover, when you compare Arrhenius, van't Hoff, Ostwald, and Planck it becomes clear that they had similar ways of working and thinking, even though they were schooled differently. We believe that some of these shared approaches might be extremely useful for training scientists today—to place them at the nexus of important emerging problems.

▷ **Hunter:** In short, we believe we've found some more of those strategies for research Imp's been after.

▷ **Imp:** Good! So at last we address the third relationship you promised us: that between nurture and genius or, looked at another way, between training and creativity.

▷ **Richter:** Frankly, I would prefer to discuss illuminations, such as the one that solved van't Hoff's problems with the carbon atom. Why may a man spend years trying to solve a problem and then suddenly, without any warning, the answer appears to him? Eh? You can place yourself at the nexus, as you call it, but what is the point if you must loiter about waiting for an unpredictable, irrational insight?

▷ **Hunter:** One thing at a time. Illuminations aren't something you just explain, any more than you just have. You need preparation. Besides which, Ariana and I haven't really gotten a handle on the matter yet.

▷ **Richter:** Ah, the crux of the matter.

▷ **Hunter:** In any case, we've got quite enough material to keep us busy this afternoon. Why not illuminations next week? Spread the burden of thinking! Give you a chance to solve one of your own problems for once, Richter.

Now, then, on to van't Hoff, Arrhenius, Ostwald, and Planck. First thing you should know is that, like van't Hoff, Arrhenius and Ostwald are intellectual heirs to Berthollet, both through student-teacher apprenticeships and in terms of the problems and theories they address. Planck is not. He's trained completely within the tradition of physical theory. That becomes important later, as we'll see. Van't Hoff, Arrhenius, and Ostwald become known as "the Ionists," for their work in inventing a thermodynamic theory of solutions based upon Arrhenius's idea that salts ionize when dissolved.[1] His is the essential insight that provides chemists with their first relatively complete understanding of how chemical reactions occur in solutions. It also allows van't Hoff to apply the physical laws describing the behavior of gases to solutions, thus giving chemists a powerful mathematical theory with which to work. While all this is going on, Planck is proposing an alternative thermodynamic theory of solutions that also postulates the dissociation of salts, but without ionization—that is, without becoming electrically charged. As we'll try to show you, Planck's oversight probably stemmed from his lack of chemical training. Another case, like Le Bel's, of limited training resulting in limited theory.

In any event, all four men make important contributions into old age. Van't Hoff invents the concept of the tetrahedral carbon atom at the age of twenty-two, and the first thermodynamic theory of solutions at the age of thirty-five; lays the basis for experimental and theoretical petrology in his forties; and provides some basic insights into enzyme kinetics during his late fifties.[2] Note that he keeps changing fields. Same with Arrhenius and Ostwald. Arrhenius invents the theory of ionic dissociation between the ages of twenty-five and twenty-nine; predicts what we now call the "greenhouse effect" when he's thirty-seven; invents the field of immunochemistry during his forties; and contributes significantly to cosmology and meteorology during his fifties.[3] Ostwald moves from mass action to general theories of chemistry in his twenties and thirties, catalysis in his forties, energetics (an odd, neopositivistic application of thermodynamics to chemistry) in his fifties, and color theory in his sixties.[4] Planck begins his career with a series of papers on classical thermodynamics and ends up helping to create quantum theory.[5]

▷ **Ariana:** The important point is that these men didn't just luck out once. They repeated their successes in a variety of fields. That suggests they knew how to increase their probability of discovering.

▷ **Hunter:** And one of their strategies was to work on problems that had been dropped by earlier researchers as being unfruitful or impossible to solve. Van't Hoff, Ostwald, and Arrhenius all worked on problems of mass action early in their careers. That makes them unusual. Historians have documented that work on mass action problems nearly died out earlier in the century.[6] For one thing, equilibria were difficult to study. Most reactions go rapidly to completion, especially ones involving electrolytes, or salts. Also, organic chemistry—the synthesis of carbon compounds—developed so quickly and offered so many opportunities that most chemists ignored inorganic chemistry.

▷ **Richter:** In short, you are saying there is cultural inheritance. These scientists are of a different intellectual lineage.

▷ **Hunter:** Yes, in terms of both codified science and science in the making. And they also compound an unusual variety of scientific specialties, which again sets them off from their colleagues.

▷ **Ariana:** Not just scientific skills! They have an unusual range of skills and interests, period. Consider Ostwald, my personal favorite.[7] He was born in Latvia in 1853 and grew up in a family of do-it-yourselfers who practiced all kinds of arts and crafts. As a teenager he concocted his own fireworks, made his own collodion for the plates of a box camera he built

Wilhelm Ostwald (standing, with violin) as a university student. (Akademie der Wissenschaften der DDR, Berlin)

himself, mixed his own paints for his experiments in painting, and even invented a new method of decalcomania—a dye transfer process for making pictures. He was to become the primary experimenter of the Ionists and the chief inventor of new techniques and apparatus. He also played the piano and the viola, dabbled in most of the sciences, and excelled at mathematics. He attended the local *Realgymnasium* and matriculated at the University of Dorpat in Tartu in 1872. Like van't Hoff, he was pushed toward engineering by his parents. He vacillated. Unable to commit himself to any particular subject, he spent most of his time with his artistic and musical friends, and he studied philosophy. Like van't Hoff, he found himself drawn to Comtian positivism, and later joined Ernst Mach in formulating neopositivism. When his university years extended beyond the usual number, his father put his foot down. Ostwald chose to sit for a degree in chemistry.

▷ **Imp:** Sounds a lot like Jacques Monod. Even as a doctoral student he couldn't decide between music and biology, and he spent most of his time at Caltech organizing the Bach Society and conducting a local orchestra instead of learning *Drosophila* genetics. He didn't know whether to follow Pasteur or Beethoven![8]

▷ **Hunter:** Not atypical. Planck was faced with a similar dilemma—really trilemma: whether to become a professional pianist, a philologist, or a physicist. When he asked his music teacher's opinion about a career in music, the man said, "If you have to ask, then you should do something else."[9] Good advice for deciding about any profession.

▷ **Richter:** Fine, but this tells me nothing about how these men manipulate the process of discovering.

▷ **Ariana:** Wrong, Richter. It tells you how they develop their particular style of research. We had to know Fleming's personality, skills, and interests to understand what he did; and the same goes for Arrhenius, van't Hoff, Ostwald, or anyone else.

▷ **Richter:** Style. I see. Now we are artists?

▷ **Ariana:** In a sense, yes—creating new ways of perceiving nature, and with just as much individual style. Absolutely. Define style as the unique confluence of problem types an individual addresses, whether these problems are constructed within or across traditions of codified science, the techniques chosen to investigate them (experimental, theoretical, and so on), and the criteria employed in defining acceptable solutions to the chosen problem. Then every individual has a unique style.[10] Compare yourself to me or Imp or Constance, Richter. Style? You bet! Very individual; very much limited to the range of our skills, interests, and abilities; very much shaped by personality and personal experience.

▷ **Constance:** Exactly David Nachmansohn's opinion. He wrote not too long ago that "character, emotions, literary and artistic experience, philosophy, and political involvements form an integral part of a personality. Since scientists are human, all these factors determine their reactions, their way of thinking, and must be essential elements in the formation of scientific ideas and views, motives, and attitudes. Knowledge alone of a special scientific field, however solid and profound, provides only the tools. What is achieved with these tools depends to a very large extent on the complex factors of personality."[11]

▷ **Ariana:** Sure. Lots of other scientists have said similar things. What Hunter and I have set out to do today is to demonstrate just how important this concept of style—the confluence of personality and knowledge—is in science.

▷ **Hunter:** Arrhenius is a beautiful case—the prototype for the hard-working, hard-drinking visionary Sondelius in *Arrowsmith*. Very different from Ostwald. In fact, when he and Ostwald first met, Ostwald wrote to his wife that Arrhenius's "substantial abilities do not lie in the realm of experiments, but in the realm of speculation. I myself am primarily an experimenter: we add up to a beautiful whole."[12] Indeed, Arrhenius had quite a different background. He's another of our scientifically and mathematically talented young men, but as far as I can tell without the artistic or literary flair of van't Hoff and Ostwald. Whether this is a matter of personality I can't say, but economics certainly played a role in limiting his experiences. Arrhenius's father was not well-off. He didn't earn enough in his position as an administrator at Uppsala University to support the family and had to hold a second job as an estate manager. Arrhenius used to do the accounts with his father and so developed an extraordinary facility with numbers. He also learned economizing and a utilitarian philosophy as a matter of course. Arrhenius's son thinks these early experiences set the pattern of his father's later work habits as a scientist: "He worked with very simple means; test tubes, glass bottles, etc. He looked at the over-use of refined apparatus as to a certain extent dangerous, the use going over into playing with instruments instead of using one's brain. Besides, from his youth he was kept to saving and economy and therefore in all branches of life was forced to the most economic way for solving problems. With very simple means he thus could obtain brilliant results."[13]

▷ **Imp:** That puts Arrhenius in the same league as Szent-Györgyi, at least with regard to the primacy of brainpower over instrument power.[14]

▷ **Constance:** Baeyer is another example. He studied with Kekulé in a tiny, cramped attic, and much of his early work was done without much money or equipment. Later, even when it became possible for him to buy elaborate apparatus, he refused to use anything more complicated than a test tube if he could help it.[15] C. T. R. Wilson was the same way. He came from a poor family—probably poorer than Arrhenius's—and made every penny count both in his private life and in the lab. It's said that he never spent more than five pounds on an invention.[16]

But the most fascinating case of frugality I've come across is that of John Strutt, the third Lord Rayleigh. Now *he* came from a rich family, but his father taught him early the value of money relative to thoughtfulness and skill. For example, the young Rayleigh got no allowance; his father paid all his bills. So Rayleigh took to wearing his clothes till they were literally threadbare, went hunting with an old-fashioned muzzle-loader rather than buy a new rifle, and could satisfy himself with a bun and beer at the local pub. Perhaps it was his way of being independent. In any case, his scientific style was commensurate. He never used anything but the simplest and crudest apparatus. Many scientific visitors were amazed that he was able to obtain *any* results, let alone first-rate ones that characterized his work.[17]

▷ **Hunter:** The same is true of Ostwald. None of the schools or universities he attended had anything like adequate equipment, yet with his skills at inventing and building things, he was never kept from doing research. If he didn't have it, he made it. In fact, he actually considered his tenuous existence on the peripheries of science a spur to innovation. He wrote in his autobiography that had he been at a major scientific center such as Berlin, he would undoubtedly have been swept up by the same fads in organic chemistry that almost every other chemist followed.[18]

▷ **Ariana:** Leading to the conclusion that one of the things that sets off the Nobel laureate from his less successful colleagues is the ability to make the best of a poor situation. "So I'm out in the boondocks. So I've got no money, little equipment, and no assistants. I'll just have to keep it simple and use my brain instead." The unsuccessful type spends all his time bemoaning the fact that he can't get any work done because he doesn't have the resources.

▷ **Richter:** *Mental* resources! Equipment does not think. Money does not invent.[19]

▷ **Constance:** Well, it can help those with ideas, can't it? But getting back to how and where people are trained, it's certainly true that a lot of the inventors I deal with aren't trained the way you'd expect. Some of the most successful don't even have engineering degrees. That's becoming so well recognized that NASA has even taken to looking actively for "outsiders."[20] And certainly some of the most startling inventions have origins in just the sort of unpromising backwater conditions you mentioned. I don't know how much you can make of such stories, but a friend of mine used to work for the Jet Propulsion Lab in Pasadena designing fault-tolerant computer systems—the space shuttle has five computers running concurrently, you know, so that if one or two fail, the whole thing doesn't come crashing to the ground. Too bad they didn't use the same idea with the O-rings. Anyway, he's originally from Czechoslovakia and told me that one of his countrymen, Anton Svoboda, invented some of the first fault-tolerant computer architectures when he was assigned by the Czech government to build computers out of rejected parts from the Soviet Union. Since he knew that one out of three components was sure to fail, he had to think of ways to build systems that would work even when many of the parts didn't. Most people would have said it couldn't be done, but of course it could, with a bit of ingenuity. In fact, I think Sloboda ended up at MIT. Anyway, there's a man who turned a bad situation into something important.[21]

▷ **Imp:** Yes, yes, yes! Why didn't we think of this before? Fault tolerance!

What's the matter with science today? We've become intolerant! We want everything planned, everything organized. No mistakes, no errors, no ambiguities. Perfection! If a student makes an error in calculation, or answers a test question wrong, he's penalized. Peer review tries to assure that nothing untrue enters the literature, nothing unlikely to succeed is funded. Don't make mistakes—that's the common refrain. Yet think of Pasteur, Fleming, and—dare I say it?—even ourselves. People aren't perfect—can't be. Nature's not perfect. That's a tenet of evolutionary theory! The imperfections, mistakes, mutations are necessary to evolution.

So apply this to science. We don't want error-free scientists creating error-free science, because the only way to do that is to repeat and repeat and repeat again what we already know. We've got to realize that most of our conjectures are failures, most of our experiments unenlightening, our research fraught with mistakes, but that we couldn't achieve the one important result without the stumbles and bumbles. As the computer types say, debugging makes a great programmer, not writing perfect code. Same here. We've got to stop planning for impossible perfection and learn how best to utilize the faulty parts we've got—us!

▷ **Hunter:** Which means a look at systems theory. If I recall correctly, von Neumann wrote a theoretical paper back in the fifties about how to design a working system from fallible parts. Svoboda probably had access to it. It might provide some clues for us, too.

▷ **Richter:** Or may I suggest evolutionary theory? A fund of diversity with selection criteria to weed out the nonadaptive.

▷ **Jenny:** Which is more or less what Pauling and Poincaré advocated, right? Try many things and select the promising ones. But isn't it true that to try many things, you have to have lots of ideas, and that means an unusually wide background, unusual resourcefulness?

▷ **Imp:** Implying we need to produce more eurokates and fewer stenokates!

▷ **Jenny:** The word war is escalating! Where *do* you get these words?

▷ **Imp:** My odd background! A stenokate is an organism adapted to a very restricted niche.

▷ **Ariana:** Same root as "stenosis" in medicine—a constriction of a blood vessel or duct.

▷ **Imp:** The snail darter, or whatever that little fish is that kept some dam from being built. Couldn't be transplanted anywhere else. Whereas a eurokate is an organism adapted to a wide range of niches, like *Homo sapiens*. Think of it this way. Imagine what would happen to most scientists if you took away the trappings of their profession: their lab equipment, their assistants, the funding agencies, and so forth. They'd die, professionally. They survive only because there's a highly developed system providing a niche for them. Now, think about Darwin, Pasteur, Almroth Wright, Fleming, Szent-Györgyi, Einstein. Somehow I think they'd do as well—nearly anyway—anyplace, anytime. They made their laboratories, rather than their laboratories them! They had the skills to invent the ideas and the techniques necessary to their science.

▷ **Hunter:** Include Ostwald in your list. He definitely made his lab, more or less from scratch. Invented more physicochemical techniques than anyone else at the time. Constantly looking for new and better ways to do things. All this bears directly on the next point Ariana and I wanted to make about the Ionists. Consider these testimonials. Ostwald on first meeting Arrhenius: "A. is still very young and wishes to understand and explain everything."[22] Ostwald in his own dissertation: "Modern chemistry is in need of reformulation."[23] Imagine a proposition like that appearing in a modern dissertation! Van't Hoff on the origins of his first book: "Young as I was, I desired to know the relations between constitution and chemical properties. The constitutional formula should be the expression of the chemical activity as a whole."[24] He directed his book at all those scientists who needed to know the principles of organic and inorganic chemistry; technical, pharmaceutical, and plant chemistry; physiology; and zoology.[25] These men were unusually broad thinkers. Science was their preserve, not just some part of chemistry. Not even chemistry as a whole. Their goal was the elaboration of what van't Hoff's friend Willem Gunning called "allgemeine Wissenschaft," general science, or what Ostwald himself called "allgemeine Chemie"—a universal chemistry that would be a coordinating discipline for all the other sciences.

▷ **Jenny:** But how do you invent such a science?

▷ **Ariana:** Through unusual training, in part. Listen to the list of subjects studied and taught by these guys: Van't Hoff studied chemistry, mineralogy, physics, and mathematics as a graduate student. Because he's at a fourth-rate place like the Utrecht Veterinary College, he ends up teaching mathematics, physics, pharmaceutical chemistry, and organic chemistry—something no first-rate university would have allowed. Then, at the age of twenty-six, he becomes full professor of chemistry, mineralogy, and geology at the new University of Amsterdam.[26] You want to know how he's able to make contributions to organic chemistry, thermodynamics, geology, and oceanography? He's taught all of them; he's done research in all of them. Ostwald, too, is able to make a wide range of contributions because he studied chemistry, physics and mathematics in grad school and taught all of these subjects as a young instructor. Arrhenius has a similarly broad background, studying the full range of sciences as an undergraduate, doing graduate work in both chemistry and physics, and publishing papers on electrochemistry and meteorology while still in school. Again, essential training for one who will make advances in physics, chemistry, and immunology. The Ionists were, in a very real sense, scientists rather than specialists.

▷ **Hunter:** I want to stress how unusual such backgrounds were even a hundred years ago. Specialization was already beginning to plague science. Specialty journals had appeared within physics and chemistry. The English physicist Oliver Lodge was calling the area between physics and chemistry a "no-man's-land"—sure to deter the average scientist, who could hardly be expected to master two separate disciplines, let alone integrate them in a new way.[27] The German physicist and physiologist Emil du Bois-Reymond, noting the same chasm, wrote in 1882 of the necessity of creating a new mathematics-based science to bridge physics and chemistry.[28] Richard Willstätter, an organic chemist, records that despite his recognition of the possibilities held out by du Bois-Reymond's vision, the repulsiveness—his word—of trying to master more than one science was too difficult and he abjured physics in college. He subsequently found himself unable to understand, let alone participate in, the exciting developments brought about by van't Hoff, Arrhenius, and Ostwald.[29]

Unfortunately, he was typical. Ostwald complained for years about the inability of most chemists to understand the simplest physicochemical equations. On the other side of the coin, it appears that most physicists—and here I'm thinking of men such as J. J. Thomson, Lord Rutherford, and Niels Bohr as typical examples—knew virtually nothing about chemistry. Planck, too, fits in this group. Where their physics met chemistry, their insights failed.

▷ **Jenny:** But it's not enough just to know about lots of things, is it? I know scholars who have degrees in several fields, but they never meld what they know into anything coherent.

▷ **Constance:** "In some heads the most diverse ideas can live together in peace because they never meet."[30] I was just about to make the same point. One thing that always amazed me about men such as Descartes, Newton, Leibniz, Darwin, and Helmoltz is their drive to synthesize everything into a harmonious whole. Helmholtz, in fact, credits his inability to memorize individual facts as a constant spur to the invention of general laws. Separate facts and individual solutions to problems never meant anything to him because he couldn't remember them. He had to find the underlying principle. Thus, when other scientists were satisfied that they'd explained some body of data, Helmholtz was still dissatisfied.[31] Poincaré wrote much the same thing about himself, too. Individual theorems never had any meaning for him—only the order of the argument.[32] He had to invent or reinvent everything for himself before he could understand it.

▷ **Hunter:** Planck's philosophy in a nutshell: "Nur nach eigener Überzeugung"—"Only when I have satisfied myself."[33] In reality most of these scientists weren't trained to be broad-thinking integrators; they trained themselves to think that way. They were autodidacts.

▷ **Constance:** Didn't Albert Michelson, Robert Millikan, and Enrico Fermi essentially teach themselves physics? I know Edison and Marconi were completely self-trained.[34]

▷ **Hunter:** Providing each with his own unique way of understanding nature—very important. Moreover, Ariana noticed as we were going through the biographies of the Ionists that each invented for himself some overarching idea to direct his research. They explicitly set out to look for coherence and unification in nature. A big idea—something like Pasteur's cosmic asymmetric force or Fleming's ultimate antibiotic that had to be there, even if it never quite materialized. Van't Hoff's holy grail, for example, was described by his student Charles van Deventer: "Whoever knows the Amsterdam laboratory knows that things do not take place there in an ordinary way. There is something mystical, something uncanny in the air. And this demonic something is the belief—one might call it superstition, if success had not so often followed it—the belief of van't Hoff that his fundamental idea, the connection between physical and chemical phenomena, is profoundly true."[35]

Ostwald subscribed to the same superstition, having been struck with a sudden illumination during his schooling that every property of a chemical should in some way be a function of every other.[36] Planck, too, wrote that the first law of thermodynamics struck him as "like a sacred commandment, which possessed absolute validity independent of man."[37] He decided to devote his life to developing the second law to the same degree of universality.

▷ **Constance:** Oh! Themata!

▷ **Richter:** Oh, of course: themata. What the hell are themata? The scientific equivalent of stigmata?

▷ **Constance:** Well, not literally. But if you define themata as distinguishing marks burned or cut into the mind, or as marks or signs indicating something unusual, I suppose they're close. Themata certainly set someone off from other people. But actually the term is employed by Gerald Holton to mean "unverifiable, unfalsifiable, and yet not-quite-arbitrary hypotheses" about how the universe works.[38] The sort of broad, aesthetic criteria—like "God doesn't play dice with the universe"—that both Ludwig Wittgenstein and Nicholas Maxwell say are so essential to science.[39] Such themata guide the pursuit of scientific work in a particular direction without dictating specific results. Such themata are, Holton says, "directly coupled neither to phenomena nor to tautological analytical statements, but to the persisting theme of an active potency principle that stands behind the whole sequence of concepts from which our idea . . . has developed." He believes that "it is the mark of a certain type of genius, in particular the nonconservatory mind"—he's writing here about Einstein—"to be 'themata-prone.'"[40]

▷ **Imp:** I see! So examples of themata would be an evolutionist's belief in gradual change or in punctuated equilibrium, right? Or the geologist's belief in uniformitarianism. You can't prove or disprove that the same forces have been operating in the same range of intensities since the creation of the universe; you assume they have since nothing else seems to make sense.

This reminds me of a passage I read recently in a speech by the geophysicist Tuzo Wilson—the first Canadian ever to be trained formally in both physics and geology, by the way. He calls himself a "global thinker." He said something quite striking: "I remember meeting a geologist who said 'Oh, I couldn't say anything about the geology of Venezuela, I've never been there.' Well, that's ridiculous to my mind. What's the use of geologists writing reports about Venezuela, if other geologists can't read them and understand them? It seemed to me that the whole earth should be understandable."[41] Isn't that what themata are really about? Global thinking? A faith that everything is held together coherently by some set of comprehensible principles?

▷ **Constance:** What you say certainly holds among inventors. Osborne Reynolds maintained an uncompromising conviction that all of engineering was unifiable.[42] Buckminster Fuller labeled himself a "deliberate comprehensivist."[43] And Thomas Hughes has remarked that one of the characteristics of "heroic inventors" such as Edison, Sperry, Tesla, and De Forest was that they addressed entire systems of inventions rather than individual ones.[44]

▷ **Hunter:** And that's where modern specialization and departmentalization are making revolutionary breakthroughs well-nigh impossible.[45] I recently heard John Wheeler give a talk in which he discussed the fact that young men and women can't address the big problems of physics because they can't afford to risk years of possibly fruitless effort. They need to get tenure. So he says he feels it's his responsibility at the age of seventy to make sure the big questions are being addressed. He'll go anywhere, talk to anybody, and ask any question if it might lead to progress, even if it makes him look like a fool. Only an "old fogey" can do that, he says. If he won't, who will?[46]

▷ **Richter:** Yet if a scientist's greatest work is usually done before the age of thirty-five, can we afford to leave the big questions to the old fogeys?

▷ **Ariana:** More to the point, if you get in the habit of thinking narrowly—if you are selected by the current system for thinking narrowly, within disciplines—can you break the habit when you become an old fogey?

▷ **Imp:** Or if you're young like me, will you stick with a scientific career if no one will let you address the big questions?

▷ **Jenny:** So, early opportunities to do independent, global, or thematic types of research are important for stimulating explorers and revolutionaries.

▷ **Hunter:** Or, put another way, you become a global thinker only if you are globally trained.

That certainly seems to have been true of the Ionists. All three had the good fortune to work under someone who recognized their potential and granted free rein to their imaginations. Not that any of them had an easy time. There is some validity to the Einstein myth of the revolutionary going unrecognized and unappreciated by his superiors. Part of the problem are the "Bonzes"—sorry to introduce yet another bizarre term. "Bonze" is a German corruption of a Chinese word for a senior Buddhist monk and is used to connote a self-important, all-knowing, dictatorial ass.[47]

▷ **Imp:** Which is not the end that should be speaking!

▷ **Hunter:** No. In some ways, Kekulé was a Bonze, telling van't Hoff that there was nothing left to do in chemistry but work out the details he'd left unfinished. Arrhenius's professors at Uppsala were much the same way: You either followed in the footsteps of Berzelius by working under Cleve or you found some variation of Ångström's ideas to endear you to Thalen. Otherwise, you left. Which is exactly what Arrhenius did.[48] And Planck's professor told him what Kekulé had told van't Hoff: Physics is dead; go do something else.[49] What a test of your commitment to science!

Not that any of these men endeared themselves to their professors, you understand. Each of them insisted on addressing some huge, "impossible" problem as a graduate student. Van't Hoff wanted to understand the structure of molecules—the big, unsolved (some people thought unsolvable) problem in organic chemistry at the time. The tetrahedral carbon atom was his solution. When he returned to Holland to stand for his degree, he was advised not to submit the idea as his dissertation since it was too controversial and theoretical to earn him his doctorate. As Lord Rayleigh observed after contemplating the fate of Waterston's lost discovery of kinetic theory, it is perhaps best that young scientists with highly novel ideas should not present them to the scientific community until they have first established a sober and sound reputation.[50] So van't Hoff submitted some humdrum experimental work, which earned him his degree without any trouble, as well as a mediocre public reputation. He spent one-and-a-half years unemployed after obtaining his doctorate, and then was able to obtain only a position at the Utrecht Veterinary College.

▷ **Jenny:** So much for Rayleigh's advice!

▷ **Hunter:** Yes and no. Arrhenius tried it the other way, with no better results. During a lecture on complex sugars and starches, his chemistry professor at Uppsala, Per Cleve, mentioned that it was "impossible" to determine their molecular weights. This is the problem Arrhenius chose to address in his doctoral research. Cleve would have none of it. Arrhenius was adamant and transferred to the Stockholm Högskola, where Erik Edlund and S. O. Pettersson fostered his radical ideas. The molecular-weight problem fell through, but the technique Arrhenius had chosen to use for his research—electrolytic conductivity—led him to develop his electrolytic theory of chemistry based on ionic dissociation. Unfortunately, the Högskola couldn't grant degrees, so back to Uppsala to defend his dissertation, followed by the faculty's decision that although his work was sufficient to earn a degree, it was not of high enough quality to permit him to continue up the academic ladder as a privatdocent. He, too, ended up unemployed for the better part of a year before Edlund and Pettersson, with the help of Ostwald, were able to convince the Swedish Royal Academy of Sciences to grant him a five-year travel stipend.[51]

Ostwald himself, as I mentioned a moment ago, undertook the total reformulation of chemistry as a grad student. He was lucky. Carl Schmidt, his mentor, recognized his genius and did everything possible to help the young man—even wrote him letters of recommendation comparing him to Helmholtz, Bunsen, and Kirchhoff. Again, to no avail. Ostwald, like van't Hoff, took to tutoring and then finally obtained a position at a secondary school in Dorpat when two preferred candidates refused it. This is where he began his very influential *Textbook of General Chemistry*.[52]

And Planck had difficulties finding a position, too. He was one of the first purely theoretical physicists. Almost everyone before him was an experimentalist. Planck didn't do experiments. He claims in his autobiography that the only reason he found a position at Kiel was that his father knew someone there.[53]

▷ **Imp:** Let's hear it for the backwaters and peripheries of science! Where else could innovators spring up and nonconformists flourish?

▷ **Ariana:** And for the people such as Edlund, Gunning, and Schmidt who foster these unusual scientists. Creative people always tackle impossibly big problems and then let their goal guide them through a process of filling in the gap between what they already know and what they need to know. My artist friend, Tom van Sant, calls this the "leap and fill" method. That's how these creative types push beyond the limits of the known. They have an amazing ability to handle the ambiguities involved. To most other people, however, they look crazy, because the initial leap is based on nothing more than stochastic aiming into the unknown guided by one or more themata. (How's that for pulling together some of our own themes?) Someone simply has to have faith in such people and give them a chance to succeed.

▷ **Richter:** Or to fall flat on their faces.

▷ **Imp:** Hey! That's their lookout.

▷ **Hunter:** But there is a prerequisite to exploration: independence. Van't Hoff obtained independence in two ways. When the University of Amsterdam was elevated to the status of a state university in 1877, Gunning, as chairman of the department, pulled strings and had the twenty-six-year-old van't Hoff appointed—are you ready for this?—ordinary (that is, full) professor. Talk about an opportunity! Furthermore, van't Hoff found that he could scrape together sufficient money for just about any research he wanted by doing analysis of milk (fat content and so forth) for a local dairy.[54]

Arrhenius ended up almost equally well-off. His five-year travel stipend allowed him to go study with anyone in Europe. Since *Wanderjahren* were common at the time, everyone he wrote to said sure, come: Ostwald; Friedrich Kohlrausch, the current expert on electrolytic dissociation; Ludwig Boltzmann, then studying gas dissociation; van't Hoff. Very important for creating the personal ties that helped knit together the common themes that were emerging in these men's research.

Both Ostwald and Planck had it a bit rougher at first. But being isolated in poor institutions had this advantage: Both were free to do pretty much as they pleased. There was no eminent Bonze to cater to. Like Einstein in his patent office, as long as they did their work competently, whatever else they did was their business.

▷ **Imp:** Not like today's constant evaluation of every aspect of one's career: publications, committee memberships, external funding, students, editorships, external colloquia, et cetera!

▷ **Hunter:** Not on our scale. But don't think the system was any utopia, either. It was a hell of a lot harder to get a professorship in the first place, and there was just as much politicking involved as there is today.[55] But once you were in, there weren't nearly as many hurdles to doing what you wanted. You got a certain amount of money with your position, and you used it in whatever way you thought would best further your research. If you impressed the Bonzes, you might get transferred to Cambridge or Paris or Berlin. If not, then you stayed out in the provinces. But you still did your work.

▷ **Richter:** Right. We can all dream. But what is this work, and why does it require knowing this jumble of personal and institutional trivia?

▷ **Ariana:** It is *not* trivia, I assure you! Look, Hunter, can we get right to van't Hoff and transformational thinking? Let's put it all together so Richter can see it.

▷ **Hunter:** Fine with me.

One of the things Ariana and I found as we went through the various discoveries made by the Ionists was something I think we all do, but without realizing it. We take information from one field and apply it to another; we adapt a skill developed for one purpose to another.

▷ **Ariana:** We actually *transform* ideas from one way of thinking to another. We take numbers and turn them into pictures, or an apparatus and turn it into a mental abstraction. "Transformational thinking," I call it. So the more different mental skills—tools of thought—you can utilize, and the more ways you can transform ideas, the better your chances of solving problems.

▷ **Richter:** Such gobbledygook I have rarely heard.

▷ **Hunter:** Now hold on, Richter. It sounds strange, but it isn't gobbledygook. We can demonstrate very precisely what we're talking about. Van't Hoff 's

Van't Hoff's laboratory in Amsterdam. (Jorissen and Reicher, 1912)

invention of the thermodynamics of solutions is one case. Before van't Hoff moves to Amsterdam, he tries to utilize his tetrahedral carbon atom to solve the Berthollet-Laplace problem of providing a physical basis for understanding chemical affinity. It's a complete failure.[56] We won't even bother you with it. Suffice it to say, the microscopic approach doesn't work. Van't Hoff makes a major switch in perspective when he gets to Amsterdam and begins to utilize his "demonic belief" in the interrelatedness of all physical and chemical properties of compounds to study their aggregative properties. He studies mass action, the dependence of reaction rates on temperature and pressure, and so forth. Eventually he reads an interesting set of articles by a pair of virtually unknown scientists, Friedrich Horstmann and Leopold Pfaundler, both of whom had studied with van't Hoff's teacher Wurtz and his colleague Regnault in Paris. Perhaps Wurtz introduced him to their papers. Like van't Hoff, both men were unusual in combining mathematics, physics, and chemistry in their training and research. Essentially, they suggested that it might be possible to extend the laws of thermodynamics from gases to aqueous solutions if one could find an appropriate analogy for each of the thermodynamic variables of temperature, volume, and pressure. The first two were simple, since they are the same for gaseous and liquid systems. More troublesome was pressure. At constant temperature, the pressure of a gas changes as a direct function of changing volume. Not so in a liquid. Indeed, most liquids are nearly incompressible. Horstmann had suggested that perhaps the phenomenon of disaggregation—or, in our terms, the dissolving of a solid in a liquid—held the key to making the appropriate analogy.[57]

At about the same time, 1884, van't Hoff came upon a paper by Eilhard Mitscherlich—the same fellow who had provided the erroneous report on the tartrates that so stimulated Pasteur. Mitscherlich had demonstrated that when crystals of various salts are placed in a barometer, they absorb enough water vapor to decrease the barometric pressure. However, the resulting affinities were so small that van't Hoff found them unbelievable.[58] He began to wonder if there might be a way to measure affinities between salts and water more directly by dissolving the salts.

▷ **Jenny:** So, two problem areas—extension of thermodynamic theory and measurement of affinity—leading to the same subproblem: how to measure the dissolution of salts in water. Right?

▷ **Hunter:** Right. Van't Hoff leaves the lab and on the way home runs into his colleague Hugo de Vries. De Vries gives van't Hoff the clue he needs to solve both problems.[59]

▷ **Constance:** That's odd. Wasn't de Vries one of the men who reinvented Mendel's laws in 1900? What does genetics have to do with thermodynamics and affinity?

▷ **Hunter:** Well, actually, the rate of temperature-induced mutation in flies can be described by the Arrhenius equation. That's H. J. Muller's work.[60]

But in general genetics has nothing to do with thermodynamics. De Vries, like most of the important scientists we've discussed, had more than one scientific interest. During the 1880s his main interest was plant physiology, particularly the study of the osmotic pressure that keeps cells plump and keeps plants from wilting.

▷ **Jenny:** And osmotic pressure is?

▷ **Hunter:** That's precisely what de Vries asked van't Hoff when they met.[61] I don't mean to evade your question, but the fact is that prior to van't Hoff's investigations, nobody knew. It was just an observable phenomenon.

The phenomenon was first described in 1748 by J. A. Nollet. Why he did these experiments I've been unable to ascertain, but he placed red wine in a glass cylinder sealed with the bladder of an animal. The cylinder was then immersed in water, and the bladder was seen to expand, often until it burst. Something in the water was passing into the cylinder by way of the bladder, increasing the pressure in the cylinder and making the bladder expand. At the same time, none of the red wine could be seen to leak into the water. The bladder was acting as a semipermeable membrane.[62]

▷ **Imp:** Maybe Nollet was trying to figure out why the body's blood, or for that matter the wine we drink, doesn't get excreted in the urine. We could do the same thing with plant cells or even egg yolks in concentrated salt solution or—hey, why not wine?

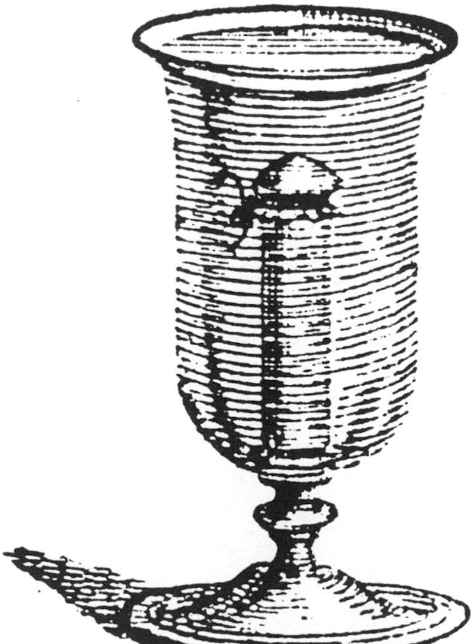

J. A. Nollet's semipermeable membrane experiment, 1748.

▷ **Jenny:** Play some other time, dear. We can take this on faith.

▷ **Hunter:** There's a continuous line of research on osmotic pressure from Nollet to de Vries accompanied by various explanations of the phenomenon. R. J. Dutrochet had suggested that the two liquids had an affinity for one another, pulling at each other through the membrane. Justus Liebig and Thomas Graham had proposed that the semipermeable membrane itself had affinity for something in one solution, but not for anything in the other, thereby creating an asymmetry of the attractive force.

De Vries himself countenanced neither view because he had discovered the phenomenon of isotonicity. He had shown that if a cylinder containing a salt solution is immersed in a different salt solution at the same concentration, then no osmotic pressure is exerted on the membrane. The two solutions are isotonic with respect to one another. Same term you sometimes see on various liquid pharmaceuticals. Since isotonicity was independent of the type of salts or sugars employed, Dutrochet's chemical affinity theory was clearly unacceptable. Furthermore, the effect was independent of the type of membrane used—animal bladder, small intestines, plant cell membranes—and therefore had nothing to do with the chemical affinity of the solutions for the membrane. So much for the Liebig-Graham theory.[63]

De Vries presented van't Hoff with the most recent and comprehensive book on the subject, Wilhelm Pfeffer's *Osmotic Investigations*, and

De Vries's observations of the effects on plant cells of (left to right) *hypotonic, isotonic, and hypertonic solutions (ca. 1880).*

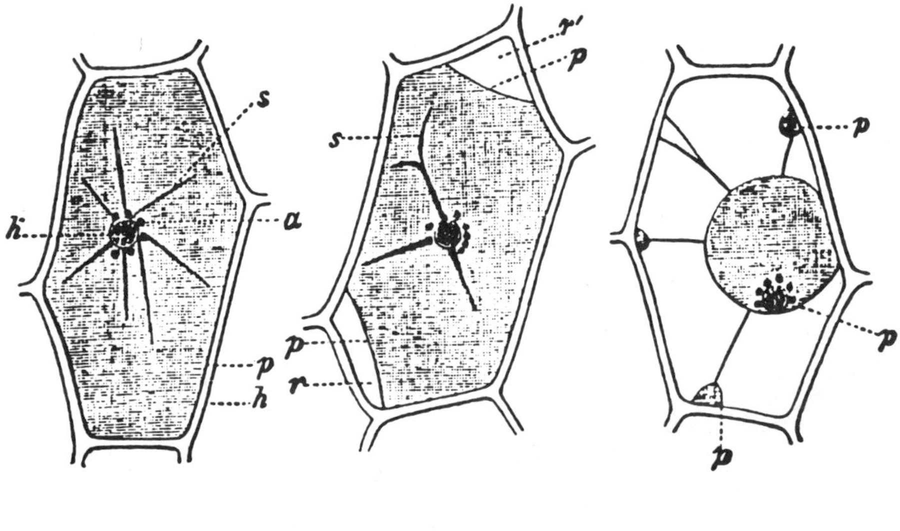

asked him for his thoughts on the matter.[64] Van't Hoff, for his part, quickly found the information he needed: a method for measuring directly the affinity of a solvent for a solute.

▷ **Jenny:** So here's Wheeler's "ask anyone anything" approach to problem solving. To solve a problem in physical chemistry, van't Hoff turns to physiology for insight. More interdisciplinary thinking. More global thinking. Chemistry is chemistry whatever its manifestation.

▷ **Hunter:** Right. Van't Hoff couldn't care less who's done the research in what field if it provides him with relevant data or techniques.

▷ **Constance:** Mendel's and Darwin's approach as well. You should see Darwin's correspondence. He wrote to anybody in the world who might provide him with information, and even challenged the class barriers of nineteenth-century England by consorting with local agriculturalists, pigeon breeders, sheep farmers, and the like in search of examples of artificial selection.

▷ **Hunter:** Van't Hoff crossed a similar academic barrier. The key to van't Hoff's research was an invention by the Polish polymath Mauritz Traube, reported in Pfeffer's book. Traube never obtained a university position but managed to persevere in his studies in his spare time in a small private laboratory. In consequence, his results, like those of Herapath and Waterston before him, were largely ignored.[65]

▷ **Imp:** Ah, yes, the old "if he were any good he'd be at Harvard" routine.

▷ **Hunter:** Well, fortunately Traube persevered, and Pfeffer, for one, was impressed. What particularly caught Pfeffer's attention was Traube's theory of how cell walls and membranes are built. In 1864, Traube had suggested that cell walls result from the precipitation of two substances at the interface between cells to form a solid, semipermeable membrane.[66] Another incorrect but useful theory for our list. Three years later, he reported two methods for creating artificial membranes by such precipitations.[67] The first involved placing globules of gelatin in tannic acid. The tannic acid reacted with the gelatin to form a membrane semipermeable to water, but not to tannic acid.

The second method involved immersing a tube full of Prussian Blue dye in a bath of ferric cyanide. A semipermeable membrane of cupric ferrocyanide was formed across the opening of the tube. Unfortunately, both the gelatin–tannic acid and the cupric ferrocyanide membranes were extremely fragile, and so useless for measuring osmotic pressures. That was not, in any case, Traube's intent.

Pfeffer was fascinated. The son of an apothecary, he was well versed in both chemistry and plant physiology.[68] He adapted Traube's experiments to measuring osmotic pressure as follows: He immersed unglazed, porous porcelain jars in a solution of copper sulfate while simultaneously filling the inside of the jar with potassium ferrocyanide. The two solutions slowly penetrated the walls of the jar, forming a semipermeable precipitate where they met. Traube's membrane, but now fixed in a matrix to give it strength. Beautifully simple.

Pfeffer then filled the jar with one solution, sealing into it a manometer for measuring pressure, and immersed the jar in a second solution to determine its osmotic pressure. Scientists now had a method for making a standardized osmometer with the capability of measuring relatively high osmotic pressures.[69]

▷ **Richter:** But the importance of a technique lies in what you do with it, not its invention.

▷ **Hunter:** Pfeffer was no dummy. He quickly realized that his results could be used to measure affinities, and that these affinities would have quite large values. Lacking the physics background to work through the implications himself, he approached Rudolf Clausius, the grand old man of thermodynamics. Clausius didn't believe Pfeffer's results, even when Pfeffer demonstrated the apparatus for him, and refused to investigate the phenomenon.[70]

▷ **Constance:** That's typical. Clausius rarely paid any attention to anyone's results but his own.[71]

▷ **Hunter:** True. He also refused to comment on Arrhenius's application of his kinetic theory to ionic dissociation in 1884.[72] The young, unknown van't Hoff, however, paid attention to both Pfeffer and Arrhenius and so pulled together what Clausius did not. For van't Hoff, Pfeffer's results were a revelation. Whereas Mitscherlich's barometric studies had resulted in a pressure of one two-hundredth of an atmosphere due to the attraction of a salt for water in a 1 percent solution, Pfeffer measured a pressure of two thirds of an atmosphere.[73] This was more in keeping with van't Hoff's expectations. But now van't Hoff was faced with our favorite phenomenon: a contradiction in data.

▷ **Ariana:** Yet not really a contradiction. As Claude Bernard advised, accept both sets of data as valid, and then determine the differences in the conditions that produce the different results. This is just what van't Hoff did. He realized that Mitscherlich had measured vapor pressure; Pfeffer, osmotic pressure. How were they related?

So, what does van't Hoff do now? Write equations? Do experiments? No. He *thinks*! He invents an imaginary system—a mental model, if you will—in which it would be possible to measure the vapor pressure and the osmotic pressure of the same solution. A physicist's variation on understanding a system so completely that one internalizes it.

▷ **Jenny:** Is that typical?

▷ **Ariana:** Probably not of all physicists. But certainly of many of the best. Listen to Einstein on his thought processes: "The words of the language, as they are written or spoken, do not seem to play any role in my mechanism of thought. The psychical entities which seem to serve as elements in thought are certain signs and more or less clear images which can be 'voluntarily' reproduced and combined . . . The above mentioned elements are, in my case, of visual and some of muscular type. Conventional words or other signs [presumably mathematical] have to be sought for laboriously only in a secondary stage."[74] Imagery, feelings—not numbers, not words. That's the key. Riding the light wave and feeling what it's like. Falling in an elevator at the speed of light and imagining the consequences.

▷ **Richter:** Yet only the mathematical part is communicable.

▷ **Ariana:** Only because we fail to develop adequate means of representing other forms of thought.

▷ **Hunter:** I agree. Even accepting your point, Richter (which Ariana and I hope to prove false in just a moment), most complex physical problems can be solved by so-called pure reason and without mathematics.[75] Indeed, Feynman and Dyson have suggested that much of Einstein's later work failed precisely because he became an equation manipulator, thereby losing touch with the mental models and visual-kinesthetic intuitions that had provided his earlier insights.[76] My physicist friends inform me similarly that the greatest difficulty in teaching relativity theory or the latest string theories and so forth is not the mathematics—there are lots of mathematically sophisticated undergraduates—but the inability of these mathematically talented students to imagine *Gedankenexperimenten* for themselves, to visualize four- and five-dimensional objects, and to translate the equations into physical situations and the physical situations into equations. What they learn in calculus or linear algebra often fails the transplantation to physics, apparently because they fail to learn the qualitative aspects of mathematical sciences along with the math. Equations don't *mean* anything to them.[77]

▷ **Ariana:** And that's where all of van't Hoff's vocational and avocational skills come together. Remember, he's highly imaginative, capable of inventing three-dimensional images of things such as atoms that he's never seen. He can convert his mental models into physical ones, as we've seen with his tetrahedral carbon atoms. His poetry has given him practice in transforming sensory impressions into words and back into images and feelings. He can converse and create as competently in the language of mathematics as in the Dutch, German, French, and English in which he composes verse; he can translate a line of mathematics into any of these other languages as well. Moreover, his peripatetic studies have introduced him to a very unusual range of physical, chemical, and physiological problems and techniques; and his thematic, global style of thought has driven him to understand how everything fits together. The amazing thing is that it does. And that's what Hunter and I want to try to convey to you now: how van't Hoff gets it all to fit together.

▷ **Hunter:** Put another way, Ariana and I believe that van't Hoff's broad experience of the creative process in its many manifestations developed an unusual sense of how the world works—habitual patterns of action and understanding that we've been calling intuition.

▷ **Ariana:** And he uses this intuition, not math, to solve his problems. Oh, sure, all of van't Hoff's thermodynamic ideas are expressed as equations in his publications, but we know from what he says, and from the images he utilizes in deriving his equations, that he didn't solve his problems using math. In consequence, we can show you the entire process we think he went through without a number or an equation.

In fact, the very first thing that happens is that the numbers disappear. They raise the problem, but they are not the problem. It doesn't matter what particular numbers Mitscherlich and Pfeffer obtained beyond the fact that their values and their methods were different. The real problem is one of relationships—relationships between experimental procedures. The data have meaning only with regard to these procedures. So that's what we have to concentrate on: what's going on inside the apparatus and how it relates to the mental manipulations you perform in trying to understand the apparatus.

▷ **Hunter:** Precisely. Everything of importance goes on in here, between the ears, not out there in the instruments. A very difficult point to get across to students.

▷ **Imp:** Hell, we teachers are half the problem. I had exactly one teacher in my whole career who insisted that we know how each technique worked, what its limitations and its theoretical foundations were, before we began interpreting data. The rest just said, "Here are the data; here's the interpretation. Make sure you understand."

▷ **Hunter:** Unfortunately, that won't suffice. Theories are always abstractions and simplifications of reality and measuring devices are never perfectly accurate. They rarely correspond exactly, so that the raw data from a measurement often aren't appropriate to testing a theory. You've got to have a model in your head, and understand its limitations as a mirror for reality. Let me give you an example relevant to van't Hoff's case: the use of a manometer, a pressure measuring device, by Regnault. Regnault, you recall, was the man who taught Horstmann and Pfaundler, the two fellows who suggested the basic thermodynamic ideas that van't Hoff used in setting up his osmotic pressure problem. Duhem, who observed Regnault, describes his actions as follows:

> *The height of the column of mercury is ascertained by an assistant and given to Regnault, who makes the proper corrections. Those corrections are made, not because Regnault believes that his assistant was mistaken—on the contrary, if he believed that, he could not make the corrections. Another reading is substituted for the one actually observed in order to bring the ideal and symbolic manometer, to which he has applied his calculations, nearer to the real manometer, on which the readings have been made. The ideal manometer is filled with an incompressible fluid, having throughout its volume an identical temperature and being subject at every point of its free surface to an atmospheric pressure independent of its height. However, this [ideal] manometer is too far removed from reality to be sufficiently accurate for the purposes of experiment. Therefore a new ideal manometer is conceived [and] all these modifications [made to it] amount to corrections.*[78]

▷ **Ariana:** Similarly, the problem van't Hoff faces is to uncover the methodological and theoretical presuppositions underlying the techniques that yield the data so that he can correctly compare them. Like Regnault, he has to invent an ideal vapor presure–osmotic pressure device that somehow embodies the physical connection between Mitscherlich's and Pfeffer's data. He needs an ideal representation of the physical relationships.

So how does he do it? By transformational thinking. He takes the problem as it's presented in its original numerical form and converts it into a picture; mentally manipulates the picture, and then transforms it yet again into an equation or words. The same thing Einstein describes. Let's see it in action.

Imagine yourself as van't Hoff, with all his propensities and skills. What do you do? You think yourself into the apparatus. How does Mitscherlich measure his vapor pressure? How does Pfeffer? You visualize what happens to the molecules inside: Mitscherlich's crystals absorbing water vapor from the gas phase, their affinity sucking some of the water molecules down into the liquid phase which otherwise would leap free of the surface through evaporation. Does van't Hoff *feel* these forces at work, as some of the other scientists we've discussed professed to do? Who knows? But certainly he imagines Pfeffer's crystals drawing water through the semipermeable membrane to satisfy the affinity of the salt molecules, increasing the pressure on one side of the membrane at the expense of the other. The systems begin to operate inside your head.

You abstract. Reality is too complicated. Physical laws must apply

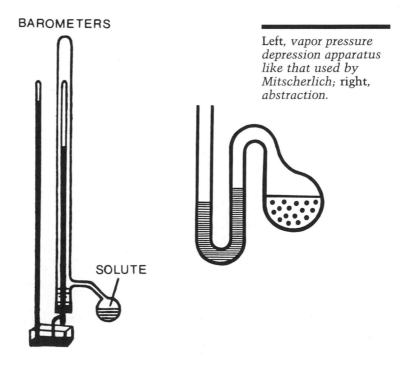

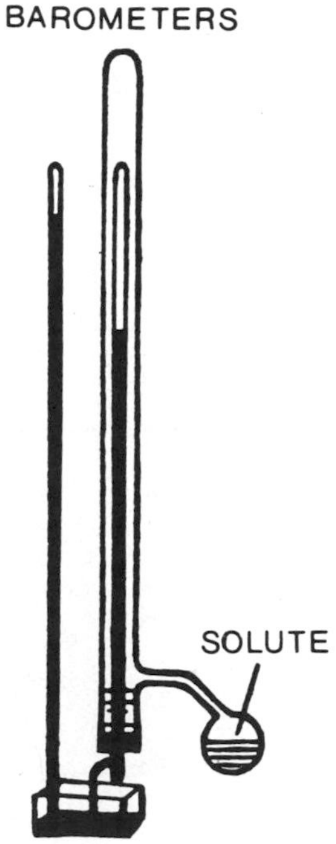

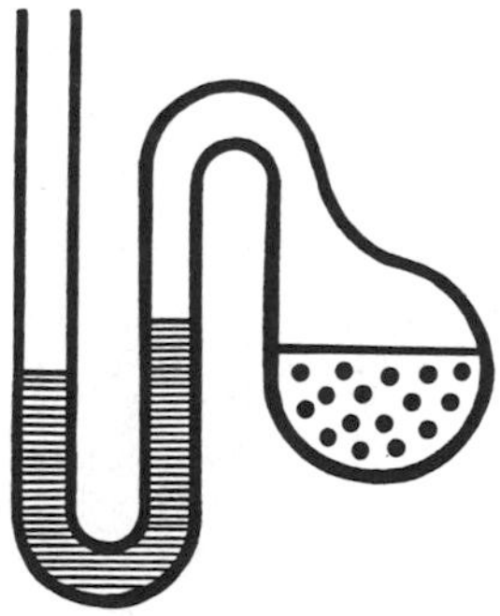

Left, *vapor pressure depression apparatus like that used by Mitscherlich;* right, *abstraction.*

to any apparatus whatever, any analogous situation. You discard all the unique and inessential elements of each system. Mitscherlich's apparatus becomes a box with solute-solvent in the lower half, water vapor in the upper half, and a pressure-measuring device such as a manometer. Pfeffer's osmometer becomes another box, this one filled with water and divided vertically by a semipermeable membrane. In the left half is the solute, unable to pass into the right. Again, a pressure measuring device is attached. Nothing left but essentials.

Next, you search for common overlaps in your abstractions. You integrate your mental images, much as Imp claims he did with the Fleming material. Overlaps become recognizable. An abstract, imaginary device emerges. Still not simple enough. You discard the measuring devices (which have nothing to do with the physical system per se) and this yields an even more schematic representation.

▷ **Richter:** All of which is hypothetical nonsense.

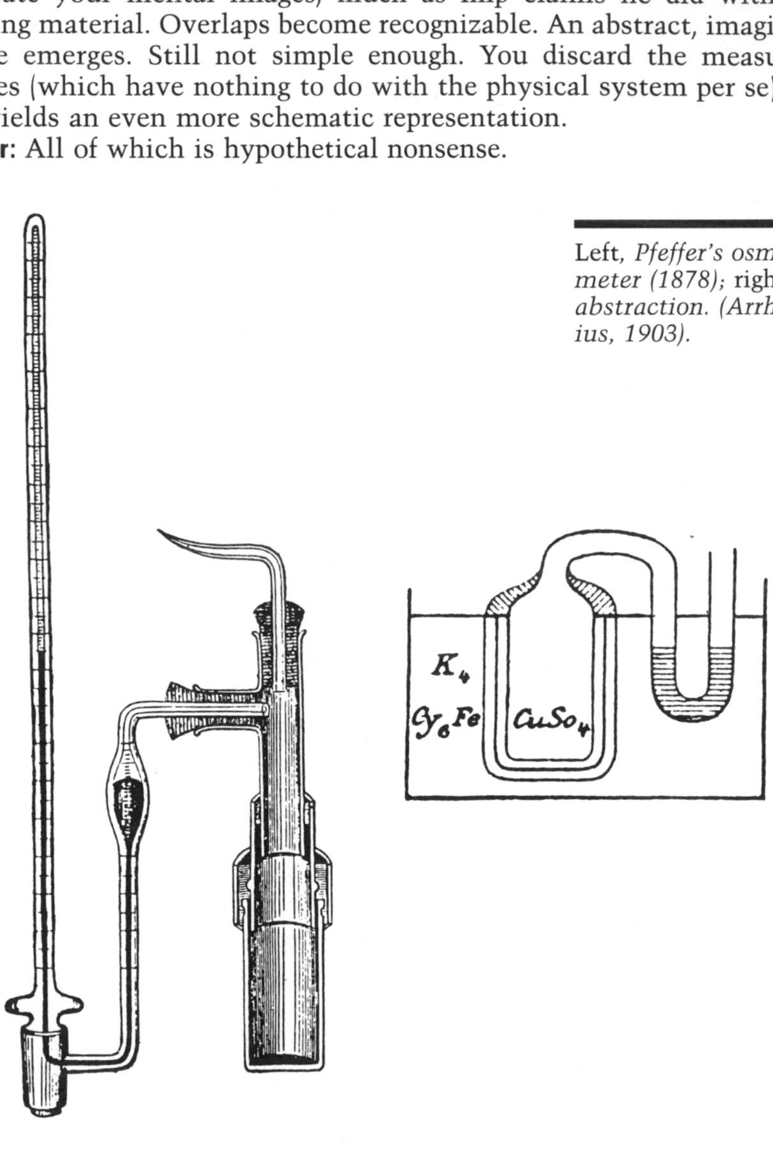

Left, *Pfeffer's osmometer (1878);* right, *abstraction. (Arrhenius, 1903).*

▷ **Hunter:** All of which yields precisely the figure that van't Hoff publishes some years later in explaining how he arrived at his conclusions.[79]

What next? Van't Hoff applies the well-known gas laws to his imaginary system, deriving corrective factors for comparing gas and solute pressure. He begins with the gas law $PV = RT$, where P is pressure, V volume, R the gas constant, T the absolute temperature in degrees Kelvin. He inserts the values of Mitscherlich's and Pfeffer's data, and compares. That's when he gets his first big surprise. Pfeffer reported that a 1 percent sugar solution exerts a pressure of 49.3 centimeters of mercury (⅔ atmosphere) at 0 degrees Celsius. The pressure in grams per square centimeter is therefore 49.3 times 1022. The volume a gram of cane sugar will occupy in a 1 percent solution is 34,200 cubic centimeters, since cane sugar has a molecular weight of 342. The temperature will be 273 degrees Kelvin. Substituting into the equation yields a value of $R =$ 84,200. $R =$ 84,700 for gases. Almost identical![80] So the osmotic pressure exerted by a solute in solution is equivalent to the gas pressure exerted by the same compound in its gas phase.

▷ **Jenny:** You lost me. Besides, I thought you weren't going to use any equations.

▷ **Hunter:** Sorry—habit. Think of it this way: Now you've seen the mathematical translation.

▷ **Ariana:** And we can still do it without any equations or numbers. Try this: Imagine the idealized osmometer–vapor pressure apparatus system at equilibrium. That is to say, there is a balance of opposing reactions—one of Pfaundler's insights. The amount of water going from right to left across the semipermeable membrane will be exactly equal to the amount going from left to right. In other words, the net amount of water crossing the membranc is zero. Physically, this means that at equilibrium, the passage of water across the membrane cannot be contributing to the osmotic pressure. We can therefore ignore the presence of the water in considering the cause of osmotic pressure. So once again we abstract. We eliminate the water from the system. What are we left with? Just the solute—the dissolved compound—floating around in our idealized box as if it were a gas!

▷ **Hunter:** Which, once again, is van't Hoff 's conclusion: The osmotic pressure, he writes, "is equal to the gas pressure the solute would exert if it were present as a gas in a volume equal to the volume of the solution."[81] The astonishing thing about all this is the incredible osmotic pressures that can potentially be generated by such a system. Ostwald calculated that a 17 percent solution of ammonia would exert 224 atmospheres of pressure against pure water at 0 degrees Celsius.[82] To imagine this, consider that an atmosphere of pressure is about fifteen pounds per square inch at sea level. We pump up car tires an additional thirty pounds per square inch, and bicycle tires to about twice that. Ostwald's talking about pressures a hundred times larger—the kinds of pressures submarines have to withstand in the deep seas. So when Clausius doubted

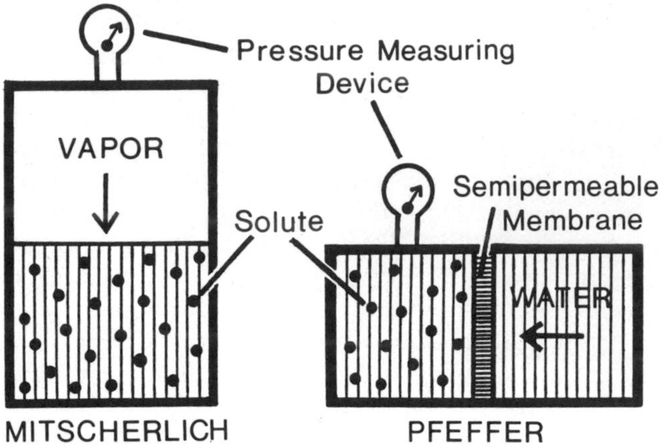

Further abstractions of Mitscherlich's and Pfeffer's apparatuses.

Pfeffer, he wasn't simply being stubborn. A sense of physical reality may have made him doubtful.

▷ **Imp:** Had he known about deep-sea fishes, however—they need to generate such high osmotic pressures.

▷ **Jenny:** Fine. But you can't really do the things you've just discussed, can you—extract the water and all that?

▷ **Hunter:** No. It's all done in your head, just as Ariana has presented it. There's nothing novel about that. Physicists have done it for centuries. Galileo imagined falling beside two bodies tied to one another and cutting them apart as they fell. Would they, as Aristotle claimed, suddenly change their acceleration, since acceleration was supposed to be a function of weight; or would they continue to fall at the same rate as before, as Galileo himself believed? The laws of thermodynamics were generated using an imaginary frictionless piston, working reversibly on a perfect gas. Nothing but a set of useful fictions that approximate reality closely enough that the sort of corrections Regnault made on his manometer measurements become possible.

In fact, van't Hoff extends these thermodynamic laws to solutions by inventing a simple osmotic pressure piston analogous to these imaginary frictionless pistons. Instead of altering the pressure and volume of a gas, however, his pistons alter the osmotic pressure and volume of solutes by manipulating their passage through semipermeable membranes. He takes his imaginary piston system through the same set of reversible operations that Clausius had done to derive the gas laws, and obtains the equivalents for solutions. The result is a set of equations for describing the thermodynamic properties of solutions that is completely analogous and fully as general as Clausius's equations for gases. A tremendous achievement that still underlies all physicochemical theory.

▷ **Ariana:** Yes, but tell them about what you said concerning van't Hoff as a poet of equations.

▷ **Hunter:** Ah, yes: "Equations properly expressed are a poem." Another saying I have over the door of my lab. I've refrained from presenting you with van't Hoff's derivations because of Ariana's and my joint purpose, which unfortunately limits me to asserting something that I wish you could appreciate for yourselves: the extraordinary beauty and simplicity with which van't Hoff produces his desired effect. His derivations of the thermodynamics laws governing solutions are truly a poem, no less than the ones he invented with words. He sketches his characters—pressure, volume, temperature; he puts them through their paces, allowing us to learn their limitations and possibilities; he brings pairs of them on stage for us to study in all their varied combinations; and then he reveals the unexpected hidden relationships that every good poem or play contains in its structure. There are no extra words, no slips of the tongue or inconsistencies to jar us. Nothing is forced. And you sit in wonder after reading it thinking how wonderfully, gloriously obvious—how beautiful—it all is.

▷ **Ariana:** Again suggesting that van't Hoff's facility in poetical composition—his fluency, his sensitivity to the symbols of language—is carried over to the language of mathematics as well.

▷ **Hunter:** Indeed. There was a story going around Princeton when I was a grad student there to the effect that one day a famous mathematician returned from a year in Europe to find his favorite pupil gone. "What's happened to so-and-so?" he inquired, and was told by one of his colleagues: "Oh, he didn't have enough imagination to be a mathematician, so he's gone off to write poetry." The slander against poetry aside, I think there's an element of truth in the connection Ariana wants to make.

But I have to tell you that van't Hoff's derivations fail at one crucial point. There's one ugly spot in his "poem" where the rhyme scheme is inconsistent. He finds that he must throw a "fudge factor" into all his equations to make them match the data, and he hasn't a clue why. Somewhere, his mental images don't correspond exactly with nature.

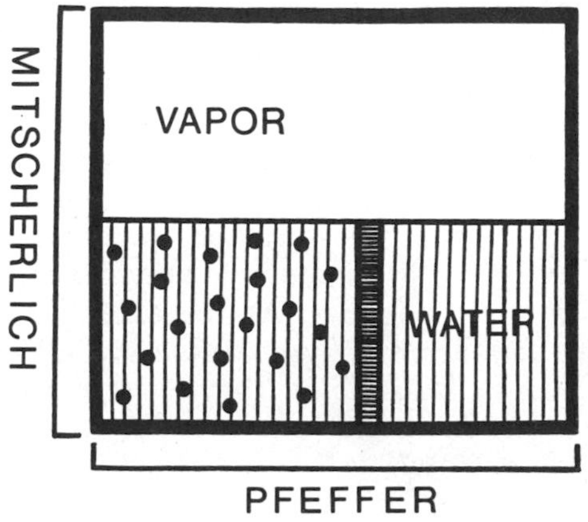

Elision of abstractions in previous figures.

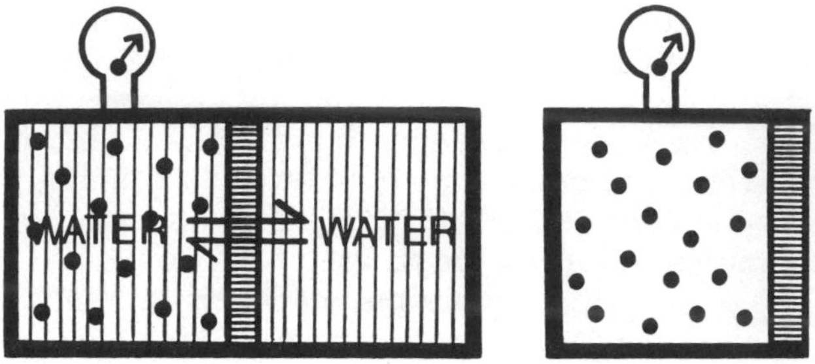

Final abstraction, showing that the pressure exerted by a solute in solution is equal to the pressure the solute would exert as a gas in the same volume.

Very frustrating. All he knows is that there appear to be more molecules of most salts in solution than one would predict according to his theory. Whether this is a flaw in the theory, a problem with the data, or evidence for some unexpected phenomenon he can't say.[83]

It's quite interesting to observe Planck's reaction to this. Remember, he's the mathematical purist of the group. He's never done an experiment in his life. He's never studied chemistry. He wants everything reduced to a set of equations whose validity exists independent of any experimental considerations. So he approaches the whole problem completely differently. In 1886 he's introduced to van't Hoff's results by Ostwald and rederives them from entropy considerations. When he reaches the point at which van't Hoff introduces his "fudge factor," Planck balks. Too ugly for him. No justification in the laws of thermodynamics. What does he do? True to his thematum that thermodynamics is sufficient to explain the universe, he boldly proclaims that the chemists must have gotten the molecular weights wrong. There have to be more molecules in solution than they suppose, or those molecules must be dissociated.[84] How the dissociation occurs, and what the chemical implications might be, Planck doesn't even venture a guess. He's simply interested in maintaining the pristine purity of the equations.

▷ **Imp:** Another example, then, of personal style consequentially affecting the route different scientists take to simultaneous discoveries, and the forms in which the various results are announced.

▷ **Hunter:** A point that becomes even more clear when we compare Planck's statement of dissociation with that of his codiscoverer Arrhenius. At about the same time that Planck is working on his rederivations of van't Hoff's equations, late 1886, Arrhenius, following up a sudden illumination of a few years before, solves the "fudge factor" problem another way. As we've documented today, his background is very different from Planck's, and so is his style. He's a realist. Atoms exist for him as chemically reactive particles. Equations have to describe what these particles are doing, what they look like, and how they act. It's not enough for him to say that there must be more particles in solution than van't Hoff expected. Why? he wonders. How?

He demonstrates that dissociation does, in fact, occur, producing more solute particles than predicted by molecular weights. Salt decomposes into sodium and chlorine, right? So the greater the dissociation into ions, the greater the value of van't Hoff's "fudge factor" for osmotic pressure. So far, he and Planck agree. In addition, Arrhenius demonstrates that these dissociated particles are electrically charged ions, and that the number of ions present in solution correlates exactly with how chemically reactive the compound is, rates of reactions, decrease in freezing point and increase of boiling points of solutions, and so forth. In short, he proposes an ionic dissociation theory of chemical activity that fully incorporates van't Hoff's thermodynamical approach.

Planck can't accept his conclusions, because they go beyond the physical principles embedded in his own derivations. They get into a controversy, in which Ostwald acts as mediator. Ostwald also provides the experimental tests of ionic dissociation. The result is the integrated, *allgemeine* chemistry that each of the Ionists had sought in the first place, and the first mathematical theory of chemistry to be integrated successfully with physics.[85]

▷ **Ariana:** And in the process these men provide a slew of examples of the sort of integrative, transformational thinking we've just illustrated with van't Hoff: Arrhenius manipulating data to find general equations and interpreting these to explain the physical behavior of chemical interactions—unlike Planck, interpreting every equation as a physicochemical process affecting real atoms and having testable chemical implications. Ostwald transforming between equations, physical models, visualized inventions, sketches, and blueprints; building the physical apparatus, performing the experiments, producing more numbers and going through the cycle again. Every new problem an occasion to invent a new technique. Numbers, images, models, data, equations, words, carrying ideas from one realm of knowledge to another. Planck the musician, turning to the work of that other great musical physicist, Helmholtz, and drawing upon his invention of the resonator—musical instruments that absorb and emit sound at a single frequency—to understand, by analogy, the physical properties of even tinier resonators, atoms: the basis of his black body theory of radiation.

Well, we can't go through all of it for you. The point is that every skill these men developed, every interest, every branch of knowledge they made their own, eventually was pulled together into an integrated whole by their drive to fulfill their thematic, global style of research. And that, to me, is why they are honored for their contributions to science, and worthy still of emulation.

▷ **Richter:** All of which is well and good. You convince me that what a person knows and what skills he acquires determine the range of possible problems he may recognize and address. That is obvious. I am willing to admit that simultaneous discoveries are not interchangeable. Fine. Perhaps you are even right to say that eliminating certain individuals might alter the course of history—how much, I still have my doubts. But the most important thing I hear is that you have raised the specter of irrationality again. You yourselves state that van't Hoff arrived at his tetrahedral carbon atom by a sudden insight; that Arrhenius invented ionic dissociation by an unexpected illumination; that Ostwald's thematum came to him in a flash of inspiration. If these are crucial events, then we still do not understand the basis of discovery and invention, do we? You place us in the position of the discoverer, but for what? To leave us hanging at the crucial point.

▷ **Ariana:** You underestimate how much we have accomplished. Remember, you yourself agreed that many discoveries do not occur by chance. Perhaps we can't explain them all—but haven't we learned some important things from the ones we have explained?

▷ **Hunter:** Obviously I think so. Yet Richter's right. We do have a problem with illumination. And it is one we haven't solved yet, I'm afraid. I'm planning to try to make sense of Arrhenius's insight this week, if I can. But as I suggested earlier this afternoon, why don't all of us take up the issue of illumination for next Saturday and see how far we can get? Especially you, Richter. It would certainly be nice to make everything coherent, rather than leave bits and pieces strewn about in our path.

▷ **Imp:** Absolutely! It would also give us a chance to rethink some things. I know I've experienced illumination, but I've certainly never tried to figure out why or how it unfolded. Let's give it a shot next week.

▼ JENNY'S NOTEBOOK: Renewing Old Knowledge

Richter and Constance left together, murmuring quietly. I gather Richter was asking for information of some sort. Hunter was on his way out the door when Ariana called him back: "I have something I must ask you about those semipermeable membranes invented by Traube."

Hunter sat down beside Ariana. "Of course."

"Gelatin, or at least some gelatin, is boiled animal tissue, especially connective tissue, right?" began Ariana. "Proteins. And Traube made his semipermeable membranes by treating gelatin with tannic acid."

"That was one of his two methods."

"Then listen to this. A few nights ago I was dipping into Paul de Kruif's *Why Keep Them Alive* looking for some anecdotes to liven up a lecture—not, I know, a source Constance would approve of, but useful in its place.[1] De Kruif describes Edward Davidson's attempts to treat burn victims during the 1920s. What's the result of burning or scalding connective tissue? Crudely stated: gelatin. What did Davidson finally hit upon as a treatment? Tannic acid![2] De Kruif described it as a miracle treatment. It sealed the wound by creating a hard blackish skin, or 'eschar,' eliminated the pain, and, according to Davidson's records, allowed dozens of patients to live who otherwise would have died."

Hunter nodded his head vigorously. "I see where you're headed. You want to know whether Davidson's tannic acid treatment created a semipermeable membrane—in essence, an artificial skin—equivalent to Traube's semipermeable membranes. Fascinating idea.[3] It's certainly testable. But you aren't thinking of reviving Davidson's treatment, are you? Isn't tannic acid toxic?"

Ariana nodded. "Supposed to be. Injected intravenously, it is. That was demonstrated during the 1940s, and was one of the major arguments used against Davidson's treatment. Several doctors who'd invented their own burn treatment started a smear campaign a few years after Davidson's death, claiming that burn patients often died of liver damage caused by the tannic acid.[4] I have my doubts. Patients dying of severe burns often have liver damage whether they're treated with tannic acid or not. Moreover, what we know of pharmacology would suggest that how you get the tannic acid and in what dose, has to make a difference—i.v. isn't topical application, and the absorption isn't the same. And besides, we drink the stuff every time we drink tea."

Imp had disappeared into his study during this conversation, and now returned holding a book. "Exactly! Lots of plants have tannins. Tree bark—that's what people used to use to tan animal skins; fruits, all sorts of things. Your talk of tanning reminded me of something: my perennial favorite scientist, Albert Szent-Györgyi. Remember I said something about Szent-Györgyi isolating vitamin C after noticing that some fruits turn brown or black when damaged and others don't? For some reason I had that mentally filed along with tanning. So when you mentioned a hard, black 'eschar' formed by tannic acid . . . Here's the passage: 'this oxidation was known to be due to the oxidation of some polyphenol. There were complex chemical mechanisms proposed for this reaction involving peroxide formation. I could show in simple experiments that all that happened was that a polyphenol was oxidized by a ferment [enzyme, we'd say] to the corresponding quinone. This, then, tanned the damaged surface, forming pigments, closing the wound, and killing the invading bacteria. This system has great survival value for the plant.'[5] So plants have been using tanning to seal wounds for eons!"

"And vitamin C is necessary to promote wound healing?" I asked.

"On the contrary," replied Imp. "Vitamin C *retards* the tanning reaction. Banana skins are very high in phenolic compounds related to adrenalin and noradrenalin—catecholamines—and very low in vitamin C. Banana skins turn black. Fruits such as lemons and limes that are high in vitamin C are low in phenolic compounds and don't turn brown or black when bruised. That's the observation that allowed Szent-Györgyi to isolate vitamin C. Lemon juice retarded the banana reaction."

"Oh," I said, "so that's why recipes recommend sprinkling cut fruit and avocados with lemon juice to prevent browning."

"Sublimity of the mundane." Imp beamed.

Hunter, however, looked uneasy. "Well, pH—the increased acidity—

probably plays a role in preventing the oxidation, too. But your mention of catecholamines brings back vague recollections of other color reactions. Don't catecholamines oxidize to form colored compounds?"

"Absolutely!" exclaimed Ariana. "Dopamine, adrenalin (otherwise known as epinephrine), norepinephrine—they're the metabolic precursors for the melanins."

"Which are?" I asked.

"The pigments that give skin its brown color," explained Ariana. "They also act as the markers of melanomas—black skin cancers. And there are some neuromelanins in your brain, too. Nobody seems to know quite why they're there, though, or even their precise chemical structures."

Imp slapped his forehead. "Yes, yes, yes! How could I be so stupid! More connections—and the whole thing sitting there in front of my nose for years! Look!" He began pacing up and down the room. "Catecholamines don't just form melanins; they also form other oxidation products that are chromophores. Norepinephrine turns pink at pH 7—body pH—after just a few minutes. One of the big mysteries is how the catecholamines in your body are prevented from oxidizing when they're stored in nerves. Complexes with ATP have been suggested; and binding to proteins.[6] First clue: Everybody who uses catecholamines in their assays or tissue cultures always adds vitamin C to prevent the oxidation. Standard procedure. Why? Norepi with vitamin C can last for days, even at pH 7. How does it work? Supposedly the vitamin C protects the norepi by oxidizing first—supposedly, mind you. Second clue: A couple of years ago, a colleague at the University of Rochester School of Medicine mentioned to me this weird observation that vitamin C seems to be stored in nerve terminals secreting dopamine and norepi. Same anti-oxidant as in plants and assays, but this time in the nerve terminals themselves. Think about it! Here we have Szent-Györgyi's polyphenol–vitamin C reduction-oxidation system all over again!"

"Hold on." Hunter frowned slightly. "If I recall, it only takes a few minutes to oxidize vitamin C in aqueous solution at pH 7. You can't take two compounds, each of which oxidizes very quickly, and have them last for hours or days. Doesn't make sense."

"Exactly!" exclaimed Imp. "It doesn't make sense. Nobody's paying attention to the reaction rates. To protect norepi from oxidation, it's got to be bound to something. Why not vitamin C? We know they're mutually protective in solution. We know they're in the same nerves. If the reactive groups on one were bound to the reactive groups on the other, neither would oxidize. They'd be a stable complex."

"Is this another of your infamous complementarity ideas?" I asked. "More little molecules sticking to one another?"

"You bet! And probably just as crazy as its predecessors," replied Imp happily.

"Okay," I continued tentatively, "but what does this have to do with tannic acid and burn treatment? Or is it a different subject?"

"No, no, no! Same subject, different angle! Can't you *see*? Global thinking! It's the same principle, even the same actors, interacting in different ways. Where's my book? Here! It's just as Szent-Györgyi writes:

> *Looking back on all this work today, I think that bananas, lemons, and men, all have basically the same system of respiration [and much else, I might add!], however different they may appear. Like the plants, we, too, have in our own oxidation-system polyphenols and ascorbic acid [vitamin C], but nature is clever enough to kill several birds with the same stone and so emphasizes in the various species, for certain ends, the one or another member of the system, as it emphasizes the polyphenols in bananas and ascorbic acid in lemons. This I would call the principle of* horizontal organization, *by which I mean that in the various species we find the same row of substances or reactions from which one may be pushed to the fore.*[7]

Now do you see? Vitamin C controls tanning reactions involving phenols in plants, and the same set of chemicals are present in the nerves of animals—the same set of reactions, the same set of controls, all adapted to a variety of uses. Maybe vitamin C even controls the rate of synthesis of melanins, for example!"

Above left, *ascorbic acid (vitamin C);* above right, *norepinephrine (noradrenaline);* below, *complex of ascorbic acid with norepinephrine.*

Hunter still wasn't convinced. "An interesting idea, Imp, my friend, but is there any actual evidence for a catecholamine–vitamin C complex? Or any of the rest?"

"Oh, come on, Hunter, you know better than that. There's never evidence of anything until you look for it! And since we can't look just at the moment, why don't we manufacture some evidence?" Imp disappeared into his study again.

I turned to Ariana and Hunter. "Well, it appears that this may take some time. How would you two like to stay for dinner?" They agreed after only the slightest persuasion. Then Imp reappeared.

"Models!" he cried happily, and sat down with a large box full of black, red, white, and blue plastic pieces of various shapes, each representing an atom. "CPK models. As Pauling says, modeling is a way of thinking.[8] So let us think! Here, Hunter, you make vitamin C while I make norepi."

And so the rest of the evening was spent eating and playing. Imp and Hunter snapped and popped and pushed and pulled and patted their molecules into various shapes, twisting and turning them to see how they might fit, and making a variety of cryptic comments like, "Isn't there some way to maximize the pi-pi overlap bonds?" or "What about this hydroxyl? We can't just leave it hanging there!" Eventually, after dozens of false starts, they created a model of a norepi–vitamin C complex that satisfied them both by "tying up" all the reactive groups on each molecule. "This should do it," proclaimed Imp. "No place for an oxidation reaction to occur."

"At least not with any facility," amended Hunter. "You realize, of course, that all we've done is to invent a hypothesis. And even if we can find the same thing happening in a test tube—say by infrared or ultraviolet spectrometry—that's still a far cry from demonstrating the existence of such a complex in a nerve."

"Sure," responded Imp, "but if they're in the same vesicles in the same cell, and if you could demonstrate corelease . . ."

Actually Hunter seemed more interested in Ariana's idea and returned to it consistently throughout the evening. "It seems to me," Ariana said at one point, "that it might be worth looking back at Davidson's work if only because it was so simple. Medicine's getting too complicated. I can't help thinking what might happen if, heaven forbid, we actually had a nuclear war or a major catastrophe like the San Francisco earthquake and fire at the turn of the century. Take away the high-tech tools of the major hospitals, and most of our medical expertise goes, too. Same distinction Imp made between stenokates and eurokates: Most doctors deliver health care, but very few invent it or could even recreate it if they had to. Scary. So why not develop some of these simple techniques for emergency use?"

Hunter agreed. "The armed forces might be interested, too. You can't treat battlefield wounds with skin transplants. No facilities. A tannic-acid salve, or something based on the same principle but less toxic—assuming it is toxic—could be just the thing." I began to understand some of our earlier discussions about the necessity for a scientist to make something of a discovery. Nobody was going to care about Ariana's insight unless it could be put to some use.

A few minutes later Hunter had another thought. "Even if Davidson's treatment is too outmoded to have much medical interest, it could still be very useful for studying the chemical reactions that create semipermeable membranes, and the mechanisms underlying semipermeability. It's not a field I'm all that familiar with, but I'd say that an insight stemming from an 1870s botanical paper and a 1920s medical treatise is unlikely to be common knowledge. And, as you say, Ariana, the beauty of it is its simplicity. You see, if we can make such membranes from something like egg albumin, which has been completely sequenced, then perhaps we can elucidate the reactive sites and gain sufficient information to create new kinds of artificial membranes—things that could be used in medicine, in industry, as molecular sieves in laboratories, or even for water purification. Embed them, as Pfeffer did, in another porous material such as porcelain or various plastics, and there are all sorts of possibilities . . ."

They left together fairly early, talking of various ways they might collaborate. Imp's still playing with his models, and will no doubt continue to do so into the early morning hours, as he always does when inspired. Altogether a very interesting and very exhausting day.

I begin to wonder how many things I know that would suddenly take on new meaning if only I could perceive the connections. I foresee a restless night.

DAY FIVE

Insight and Oversight

I go on thinking about my problems all the time, and my brain must be going on thinking about them even when I sleep because I usually get the answer to my problems ready-made at the moment I wake up, and sometimes in the middle of the night. My brain must do as the laxative that was advertised by saying: "While you sleep it does the work."

—Albert Szent-Györgyi (1963)

► IMP'S JOURNAL: Illumination

Flash of insight. Yes, let's see—four years ago now.

Started innocuously enough when I approached Gustav about amino acid pairing. Refreshing chat. Said if it hasn't been tried, let's try it. Turned out he had his own reasons for being interested. He'd made peptides for testing EAE [Ed. note—experimental autoimmune encephalomyelitis, an animal model for human diseases such as multiple sclerosis, in which the body's immune system attacks itself]. Textbooks all said EAE caused by autoimmune reaction to myelin basic protein. But Gustav and his colleagues found that no matter how much myelin basic protein they injected into an animal, no EAE. In fact, opposite: actually suppressed immunological reaction at high doses. Why?

Gustav informs me that textbooks often leave out key ingredient: so-called adjuvant. Everyone in field knows you need adjuvant, but ignore when explaining disease induction. So what's an adjuvant, I ask. "An immunopotentiator." Christ. Just like Molière's doctor. Drugs make you sleep because they have somnogenic actions. Adjuvants increase immunological responsiveness because they're immunopotentiators. When are we gonna learn?

Turns out adjuvant in this case is bacterial cell walls or their chemical components. Gustav mentions that someone's found the smallest active adjuvant component for EAE induction—muramyl dipeptide—just as he's found the smallest active peptide fragment of myelin basic protein. Would I be interested in seeing if they pair? He thinks the active inducer might be a complex of adjuvant and peptide. Very unlikely it'll fit my theory, but what the hell. I try it. Sure enough, doesn't fit my theory. But I do manage to invent a model of a complex that explains why each known alteration of peptide or adjuvant is active or not.

Model's fine. But how's adjuvant-peptide complex active when peptide alone isn't? Try every idea imaginable. Prevents enzymatic degradation of peptide so more's available for stimulating immunological response. (Then why doesn't a high dose of protein elicit EAE? More you inject, more there is available to the immune system.) Elicits an immunological response to complex. (Then why is autoimmunity directed only at myelin basic protein? Is it? Well, there sure isn't bacterial cell-wall material bound to every myelin basic protein molecule in a normal cell. How about some chemical analogue?) Back and forth for weeks. Model works. Explains available data. But we can't figure out why! Frustration! But excellent education for me. Never thought about autoimmunity before.

The big day. I can still relive it. Sitting in a chair, facing the floor-to-ceiling window as the sun set over the ocean. Leaning back, listening to Gustav go over arguments yet again. Bored. Not getting anywhere. Drifting off, aware only of golden light streaming in, making room too warm. Tired. Eyes closed. Red haze of inside of eyelids. Gustav still droning on. Were his words evoking images in my mind? Don't know. But something welled up from previous conversation. "Didn't somebody say that the adjuvant material alone induces an immune response?" I'd asked. "Of course. We induce EAE in Lewis rats with *Mycobacteria tuberculosis*. Same organism that causes the disease tuberculosis. Everybody knows that." Obviously unimportant then. But in mind's eye saw adjuvant and antigen swirling around, combining, dissociating. Each separately inducing an immune response; both together—

Then it hit me. Everything fell into place. Physical wave of relief. Tension of past months gone. Tremendous excitement! Didn't hear Gustav at all now though he was still talking. Totally within myself, in warm red light of knowing! This was it!

The adjuvant wasn't an adjuvant! It elicited immunological response of its own. Makes it antigen in own right. So not myelin basic protein that induced EAE. Not even our complex! It's *pair* of antigens each eliciting own immunological response! Complex provides key for how and why pair induced autoimmune response. Reasoning as follows (a posteriori—at the time, simply fit pattern of criteria defined by problem instantaneously. No logical, verbal reasoning involved, just images.) (1) The two antigens complementary in chemical structure (demonstrated by binding). (2) Each antigen elicits own antibody (or whatever on T-cells). (3) Antibody is chemically and structurally complementary to antigen (known since Ehrlich, ca. 1900). (4) Therefore—key point—the antibodies elicited by a pair of chemically complementary antigens will themselves be complementary. One antibody will attack other. Same thing for binding proteins on T-cells. Autoimmunity resulting from the immune system actually fighting itself! A whole new way of thinking about autoimmunity.[1]

Analysis: We'd been trapped by previous dogma into thinking about a single cause for a single disease. Trapped by language into thinking adjuvants immunologically different (more general, less important) than antigens. Trapped by own model of complex into thinking about complex as single entity rather than as a result of complementarity. Whole series of hidden assumptions directing us to dead ends. Us and everybody else in field.

Of course, illumination only beginning. Verified immunological response to myelin basic protein not cross-reactive with adjuvant. Other

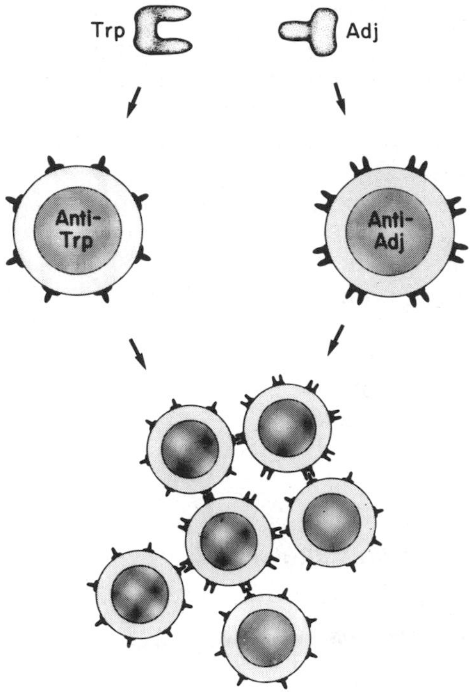

Imp's illumination concerning complementarity between the two antigens (tryptophan peptide and adjuvant peptide) that cause experimental allergic encephalomyelitis. Each antigen elicits a complementary "antibody" on a lymphocyte, and the resulting lymphocytes are in turn complementary. These complementary lymphocytes attack one another as well as the antigens. (Westall and Root-Bernstein, 1983)

experimental autoimmune diseases all require dual-antigen induction. Demonstrated binding actually occurs as per model. Most surprisingly, found theory explains related phenomena—immune complexes and perivascular cuffing. Hadn't even included those in prior considerations. Another instance of misdefining problem by too-narrow focus.

And then a missed discovery. Almost. Stared us in face, but wouldn't acknowledge it. Again, unquestioned preconceptions. Had reviewed lit on myelin basic protein. Found refs to serotonin binding site on it. Did work to demonstrate structure of site. Model for serotonin binding site on myelin basic protein same as muramyl dipeptide (adjuvant) binding site. Implication: muramyl dipeptide should bind to serotonin binding sites, and therefore act like serotonin. Remember thinking: But how could a bacterial cell-wall breakdown product act like a neurotransmitter? Different sizes and molecular weights. Different shapes. Nonsense. Shrugged it off. Should've known better. LSD binds to serotonin sites, too, and it doesn't look much like serotonin! (New idea: Is that why bacterial infections can cause hallucinations during high fevers? Serotonin controls temp. Muramyl dipeptides cause fevers.)

Then van't Hoff–like event. For him, Wislicenus's paper on different lactic acid forms provides key to problem. For us, Pappenheimer, Karnovsky, and Krueger announcing that sleep factor in mammalian urine is muramyl dipeptide.[2] Sleep controlled by serotonin pathway. Would explain why many bacterial infections associated with drowsiness. No time to lose. Gustav and I send letter to *Lancet* suggesting connection.[3] Silverman and Karnovsky verify by demonstrating competition between muramyl dipeptide and serotonin for same receptors on macrophages. We do chemical studies of binding.[4] Bingo! Still following that one up.

That's key. Too easy to get trapped, invent tacit rules for self about what's acceptable, what isn't. No basis for most. Just common sense or popular usage, past experience, structure of language. But discovery's surprise. Illumination's surprising oneself. You either have to relinquish tacit rules—is that what drowsing or sleeping does?—or be kicked in the pants so you'll turn around to see what's standing right behind you. Thinking something is impossible should be key that something important may lurk behind our preconceptions.

But can you deliberately break such preconceptions?

▼ JENNY'S NOTEBOOK: Patterns on Paper

Constance. Now why do I have this idea that she's too pedantic to be creative? All those boxes of notecards and lists of quotations? Her personality? Her meekness? (I'm one to talk!) Or do I have this preconception that finding information or pulling it together is less creative than, say, inventing a new theory? Well, she surprised us again. Walked in on Saturday with a couple of books and a stack of photocopies, which she flopped on the table. "Oh, good, you're already here, Ariana. I thought you might like a look at these. What do you think?"

"Are they all valid?" responded Ariana, smiling broadly.

"Most of them," replied Constance.

"Come here and take a look at this one." Ariana beckoned me over. I was already sidling up to peer over her shoulder. "Look familiar?" Sure enough, there were the weird, loopy form of the periodic table I'd seen in Annie Besant's book at the County Art Museum a couple of months earlier, but this time attributed to a scientist named Crookes. Same guy who invented the Crookes tube used by Roentgen to discover X-rays.

Constance proceeded to give us a guided tour of her finds: more than 150 invalid and some 400 valid versions of the periodic table of elements, over twenty of them by Mendeleev himself; some as recent as 1980; most summarized in a book by a modern inventor of such tables, Edward Mazurs.[1] "Just look at the creative imagination being displayed here!" cried Ariana. "A given set of elements—pun intended—that must appear in a predetermined order and display certain periodicities, and look! It's wonderful!"

"So much for paradigms that you must accept or be kicked out of the field," added Imp, who'd emerged from the kitchen with Richter.[2]

"Or might you consider them as different illustrations of the same paradigm?" inquired Constance.

"Doesn't matter," Ariana responded. "Just look at how these figures give the lie to the old saw that if you give ten painters the same scene to paint they'll produce ten different paintings; but give ten scientists the same problem to solve and they'll all reach the same solution. *Now* try to tell me that scientists aren't as imaginative in their work and as diverse in their solutions as any group of artists!"[3]

"With pleasure." Richter looked at Constance's cache quietly.

"Now, aren't these the kind of thing you were talking about last week?" Constance asked. "You wondered whether, if neither proof or disproof were possible, several—in this case many—hypotheses or theories might coexist.

Richter perused the various figures carefully, noting dates and names, making comparisons. He refused to be drawn by Ariana's insistent remarks about aesthetics, creativity, visual thinking and pattern forming. He responded noncommittally and sat down to our session looking very thoughtful.

◀ TRANSCRIPT: Insight and Oversight

▷ **Imp:** Everybody ready? Today we illuminate illuminations, eulogize the Eureka act, and hope that hindsight will give us insight into the foresight of inspiration!

▷ **Richter:** Your lightheartedness is out of place. The issue is quite serious.

Upon consideration I find two—make it three—problems. First, I personally have never had an "illumination," whatever that is supposed to be. Nor have a number of my colleagues. I find that most solutions occur after tedious, step-by-step grinding it through. Though you say that van't Hoff, Ostwald, and Arrhenius each had an illumination, we had recourse to no such vague term in our previous discussions of Berthollet, Pasteur, or Fleming. Therefore, I question the utility of the concept and its place in Constance's model. Second, a friend—you know Normanson—admitted to having had an illumination, but asserts it was wrong.

▷ **Constance:** Sir John Eccles's experience, too.

▷ **Imp:** Well, you have to be a little careful about Eccles's story. I looked into that myself. What really happened was that he had a flash of insight, performed some experiments, and confirmed it. He published his results. Then, several years later, he reported that he'd been wrong. That's the part of the story he likes to relate. Then, a few years after that, he realized that his first idea was actually valid under some conditions and not others.[1] A problem of boundary conditions. The point is that, right or wrong, his insight got him moving in the right direction. Surely if there's one thing we've learned from these sessions it's that ideas shouldn't be judged by their correctness but by their ability to catapult the researcher into unexplored territory.

▷ **Richter:** To resume—third, my impression is that these rare insights come unannounced and uncontrollably. What is the point in discussing something we cannot predict or manipulate?

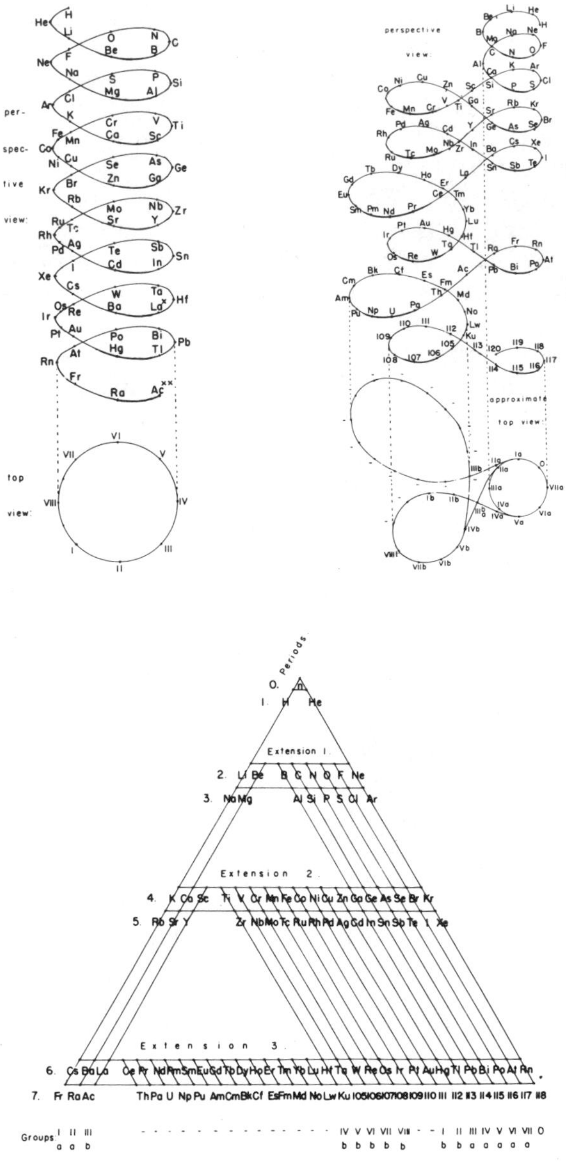

Five of some 450 ways the periodic table of elements has been portrayed in the past 120 years. (Mazurs, 1959/1977. Copyright University of Alabama Press. By permission)

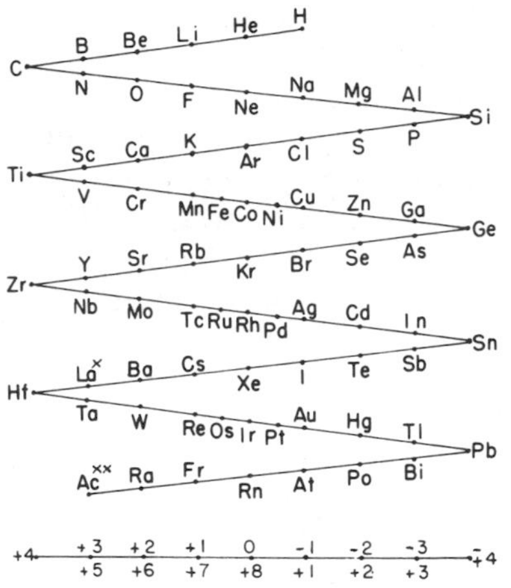

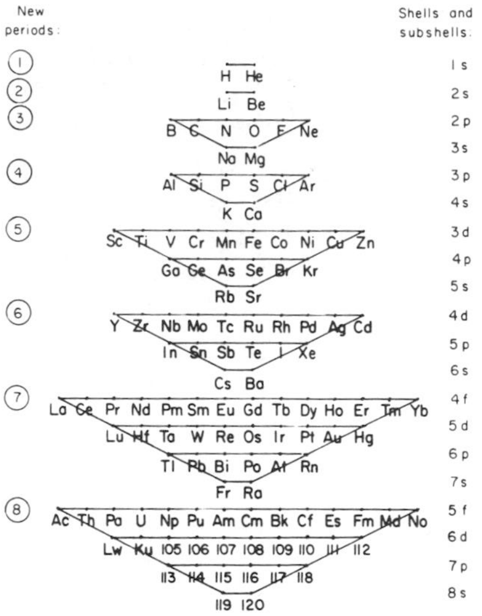

▷ **Imp:** For the same reason geneticists study mutations: to try to determine the conditions under which they arise and how to create such conditions. To move from pure randomness to probability.

▷ **Richter:** Perhaps.

▷ **Jenny:** That would be great if we can do it. But I've got a different sort of problem. I looked through a few of the books Constance mentioned last week, and I'm not clear what we are calling an illumination. From what I gather, the term was introduced by Graham Wallas in his *Art of Thinking* to refer to the sudden resolution of a problem achieved after prolonged preparation.[2] He writes as if all important problems are solved in this way. So does Arthur Koestler, who calls it the "Eureka act"—Archimedes jumping out of the bath shouting, "Eureka! I've found it!"[3] Then I came across other people calling it a *Geistesblitz* or "flash of insight," but they make it sound like a rare, chance event.[4] And even more confusing, a lot of the examples are dreams, daydreams, or just unexpected intuitions. So what are we including in our discussion?

▷ **Ariana:** How about everything? Why assume that illuminations, dreams, whatever, are different from any other kind of discovery process? Why not think globally and try to perceive these things as particular manifestations of the general process of discovering?

▷ **Constance:** How would that work?

▷ **Ariana:** It's a matter of policy. We can subdivide and subdivide until we have an infinitely long list of all the different ways of making discoveries, or we can try to understand how the diversity of ways of discovering results from some underlying unity. Anyone can learn to perceive differences, but one definition of genius is the finding of useful connections between things that other people overlook.

▷ **Hunter:** Yes, Ulam has said that a good mathematician perceives analogies between things; a great mathematician perceives analogies between analogies. I do think Ariana's proposal is the way to proceed. But let me point out another hidden assumption: Almost everyone I've read on the subject believes that illuminations or dreams or whatever are, as Koestler says, "acts"—events, localizable in time. Do we really want to think about them that way?

▷ **Constance:** Well, I certainly don't. I mean, I put illumination in as a step in the process, but I didn't necessarily mean it to be discrete. And, at least from my point of view, it couldn't occur if a person didn't go through all the previous parts of the process.

But look, I can answer some of the factual questions concerning illumination right now. As Richter says, not everyone experiences them, but most people do. That's why I put it in my model. Two chemists, Platt and Baker, did a survey in which 83 percent of their colleagues claimed to gain insights from unconscious intuitions. Seven percent claimed that these inspirations were always right, whereas most chemists found that between 10 and 90 percent of their intuitions were wrong.[5]

▷ **Richter:** So much for reliability.

▷ **Hunter:** I don't know. I wouldn't mind having a technique for solving important problems that worked only 10 percent of the time. That's probably better than my current average!

▷ **Constance:** To continue, of sixty-nine mathematicians who responded to a questionnaire by Maillet, only two had solved a mathematical problem in a night dream, or knew of someone else who claimed to have done so; five reported working out trivial arguments; and the rest never solved anything in their sleep.[6]

▷ **Richter:** Nor would one expect them to. Whether one reads Mach or Crick, they agree: Dreams are the waste products of the mind.[7]

▷ **Imp:** Now, wait a minute! What about Otto Loewi? He was trying to figure out how to demonstrate that some chemical released from nerves helped to control the rate of the heartbeat. He'd hypothesized the existence of such a substance in 1903. Then, one night during 1920, he awoke from a dream convinced he'd invented a way to test his idea, wrote some notes concerning the experimental protocol, and went back to sleep. When he awoke, he found his notes illegible, and he was unable to recall his idea. The next night, he had the same dream, awoke, dressed, went straight to the lab, and performed the experiment—which was successful. How can you call that the refuse of the mind?[8]

▷ **Constance:** Judson reports another scientist having the same experience a few years ago.[9] And Bernd Matthias, who discovered more superconducting elements than anyone else, says he discovered many of them in his sleep.[10]

▷ **Richter:** So? I am counting the instances on my fingers.

▷ **Constance:** But, Richter, this is like the chance discovery issue. I mean, how many examples do we need to convince you that it's not unusual?

Kekulé invents his benzene ring in a moment of reverie sitting in front of a fire. Van't Hoff, Arrhenius, and Ostwald we've already mentioned. Darwin realizes the importance of diversity in a flash of insight during a carriage ride in 1844. Wallace, suffering from a high fever in 1858, suddenly recalls reading Malthus a dozen years earlier, and realizes that selection would cause the fittest to survive. Ampère, Gauss, Poincaré—I've got a dozen others—all report similar unexpected insights during periods of relaxation or sleep.[11] In fact, Robert R. Wilson thinks this sudden emergence of a fully formed idea from the subconscious is a universal phenomenon: "It's like throwing up when you're sick," he says. "You know it's going to happen long before it does, and when it does happen there's a satisfaction—almost a pleasure."[12]

▷ **Richter:** So we return to the subconscious.

▷ **Constance:** But that's not my point.

▷ **Richter:** It is mine.

▷ **Constance:** The question is how typical this is. Most of the eminent scientists I've studied report several instances of illuminations, whereas most of the people interviewed by Platt and Baker and Maillet were unknowns.[13]

▷ **Imp:** Suggesting that the most creative scientists are those who have developed tricks for paying attention to ways of thinking other scientists ignore.

▷ **Hunter:** Exactly what I was about to say. One thing we're ignoring here is that in every instance Constance mentioned, the scientist spent months or even years working on the problem solved by the illumination. Several of them report using some basic technique to increase the probability of such insights. Lipscomb, for example, says that he worked late at night in order to prime his brain to think about the problem as he slept.[14]

Pauling does the same thing. He says he's trained his brain to have illuminations. He habitually thinks about the problem he wants to solve as he lies in bed preparing for sleep, reviewing his unsuccessful daytime attacks. This primes the pump, so to speak. Here's the problem; here's what doesn't work; the solution must lie in this increasingly well-defined area. He does this every night for several weeks, impressing on his mind instructions to filter all possible solutions he may invent during his dreams and to bring any pertinent ones to his attention. Presumably these instructions contain a set of criteria for recognizing viable solutions. Pauling doesn't elaborate. Finally, he stops working on the problem—he gives up.[15] That's when the solution emerges, says Pauling:

> *Some weeks or months might go by, and then, suddenly, an idea that represented a solution to the problem or the germ of a solution to the problem would burst into my unconscious.*
>
> *I think that after this training, the subconscious examined many ideas that entered my mind, and rejected those that had no interest in relation to the problem. Finally, after tens of hundreds of thousands of ideas had been examined in this way and rejected, another*

idea came along that was recognized by the unconscious as having some significant relation to the problem, and this idea and its relation to the problem were brought into the consciousness.[16]

▷ **Constance:** Poincaré, you know, said very much the same thing about his unconscious process of intuition. But you left out something else both Poincaré and Pauling said. You'll like this, Ariana. They both state that the selection of a plausible answer is actually an aesthetic decision. For example, Pauling says that whereas some scientists ask, "What conclusions . . . are we forced to accept by these results of experiment and observation?" he asks instead, "What ideas"—note the plural—"about this question, as general and as aesthetically satisfying as possible, can we have that are not eliminated by these results of experiment and observation?"[17] Poincaré actually equates aesthetic reasoning with subconscious thought, placing it above conscious reasoning as a means of selection and invention.[18]

▷ **Richter:** Here we go again.

▷ **Constance:** I must say, however, that Poincaré raises an objection to his own description of illumination that we haven't considered. Can we really believe, he asks, that the subliminal mind makes every possible combination—Pauling's tens or even hundreds of thousands of ideas—tests them, compares them all for their relative beauty, and then brings only the best solution to the awareness of conscious thought? All without us knowing it?[19]

▷ **Imp:** Not only unbelievable, but contrary to what we know about illuminations. Having had one, I'll speak from experience. I don't think your unconscious sits and permutes everything under the sun. My experience is that your conscious mind does that, and it reaches a dead end. You can't think of any other way to attack the problem.

The beginning of all illuminations is *giving up*. You carry the problem and the criteria for its solution around like a ball and chain, but you stop trying to free yourself from it. When you stumble on the key, then you open the lock. And usually the key sits right in front of you all the time, but you overlook it.

▷ **Ariana:** Letting go. That gets us somewhere, I think. What do all types of illuminations, sudden insights, dreams, and so forth, have in common? They all occur during periods of relaxation, right? During illnesses, sleep, dozing, taking a walk, going on vacation. Getting away from work. Thinking about something else. That's an important clue. According to most reports, the individual is not trying to solve his or her problem, but is thinking about something else when the solution emerges.

▷ **Richter:** Or not thinking, more likely.

▷ **Ariana:** So, hypothesis—not entirely original, I admit: What if illumination results simply from an absence of interference by the logical faculties?[20] Please! I'm not finished, Richter.

This gets us back to our notion of discovery as surprise. The problem of illuminations reduces to one simple question, then: How can you surprise yourself?

▷ **Richter:** By being schizophrenic!

▷ **Ariana:** Which is essentially what the unconscious-conscious distinction assumes. I find the distinction intellectually repugnant. Let's try another possibility related to the "eye of the mind." You perceive what you expect to perceive. You try to fit everything you observe into prefabricated logical frameworks, usually acquired through studies of codified science. But what happens when the logical framework is insufficient? One of two things: You reject the observation as nonsensical, or you reject the consequences of the observation as impossible. Or perhaps you never even conceive of the consequences because they cannot be imagined within your set of codified rules.

▷ **Constance:** Poincaré gave an example of that in his discussion of the mathematical style of Félix Klein:

> *He is studying one of the most abstract questions of the theory of functions: to determine whether on a given Riemann surface there always exists a function admitting of given singularities. What does the celebrated German geometer do? He replaces his Riemann surface by a metallic surface whose electrical conductivity varies according to certain laws. He connects two of its points with two poles of a battery. The current, says he, must pass, and the distribution of this current on the surface will define a function whose singularities will be precisely those called for by the enunciation [of the problem].*
>
> *Doubtless Professor Klein well knows he has given only a sketch; nevertheless he has not hesitated to publish it; and he would probably believe he finds in it, if not a rigorous demonstration, at least a kind of moral certainty. A logician would have rejected with horror such a conception, or rather he would not have had to reject it, because in his mind it would never have originated.*[21]

▷ **Jenny:** So what you're saying, Ariana, is that logical rules limit the possible ways an individual will conceive of combining or interpreting data.

▷ **Ariana:** Right. They are patterns. Within any pattern, the possibilities are limited and knowable in advance. The only way to surprise yourself is to consider something impossible.

▷ **Hunter:** That's what Arrhenius said in his Nobel lecture. "Just those things that are considered impossible are the most important for the advancement of science."[22] When accepted rules are abandoned, the unexpected becomes likely.

▷ **Ariana:** That's it! The Fleming-Lorenz-Delbrück limited-sloppiness approach adapted to explanations. And that's why illuminations occur only when you're *not* sitting at your desk pounding away at the problem, *not* reading carefully articulated logical accounts of research, *not* experimenting. If the logical rules you know could work, then you'd have solved the problem in a straightforward manner. But, as Imp says, illuminations occur only when the expected avenues don't lead to your goal. You need an unexpected path. So you have to abandon the concourses—the paradigms, the codified patterns of thought. Logical strictures have to be discarded so that you can consider an otherwise impossible solution.

▷ **Jenny:** I'll say this: Your description certainly fits Snow's description of an illumination in *The Search*. The scientist is stalemated. He quits working, goes home and has a chat with his wife. Then: "I started. My thoughts had stopped going back upon themselves. As I had been watching Audrey's eyes, an idea had flashed through the mist, quite unreasonable, illogically."

▷ **Richter:** Note.

▷ **Jenny:** "It had no bearing at all on any of the hopeless attempts I had been making; I had explored every way, I thought, but this was new; and, too agitated to say even to myself that I believed it, I took out some paper and tried to work it out."[23]

Mitchell Wilson gave a similar example in *A Far Meridian*.[24]

▷ **Richter:** Look, this is all very well, but you forget something. You are supposed to be *explaining* illumination, not describing it. Give me a model. Tell me how it works.

▷ **Constance:** Well, what about Koestler's and Rothenberg's "bisociation" kind of idea?[25] They'd probably shudder at being lumped together, but I don't really see much difference in their ideas on the subject. Essentially, they both claim that you bring two apparently contradictory ideas into conjunction, and create a synthesis. A kind of Hegelian dialecticism, I suppose. Rothenberg calls it "Janusian thinking," after the Roman god who has two faces looking in opposite directions. He presents an interpretation of Einstein's research as an example, though I sort of doubt its accuracy since Rothenberg isn't trained as either a physicist or a historian of the subject.

▷ **Richter:** I'm happy to note that I am not the only skeptic here. But surely you realize this bisociation or Janusian thinking or whatever name you call it is nonsense. All I have to do is take any two contradictory ideas and show they are compatible, and—flash!—I am a genius. I am pleased to hear it is so simple. Let us all go home and think antithetically.

▷ **Ariana:** Oh, come on, Richter. Don't be ridiculous!

▷ **Richter:** Just playing "Implications," my dear. Did not Constance tell us of Poincaré's dictum, "To invent is to choose"? It is not sufficient merely to put things together in a new way. There are an infinite number of ways of putting ideas together that no one has tried before. Nor should anyone bother. The object of science is to put things together in a new and meaningful way. By illumination or otherwise.[26]

▷ **Imp:** Which is one reason that dialectical thinking is so useful, Richter—because it stresses process rather than product, the ability to synthesize links between apparently antithetical ideas to yield surprise. Read J. B. S. Haldane or Richard Lewontin.[27]

▷ **Richter:** I still fail to see how one does that in practice. Perhaps it appears that way in retrospect. But does the scientist really sit down and say, "These two ideas or observations are antithetical. I will attempt to find a perspective from which they are synthetic"?

▷ **Hunter:** Absolutely. As you yourself have repeatedly pointed out, paradoxes, contradictions, and anomalies tell us where the next discovery lies. Why? Because we know that they must fit, but they do not. And sometimes just searching for another perspective does solve the problem.

▷ **Constance:** Which suggests that Kuhn's model of illumination as a gestalt shift might be useful.[28]

▷ **Imp:** No, absolutely not. I've been thinking about this since the first session. Gestalt shifts occur spontaneously. You don't generate a set of logical rules that prevent you from perceiving the other figure. You perceive one and then the other, back and forth. And this flip-flopping occurs only when you're actually staring at the picture. Illuminations occur when you *aren't* looking at the picture, and once you've got the new figure in your head, you can't go back to the old one. It doesn't work anymore.

▷ **Hunter:** Precisely—the anomaly again. Kuhn himself recognizes the central role of the anomaly as a motivator for theoretical change, yet he ignores it when he gets to his gestalt stuff. That's sloppy. The reason for inventing a new theory is always to explain something, or to make a series of ideas or observations coherent. There's got to be a set of incommensurables involved somewhere—internal contradictions or inconsistencies, thematic discords, something. So there can't be just one line that describes equally well two different figures. There's one line that describes one figure, but leaves out a point, and there's another line that describes a different figure that incorporates all the points. Or the figure that the line describes becomes ugly to the point of internal inconsistency. If this weren't true, then how could Kuhn claim that once a new paradigm is invented, anyone adhering to an old one is cast out of the profession? If caloric theory is really equivalent to the kinetic theory of heat, then what's the advantage of one over the other?

▷ **Constance:** Well, I think you're being a bit too literal, Hunter. I mean, if one has two images, say a duck and a rabbit, the observed differences predict other differences. Further experimentation can add the data necessary for choosing one image over the other. Does the animal fly or hop?

▷ **Imp:** No, I'm with Hunter. You still have the problem that if the same data define two equivalent theories, then why does one have to be invented? In a gestalt shift, both images already exist. In science, the new one doesn't until it's invented. Look at all those periodic tables you showed us. There's nothing intrinsic to the data that determines or even suggests any particular pattern. Periodicities, yes; but not specific patterns. Unlike the gestalt image, the second pattern isn't there in the first. No, let me correct myself: The other *patterns* aren't there in the first place—emphasizing another point that needs to be made explicitly:

The world isn't either/or; it's incredibly variegated. A gestalt model can't account for that.

▷ **Hunter:** Then allow me to make a suggestion. One thing we've all ignored so far is Poincaré's emphasis on the role of preparation in the process of illumination. As far as I can tell, nobody ever had a flash of inspiration without first spending a great deal of time searching for solutions. Perhaps the key lies there, in what the person already knows. Arrhenius's route to ionic dissociation is quite revealing on this point.

▷ **Imp:** Objections, anyone? Then go ahead, Hunter.

▷ **Hunter:** Right. Give me a minute to get my notes together.

Arrhenius's illumination occurs on May 17, 1883.[29] The fact that he could remember that date years later indicates how exciting his insight was for him. For most of the previous year, he's been sitting in Edlund's lab in Stockholm measuring how much electrical current salt solutions of different types will conduct. The work is boringly repetitive, especially for someone like Arrhenius who gets his thrills by thinking his way through problems rather than by experimenting. You may recall from last week that he undertakes this research in an attempt to solve the "impossible" problem of the molecular weights of complex sugars. He fails, but realizes that different types of molecules—sugars, salts, alcohols, and so on—each have characteristic electrolytic conductivities. Other than some recent work by Friedrich Kohlrausch and Rudolf Lenz, there isn't much information on the subject, so he takes it up with a vengeance, especially since he's convinced that an understanding of the electrical nature of chemical combination will yield new and basic insights into all of chemistry.[30]

▷ **Imp:** So Arrhenius is young, and he's new to the field.

▷ **Hunter:** As are van't Hoff and Ostwald when they have their insights as well. That's fairly typical, though not completely.

▷ **Constance:** Kekulé had his illumination while taking a break from writing a general textbook, so maybe they occur mainly when the scientist is trying to pull everything together coherently.[31] Think about Darwin and Wallace.

▷ **Hunter:** That's an interesting point. Certainly Arrhenius was trying to pull everything together when he had his. But a little more background first.

Arrhenius has an advantage over Kohlrausch and Lenz in performing his research: He's using a measuring apparatus invented by Edlund. Edlund's apparatus is a magnitude or so more accurate than Kohlrausch's, and allows Arrhenius to study the conductivity of much more dilute solutions.[32]

He knows from Kohlrausch and Lenz that each compound has a more or less unique conductivity. The question is how accurately these can be distinguished. By the time he's experimented for a few months, he knows that at extreme dilutions they can't be. They fall into groups. Why?

Svante Arrhenius as a student, 1876. (Universitetsbibliotek, Stockholm)

As in all experimentation, there comes a point at which further data are useless. One needs to understand what the data mean. Arrhenius reached this point at the beginning of May. "I have experimented enough. Now I must think," he says to himself.[33] He decides to go home to Uppsala for a vacation. We find him sitting upstairs in his bedroom in his parent's house, past midnight, thinking furiously. He's just had his first insight into the internal workings of salt solutions.

Had you talked with him a few days earlier, however, you would have found that he'd reached an impasse. Part of his data makes sense, part doesn't. What makes sense is that at low concentrations, the more salt one has in solution, the more electricity the solution conducts. Nothing surprising there. Again, as one might expect, as the amount of salt drops to zero, the conductivity drops to zero. That's fine, too. The problem is that when the concentration rises above a certain level, the conductivity peaks and then begins to decrease This is odd, especially as Faraday and other electrochemists had proclaimed that the amount of electricity carried by each ion in solution—that is, by the charged molecules—is identical.[34] If the salt carries the electrical charge—that is to say, acts as the ions—then why should adding more salt decrease the conductivity? If the salt isn't the ionized part of the solution, then why is conductivity proportional to salt concentration over a wide range of dilutions? Arrhenius decides to look at the amount of electricity carried per molecule of salt.

Erik Edlund, ca. 1880. (Universitetsbibliotek, Stockholm)

Here's where he gets his biggest surprise. He expects every salt molecule to carry the same amount of electricity, at least at high dilutions. It doesn't. The molecular conductivity—that is, the amount of electricity carried by each molecule—actually appears to rise to a maximum at minimum concentrations; and that maximum is two or even three times the molecular conductivity he'd been led to expect by his predecessors. In other words, the less salt you have, the better a conductor it becomes, up to a point at which every salt molecule appears to be carrying twice the charge you'd expect for its molecular weight. Arrhenius is, to say the least, surprised. He's also confused.

▷ **Jenny:** As am I. I'm afraid I'm not following this very well.

▷ **Hunter:** Then let me show you the various models Arrhenius had available for explaining electrolytic phenomena. Each has imbedded in it a set of assumptions that become obvious only in light of a subsequent experiment or model. As with most of my previous examples, the answer derives ultimately from Berthollet, so that's where I'll start. The first mechanical model of electrolysis—the pulling apart of chemicals by electrical force—was given by Berthollet's student Grotthus in 1805, and it underpinned all subsequent models until Arrhenius's.[35] Grotthus proposed that anions and cations alternate to form a chain from one electrode to the other. The difficulty with his model is that once the compound is pulled apart, all the remaining molecules must turn over to avoid the repulsion of like charges. Do you see what I mean?

Humphrey Davy published a very similar model in 1807, which Berzelius amended a few years later. Unfortunately, these men failed to account for the placement of the ions and ended up with chains of positive ions and chains of negative ions, which are impossible. Like charges repel. Not until Hittorf invented a new model in 1858 was the problem of charges solved. In Hittorf's model, the positive and negative ions exchange with one another all the way down a line of molecules to release a positive ion at one end and a negative one at the other.[36] None of these models, however, attempts to explain how concentration affects conductivity. They're useless for Arrhenius.

There's only one available model that accounts for concentration effects: the so-called Clausius-Williamson hypothesis.[37] Both Clausius and Alexander Williamson had explained various chemical reactions in solutions according to the kinetic theory. As the molecules move about in solution, they occasionally bump into one another with sufficient energy to cause a momentary dissociation, allowing an exchange of parts.

Otto Petersson's laboratory during the 1880s, when Arrhenius was working on his dissertation. (Stadsmuseum, Stockholm)

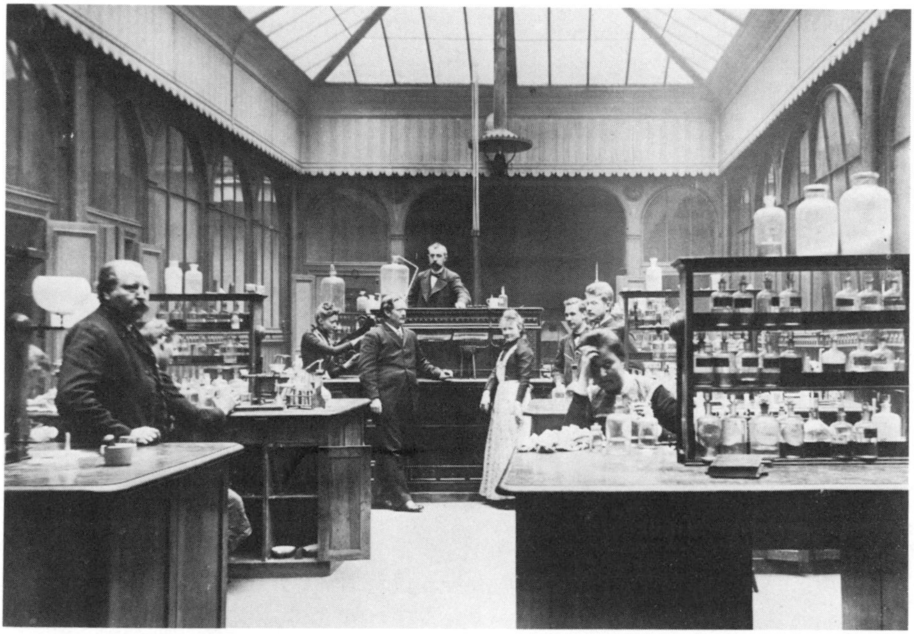

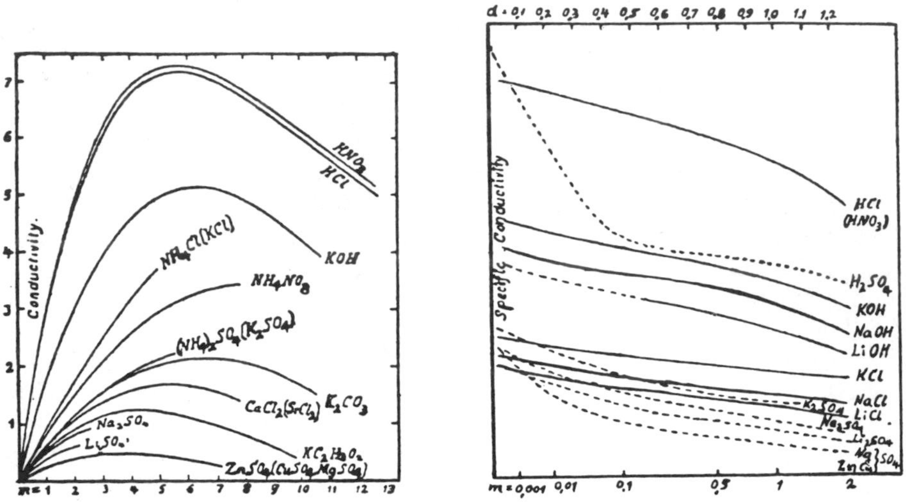

Left, *electrolytic conductivity versus concentration;* right, *specific (or molecular) conductivity versus concentration. (Kohlrausch, 1888)*

Thus, if ionized salt molecules were randomly to bump into one another, the resulting exchange of parts would convey the electrical charge from one electrode to the other. Not in a straight line, as in the Grotthus-Hittorf mechanical models, but by a path resembling a random walk.

This model has two distinct advantages over its predecessors. First, it explains how even the smallest electrical current results in electrolysis. If the molecules are already dissociating and recomposing as a result of their kinetic energy, then even the smallest force would prevent some of the reassociation. Furthermore, the model explains how conductivity decreases with concentration. The fewer salt molecules in solution, the fewer the collisions between them, and the less the conductivity.

▷ **Imp:** But that doesn't explain why conductivity decreases at high concentrations.

▷ **Richter:** And it predicts that the molecular conductivity should also decrease, not increase as Arrhenius observed, yes?

▷ **Hunter:** Exactly Arrhenius's problem. If the electrical charge is carried by the ions bumping into one another randomly, then as the number of ions in solution approaches zero, so will the conductivity—specific or otherwise. And, as Imp says, there's no reason for the conductivity to decrease as concentration increases. So—what's the problem?

▷ **Ariana:** Let me guess. Something's been left out. Arrhenius needs to perceive what everyone else overlooks.

▷ **Jenny:** I know! The solvent! Just like last week with van't Hoff.

▷ **Hunter:** Why are you an historian? I've had dozens of students with less natural aptitude than you.

▷ **Jenny:** But hopefully more interest. Besides, what makes you think I don't need the same aptitudes to be a historian?

▷ **Hunter:** Sorry. My scientific chauvinism is showing.

But you're right. The models ignore the solvent. That's Arrhenius's insight. He's read Berthollet and some of his more recent followers, notably the Norwegians Cato Guldberg and Peter Waage.[38] They clearly state that you can't ignore the affinity of the solution for the solute. In a flash of insight, you recognize that not only do the ions bump into one another, they also bump into the solvent molecules.

▷ **Imp:** In other words, the solvent—water—is an electrolyte, too.

▷ **Hunter:** That's it. A very poor one, but still, for Arrhenius, an electrolyte. So what happens? Essentially, Arrhenius turns things on their heads. For Clausius and Williamson, the electric current is carried by ions bumping into ions and exchanging parts. For Arrhenius, the electric current is carried by ions bumping into water molecules to form ionic hydrates. He can now introduce mass action effects to explain the previously anomalous behavior of salts at extreme concentration and extreme dilution. At extreme concentration, the ionic molecules will spend a significant time recombining with like molecules, and therefore carry the current in circles. At extreme dilution, most of the salt molecules will be dissociated to form hydrates with the water. The fewer salt molecules, the more likely they will form such hydrates.

The result will be two ions, one negatively charged and one posi-

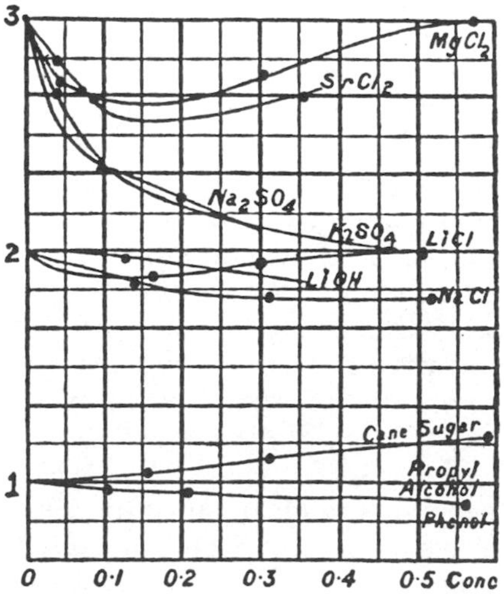

Raoult's data for specific (or molecular) freezing point depression. (Arrhenius, 1903)

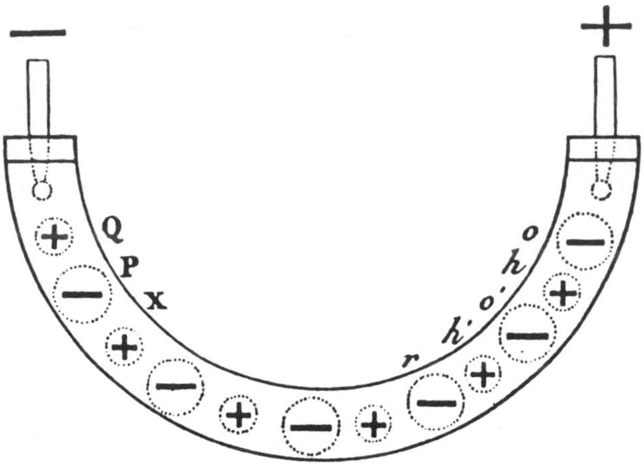

Grotthus's model of electrolysis, ca. 1806.

tively charged, for each molecule of salt dissociated. Since specific conductivity increases as salt concentration decreases, he concludes that the active molecules in terms of conductivity must be in this hydrated form. Thus, for Arrhenius, ammonia is a nonconductor until it forms a complex with water:

$$NH_3 + H_2O \rightleftharpoons NH_4^+ + OH^-$$

He also recognizes immediately that his insight has chemical implications. For instance, Berthelot had reported that ammonia is chemically unreactive in its pure, unhydrated form—it won't combine with anything else, and it won't conduct electricity.[39]

▷ **Jenny:** You said something about that with regard to Lavoisier and Berthollet, didn't you? That a pure acid or base is unreactive until it's put in water?

▷ **Hunter:** Certainly in nineteenth-century terms. No chemist since G. N. Lewis would accept that formulation, but it's close enough for our purposes here. The point is that ammonia in water is chemically very reactive, and the molecular reactivity increases as the solution becomes more dilute, just as the molecular conductivity does. So Arrhenius suggests that his electrolytically active molecules are also the chemically active ones. The result is an electrochemical theory of chemical reactions that eventually links, among other things, studies of electrolysis, chemical affinity, reaction rates, colligative properties, and mass action. Arrhenius claims he stayed up most of the night outlining the various applications his idea might have.[40]

▷ **Imp:** I'll bet he did!

▷ **Hunter:** How far he actually got remains unclear, since his dissertation is quite confusing on the issue of what his active and inactive molecules actually are, but by 1887 he states the concept of ionic dissociation in very much its modern form—dissociation of salts into free ions in solution—thereby solving van't Hoff's problem of why salt solutions appeared to have more molecules present than could be accounted for by molecular weight. They had dissociated.[41]

▷ **Constance:** But wait a minute. I thought you were going to take us through an illumination—you know, a sudden insight.

▷ **Hunter:** I just did.

▷ **Constance:** Well, I'm sorry, but I don't see it. I mean, it seems structurally no different from the other examples of discovering we've discussed.

▷ **Hunter:** Right. Except that instead of occurring over a period of months or years, it happened in a few hours late one memorable night in someone's bedroom. Sounds almost sexy, doesn't it?

I think it's this romantic version of illumination that's fooled people. We've made illumination into a different sort of discovery. But it isn't; it's a step in the process by which *all* discoveries are made. I think Constance is right on that point. Sometimes discoveries involve a series of small, almost unnoticed insights—minor surprises accumulated from many passes through the discovery process that add up to a major revision of thought or practice over many years. Other times, the pressures build as the research hits one major snag, and then the pent-up energy is released in a sudden lurch, and we identify that point in the process as the epicenter of the discovery. So, in essence, we ourselves have been trapped by the terminology—indeed, by Richter's question concerning the universality of illumination—into thinking of it as separate when it isn't.

No offense meant, Richter. You yourself pointed out that some problems are artifactual and can't be solved until the false assumption is exorcised. And you made the point that artifactual problems are much more common than we care to admit. So we tested your argument—and our own assumptions—and found them wanting. Good. Now we have a new way of looking at insight.

▷ **Richter:** Do we? I am not convinced yet. You claim to have solved the puzzle of illumination. All you have really done is reformulate the ques-

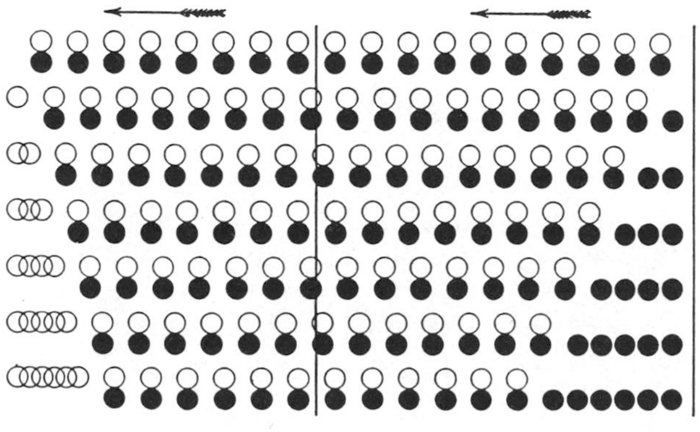

Hittorf's model of electrolysis, 1858.

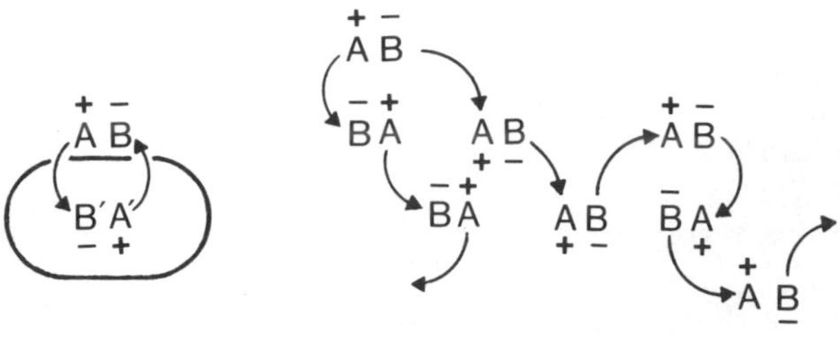

Arrhenius's kinetic model of electrolysis based on the Clausius-Williamson hypothesis. (After Arrhenius, 1884)

tion and its boundary conditions. Fine. Now we need Imp's research and rethinking. How does the process work? Where do the assumptions come from? Where is your model?

▷ **Ariana:** First, I think we'd better get straight exactly what the elements of illumination are, so we won't disagree over the outline of the subject.

The key to illuminations, as to all discoveries, seems to be the anomaly or unsolved problem. Tremendous effort (in the cases of Darwin and Poincaré, for example, years of effort) may precede the solution. During this preparation period all the obvious solutions are essayed, fruitlessly. The researcher begins to go in circles and finally lets the problem go. He or she sleeps, goes on vacation, takes a walk, goes off somewhere to get away from it all.

Now what happens? This is crucial. As far as I can tell, all of Constance's examples are characterized by one of two things: either something forgotten is suddenly remembered, or an untried avenue is explored. Why these two? Because we compartmentalize things, build taxonomies and hierarchies. We hold preconceptions, probably picked up through paradigmatic teaching or thematic preference, of what are acceptable solutions. Somehow these rules, patterns, preconceptions—these compartmentalizations—have to be broken down so that previously disparate ideas can meet.

▷ **Richter:** Consciously? Or unconsciously?

▷ **Ariana:** What's the difference? Is driving a car or riding a bicycle a conscious act? When you're first learning, absolutely! Probably interferes with actually doing it. After you've practiced, it comes naturally—but that's another misleading term. No human being knows from birth how to ride a bike or drive a car. The "naturalness" comes only when the pattern of responses has become so imbedded that there is nothing novel in them anymore.

In the same way, you learn to invent consciously, but after you've learned it becomes habitual. And that explains the facility that someone like Pauling acquires with practice. Sure, he can prime himself every night and then wake up one day with an answer out of nowhere—he's no more aware of what he does than I am of how I play the cello or manage to hit a tennis ball. But that doesn't make the process any more mysterious, or any less amenable to being taught.

▷ **Richter:** And yet you persist in circumventing the issue of what is to be taught and how. I can only conclude you really do not know.

▷ **Hunter:** Well, we told you last week we were having problems understanding illumination, at least in the sense you seem to mean. We can describe the process. We can't explain how it actually occurs.

▷ **Jenny:** What about Imp's stereo vision analogy for how he suddenly understood Fleming's lysozyme work?

▷ **Richter:** No. Insufficient. Oh, it has some of the elements—connecting information in a pattern, a surprising result, an added dimension of information—but it's too much like Koestler's bisociation. Put two unlikely things together and you get an invention.

▷ **Imp:** Now hold on, Richter. There are only a few combinations that will yield that jump from 2-D to 3-D, so it's certainly more rigorous than the Koestler-Rothenberg kind of thing.

▷ **Richter:** Granted. But it certainly doesn't allow for a sudden shift in meaning. There's only one way to make the image. In that respect Kuhn's gestalt shift is a better model. At least it shows how a data set can mean one thing from one perspective and something else from another. None of the models, however, allows for the single most important element of invention that we've discussed so far, the amazing diversity of possible solutions scientists generate. It comes to this: You must explain not only how one gets from one image to another, but why a diversity of images is possible at one instant in time, and why none of these indicate the possibility of the new image that suddenly emerges.

▷ **Jenny:** Maybe they do but we don't perceive it.

▷ **Ariana:** Look, Richter. At least we've gone somewhat further than the existing literature in redefining the problem. According to you that's more than half the battle. Unless you've got the solution yourself, lighten up, okay?

▷ **Imp:** Yes. Despite your earlier admonition against my—shall I admit it?—indiscreet silliness, surely what we've just discussed indicates a need for relaxed playfulness rather than more serious analysis. What we need are more puns, more fun, more games!

▷ **Jenny:** New ways to surprise ourselves. In the meantime, how about some coffee to stimulate our brains?

▼ JENNY'S NOTEBOOK: Modeling Illumination

As with so much scientific research, the direct approach often gave no results; one had to get at it, so to speak, from behind.

—Naomi Mitchison, novelist

My father, J. J., often maintained that a certain amount of interruption was good and almost necessary.

—G. P. Thomson, physicist

We all agreed that we'd reached a dead end on the topic of illumination, but no one really wanted to let it go. Everyone got up and stretched and a few of us yawned—perhaps, as dogs do, out of frustration. Constance offered to help me get the coffee (still a woman's job). Imp, Richter, and Hunter discussed the merits and demerits of a recent paper claiming that the complementary strand of the DNA coding for some hormone contains a peptide mimicking its receptor protein.[1] Something to do with how we understand the genetic code. Ariana, who was the only one who really knew anything about hormones, half-listened and, as always, doodled—that is, if you can call it doodling. The drawings and caricatures she whips off in a couple minutes are far better than anything I could turn out in a month. Some people just think in lines, she says. Some people just have talent, I say.

That's when it happened. I was setting down coffee for her as she drew a random array of dots and began connecting them with lines. It's an exercise I've seen her use when she can't think of anything specific to draw. She pulls the most amazing figures out of the dots. Just like those crazy periodic patterns scientists have pulled out of the data on elements, I thought. And then it hit me. "I've got it!" I exclaimed. I was really excited. "I've got a model for illumination! I think . . . maybe I can even take you through the process itself. Sit down, everybody." They looked at me like I was crazy. After all, I wasn't the person they'd have voted most likely to solve the problem. But that's part of the surprise of discovering, too.

I ripped a piece of paper off my pad. "All of you've probably seen this before, but I guarantee you haven't seen it used this way. The familiar suddenly perceived as unfamiliar. Conscious elaboration of diverse answers. Unconscious assumption of logical rules that precludes perceiving the correct answer. Surprise. The sublimity of the mundane. Elements of the gestalt switch. All those periodic tables Constance showed us earlier this afternoon. Roll it all together and here's what you get!"

I briefly explained how my own illumination had occurred while watching Ariana doodle. "I suddenly remembered a mathematical puzzle Ariana had shown me a couple months ago. I don't know what it's called, but you put nine dots in a square like this, and try to attach them with five straight lines."

"Of course!" cried Ariana. "Now, why didn't I think of that?"

"Because, like the rest of us," replied Imp, "you've been trying to get at the answer directly, without detours or glances to the side."

"Whereas," I added, "I wasn't actually trying to solve the problem at all. I was simply aware of it and of all the criteria necessary to recognize a solution. But I had no expectations to fulfill or preconceptions to overcome."

"Or," suggested Hunter, "perhaps because the puzzle is so familiar to you, Ariana, that you don't consciously think about it anymore. It's still something quite novel, fascinating, and problematic to Jenny."

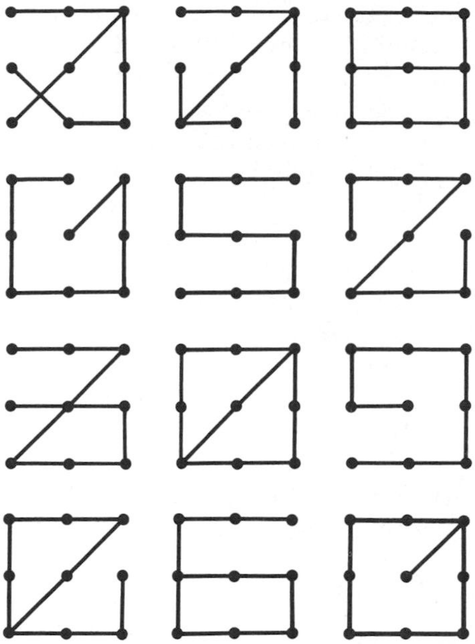

Five-line solutions to the nine-dot problem.

"Fine," interrupted Richter. "But the object is to solve the puzzle with four lines, not five. And there are other rules. You may not lift your pencil or pen off the paper or go over any line twice."

"Look," responded an annoyed Ariana, "who had the insight, you or Jenny? Let Jenny dictate the rules, please." Richter sighed and waved his acquiescence. If we insisted on doing it wrong, it wasn't his fault if nothing came of it.

I continued. "Forget the rules Richter just enumerated. There's a reason I want you to connect the dots with five lines first. Do it any way you can imagine. Lift your pen. Go over lines twice. Do whatever comes to mind. And if you already know the point to which this is leading, pretend you don't—because it isn't what you expect. That's one of the characteristics of illumination, right? Encountering the unexpected." All of us but Richter began drawing. A minute or two later, I called a halt.

"Okay, that's enough. Let's see what we've got. It looks like we've come up with about a dozen possible solutions, some by lifting the pen off the paper, some by never doing so, and so forth. There are probably more, for all I know. But it doesn't matter. What struck me a few minutes ago is that we've seen this variety of solutions before—with the periodic tables."

"Of course." Constance nodded. "And with the benzene rings I dug up."

"Not to mention the diverse models of DNA structure and the various rearrangements of the genetic code I've been playing with," added Imp.

Ariana reminded us of the many attempts to portray Saturn she'd shown us the evening we discussed Herschel and the art of perceiving.

"Multiple solutions," I emphasized. "General consensus about the nature of the solution but specific dissent about its details. This game is exactly analogous to what all of you keep telling me scientists do with data. Even given stringent rules about how many connections you can make and how they can be made, look at the possibilities! And just as with choices between the benzene models or the DNA models, it all comes down to what criteria of choice you prefer. For example, you might interpret an injunction against lifting your pen off the paper as a desire for coherence. Imp's always talking about minimizing your hypotheses, and making sure that what we know about, say, serotonin receptors corresponds with what we know about serotonin antagonists. It's all got to fit. We have the same problem in history. Imagine what would happen if we were allowed to explain every fact with a separate hypothesis—we'd never get anywhere. Nothing would ever connect. No principles would emerge. So you might insist on the greatest coherence possible."

Hunter nodded his agreement. "Or, if you're dealing with equations, you might want all your equations to be symmetrical with respect to physical operations—or asymmetrical, if you're dealing with reactions irreversible in time."

"In other words," I continued, "my model explains why scientists utilize aesthetic criteria." I smiled at Ariana as I said this, and awaited Richter's bark. It didn't come. "What? No objection, Richter?"

"I am restraining myself. To a purpose. I am not sure I would call coherence—you might add internal consistency in applying one's rules, simplicity, and so on—an aesthetic criterion. Logical, perhaps. But since I too am puzzled by the diversity of answers scientists invent, I won't quibble at present. If you have thoughts on the matter, please proceed."

"That's progress already. Okay. Think about the possibilities. If I can't lift my pen, for example, I have to eliminate the top three models. If it's got to be symmetric, another batch goes. If the lines can't cross, still other solutions are eliminated, right? So for any set of data there will be a variety of possible solutions, and the only way to discriminate between them is by the kind of extrascientific aesthetic criteria that Poincaré and Pauling relied on. Each rule we invent for the game represents one of these aesthetic criteria. The object isn't to find the right answer but, as Pauling said, to eliminate all the impossible answers."

Ariana was obviously excited. "And like Holton's themata, there's no logical way to justify these criteria except to say that we prefer simple solutions to complex ones, or symmetrical ones to asymmetrical ones. They guide our stochastic search into the unknown while simultaneously delimiting it."

"Which suggests a role for culture in forming these aesthetic rules," I went on. "There's no *a priori* reason an American should prefer European music to Japanese music except that it's what we're used to. In fact, Americans raised in Japan, or descended from Japanese forebears, often retain a preference for Japanese music. That means every culture, every language, every method for analysis contains not only certain potentials, but also certain limitations. Constance's point about Fleming's runny bottom–runny nose pun, for example.

"Okay, next step: invent a new aesthetic challenge. Try to connect the nine dots using only four straight lines without lifting your pen off the paper. In other words, try to solve the problem as coherently as possible." This time Richter absentmindedly sketched the answer.

Constance, however, was puzzled. After a while she asked, "Are you sure this is possible?"

"Absolutely," Ariana assured her.

"Oh, of course," said Imp a few seconds later. "I knew there was a trick."

"No. Not a trick!" exclaimed Ariana. "That's the problem with discussions of discovering—we use too many meaningless words. You don't solve it with a trick; you solve it by realizing that as a result of prior practice you've invented a tacit rule for which there's no justification but prior practice."

"Which is?" Constance asked.

Ariana smiled. "What shape do the nine dots define?"

"A box," Constance replied. "Isn't that obvious?"

"So obvious, you have to work at it to realize that it isn't necessarily true. The obvious answer is often the wrong answer in science. Remember Imp's repeated admonitions about the problem of determining how many pieces there are to your puzzle? Are the dots the complete figure, or just part of a large figure?"

"Oh, I see," said Constance a moment later. "You can go outside the box!"

"Surprise!" Ariana laughed. "You see what happened, of course. You succeeded in solving the five-line problem working within the box, so you assumed that you could solve the four-line problem utilizing the same criteria. But you can't. Jenny's absolutely right: That's what happens to scientists all the time. We learn paradigms of problem solving

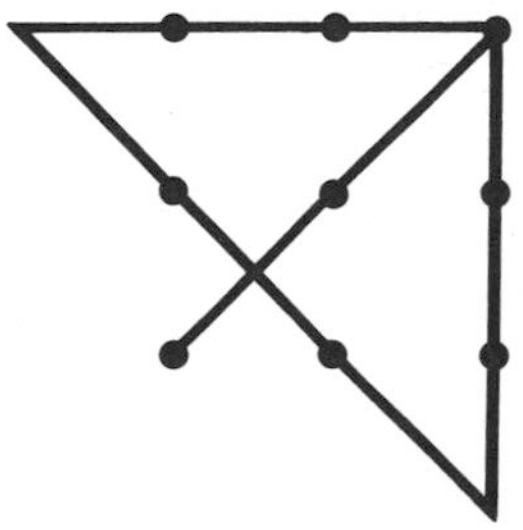

Coherent four-line solution.

and try our damnedest to make them work, even on problems where they can't. Not until we relinquish this acquired set of logical operations by questioning the unstated assumptions underlying them can we solve the recalcitrant problems—and surprise ourselves in doing so."

"Which is why this is such a good model for illumination," I added. "It explains why scientists so often report abandoning a problem after many futile attempts to solve it and then suddenly perceiving the answer in a moment of relaxation. As long as you consciously employ the paradigm you expect to work, you lock yourself inside a logical box that's inappropriate to your problem. Not until you abandon your preset notions of what you think the answer ought to be—not until you relinquish your control over how to achieve the answer—can your mind do what it does in Imp's stereo vision analogy: allow the pattern to form by itself."

"Well, I'm not sure I'd go so far as to relegate the problem solving totally to the unconscious mind," Ariana objected. "Not only do we have Pauling's testimony that he programs his brain to do this sort of thinking, but it seems likely to me that we could learn to do such problem solving quite consciously by playing the sorts of games Imp advocates. Once you hit a snag, stop trying to solve the problem itself and start examining your preconceptions and assumptions one at a time to see what happens if you abandon or alter them. Look around, as you did, for apparently unconnected things that might provide useful analogies. You see, it's not the data that create problems of comprehension, but what's inside your mind that you take for granted."

"Interesting," said Hunter. "That might explain several things. For example, Murray Gell-Mann tells a story about giving a lecture on why conservation of isotopic spin failed to explain the long lifetime of baryons—ignore the jargon. The point is, everyone who had addressed the problem assumed that the values of the isotopic spin, I, had to be multiples of ½. So just as he was about to say, 'Assume $I = 5/2$' to illustrate why such values don't work, for some reason he said 'Assume $I = 1$.' It was unthinkable! But before he could point out how ludicrous such a suggestion would be, he realized that this value solved the problem. Nobody had ever considered whole number values before because the 'superstition,' as he calls it, held that whole number values were unthinkable.[2] Yet, as you say, Ariana, anyone who had sat down and systematically assumed the 'impossible' or just played 'What would happen if . . .'—well, they would have gotten there, too.

"The same line of reasoning might also explain why so many important discoveries are made by scientists first entering a field, novices like Jenny, who's trying to think about scientific reasoning seriously for the first time. The novice's thinking hasn't become habitual yet. Everything's new and everything needs to be considered consciously, including questions of why this and not that."

"Suggesting," continued Imp, "that those most likely to rethink a field are those who start with different preconceptions about it, or who learn it from scratch their own way—the autodidacts we keep encountering. Speaking of which, you realize there's another set of possible answers to the five-line problem if we go outside the box? Look." Imp drew three more five-line solutions.

"Research and rethinking." I repeated the litany. "You see now why I was so excited when I perceived this puzzle as an analogy. Everything suddenly fell into place."

"Sure," said Constance. "I can understand that. But I'm not entirely clear what the analogy is. I mean, I have a sense of the general idea, but—"

"Then let's spell it out," I responded. "Let me begin by proposing some definitions. Let the dots be data points; and the lines connecting them, predicted lines of evidence that connect these points. The pattern formed by the lines then represents the hypothesis. You test the hypothesis by searching for the data points predicted by the lines. In essence, these lines tell you where to look for the evidence and how it should connect with existing evidence. Does that make sense?"

Constance turned to Hunter. "What was it van't Hoff said? 'The fact is the basis . . .'?"

"The Fact is the basis, the foundation; Imagination is the building material; the Hypothesis, the ground plan to be tested; Truth or Reality, the building."[3]

"Perhaps that's what I had in the back of my mind," I said. I wondered just how much of this process was conscious. The problem and the criteria for its solution were conscious, but were all of the data? Or was I just really testing it now? No, because I'd known immediately that most of the data fit. I was elaborating the theory and testing it at the same time, as in the cycles of Constance's model. "Let's reinterpret our various solutions to the five-line problem in light of these definitions. Given any of the fifteen solutions we've got, which one do you choose? Clearly, as long as only nine data points exist, your choice will be determined by aesthetics, right, Ariana? Further testing will eliminate only a few possibilities because many of the patterns predict the exis-

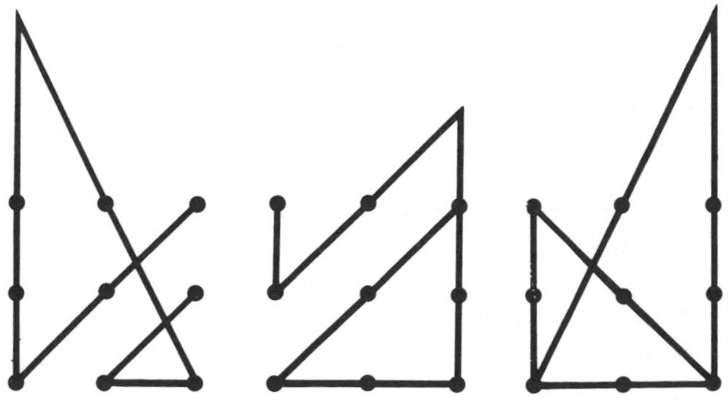

New five-line solutions suggested by four-line solution.

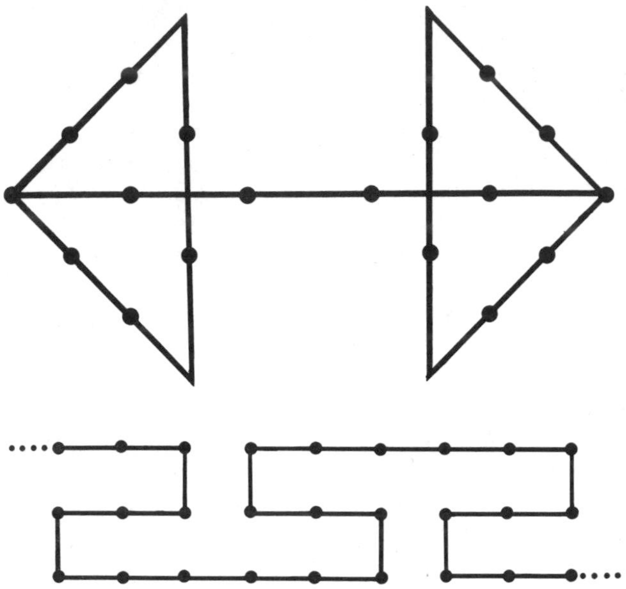

Limitations on dot-set linkages imposed by various solutions.

tence of data along the same lines. In fact, until you have elaborated all of the possible solutions within a given set of criteria, you can't devise adequate tests to differentiate among alternative explanations."

Imp was too impatient to wait any longer. "And even then, you may fail to invent some of the possible solutions by failing to check your assumptions. We thought we had most of the five-line solutions until the four-line solution made us go back and invent new ones.

"But you see what this does to our standard descriptions of scientific method, don't you? It shoots them to hell! Induction can't solve a problem or even tell you how many possible solutions there may be. Deduction isn't sufficient because it tells you what may lie between existing data points only after you've got a pattern. If you have the wrong pattern, your deduction is useless. And how do you invent the patterns in the first place? Falsification doesn't work, because many patterns predict the existence of the same data, and none may predict the existence of the data needed to falsify the whole class of solutions. Which also suggests what's wrong with Feyerabend's anarchist approach to science. Try anything, he advises. Hell, look at the number of points that are predicted by *no* pattern. You could observe billions of things without once hitting upon anything resembling a series or a pattern!"

"And," added Ariana excitedly, "it explains why guys like Kelvin, Michelson, Kekulé, Burnet, and whoever else could claim that their field had nearly reached closure. Within any given pattern—that is within any given set of conceptual assumptions—it *is* possible to get closure. But closure simply means that that particular approach to the field is played out and a new set of assumptions is needed to reopen it. Einstein's sort of thinking. It's all in the assumptions!"

"On the other hand," Hunter cautioned, "Jenny's model also incorporates a lot of previous accounts of theory building and choice, so let's not claim more novelty for Jenny than is appropriate. Here are some things I've thought of at random. Brillouin stated a number of years ago that new theories will always overlap old ones so as to preserve the relations between observable quantities—Jenny's data points—while disagreeing solely upon nonobservables, the specific pattern of lines. Therefore, one of the things a new theory must do is produce proof that it accounts for observables as well as or better than existing theories, and that all other points of difference pertain to nonobservables—assumptions or aesthetic criteria, for example.[4]

"In keeping with this methodological imperative, I'd also say that Jenny's model adequately integrates Kuhn's concept of normal science—the sort of confirmatory or developmental science one does to demonstrate that the predicted line between two data points exists. At the same time, the dot model provides a more than adequate alternative to Kuhn's gestalt shift model of sudden paradigm shifts. Moreover, it allows a multiplicity of answers, not just two, to exist concurrently. As Poincaré and Duhem argued, any limited set of data will have many possible solutions. And Jenny also integrates into her model another observable that's essential to Kuhn's description of science, but which he was unable to fit into his gestalt analogy—the existence of the anomaly that provokes the paradigm shift. In this instance, I presume it would be the discovery of a point lying outside the boundaries of the box. Say one at the corner of the triangle."

Ariana concurred. "Jenny's model also allows people to disagree on what's the best answer, based on aesthetic and sociological criteria. If a scientist already has a model of the world that works for her, then she's not going to change it just because someone can incorporate another point that lies outside that model. Right? Which gets us back to what Constance told us about Helmholtz, that he was often dissatisfied with particular solutions because he couldn't see how the particular solutions exemplified some more general principle. That's what's missing from the model so far. We haven't incorporated the idea of global thinking yet—yet, I repeat, for it's quite possible to do so. You add a criterium to your search for solutions that says something like: 'The best solution will be one that demonstrates connections between as many *sets* of data points as possible.'"

"Meaning exactly what in terms of the figures we've drawn here?" asked Constance.

"Well, one of the five-line solutions is a closed box. This would be equivalent to saying, 'Everything of importance lies within these boundaries.' On the other hand, the S-shaped five-line solution can be linked to other S-shaped five-line solutions to form a long chain. It offers paradigmatic possibilities, applications to other data sets."

"Or," interjected Imp, "put another way: The box is a particular solution, but the S-shape is algorithmic—it applies to all such data sets."

"The four-line solution has only one connection point," Ariana continued. "Put two four-line solutions together, and the system becomes closed—very limited applications. The same limitation I see with bisociation and Janusian thinking. It's not a matter of putting two ideas together or melding opposites; it's finding patterns linking as many data points or data sets as possible."

"Yet that is just where your aesthetics gets you in trouble," warned Richter. "The four-line solution may be simpler, but some of the five-line solutions provide greater possibilities for integration. You end up with conflict between criteria. Which do you adopt?"

"Isn't that exactly what many scientific controversies are about?" queried Hunter. "Not about the data, but about what criteria we use to evaluate the data."

Ariana thought for a moment. "In this case, perhaps neither set. As Imp said about trying to understand Fleming's penicillin work, when none of the available theories suffices, it's time to invent a new one. If the object of science is to make coherent as much knowledge as possible, then when simplicity and connectedness come into conflict, it's time to reexamine your assumptions yet again. Search for an even more elegant solution to the problem: one that requires fewer independent data sets or variables, one that integrates even more data sets and theories. The sort of thing I hear the physicists are trying to accomplish by inventing a unified field theory. And we can model that sort of scientific change, too. First, to get from five to four lines, you have to realize you can go outside the box. There's a three-line solution, too, that can integrate an almost infinite set of boxes."

"Obvious, at this point," said Richter. "Question other assumptions. Say points are not geometrically dimensionless. The lines pass through the points at slight angles that will meet before infinity."

"Or," suggested Imp, "assume a non-Euclidean geometry in which lines meet at infinity. Then the lines can be parallel."

Constance shook her head. "But that's not a real answer. You can't actually do it."

"That's what's fun about science, Constance," Hunter said. "You let your imagination get carried away, and every once in a while you find that you've come up with something eminently real. Einstein's thought experiments, for example. Speaking of which, although I haven't been able to imagine a two-line solution, there is a one-line solution to the puzzle."

"Let me guess," said Imp. "You imagine the nine points on an infinitely large sphere or tube intersected by a single line in a spiral around the sphere. Since the sphere is infinitely large, the line will appear to be straight, and so a single line will pass through all nine—for that matter, an infinite number of points."

"Or you plot the square in *n*-dimensions and send a circle through them," rejoined Hunter. As Piaggio wrote many years ago:

'If the path of a planet you'd trace, / You've Christoffel's weird symbols to face, / For an orbit, you see, / Is as straight as can be / On a surface in quintuple space.'[5] Another limerick for your collection, Jenny."

Ariana wasn't impressed. "No need to be so fancy," she said. "There's a physically realizable one-line solution suggested by Paul McCready—the guy who invented the Gossamer Albatross and the flying pterodactyl model for the Smithsonian. He pointed out that if you crumple up a piece of paper enough times, each time spearing it with a pencil, eventually you'll pass the pencil through all nine dots. Probabilistic thinking."[6]

I suddenly had another insight. "Hey! Then there has to be a no-line solution."

"Impossible," grunted Richter.

"No, no, I'm serious," I countered. "As long as we're playing games with space—questioning assumptions, inventing new rules, or whatever you want to call it—why not just tesseract!"

"How in god's name does one tesseract?" asked Richter.

"You read Madeleine L'Engle's *A Wrinkle in Time.* A children's science fiction story. I know that's not considered valid scientific fare, but an idea's an idea, regardless of where it comes from, right? L'Engle imagines a universe in which the shortest distance between two points is not a line, but the folding of space so as to bring the two points together. So why not simply fold up the paper, origami style, to bring all the points together at one point?" It was surprisingly easy to do.

"Yes," said Richter, eyeing me thoughtfully. "Well, I can see no objection. But what has this to do with discovering?"

"But, Richter, this *is* discovering!" admonished Ariana. "This is the systematic consideration of alternative assumptions to elaborate all possible hypotheses that we've been talking about all these weeks.

"Or do you mean, what does this have to do with discovering in biology rather than in mathematics or physics? You haven't read Bronowski, have you? He says the only difference between biological and physical theories is that the physicists have more fun.[7] Since physical phenomena are well established, physicists can play with conceptual underpinnings and assumptions. Their concepts can become unrealistic in the very way our solutions to the nine-dot problem became progressively less realistic—until, suddenly, Jenny popped up with the simplest, most realistic model of all: a piece of folded paper. Or "string theory" perhaps, whatever that means to a physicist. Pattern forming is pattern forming in any discipline. But you and I as biologists spend all our time trying to figure out what the points in the pattern are. We haven't yet begun to play with the conceptual underpinnings, save in evolutionary theory and a few other areas. And perhaps that's why our vision is still so myopic. Not enough imaginative possibilities to test our mettle and enlarge our horizons."

"Long live imagination!" shouted Imp.

Richter actually nodded agreement. "We do have far fewer principles—or at least principles worthy of the name—than we could use. Some conceptual subversion would not be amiss."[8]

Constance looked pained. "Getting back to the point about differences between physics and biology, I really can't agree that the level of conceptual play is the only difference between the biological and physical sciences. There are others—historicity, for example, and teleonomy—which we might want to consider at some point. Though the process may be the same for all scientists, what they need to bring to it is very different. And their historical development is certainly different."

"Now, wait," I said, trying to regain control of the conversation. "We haven't finished with the model. There's at least one other aspect of the nine-dot puzzle that's typical of scientific discovering and especially of illumination. You either perceive the entire answer to a problem or you don't. As Diderot said, there's no such thing as part of a thought. Extrapolating from the five-line solutions doesn't yield a four-line solution. Nor does any finagling with the four-line solution yield the solutions with three, one, or zero lines. So you can't proceed step by step. At some point you simply have to rethink the whole thing, starting with your assumptions, or you get stuck in the box forever. It's all-or-nothing, like Kuhn's gestalt model and Imp's stereo perception model."

"In fact," said Imp, "you could combine your model with mine, Jen. Assume the points aren't necessarily in a plane, but can exist in three or even *n* dimensions. Then in some cases a new solution will result only if disparate sets of points—or different patterns—are juxtaposed in just the right positions to yield that sudden leap into the next dimension. That's it—I just figured out what we've been doing!" I suddenly realized he wasn't going to let me have this insight all to myself. "Botryology! A word invented by I. J. Good, meaning the science of defining and finding 'logical clumps' in sets of objects.[9] Science is a botryological game."

Richter snorted. "Or a taxonomic one. Why invent a new word when a perfectly good one exists already?"

"To emphasize the fact that the taxons are not external, observable properties of objects, but logical abstractions concerning them," replied Imp. "The message, the logical clump, you invent is dependent upon the assumptions you're willing to entertain. And I think you'll agree that there's no *a priori* reason for preferring one set of assumptions over another, until we invent ways to distinguish between them."

"So we may conclude that science is the elaboration of all imaginable explanations of the world, successively constrained by an ever-increasing body of tests," Ariana suggested.

"In which the elaboration of possibilities represents science in the making, and the body of tests, theories, and techniques represents codified science," Imp added.

"And the individual personality the coordinator-permuter-inventor," I proposed.

"I presume, then," said Constance doubtfully, "that the social context sets the limits on acceptable problems and aesthetic criteria? But now I'm confused. Which of these analogies or models is right? I mean, we've got at least two now: Imp's stereo vision model, which he used to explain his Fleming insight, and Jenny's mathematical puzzle model. Which one are we going to use?"

"Both of them," replied Imp, "and neither of them. That's the whole point. We don't want *the* answer; we want all the possible answers. This is when the playing begins! The worst thing that can happen to an emerging science is to reach a premature consensus.[10] It's too easy to stop doing research when you've hit on the first plausible answer. That's what happened with Kuhn and gestalt shifts, and Koestler and Rothenberg with their bisociation and Janusian thinking. On the contrary, the first answer should be taken as a sign that many other possible answers also exist.

"So, good: We've finally come up with a way to model illumination and the surprise of discovery. An answer's only a beginning." I was beginning to get angry. Hadn't I accomplished anything? "Now we want to imagine every conceivable way to think about the process, so that we account for all the important aspects and successively define the area of overlap within which the correct answer must lie.[11] And we want to portray our answer in as many ways as possible so that people of very diverse interests and backgrounds may use it. Richter, I know, isn't too keen on pictures, so why insist on a visual model if an adequate verbal or mathematical one can be invented?"

"Maxwell's philosophy exactly," said Hunter. "For him a physical illustration—by which he meant a mechanical model or a diagram—was much more robust and vivid than any set of equations. J. Willard Gibbs was the same way. Yet both of them recognized that for other scientists the reverse was true. They therefore maintained that scientific results should be expressed in as many ways as possible—verbally, graphically, and mathematically."[12]

"Emphasizing once again the point we made last week that all of these expressions can be transformed into the others," added Ariana.

Hunter continued. "It's simply a matter of recognizing that people have different styles: logical, analytical, algebraic on the one hand, and visual, modeling, geometrical on the other.[13] People of each type must be addressed in the style they best understand.

"I also agree with Imp's point that fuzziness is an important part of new theories. De Broglie warned us that the more precise and rigid a concept becomes, the more limited its sphere of application; and Eddington even suggested that we should avoid final answers, for once an answer is codified there's nothing further to be learned by investigating along those lines.[14] The field approaches senescence when the box becomes too rigid."

I'm not sure how much of this I heard at the time (thank heavens the tape recorder was left on). I was too mad. I was proud of my contribution, and Imp was making hash of it. I couldn't wait for things to end so I could get him alone.

"Hmmm," said Constance. "That explains something that always struck me as paradoxical. Kuhn has been pilloried for years for his failure to define what he meant by 'paradigm.' Someone counted up the various ways he used the word in his *Structure of Scientific Revolutions* and came up with something like a hundred and fifty. On the other hand, I've always thought that had he been too specific, no one would have been able to use the book. It offered the potential for each reader to select from the range of meanings those most useful to her or him, and then to connect them to other ideas in ways unforeseen by Kuhn himself. In fact, Gustave Nossal once commented that every new theory is sufficiently fuzzy to allow an intelligent person to pick any number of holes in it, but that such critical actions are detrimental to progress."[15]

"Clearly," agreed Ariana. "We all know it's easier to criticize than to invent, so we've got to learn to judge novelty not by its faults but by its possibilities."

"Which brings to mind Benjamin Franklin's reply to a critic of one of his new ideas. 'What good is it?' the man asked. 'What good is a baby?' Franklin responded."[16]

The conversation penetrated my silent fuming and I realized that Imp wasn't really to blame. I was jealous of my idea the way a mother is jealous of a newborn. Yet the greatest praise anyone could offer was to want to help it mature. All too soon, it takes on a life of its own, over which the progenitor has no control. Misunderstandings, misapplications, transformations are part of the process too. I was still the originator, and everyone knew it. As Hunter said, a discovery isn't a discovery until it's used by somebody. My anger evaporated. "So where do we go from here?" I asked.

Constance made a tentative motion. "Could I add one more thing? I just want to suggest that your illumination model might be useful for explaining theory and data reception, too. The basic idea is that people who already have a pattern in their heads won't want to replace it with another. Tacit rules or patterns of expectation prevent them from perceiving or perhaps even considering alternatives. Studies of the reception of new theories—Darwinian natural selection, the continental drift theory in geology, atomism in chemistry, for example—have concluded that all novel ideas and results meet what Bernard Barber calls 'resistance to knowledge.'[17]

"A lot of this resistance seems to arise from conflicting themata and preconceptions. For example, M. J. Mahoney sent out a set of seventy-five anonymous manuscripts to psychologists. Every manuscript had an identical introduction, methods section, and bibliography. The only differences were in the results and conclusions, which were adjusted to be congruent with or contradictory to the referee's known theoretical stance. Mahoney found that congruent results were almost always recommended for publication, but antithetical results rarely were.[18] Similarly, after studying the differences in the way relativity theory was received initially in Germany and Great Britain, Paul Forman concluded that 'reponses were . . . conditioned largely by prior expectations, predilections, and prejudices . . . people, and physicists, not only tend to find what they are looking for, but also fail to recognize what they are not prepared to see.'"[19]

Imp interrupted again. "Which is why successful frauds always involve the presentation of expected results to a breathlessly anticipatory community of previously committed believers. They don't see the flaws because they don't look for them. The results fit the pattern of their expectation. Nobody's ever pulled off a fraud involving data or theories contradicting the expectations of the scientific community. How could they? Nobody believes them anyway! So the reviewers have it backward. They should distrust most the results that fit their expectations and consider most interesting the ones that don't."

"My point throughout our discussions," Richter remarked.

Imp continued. "But look, let's not get into peer review now or we'll be here all day. It's already getting late—in fact, too late to have our usual second discussion. But Jenny's right: Where *do* we want to go from here? Peer review? Reception of scientific novelty? Fraud? That's another big topic. Personally, I'd like to address some of the literature that's coming out of the artificial intelligence labs on the possibility of discovery machines. Wild stuff—just think what having a good model of the process might mean for AI research."

"Or how about whether the process we've described for scientific creativity is similar or dissimilar to the creative process in general? Learning to utilize the arts to teach non-discipline-bound skills such as pattern forming, aesthetics, visual thinking. There are tremendous implications for education and cognitive psychology." That was Ariana, of course.

"Whoa!" I interjected. "Let's not get carried away! I realize that the sign of a good research project is that it spins off more problems than it solves, but we can't impose too much on our friends, Imp, dear. Some of us have to start thinking about preparing lectures for fall term in a couple of weeks and, as much fun as this has been, it can't go on forever. I can't, at any rate."

"Nor I," added Hunter. "Unfortunately. Perhaps next summer, another session. In the meantime, may I make a suggestion? It seems to me we need to do a bit of codification ourselves, pull things together coherently. So how about each of us writing a report summarizing the most important things we've gotten out of these discussions and drawing out whatever practical implications we can? For example, I've been thinking about science policy, trying to imagine reasonable ways of encouraging exploratory research without revolutionizing the structure of science, and without unduly challenging Big Science. I've drafted a little paper I'm calling 'Obstacles and Inducements to Exploratory Research,' which addresses some of the relevant questions. If you'd like, I could circulate it."

"Send it to all of us," Richter said. "And then allow me to reciprocate. You have all had your say about how the process of discovering works. Now let us see if you have the courage of your convictions. You say we should elaborate all the possible ways of understanding our material? Good. Because for the last few sessions, I have sat here putting things together *my* way—a repatterning of the data reminiscent of Jenny's little exercise today. Now I, too, would appreciate a chance to present my case. Not as we've been doing, with dozens of interruptions, but as a unified report. My view, coherently presented, take it or leave it."

"Why not?" rejoined Imp. "I'd enjoy a chance to pull everything together from my perspective, too. Get at all those tactics and strategies for increasing the probability of discovery and summarize them in a little 'Manual of Discovering.' After all, that's why I invited you all here."

Ariana was for it as well. "I'd appreciate a chance to present a coherent case for the importance of personality and my various 'tools of thought' as determinants of what a scientist is capable of doing and perceiving, and present the educational imperatives that recognition of these factors would demand."

I begged off. "I'll pass, thanks—on a report anyway. I'll just try to put all the transcripts, notes, and slides and sketches in order for the rest of you. After all, documents are what I'm most comfortable with. What about you, Constance?"

"Well, I'm very busy at the moment. I have several complicated patents I'm trying to get finalized and filed in the next month. This project of yours has already taken up more of my time than I'd expected—not that I haven't enjoyed it. What I'm trying to say is that I can certainly try to prepare something on the role, or possible roles, of history and philosophy of science as scientific research tools. Seems to me we've run across lots of examples of people using old data, old theories, philosophical criteria, and things like that—more than I expected, anyway—so maybe the professionalization of these fields has created a schism that needs to be healed. Well, anyway—sure."

"Great! Then that settles it." Imp beamed. "I'm excited! Is three weeks sufficient? I've got a couple of conferences to attend after that. So, final session—till next summer anyway—three Saturdays from today. Just remember: We've all got to stick our necks out a bit. Push the data to their limits. Be daring!"

"You asked for it," Richter warned.

DAY SIX

Complementary Perspectives

The Mexican sierra [fish] has "XVII–15–IX" spines in the dorsal fin. These can easily be counted . . . We could, if we wished, describe the sierra thus: "D. XVII–15–IX; A. II–15–IX," but also we could see the fish alive and swimming, feel it plunge against the lines, drag it threshing over the rail, and even finally eat it. And there is no reason why either approach should be inaccurate. Spine-count description need not suffer because another approach is also used. Perhaps, out of the two approaches we thought there might emerge a picture more complete and even more accurate than either alone could produce.

—John Steinbeck, novelist, with Edward Ricketts, marine biologist (1941)

As a biologist you must have an idea of all levels [of order], so that you know where you are, and what you are talking about. If there was a creator, he was not a quantum mechanician, nor was he a macromolecular chemist or physiologist—he was all of these, and knew all of it. In the same way, we must know a little bit of all of it.

—Albert Szent-Györgyi, physiologist (1966)

►

JENNY'S NOTEBOOK: Asymmetry and Fault Tolerance

It seems strange not to have everyone gathering here on Saturdays anymore. One week till we meet formally again—though we've seen everyone in between, except Constance. She really must be busy.

Nothing too exciting to record. Richter dropped by a couple nights ago for an hour. He's keeping the contents of his report under wraps. Meanwhile, it seems he's had some wonderful scientific ideas—he actually looked happy—though Imp says you can never tell how these things will pan out. One's a way of thinking about organisms as fault-tolerant systems. Got the idea listening to Constance's story of the guy who invented fault-tolerant computer systems. Seems the idea has some merit. Then, while reviewing the Pasteur material, he apparently realized that although crystallographers have catalogued all the possible forms of symmetry, no one—at least as far as he could find in a preliminary search of the literature—has ever tried to systematize all the possible forms of *asymmetry*.

He and Imp had a big argument about whether it's a moot point or not. (Imp's a bit testy these days.) Imp seems to think that if you know the forms of symmetry, you know the forms of asymmetry—besides which, fractals (whatever they are) will solve all the problems of form, so why waste your time? But Richter, who knows a good deal more about mathematics than Imp, doesn't buy it. He pointed out that all symmetrical objects can be decomposed into asymmetric ones, so that instead of symmetry being the underlying basis of the universe, perhaps asymmetry is. We see so much symmetry for much the same reason that chemical reactions usually yield symmetrical compounds—you've synthesized aggregates of asymmetric subunits whose collective property is symmetry. Mentioned some example having to do with aperiodic tilings of polyominoes that I didn't understand. Then he gave Imp and me a lecture on some new kinds of math called catastrophy theory and fuzzy sets, dealing with events and observations that don't fit the standard symmetry patterns and Platonic assumptions of classical math. New themata promising to open up new universes to our imagination.

All sounds strange to me, but I found the argument interesting because it reminded me of eighteenth-century philosophers arguing about the perfectibility of nature. Is it all symmetrical and harmonious—

a clockwork universe—or are we on the verge of inventing a new aesthetic for ourselves that proposes asymmetry and imperfection as the natural order and beauty of the world? Figures Richter would take the latter position. Can't wait to see what Ariana's response will be!

N. B. Speaking of responses, I've been thinking about my dot model. It really works visually for me, as an expression of everything essential to the scientific discovering. Working through it brings me understanding of the examples I'd simply heard about or known of before. So I don't care what Imp says: You may have to elaborate all possible solutions, but you've also got to choose among them. Let others select what scheme they will: I choose mine!

◀ TRANSCRIPT: Reports

▷ **Imp:** Ah, Richter! Late again. But you always were one to prepare to the last minute.

▷ **Richter:** You would not have me suppress a thought to satisfy a social convention.

▷ **Imp:** Never! Ideas first, by all means.

Here's the plan. Each of us will present his or her report without interruption, as you stipulated, Richter. Comments and questions will be permitted after each presentation. Let's keep them brief—this is going to be a long day, no matter how exciting we make it.

As to order: Ariana has volunteered to go first, followed by Constance. They've promised to discuss who discovers and what educational, avocational, thematical, philosophical, and historical traits set revolutionary scientists off from their colleagues. Then you, Richter, must deliver on your promise to turn everything we've done to this point on its head. Prove to us that the master skeptic can be an inventor. Give us that unifying framework we've all been stuggling toward. Next, Hunter with his report on institutional and social obstacles and inducements to pioneering research. How can we help those who want to eschew Big Science and organized programs for individualistic and even eccentric exploring? And finally, in the place of honor—but in truth, to make sure we don't take ourselves too seriously—my own semihumorous report on personal strategies for living in the wildernesses of the unknown.

Ariana?

▲ ARIANA'S REPORT: Who Discovers, How? Personality and the Art of Science

> *The particular kinds of sensibilities required by a scientist are more complicated. Begin with his intense awareness of words and their meanings. While the poet's affinity for words makes him sensitive to their sound, emotion, and rhythm, the scientist uses them as instruments of precision. He must be capable of inventing new words to express new physical concepts. He must be able to reason verbally*

by analogy—to explain how this thing is like that thing, and to be able to fit the many resemblances into one single generalization that covers them all.

The scientist must also think graphically, in terms of dynamic models, three-dimensional arrangements in space . . . Scientists keep these three-dimensional pictures in mind as vividly as if they were actually seeing them. Formulas and equations printed on a two-dimensional page have three-dimensional meaning, and the scientist must be able to read three dimensions to "see the picture" at once . . . Unless a man has some kind of spatial imagination along with his verbal sensibility, he will always be—as far as science goes—in the role of the tone-deaf stuggling with a course in music appreciation. On the other hand, the possessor of both verbal and spatial sensibility will rather quickly be bored if asked to limit his imagination to only the verbal domain, in the case of the humanities; or to only the spatial domain, in the case of the graphic arts.

A man accustomed to working at the peak of his powers has no patience with anything that calls on him to work at only half load. With this dual sensibility then, the true scientist would find it difficult to be like everyone else even if he wanted to be.[1]

—Mitchell Wilson, physicist, inventor, novelist, historian of technology

My message is simple. What you accomplish is determined as much by who you are as by what you know. Mastery of facts and techniques alone does not make a scientist. The difference between a technician and a discoverer is imagination. The only way to develop this trait is to practice using it. In short, to be creative, you must practice being creative. You must master what I've come to call "tools of thought"—analogizing, pattern forming, pattern recognition, visual thinking (not just in three, but in four or even *n* dimensions), modeling (both mental and physical), playacting (in the sense of becoming the thing you study), kinesthetic thinking (feeling how a system functions), manual manipulation, and aesthetics (the sense of beauty in nature, be it expressed in images, sounds, equations, or words). The mind and senses alike must be trained to perceive and to imagine; the "eye of the mind" and the "eyes of the forehead" must be trained equally and in tandem. I need hardly point out that few, if any, of these tools of thought are in our standard science curricula.

Tools of thought, in and of themselves, are useless to a scientist until linked by "transformational thinking"—that is, the ability to translate a problem expressed in one form (such as numbers) into another form more amenable to problem solving (words, perhaps, or mental images); to mentally manipulate these words, images, or models to solve the problem; and then to translate this solution into yet another form (such as an equation or diagram or experimental protocol) that can be communicated to other scientists. Clearly, such transformational thinking requires practice, more than we now give students at any level of their education. And, clearly, transformational thinking is impossible in the absence of "correlative talents"—the interactive mental skills prerequisite to transformational thinking. Knowing how to invent poetical analogies, to draw what is observed, to model what is imagined, and to solve mathematical problems will not help a scientist unless she understands how these skills relate to one another and can use them to good effect.

Thus, I claim that what a person accomplishes is a manifestation of her or his personality—that is to say, the accumulation of talents, skills, experiences, desires, and goals that define the individual. Let me aver at the outset that this idea of the who and the how of science being inseparable is not new. Paul Bert wrote of Claude Bernard, a century ago, that "Nothing in his pure and harmonious life was turned aside from its chief aim. Enamored of literature, art and philosophy, Claude Bernard as a physiologist lost nothing by these noble passions; on the contrary, they all helped in developing the science with which he identified himself, and of which he is the highest and most complete embodiment."[2] Duhem, Sarton, Kubie, Nachmansohn, and Holton have generalized the principle to all scientists of genius.[3] Holton, for example, glosses Einstein's remark "I am a little piece of nature" with the comment that "there is a mutual mapping of the mind and lifestyle of this scientist, and the laws of nature . . . of the style of thinking and acting of the genial scientist on the one hand, and the chief unresolved problems of contemporary science on the other."[4]

How does this mutual mapping of mind and nature come about? Santiago Ramon y Cajal suggested that the scientist's personal style may be laid in the earliest years of life: "For my part, I have always believed that the games of children are an absolutely essential preparation for life; thanks to them the infantile brain hastens its development, receiving, according to the hobbies preferred and the amusements carried on, a definite moral and intellectual stamp upon which the future will largely depend."[5] Wittgenstein and Holton believe similarly that the foundations of the themata that characterize all great investigators are laid in our kindergartens.[6] That is, after all, where we first encounter our notions of what constitutes knowledge, how it is ordered and divided, and of what beauty and truth consist. In this sense, the curriculum, from kindergarten through graduate school, is but a reification of how a society construes and constitutes knowledge. Tacit in these curricular arrangements are lessons about what skills and ideas are deemed valuable and whether knowledge is unifiable or irreparably fragmented. These lessons are carried into practice by students who learn—or don't learn—the tools of thought, themata, and patterns of action that will determine their later accomplishments.

To understand the great investigator, then—and to train the great investigators of the future—requires an understanding not only of her scientific training, but of the education, skills, philosophy, ethics, games, hobbies, and passions that make the human being. Notwithstanding a few studies along the lines of the Guthries' *Contributions to the Study of Precocity in Children* and the Goertzels' *Cradles of Eminence*, we know next to nothing about the childhood and adolescence of most famous scientists.[7] As two disgruntled researchers commented, "We know what they did, but not what they were."[8] Worse, if we do know what they were, we ignore it. The nonscientific life of the scientist has been deemed by most biographers and historians of science—and certainly by philosophers and educators—irrelevant to understanding or teaching science.[9] Too often science is portrayed as objective and therefore free of both context and personality. This view is wrong. We must understand the scientist as a human being first if we are to understand the science she creates. Only then can we devise the new and appropriate education for imaginative scientists that Hunter has repeatedly urged upon us.

What are the personal and educational characteristics that differentiate eminent scientists from those who fail to rate even a footnote in history? Let's begin with the so-called average scientist. Most of us, I suspect, grew up with the stereotypical image of the scientist as unathletic, totally cerebral, incapable of action, bedecked with thick glasses, surrounded by books, and unable to communicate with any normal human being. Unfortunately there are enough scientists who fit this stereotype to keep it alive. Various psychological studies do show that, in general, scientists—actually male scientists (the psychology of female scientists has largely been overlooked)—are characterized by a need for autonomy, a narrow devotion to work, a drive toward rational control, a dislike for ambiguity, a respect for impartiality and masculinity, an avoidance of emotion, a correlated ignorance or rejection of the fine arts, poetry, and literature, a lack of social interests, a passivity in home life, a preference to work with things rather than people, and an inability to make up stories or to imagine fictional situations.[10]

However, the highest achievers among scientists do not fit the common mold. Several studies have noted that the most creative scientists and engineers manifest an unusually broad set of intellectual interests as youngsters, and are often very interested in art, music, and literature as adults.[11] Eiduson, for example, noted that the highest achievers in her group of scientists—and of the forty scientists she began studying in 1958, four have received Nobel Prizes, two others were nominated for that award numerous times, and one became a presidential science advisor—had a different profile from their less successful colleagues. These unusual "gentlemen of science," as she calls them, "seem to feel that they can extend themselves emotionally, not to mention physically, into the larger intellectual sphere of humanity without actually withdrawing time, energies and efforts from their work."[12] They participate in art, music, literature, politics, and social issues almost as fully as they do in the sciences. They are energetic men in both mind and body. They contribute not only to science but also to the integration of human knowledge broadly conceived. They are not typical scientists, and Eiduson comments that they perceive their function differently: "These men feel that science can be practiced differently without deleterious effects on the kind of progress it makes, or on the rate at which it progresses, or on the genuine inner intellectual motivations of the man . . ."[13] They believe in, and are themselves, mavericks.

The history of science yields a portrait of eminent scientists more compatible with Eiduson's "gentlemen of science" than the psychological portrait we have of the average scientist of today. For example, in 1893, Ramon y Cajal asked himself, "Who discovers?" The students who are narrowly focused, high achievers in scientific subjects alone? The monomaniacal prodigies?

> *A good deal more worthy of preference by the clear-sighted teacher will be those students who are somewhat headstrong, contemptuous of first place, insensible to the inducements of vanity, and who being endowed with an abundance of restless imagination, spend their energy in the pursuit of literature, art, philosophy, and all the recreations of mind and body. To him who observes them from afar, it appears as though they are scattering and dissipating their energies, while in reality they are channeling and strengthening them . . . of intellectual attributes there is no need to speak, for I assume that the newcomer to laboratory work is endowed with a normal mentality, a not inappreciable amount of imagination, and, above all, that harmonious coordination of faculties which is worth more than brilliant but erratic and unbalanced mental gifts . . . The investigator should possess something of this happy combination of attributes: an artistic temperament which impels him to search for, and have the admiration of, the number, beauty, and harmony of things.*[14]

Ramon y Cajal's answer is, of course, an expression of personal opinion—more: a description of himself. Yet he summarizes in a single paragraph most of the attributes discussed by Francis Galton a few years earlier in his influential work, *English Men of Science.*[15] Galton's study of the most eminent Fellows of the Royal Society suggested that the highly successful scientist had a practical nature, was persistent, independent, and curious, and was energetic both mentally and physically, as well as being mechanically and musically inclined. P. J. Möbius, grandson of the famous mathematician, presented a similar portrait of eminent mathematicians in his study.[16] And Wilhelm Ostwald depicted the scientific genius in the same way in his classic *Grosse Männer.*[17] All of the men he discusses—Davy, Faraday, Liebig, and so on—had unusually broad aptitudes and experiences which contributed to the range and depth of their science.

More limited in scope, but of greater relevance to my subject of tools of thought, is a study of famous scientists by van't Hoff. Hunter mentioned that as a teenager van't Hoff began reading biographies in search of the patterns of action that characterized the successful scientist. When he was appointed full professor at the University of Amsterdam in 1878, his first duty was to give an inaugural address, which he entitled "Imagination in Science."[18] In the lecture, van't Hoff asserted that "imagination plays a role both in the ability to do scientific research as well as in the urge to exploit this capability." This conviction, combined with his own musical and poetic proclivities, prompted him "to investigate whether or not this [imaginative] ability also manifests itself in famous scientists in ways other than their researches. A study of more than two hundred biographies showed that this was indeed the case, and in large measure."[19] Copernicus was a painter who also translated poetry. Galileo had intended to be an artist as a teenager, and wrote poetry throughout his life. Newton, too, painted and wrote poetry. Kepler was a musician and composer, as were Lacepède and the Herschels. Humphrey Davy wrote poetry praised by Coleridge. And so forth. Altogether, about 25 percent of his sample, including almost without exception the most important theoreticians and conceptualizers of the early modern period, engaged in nonscientific forms of creativity.

We can add to van't Hoff 's list from the men we have studied in these sessions. We've heard that Louis Pasteur was a gifted painter as a youth. August Laurent, the chemist who influenced Pasteur's first experiments, was likewise an artist and musician. Van't Hoff 's teacher Wurtz was artistic, and Kekulé was trained as an architect. Van't Hoff himself could hardly have imagined or worked out the consequences of the tetrahedral carbon atom without practiced skills of mental imagery. Are these men typical of the most creative and successful scientists? Are unusually important scientists unusual also in their personal skills, interests, and activities? If so, how do eminent scientists come by these traits, and how are these traits correlated with their success in day-to-day research?

To answer these questions, I began a sort of epidemiological study to determine the frequency of occurrence of artistic and inventive proclivities among eminent scientists. The tables I'm handing out display the results of my rather haphazard and incomplete search.

Note first of all that I have limited myself to nineteenth- and twentieth-century scientists and engineers. One could easily compile an equivalent list for Renaissance and Enlightenment figures, but that would hardly be surprising. Knowledge was not then divided into thousands of specialties as it is today. The point I want to make is that the Renaissance man of old—and now the new Renaissance woman—is alive and well.

I also want to stress that these are, without doubt, very incomplete lists. I have, for example, hardly investigated geologists, although I am assured by my geologist friends that virtually every geologist of note was a skilled draftsman, artist, and craftsman. After all, the geologist uses surface features of the landscape to imagine the three-dimensional structure underneath, and then invents a succession of processes that could

Artistic Proclivities among Eminent Scientists and Inventors (*Nobel Prize)

Scientist	Profession	Reference
Painters, Etchers, and Sketchers		
*Adrian, Edgar (1889–1977)	Physiologist	Haldane, 1961
Agassiz, Louis (1807–1873)	Biologist	Lurie, 1960
Argand, Emile (1879–1940)	Geologist	DSB[a]
Austin, James H. (1925–)	Neurologist	Austin, 1978
*Banting, Frederick (1891–1941)	Physiologist	Jackson, 1943
Barcroft, Joseph (1872–1947)	Physiologist	Franklin, 1953
Belar, Karl (1895–?)	Cell biologist	Goldschmidt, 1956
Bell, Charles (1774–1842)	Anatomist	DSB
Best, Charles H. (1899–1978)	Physiologist	Parergon, 1947
Bezold, Wilhelm von (1837–1907)	Meteorologist	Albers, 1975
Bolyai, F. (1775–1856)	Mathematician	DSB
*Boveri, Theodor (1862–1915)	Cell biologist	Baltzer, 1967
*Bragg, William (1862–1942)	Physicist	Personal comm.[b]
*Bragg, Lawrence (1890–1971)	Physicist	Jeffreys, 1960
Bright, Richard (1789–1858)	Physician	Chance, 1940
Brockedon, William (1787–1854)	Inventor	Wilkinson, 1971
Brown, Rachel F. (1898–1975)	Chemist	Baldwin, 1981
Brücke, Ernst (1819–1892)	Physiologist	Cranefield, 1966
Busemann, Herbert (1905–)	Mathematician	Dembart, 1985
Butschli, Otto (1848–1920)	Biologist	Goldschmidt, 1953
Carothers, Wallace H. (1896–1937)	Chemist	Smith et al., 1985
Carver, George W. (1864–1943)	Inventor	Holt, 1943
Cayley, Arthur (1821–1895)	Mathematician	Bell, 1937
Charcot, Jean Martin (1825–1893)	Neurologist	Parergon, 1947
Chun, Carl (1852–1914)	Zoologist	Goldschmidt, 1956
Cushing, Harvey (1869–1939)	Surgeon	Cushing, 1944
Cuvier, Georges (1769–1832)	Anatomist	Negrin, 1977
Dana, James Dwight (1813–1895)	Geologist	Viola et al., 1986
Darlington, C. D. (1903–1981)	Biologist	D. Lewis, 1982
Doflein, Franz T. (1873–1924)	Protozoologist	Goldschmidt, 1956
Donders, F. C. (1818–1889)	Biologist	Bowman, 1891
Du Bois-Reymond, Emil (1818–1896)	Physiologist, physicist	Cranefield, 1966
Duhem, Pierre (1861–1916)	Physicist	Taton, 1957
Dujardin, Felix (1801–1860)	Biologist	Snyder, 1940
Edgerton, Alfred (1886–1959)	Engineer	Hill, 1960
*Euler-Chelpin, H. von (1873–1964)	Biochemist	DSB
Faraday, Michael (1791–1867)	Physicist, inventor	Williams, 1965
*Feynman, Richard (1918–1988)	Physicist	Grobel, 1986
*Fleming, Alexander (1881–1955)	Bacteriologist	Maurois, 1959
*Florey, Howard (1898–1968)	Chemist	Williams, 1984
Franklin, Rosalind (1920–1958)	Crystallographer	Sayre, 1975
Frisch, Otto (1904–1979)	Physicist	Frisch, 1979
Fulton, Robert (1765–1815)	Inventor	Hindle, 1981

Scientist	Profession	Reference
Gamow, George (1904–1968)	Physicist	Gamow, 1966
Geddes, Patrick (1854–1932)	Biologist	Geddes, 1895
Goodrich, Edwin S. (1868–1946)	Anatomist	DSB
Gould, John (1804–1881)	Zoologist	Ritterbush, 1968
*Guillemin, Roger (1924–)	Physiologist	Personal comm.
Haden, Francis Seymour (1818–1910)	Physician	Zigrosser, 1955
Haeckel, Ernst (1834–1919)	Biologist	Haeckel, 1899–1904)
Henle, Friedrich G. J. (1809–1885)	Anatomist, physiologist	DSB
Henry, Charles (1859–1926)	Physical chemist	Arguelles, 1972
Herschel, John (1792–1871)	Astronomer	DSB
*Hinshelwood, Cyril (1897–1967)	Physical chemist	R. V. Jones, 1971
His, Wilhelm (1831–1904)	Anatomist	Parergon, 1947
Hodgkin, Thomas (1798–1866)	Physician	Parergon, 1947
Hooker, Joseph D. (1817–1911)	Botanist	Hilts, 1975
Huxley, Thomas H. (1825–1895)	Biologist	J. Huxley, 1935
Jenkin, Fleeming (1833–1885)	Engineer	Hilts, 1975
Jung, Carl G. (1875–1961)	Psychologist	Jung, 1979
Kundt, August (1839–1894)	Physicist	Cahan, 1987
Laënnec, R.-T.-H. (1781–1826)	Physician	Lyons et al., 1978
Laurent, August (1808–1853)	Chemist	Partington, 1964
Lister, Joseph J. (1786–1869)	Physicist	Godlee, 1917
Lister, Joseph Lord (1824–1912)	Physician	Godlee, 1917
*Loewi, Otto (1873–1961)	Physiologist	Lembeck and Giere, 1968
*Lorenz, Konrad (1903–)	Ethologist	Lorenz, 1952
Lyell, Charles (1797–1875)	Geologist	Bailey, 1962
*Michelson, Albert A. (1852–1931)	Physicist	Livingston, 1973
Morse, Samuel (1791–1872)	Inventor	Hindle, 1981
Mueller, Johann H. J. (1809–1875)	Physicist	DSB
*Muller, H. J. (1890–1967)	Geneticist	Carlson, 1981
*Ostwald, Wilhelm (1853–1932)	Physical chemist	Walden, 1904
Paget, James (1814–1895)	Surgeon	Paget, 1901
Parker, William (1823–1890)	Anatomist	Hilts, 1975
Pasteur, Louis (1822–1895)	Physicist, biologist	Wrotnowska, n.d.
Penrose, Roger (1931–)	Mathematician	Liversidge, 1986
*Racker, Efraim (1913–)	Biochemist	Personal comm.
*Ramon y Cajal, S. (1852–1934)	Neuroanatomist	Ramon y Cajal, 1937
*Richards, T. W. (1868–1928)	Physical chemist	Hartley, 1929
Roebling, John A. (1806–1869)	Engineer	Hindle, 1984
Rood, Ogden (1831–1902)	Physicist	DSB
Rumford, Count (1753–1814)	Engineer, physicist	G. Thomson, 1961
Runge, F. F. (1795–1867)	Chemist	Ritterbush, 1968
Sachs, Julius von (1832–1897)	Biologist	Nemec, 1953
Salzmann, Maximillian (1862–?)	Ophthalmologist	Sugar and Foster, 1981
Samuel, David (1922–)	Physical chemist	Personal comm.

Scientist	Profession	Reference
Schuster, Arthur (1851–1934)	Physicist	Crowther, 1968
Schwartz, Ludwig (1822–1894)	Astronomer	Moebius, 1900
Sharpey-Schaefer, E. A. (1850–1935)	Physiologist	RCP archives
Varley, Cornelius (1781–1873)	Inventor	Edgerton, 1986
Williamson, William (1816–1895)	Biologist	Hilts, 1975
Woehler, Friedrich (1800–1882)	Chemist	Jaffe, 1957
Wood, Robert Williams (1868–1955)	Physicist	DSB
Wurtz, Charles Adolphe (1817–1884)	Chemist	DSB
Young, Thomas (1773–1829)	Physicist	Peacock, 1855
Sculptors		
*Holley, Robert (1922–)	Biochemist	Personal comm.
*Luria, Salvadore (1912–)	Virologist	Luria, 1984
Malina, Frank (1912–1981)	Engineer	*Leonardo*
Maxwell, James Clerk (1831–1879)	Physicist	Personal comm.
Richer, Paul M. L. P. (1849–1933)	Pathologist	Monro, 1951
Taylor, C. Fayette (1894–)	Engineer	C. F. Taylor, 1987
Urbain, Georges	Physicist	Jaffe, 1957
Wilson, Robert R. (1936–)	Physicist	Crypton, 1986
Drafters		
*Alvarez, Luis W. (1911–)	Physicist	Alvarez, 1987
*Beadle, George (1908–)	Biologist	Personal comm.
Brunel, Marc Isambard (1806–1859)	Engineer	Ferguson, 1977
Fitch, John (1743–1798)	Inventor	Hindle, 1981
Klieneberger-Nobel, Emmy (1892–)	Bacteriologist	Klieneberger-Nobel, 1980
Latrobe, Benjamin H. (1764–1820)	Inventor	Hindle, 1981
*Pauling, Linus (1901–)	Physical chemist	Personal comm.
*Ramsay, William (1852–1916)	Physical chemist	Travers, 1956
Sachs, Julius von (1832–1897)	Biologist	Ritterbush, 1968
Schulze, F. E. (1815–1873)	Biologist	Goldschmidt, 1956
Thornton, William (1761–1828)	Inventor	Hindle, 1981
Tyndall, John (1820–1893)	Physicist	Crowther, 1968
Watt, James (1736–1819)	Inventor	Baynes and Pugh, 1981
Trained in Architecture		
Dewar, James (1842–1923)	Physicist	Crowther, 1968
Hahn, Otto (1879–1968)	Chemist	Spence, 1970
Kekulé, August (1829–1896)	Chemist	DSB

Scientist	Profession	Reference
Latrobe, Benjamin H. (1764–1820)	Inventor	Hindle, 1981
Lobachevski, Nikolai A. (1793–1856)	Mathematician	Bell, 1937
Schuster, Arthur (1851–1934)	Physicist	Crowther, 1968
Thornton, William (1761–1828)	Inventor	Hindle, 1981
Tyndall, John (1820–1893)	Physicist	Brock et al., 1981
Photographers		
Edgerton, Harold (1903–)	Electrical engineer	MIT archives
Ehrenfest, Paul (1880–1933)	Physicist	Clark, 1984
Emerson, Gladys A. (1903–)	Biochemist	Yost, 1959
*Florey, Howard (1898–1968)	Chemist	Personal comm.
Gadowsky, Leopold, Jr. (1900–1983)	Inventor	Hodges, 1987
Goldschmidt, Richard (1878–1958)	Biologist	Goldschmidt, 1956
Herschel, John (1792–1871)	Astronomer	DSB
Klieneberger-Nobel, Emmy (1892–)	Bacteriologist	Klieneberger-Nobel, 1980
*Koch, Robert (1843–1910)	Bacteriologist	Lagrange, 1938
*Lippmann, Gabriel (1845–1921)	Physicist	DSB
Mannes, Leopold (1899–1964)	Inventor	Hodges, 1987
Maxwell, James Clerk (1831–1879)	Physicist	Sherman, 1981
*Ostwald, Wilhelm (1853–1932)	Physical chemist	W. Ostwald, 1926/1927
Ramon y Cajal, S. (1852–1934)	Neuroanatomist	Ramon y Cajal, 1937
Rayleigh, Lord (1842–1919)	Physicist	Crowther, 1968
*Roentgen, Wilhelm (1845–1923)	Physicist	Nitske, 1971
Roscoe, Henry (1833–1915)	Chemist	Roscoe, 1906
Rutherford, Ernest (1871–1937)	Physicist	Wilson, 1983
Sharpey-Schaefer, E. A. (1850–1935)	Physician	Personal comm.
Weavers and Cloth Workers		
Bezold, Wilhelm von (1837–1907)	Meteorologist	Albers, 1975
Hogg, Helen (1905–)	Astronomer	Yost, 1959
Klieneberger-Nobel, Emmy (1892–)	Bacteriologist	Klieneberger-Nobel, 1980
*Schrödinger, Erwin (1887–1961)	Physicist	Wessels, 1983
Woodworkers, Metalworkers, and Practicers of Related Crafts		
*Alvarez, Luis (1911–1988)	Physicist	Alvarez, 1987
Ancker-Johnson, Betsy (1929–)	Physicist	Ancker-Johnson, 1973
Barcroft, Joseph (1872–1947)	Physiologist	Franklin, 1953
Bayliss, William M. (1860–1924)	Physiologist	Bayliss, 1961
*Békésy, Georg von (1899–1972)	Physiologist	Ratliff, 1974

Scientist	Profession	Reference
Cannon, Walter B. (1871–1945)	Physiologist	Cannon, 1945
Carpenter, William (1813–1885)	Physiologist	Hilts, 1975
Evans, John (1823–1908)	Geologist	Hilts, 1975
Faraday, Michael (1791–1867)	Physicist	Williams, 1965
Fergusson, William (1808–1877)	Anatomist	Hilts, 1975
Fitch, John (1743–1798)	Inventor	Hindle, 1981
Franklin, Rosalind (1920–1958)	Crystallographer	Sayre, 1975
Gibbs, J. Willard (1847–1903)	Physicist	Wheeler et al., 1947
Gray, John Edward (1800–1875)	Zoologist	Hilts, 1975
Grove, William (1811–1896)	Physicist	Hilts, 1975
Henslow, John (1796–1861)	Botanist, geologist	Hilts, 1975
Hill, Rowland (1795–1879)	Statistician	Hilts, 1975
Huxley, Thomas Henry (1825–1895)	Biologist	Hilts, 1975
Jevons, William S. (1835–1882)	Statistician	Hilts, 1975
Kettering, Charles F. (1876–1958)	Inventor	Boyd, 1957
Kundt, August (1839–1894)	Physicist	Cahan, 1987
Liebig, Justus (1803–1873)	Chemist	Willstätter, 1965
Lucas, Keith (1879–1916)	Physiologist	Thomson, 1937
Mauchly, John Williams (1907–1980)	Meteorologist, computer scientist	Stern, 1980
*McClintock, Barbara (1902–1983)	Geneticist	Keller, 1983
*Ostwald, Wilhelm (1853–1932)	Physical chemist	W. Ostwald, 1919
Pasteur, Louis (1822–1895)	Physician, immunologist	Vallery-Radot, 1912
Payne-Gaposchkin, C. (1900–1979)	Astronomer	Haramundanis, 1984
*Ramsay, William (1852–1916)	Physical chemist	Travers, 1956
Reynolds, Osborne (1842–1912)	Physicist	Crowther, 1968
Sanderson, John Burdon (1828–1905)	Physiologist	Hilts, 1975
Wheatstone, Charles (1802–1875)	Physicist	DSB
Wheeler, John A. (1911–)	Physicist	Bernstein, 1985
String Instrument Makers		
Apgar, Virginia (1909–1974)	Physician	Speert, 1980
Dewar, James (1842–1923)	Physicist	Crowther, 1968
Fergusson, James (1808–1877)	Anatomist	Hilts, 1975
Koenig, Karl R. (1832–1901)	Physicist	DSB
Nagyvary, Joseph (1936–)	Biochemist	Stewart, 1984
Wheatstone, Charles (1802–1875)	Physicist	DSB
Musicians		
*Alvarez, Luis (1911–1988)	Physicist	Alvarez, 1987
Apgar, Virginia (1909–1974)	Physician	Speert, 1980
Auenbrugger, J. L. (1722–1809)	Physician	DSB
Avery, Oswald T. (1877–1955)	Microbiologist	Dubos, 1976
*Békésy, Georg von (1889–1972)	Physiologist	Ratliff, 1974

Scientist	Profession	Reference
Beltrami, Eugenio (1835–1899)	Mathematician	DSB
Berson, Solomon (1918–1972)	Immunologist	Overbye, 1982
Boltzmann, Ludwig (1844–1906)	Physicist	Broda, 1983
Bolyai, Janos (1802–1860)	Mathematician	DSB
Born, Max (1882–1970)	Physicist	Infeld, 1941
Bowditch, Nathaniel (1773–1838)	Mathematician	Marmelszadt, 1946
Butschli, Otto (1848–1920)	Biologist	Goldschmidt, 1956
Cannon, Walter B. (1871–1945)	Physiologist	Cannon, 1945
Carothers, Wallace H. (1896–1937)	Chemist	Smith et al., 1985
Carpenter, William D. (1813–1885)	Physiologist	Hilts, 1975
Carruthers, Peter A. (1935–)	Physicist	Broad, 1984
*Chain, Ernst (1906–1979)	Chemist	Clark, 1985
Curtiss, John H. (1909–1977)	Mathematician	Todd, 1980
*De Broglie, Louis de (1892–)	Physicist	Gamow, 1966
Dewar, James (1842–1923)	Physicist	Crowther, 1968
Donder, F. C. (1818–1889)	Biologist	Bowman, 1891
Dresselhaus, Mildred S. (1930–)	Electrical engineer	Dresselhaus, 1973
Ehrenfest, Paul (1880–1933)	Physicist	Clark, 1984
*Eigen, Manfred (1927–)	Chemist	Recorded on Polydor label
*Einstein, Albert (1879–1955)	Physicist	Clark, 1984
Emerson, Gladys A. (1903–)	Biochemist	Yost, 1959
Frieman, Edward (1926–)	Physicist	Personal comm.
Frisch, Otto (1904–1979)	Physicist	Frisch, 1979; Peierls, 1981
Fuchs, Klaus (1911–)	Physicist	Frisch, 1979
Gagnebin, Ilie (1891–1949)	Geologist	DSB
Godowsky, Leopold, Jr. (1900–1983)	Inventor	Hodges, 1987
Goldschmidt, R. B. (1878–1958)	Geneticist	Goldschmidt, 1960
Hall, Charles Martin (1863–1914)	Metallurgist	Garrett, 1963
Heidelberger, Michael (1888–?)	Chemist	Heidelberger, 1977
*Heisenberg, Werner (1901–1976)	Physicist	Personal comm.
Helmholtz, Hermann von (1821–1894)	Physicist	J. Thomson, 1937
Henle, Friedrich G. J. (1809–1885)	Anatomist, physician	DSB
Herschel, Caroline (1750–1848)	Astronomer	DSB
Herschel, William (1738–1822)	Astronomer	DSB
Hertwig, Richard (1850–1937)	Biologist	Goldschmidt, 1956
*Herzberg, Gerhard (1904–)	Chemist	Personal comm.
Jeans, James (1877–1946)	Physicist	Personal comm.
Jenner, Edward (1749–1823)	Physician	Marmelszadt, 1946
Jevons, William S. (1835–1882)	Statistician	Hilts, 1975

Scientist	Profession	Reference
Kennedy, Alexander (1847–1928)	Engineer	Hill, 1960
Koenig, Karl Rudolf (1832–1901)	Physicist	DSB
Kronecker, Leopold (1823–1891)	Mathematician	Bell, 1937
Laënnec, R.-T.-H. (1781–1826)	Physician	Marmelszadt, 1946
Lashley, Karl (1890–1958)	Psychologist	Personal comm.
Laurent, August (1808–1853)	Chemist	Partington, 1964
Loeb, Jacques (1859–1925)	Biologist	Marmelszadt, 1946
Longuet-Higgens, C. (1923–)	Chemist	Personal comm.
Lovelace, Ada (1815–1852)	Mathematician	Huskey, 1980
Ludwig, Carl (1816–1895)	Physiologist	Marmelszadt, 1946
Lyell, Charles (1797–1875)	Geologist	Bailey, 1962
Mach, Ernst (1838–1916)	Physicist	Mach, 1943
Mannes, Leopold (1899–1964)	Inventor	Hodges, 1987
*Marconi, Guiglielmo (1874–1937)	Inventor	Jolly, 1972
Mauchly, John William (1907–1980)	Meteorologist, computer scientist	Stern, 1980
*McClintock, Barbara (1902–1987)	Geneticist	Keller, 1983
Meinong, Alexius (1853–1920)	Psychologist	Personal comm.
Meitner, Lise (1879–1968)	Physicist	Frisch, 1970
*Michelson, Albert A. (1852–1931)	Physicist	Livingston, 1975
*Monod, Jacques (1910–1976)	Biologist	Judson, 1979
Nagyvary, Joseph (1936–)	Biochemist	Stewart, 1984
Nevanlinna, Rolf (1895–1980)	Mathematician	Lehto, 1980
*Ostwald, Wilhelm (1853–1932)	Physical chemist	W. Ostwald, 1926/1927
Owen, Richard (1804–1892)	Anatomist	Hilts, 1975
Paget, James (1814–1899)	Physician	Hilts, 1975
Payne-Gaposchkin, C. (1900–1979)	Astronomer	Haramundanis, 1984
*Planck, Max (1858–1947)	Physicist	DSB
Pope, William Jackson (1870–1939)	Chemist	Read, 1947
Priestley, J. H. (1883–1944)	Botanist	Armytage, 1957
Priestley, Joseph (1733–1804)	Chemist	Jaffe, 1957
*Ross, Ronald (1857–1932)	Biologist	Megroz, 1931
Schick, Bela (1877–1967)	Microbiologist	Riedman, 1960
Schultze, Max (1825–1874)	Biologist	Snyder, 1940
Smith, Homer (1895–1962)	Physiologist	Smith, 1953
Smullyan, Raymond (1919–)	Mathematician	Johnson, 1987
Sommerfeld, Arnold (1868–1951)	Physicist	Personal comm.
Squires, Arthur M. (1916–)	Chemical engineer	Bowser, 1987
Svoboda, Antonin (1907–1980)	Electrical engineer	Oblonsky, 1980
Sylvester, Joseph J. (1814–1897)	Mathematician	Bell, 1937
Teller, Edward (1908–)	Physicist	Frisch, 1979
*Theorell, Axel Hugo (1903–1982)	Physiologist	Dalziel, 1982
Thirring, Walter (1927–)	Physicist	Personal comm.
Urbain, Georges (1872–1939)	Physicist	Jaffe, 1957
Urban, Paul (1905–)	Physicist	Personal comm.

Scientist	Profession	Reference
*Van't Hoff, J. H. (1852–1911)	Physical chemist	Cohen, 1912
Voigt, Voldemar (1850–1919)	Physicist	Voight, 1911
Weisskopf, Victor (1909–)	Physicist	Cole, 1983
Wilson, Edmund B. (1856–1939)	Biologist	Muller, 1943
Young, Thomas (1773–1829)	Physicist, inventor	Peacock, 1855
Composers		
Auenbrugger, J. L. (1722–1809)	Physician	Marmelszadt, 1946
Billroth, Theodor (1829–1894)	Surgeon	Kern, 1982
Bing, Richard (1909–)	Cardiologist	Recorded on Distar label
Borodin, Aleksandr P. (1833–1887)	Chemist	DSB
Herschel, William (1738–1822)	Astronomer	DSB
Lacepede, B.-G.-E. (1756–1825)	Zoologist	DSB
Meinong, Alexius (1853–1920)	Psychologist	Personal comm.
*Michelson, Albert A. (1852–1931)	Physicist	Livingston, 1973
*Ross, Ronald (1857–1932)	Biologist	Megroz, 1931
Schick, Bela (1877–1967)	Microbiologist	Riedman, 1960
Thirring, Walter (1927–)	Physicist	Personal comm.
Poets		
Andrade, E. N. (1887–1971)	Physicist	Church and Buzman, 1945
Boltzmann, Ludwig (1844–1906)	Physicist	Broda, 1955
Bolyai, F. Wolfgang (1775–1856)	Mathematician	Moebius, 1900
Brain, Lord Russell (1895–1966)	Physiologist	Sergeant, 1980; Brain, 1962
Bronowski, Jacob (1908–1974)	Mathematician	Church and Buzman, 1945
Brucke, Ernst (1819–1892)	Physiologist	Cranefield, 1966
Burton, Alan C. (1904–)	Physiologist	Burton, 1975
Carver, George W. (1864–1943)	Inventor	Holt, 1943
Cauchy, Augustin-Louis (1789–1857)	Mathematician	Bell, 1937
Curie, Marie (1867–1934)	Physical chemist	Curie, 1940
Davy, Humphry (1778–1829)	Chemist	Davy, 1840
Dewar, James (1842–1923)	Physicist	Crowther, 1968
Eiseley, Loren (1907–1977)	Anthropologist	Gordon, 1985
Fourier, Joseph (1768–1830)	Physicist	Herival, 1975
Goethe, Johann Wolfgang von (1749–1832)	Naturalist	Gordon, 1985
*Haber, Fritz (1868–1934)	Chemist	Willstätter, 1965
Hahn, Otto (1878–1968)	Physical chemist	Spence, 1970
Haldane, J. B. S. (1892–1964)	Geneticist	Archives, Edinburgh
Hamilton, William R. (1805–1865)	Mathematician	Bell, 1937
Herschel, John (1792–1871)	Astronomer	DSB

Scientist	Profession	Reference
*Hill, Archibald V. (1886–1977)	Biologist	Hill, 1960
*Hoffmann, Roald (1937–)	Chemist	Gordon, 1985
Holub, Miroslav (1923–)	Immunologist	Gordon, 1985
Huxley, Julian S. (1887–1975)	Biologist	Church and Buzman, 1945
Kennedy, Alexander (1847–1928)	Engineer	Hill, 1960
Konner, Melvin (1946–)	Anthropologist	Personal comm.
Kovalevskaia, Sofia (1850–1891)	Mathematician	Koblitz, 1983
Laënnec, R.-T.-H. (1781–1826)	Physician	Marmelszadt, 1946
Lalande, J.-J. L. de (1732–1807)	Mathematician	Van't Hoff, 1878
Lightman, Alan (1948–)	Astronomer	Gordon, 1985
Malus, Etienne Louis (1775–1812)	Physicist	Van't Hoff, 1878
Mead, Margaret (1901–1978)	Anthropologist	Mead, 1972
Mendel, Gregor (1822–1884)	Geneticist	Iltis, 1932
Minkowski, Hermann (1864–1909)	Physicist	Santillana, 1955
Mueller, S. H. (1798–1856)	Mathematician	Moebius, 1900
*Muller, H. J. (1890–1967)	Geneticist	Carlson, 1981
*Nernst, Walther (1864–1941)	Physical chemist	Hiebert, 1983
Odling, William (1829–1921)	Chemist	DSB
Oppenheimer, Robert (1904–1967)	Physicist	Gordon, 1985
Pettenkoffer, Max von (1818–1901)	Chemist, epidemiologist	DSB
Pringsheim, Alfred (1859–1917)	Biologist	Willstätter, 1965
*Ramsay, William (1852–1916)	Physical chemist	Travers, 1956
Rankine, William J. M. (1820–1872)	Engineer, physicist	Rankine, 1874
*Richet, Charles (1850–1935)	Physiologist	DSB
Robb, Alfred A. (1873–1936)	Physicist	Rayleigh, 1943
*Ross, Ronald (1857–1932)	Biologist	Megroz, 1931
Schimper, Karl F. (1803–1867)	Biologist, geologist	DSB
Schrödinger, Erwin (1887–1961)	Physicist	Personal comm.
*Sherrington, Charles (1857–1952)	Physiologist	Macfarlane, 1984
Sylvester, James J. (1814–1897)	Mathematician	DSB
Todd, Olga Taussky (ca. 1905–)	Mathematician	Dick, 1981; personal comm.
*Van't Hoff, J. H. (1852–1911)	Physical chemist	Cohen, 1912
*Waksman, Selman A. (1888–1973)	Bacteriologist	Waksman, 1954
*Willstätter, Richard (1872–1942)	Chemist	Willstätter, 1965
Wilson, Harold A. (1874–1964)	Physicist	Rayleigh, 1943
Wood, Robert Williams (1868–1955)	Physicist	DSB
Dramatists		
Bernard, Claude (1813–1878)	Physiologist	DSB
Butschli, Otto (1848–1920)	Biologist	Goldschmidt, 1956

Scientist	Profession	Reference
Gagnebin, Elie (1891–1949)	Geologist	DSB
Goldschmidt, Richard (1878–1958)	Biologist	Goldschmidt, 1956
*Haber, Fritz (1868–1934)	Chemist	Goran, 1967
Kovalevskaia, Sofia (1850–1891)	Mathematician	Koblitz, 1983
*Richet, Charles (1850–1935)	Physiologist	DSB
Writers of Fiction		
Bell, Eric T. (1883–1960)	Mathematician	Pseudonym John Taine
Benford, Gregory (1941–)	Astrophysicist	*Artifact*, 1985; etc.
Chargaff, Erwin (1905–)	Biochemist	Chargaff, 1978
Haldane, J. B. S. (1892–1964)	Geneticist	J. B. S. Haldane, 1937; 1976
Hoyle, Fred (1915–)	Astrophysicist	Asimov, 1985; Hoyle, 1966
Kovalevskaia, Sofia (1850–1891)	Mathematician	Koblitz, 1983
Lavoisier, Antoine L. (1743–1794)	Chemist	DSB
McConnell, James V. (1925–)	Biologist	Asimov, 1985
McMahon, Thomas (1943–)	Biomechanic	*McKay's Bees*, 1979; etc.
Mead, Margaret (1901–1978)	Anthropologist	Mead, 1972
Mitchell, S. Weir (1829–1914)	Physician	Zigrosser, 1955
Pierce, John R. (1910–)	Engineer	Pseudonym J. J. Coupling
Richter, Charles (1900–)	Geologist	Personal comm.
Robertson, T. B. (1884–1930)	Chemist	Robertson, 1931
Rosen, Sidney (1916–)	Astronomer	*Death and Blintzes*, 1985
Sagan, Carl (1934–)	Astrophysicist	*Contact*, 1985
Skinner, B. F. (1904–)	Psychologist	*Walden Two*, 1948
Smith, Homer W. (1895–1962)	Physiologist	Smith, 1935
Snow, C. P. (1905–1980)	Physicist	Snow, 1934/1958; etc.
Stewart, Alfred W. (1880–1947)	Chemist	Pseudonym J. J. Connington
Szilard, Leo (1898–1964)	Physicist	Szilard, 1961
Van Stratten, Florence (1913–)	Meteorologist	Yost, 1959
Vogt, Karl (1817–1895)	Physiologist	*Biographisches Lexikon*
Weiss, Paul A. (1898–)	Biologist	Weiss, 1964
Wiener, Norbert (1894–1964)	Cyberneticist	Wiener, 1959; MIT archives
Wilson, Mitchell (1913–)	Physicist	Wilson, 1969
Wood, Robert Williams (1868–1955)	Physicist	DSB

a. *Dictionary of Scientific Biography.*
b. Information from the scientist in question or from a third party.

yield such a structure over time. One must master many mental skills to do this well.

More important is the problem of locating evidence of artistic talents. Bernal once commented that it could kill a scientific career to admit that one actually composed poetry or painted at a professional level.[20] And heaven forbid that one confess to wasting time practicing the piano every day! For example, only close family members seem to have known of William Bragg's paintings, Lise Meitner's piano playing, or Schrödinger's tapestry weaving while they were alive. I'm therefore hesitant to assert that any particular individual was devoid of artistic talents without a great deal of research, or without comments by the individual. I have appended a list of such scientists. You will notice, however, that even these nonartists are unusually broadly trained within science and sufficiently literate to write popular books about their subject. But enough. Peruse.

So much for stereotypes. I can hardly imagine a more interesting group of people to gather together than contemporary Nobel laureates in science, their equally eminent peers, and their famous nineteenth century forebears. Many, as artists and writers themselves, would be conversant about the latest trends in the arts and literature. Their paintings and sculpture would fill a number of galleries, and their best poems several fine volumes. Their novels cover every genre from social commentary, mysteries, and love stories to science fiction. One would find several good jazz musicians among them, and an entire orchestra could be assembled from their ranks to perform a whole season of compositions by colleagues such as Billroth, Borodin, Herschel, Meinong, Michelson, and Ross. They could perform for us (as they sometimes performed for

Nonartistic Eminent Scientists (*Nobel Prize)

Scientist	Profession
*Arrhenius, Svante (1859–1927)	Physical chemist
*Bohr, Niels (1885–1962)	Physicist
Darwin, Charles (1809–1882)	Naturalist
*Ehrlich, Paul (1854–1915)	Physiologist
*Einthoven, Willem (1860–1927)	Physiologist
*Fischer, Emil (1852–1919)	Chemist
Haldane, J. S. (1860–1936)	Biologist
*Hevesy, George de (1885–1966)	Chemist
Poisson, S. D. (1781–1840)	Mathematician
*Rabi, I. I. (1898–1988)	Physicist
*Rutherford, Ernest (1871–1937)	Physicist
*Thomson, J. J. (1856–1940)	Physicist
*Yalow, Rosalyn (1921–)	Medical physicist

each other) semicomic plays about science as well as serious works written by colleagues including Claude Bernard, Sofia Kovalevskaia, and Charles Richet. We would also find among their ranks extraordinary athletes, the inventors of many of our most popular puzzles and games, and even a number of magicians.

Mention of games, puzzles, and magic, by the way, leads me to a brief and relevant aside. The most creative scientists not only explore their world thoroughly; many seem literally to *play* with it. Remember Feynman generating a problem from watching a student throw a plate in a cafeteria. Think of Fleming's addiction to games, or Lorenz's "mad" method of inquiry. This playfulness is related to another characteristic that cropped up in the descriptions of many of the men (though not the women, I must say) on my lists: They were described by friends or biographers as "childlike" or even "immature." These include Faraday, Galton, Ehrlich, Einstein, Feynman, Murray Gell-Mann, Jim Watson, Carlton Gajdusek, and many others. C. D. Darlington even suggested that "great discoveries in science often *need* a child-like character in the discoverer . . . a certain lack of emotional development common in the fraternity of pioneers and explorers."[21] I prefer T. H. Huxley's opinion—that these men retain the ability to look upon the world with the all-seeing eyes of a child, and with the humility and excitement that comes with naked wonder. "Sit down before fact as a little child," Huxley advised, "and be prepared to give up every preconceived notion."[22] "I like to see things simple," agrees Szent-Györgyi, "a bit infantile, without much sophistication, and to wonder about the simple things. People often fail to see that something is a miracle if they see it often. To me the greatest and most exciting miracles are what I see around me every day."[23] "To know how to wonder and question is the first step of the mind towards discovery," said Pasteur.[24] Such scientists recognize, in Newton's words, that they are no more than children playing by the edge of a vast, unknown sea, and they are willing to admit (as too many of their more pompous and less able colleagues are not) that all but a tiny part of the universe is beyond their ken. It is not sophistication that is needed to ask deep questions, but the utmost simplicity and humility. "Knowledge is proud that he has learned so much; Wisdom is humble that he knows no more."[25]

In short, the most creative scientists retain a childlike curiosity about the universe which they indulge by exploring and playing with as wide a range of ideas and creative techniques as possible. They learn to recreate the insights of others and thereby become adept at finding their way through the mazelike process of invention Constance has sketched for us. They learn to perceive as well as to manipulate the physical world, and to recreate it imaginatively in their minds. They develop their global vision of science by recognizing common patterns in a diversity of phenomena experienced directly and personally. These experiences form their personality, which in turn forms the matrix within which their science will be created. Throughout the process, they develop the tools of thought that will define the style of their research. Let me trace some of the connections.

Without doubt, the best study of the development of inventive style influenced by training in the arts is Brooke Hindle's study of the creative processes of Samuel Morse and Robert Fulton, *Emulation and Invention*.[26] Both were acclaimed artists prior to turning to inventing. Eugene Ferguson has argued in a broader study of the subject that "thinking with pictures" or utilizing the "eye of the mind" has been an essential aspect of technological thinking for hundreds of years.[27] Arthur I. Miller has argued that such thinking is essential for modern physics.[28] Norma Emerton has studied the reinterpretation of form as a valid scientific concept.[29] Howard Gruber has written on the role of mental images, such as branching tree structures and bushes, as guides to Darwin's thought; Shirley Roe on ways in which mental imagery and concepts can cause different scientists to observe very different things in the same embryological preparations; and Geoffrey Lapage and Philip Ritterbush on the broader effects of the arts on the development of the biological sciences.[30] Essentially, the message of these studies is what Ramon y Cajal proposed: The arts teach useful observational and conceptual ways of thinking.

> *If our study is concerned with an object related to [science], observation will be accompanied by sketching; for aside from other advantages, the act of depicting something disciplines and strengthens the attention, obliging us to cover the whole of the phenomenon studied and preventing, therefore, details from escaping our attention which are frequently unnoticed in ordinary observation . . . It is not without reason that all great observers are skillful in sketching.*[31]

Many similar opinions could be added.[32] My point is that the practice of the fine arts can be a source of scientifically useful skills such as observation, abstraction, pattern recognition, and pattern formation.[33] This much, I hope, is already clear from our previous sessions.

René Taton reminds us, however, that the experimentalist must be more than an unusually keen observer. He or she must also be an engineer, craftsman, or precision worker comfortable with the invention, assembly, and handling of the most delicate apparatus.[34] While some of these skills can be developed through the fine arts, just as useful to training "thinking hands" are crafts such as woodworking, metalwork, glassblowing, silversmithing, electronics, and so forth. Thus we find the physiologist Walter Cannon writing that his experimental ingenuity was directly attributable to his boyhood experience with carpenter's equipment, and the biographers of many of the scientists on my list—Joseph Barcroft, von Békésy, August Kundt, Ostwald, Charles Wheatstone—attributing their success as inventors to diverse experiences with arts and crafts. Rosalind Franklin, Barbara McClintock, Cecilia Payne-Gaposchkin, Robert R. Wilson, and John Wheeler are only a few of the scientists I've found for whom the machine shop was as familiar as the laboratory. Nor is this kind of skill unusual. Galton found that many of his eminent scientists were skilled at various crafts, and one of the few consistent correlates to research success in chemistry and engineering, according to a survey by D. W. Taylor, was knowledge of the tools of carpentry, glasswork, metallurgy, or electronics.[35]

As you see, I believe in skill transference. Nonetheless, I must not overstate my case. Not all scientists are experimentalists, and not all visual thinkers are artists or craftsmen. There are other ways of learning these skills, including curricula based specifically upon visual thinking such as those advocated by Johann Pestalozzi during the nineteenth century and by Rudolf Arnheim today.[36] The efficacy of such curricula have been demonstrated in the case of Einstein. After failing the entrance examination to the Zurich Polytechnic Institute at the age of fifteen, Einstein matriculated at the Kanton Schule at Aarau, Switzerland. The Kanton Schule was founded in 1802 to embody Pestalozzi's principles of teaching students the "ABCs of visual understanding" ahead of and in preference to verbal and mathematical thinking. Although diluted over time, Pestalozzi's philosophy still directed the school when Einstein arrived in 1895. As Holton points out, it was the turning point in Einstein's education. In that year the largely nonverbal and highly visual Einstein invented his first *Gedankenexperiment* and encountered teachers who could appreciate and develop his visual-kinesthetic style of thinking.[37]

The importance of such a confluence of natural ability and formalized curriculum cannot be underestimated. It emphasizes the point we made at our last session, that scientific results should be communicated in as many different ways as possible so as to appeal to minds of different stylistic bents. Might not our curricular emphasis on words and equations serve to select out the Einsteins and van't Hoffs? Are those who think in words or equations unable to comprehend those who think in pictures and feelings? If word- and equation-thinkers predominate among scientists, do they create curricula satisfactory to themselves but incomprehensible to other types of thinkers?

I believe the answer to each question is affirmative. The psychologist Anne Roe recounts a personal instance of such incomprehension. When a scientist (possibly her husband, George Gaylord Simpson) in one of her studies described the ability to watch the history of the evolution of plants by conjuring up an image of a forest and watching it evolve through the geological ages—"practically a home movie," Roe writes—she was "floored." Other scientists related similar tales to her. "I must admit," she wrote, "that for some time I was skeptical that I was hearing what I thought I was hearing. Their minds did not seem to be working the way mine does."[38] Roe was intrigued and studied the phenomenon further. She reported in 1951 that among a group of sixty-four scientists, all of whom were members of the National Academy of Sciences or the

American Philosophical Society, twenty-four used concrete, three-dimensional images in their thinking; eight used geometrical images; eight manipulated visualized symbols; five worked with equations; twenty-one used verbal thought, four utilized kinesthetic feelings; and thirty-four found that most of their thoughts were imageless. (Obviously some of the scientists utilized more than one mode of thought).[39] More recently, Roger Shepard and Vera John-Steiner have documented and discussed further aspects of visual thinking in scientific work.[40] Clearly, then, there are many ways of thinking, and visual and kinesthetic thinking must be considered equally valid and useful—and therefore necessary to scientific and engineering education—as verbal and symbolic thought.[41] In consequence, our curricula must be reformed to stimulate these diverse ways of thinking.

There is more to the connection between the arts and sciences than the development of visual, observational, and manipulative skills, however. There is also the matter of learning to appreciate beauty. Consider the following passage by Ramon y Cajal concerning his motivations for his research:

> *It is an actual fact that, leaving aside the flatteries of self-love, the garden of neurology holds out to the investigator captivating spectacles and incomparable artistic emotions. In it, my aesthetic instincts found full satisfaction at last. Like the entomologist in search of brightly coloured butterflies, my attention hunted, in the flower garden of the gray matter, cells with delicate and elegant forms, the mysterious butterflies of the soul.*[42]

I could go on for hours quoting similar passages from other scientists. C. T. R. Wilson wrote in his Nobel lecture that the origins of his cloud chamber for recording the tracks of ionized particles had more to do with aesthetics than with science:

> *In September 1894 I spent a few weeks in the observatory which then existed on the summit of Ben Nevis, the highest of the Scottish hills. The wonderful optical phenomena shown when the sun shone on the clouds surrounding the hill top, and especially the coloured rings surrounding the sun (coronas) or surrounding the shadow cast by the hill top or observer on mist or cloud (glories) greatly excited my interest, and made me wish to imitate them in the laboratory.*[43]

Konrad Lorenz, the ethologist, artist, and poet: "He who has once seen the intimate beauty of nature cannot tear himself away from it again. He must become either a poet or a naturalist and, if his eyes are keen and his powers of observation sharp enough, he may well become both."[44] Poincaré: "The scientist does not study nature because it is useful; he studies it because he delights in it, and he delights in it because it is beautiful . . . Intellectual beauty is what makes intelligence sure and strong."[45] Duhem, Hardy, Dirac, Beveridge, David Hilbert, Herman Weyl, Werner Heisenberg, Chandrasekhar, William Lipscomb, Chen Ning Yang,

Donald Cram, and many others have written virtually the same thing. Rather than entering into a full discussion of the subject here, I will simply refer you to Judith Wechsler's book *On Aesthetics in Science*; the proceedings of the sixteenth Nobel conference, edited by Dean Curtin and published as *The Aesthetic Dimension of Science*; and C. H. Waddington's *Behind Appearance*.[46] These books amply document the ways in which the search for beautiful patterns, for simplicity, elegance, and harmony, are just as much a part of the sciences as they are of the arts.

Again, what is lacking in these accounts are the origins of scientific aesthetics. How similar or different are they from artistic aesthetics? Personally, I believe they are two sides of the same coin. Georg von Békésy, the great authority on hearing, also seemed to think they were virtually identical. He grew up surrounded by artists and musicians, and almost became a pianist himself. Even in his laboratory, he devoted many hours each day to the study of art and archeology. Floyd Ratliff explains that the choice was deliberately made to enhance his science:

> *It was based upon his desire to do everything well. His first idea about how to excel as a scientist was simply to work hard and long hours, but he realized that his colleagues were working just as hard and just as long. So he decided instead to follow the old rule: Sleep eight hours, work eight hours, and rest eight hours. But Békésy put a "Hungarian twist" on this, too. There are many ways to rest, and he reasoned that perhaps he could rest in some way that would improve his judgment, and thus improve his work. The study of art, in which he already had a strong interest, seemed to offer this possibility . . . By turning his attention daily from science to art, Békésy refreshed his mind and sharpened its faculties. For example, he was always much concerned about the quality of his own work and of the work of others that he was studying. But how can one recognize quality? He asked this question of practically everyone he knew. Finally, as his interest in art and in the collecting of art objects gradually developed, he put the question to an art dealer . . . The answer was: "There is only one solution—to constantly compare, and compare, and compare." . . . This was the basic method of assessing quality which Békésy ever afterwards applied both in art and in science . . . In science, this method of constantly comparing was—for Békésy at least—an almost certain guarantee of high quality work over a long period of time.*[47]

Beauty, harmony, consistency, insight, elegance, sublimity of the mundane—may they not be one and the same for every discipline? And if so, may not a developed sense of aesthetics be one of the most important traits a scientist can possess? I believe so.

But we must not limit our study of the origins of tools of thought to the visual and plastic arts and the various crafts. Like Békésy, many scientists have also been musicians; music, too, plays its role in a scientific education. The connection between musical proclivity and mathematical talent, for example, is well known, but not, I think, properly understood. Ludwig Boltzmann—who, like Max Planck, Hermann von Helmholtz, and many other mathematical physicists, was an accomplished musician—once compared mathematical style to musical style. The mathematician composes in symbols according to rules just as clear-cut as the harmonic and rhythmic patterns that underlie music. In both cases, the voice of the composer shines through. Thus, Boltzmann wrote, "even as a musician can recognize his Mozart, Beethoven, or Schubert after hearing the first few bars, so can a mathematician recognize his Cauchy, Gauss, Jacobi, Helmholtz or Kirchhoff after the first few pages. The French writers reveal themselves by their extreme formal elegance, while the English, especially Maxwell, reveal themselves by their dramatic sense." In an oft-quoted passage, he goes on to descibe Maxwell's work as a piece of music:

> *Who, for example, is not familiar with Maxwell's memoirs on his dynamical theory of gases? . . . The variations of the velocities are, at first, developed majestically: then from one side enter the equations of state: and from the other side, the equations of motion in a central field. Ever higher soars the chaos of formulae. Suddenly, we hear, as from kettle drums, the four beats "Put* N = 5.*" The evil spirit* V *(the relative velocity of the two molecules) vanishes: and, even as in music a hitherto dominating figure in the bass is suddenly silenced, that which had seemed insuperable has been overcome as if by a stroke of magic . . . This is not the time to ask why this or that substitution. If you are not swept along with the development, lay aside the paper. Maxwell does not write programme music with explanatory notes . . . One result after another follows in quick succession till at last, as the unexpected climax, we arrive at the conditions for thermal equilibrium together with the expressions for the transport coefficients. The curtain then falls!*[48]

It is easy—too easy—to pass this description off as a mere analogy used by Boltzmann to describe for laymen what he feels when he does mathematics. We must not ignore such feelings. They are part of research. In consequence, I don't believe that Boltzmann's passage is an analogy at all; I think it's a description of what Boltzmann actually experienced. I believe he actually heard music when he read fine mathematical treatises. Nor would he be alone. Listen to the mathematicians Philip Davis and Reuben Hersch (speaking of their experiences in the singular):

> *As I was working along with the analytic material, I found it was accompanied by the recollection or the mixed debris of dozens of pictures of this type that I had seen in various books, together with inchoate but repetitious nonmathematical thought and musical themes . . . Something then came up in my calendar which prevented me from pursuing this material for several years . . . It required several weeks of work and review to warm up the material. After the time I found to my surprise that what appeared to be the original mathematical imagery and the melody returned, and I pursued the task to a successful completion.*[49]

Similarly, Rolf Nevanlinna, a Finnish mathematician who was also a concert-caliber violinist and chairman of the Sibelius Academy, wrote that "Music has been a constant companion throughout my life. In a mysterious way, which I find hard to analyse, it has been a continual accompaniment to my research."[50] In the same vein, Charles Martin Hall, the inventor of the electrolytic process for purifying aluminum, was said to be a gifted pianist who would rush to the piano whenever he encountered an intractable problem. Even "while playing with such charm and feeling, he was thinking steadily of his work, and thinking the more clearly because of it."[51] The mathematician Lagrange, too, was said to work best to the sound of music.[52]

Now it seems that what we have here is something very similar to synesthesia, in which one sensory experience provokes another, as when one "tastes" a color. What Davis and Hersh describe, and what I believe Boltzmann, Hall, and Nevanlinna too experienced, might be called by analogy "synscientia"—knowing in several different ways at once. Synscientia seems to me to be the basis of transformational thinking, essential to the ability to conceive of an object or idea interchangeably or concurrently in visual, verbal, mathematical, kinesthetic, or musical ways.

You may object that music is not a way of knowing, but I am not alone in asserting that it is. As Douglas Hofstadter has explained in *Gödel, Escher, Bach*, there are logical relations between pure mathematics and music and between mathematics, music, and art.[53] Indeed, the mathematician, musician, and poet Joseph Sylvester, noting these similarities, asked:

> *May not Music be described as the Mathematic of sense, Mathematic as Music of the reason? the soul of each the same! Thus the musician* feels *Mathematic, the mathematician* thinks *Music—Music the dream, Mathematic the working life—each to receive its consummation from the other when the human intelligence, elevated to its perfect type, shall shine forth glorified in some future Mozart-Dirichlet or Beethoven-Gauss—a union already not indistinctly foreshadowed in the genius and labors of a Helmholtz!*[54]

As always I hear Richter growling that this is all a dream. Yet I think not. Jamie Kassler has written extensively on the history of musical models and analogies in science, and gives dozens of examples of the fertility of the combination.[55] I have found more examples in every field from medicine to physics.[56] And I find that music has a number of properties that continue to make it scientifically interesting. The crux of the matter is that most forms of mathematics are inaccurate when applied to describing simultaneously the interactions of more than two and less than a statistically significant number of objects. For example, complete solutions to partial differential equations for biochemical systems as simple as the Krebs cycle cannot be achieved even by modern computers. Verbal forms, on the other hand, are limited to a single voice speaking at a time. That is all the brain can interpret. The multiple themes of novels are, at best, woven like separate threads that appear and disappear in the warp and woof of the story's fabric, occasionally intersecting at a crucial phrase or event. Connections are made by the reader not, as the computer hacks say, in "real time," but after the fact—something we've all experienced in these discussions, I'm sure.

But with music, we can hear many voices speaking simultaneously, and carry many themes continuously and concurrently, summing their harmonies and dissonances to create meanings and effects that no single instrumental voice could manage. Imagine, for example, trying to create the effect of a symphony while allowing only one instrument to play at a time—clearly impossible. And yet many natural systems, such as those involved in physiological homeostasis, embryogenesis, ecological balance, weather—involve multitudes of factors interacting continuously and collectively that cannot be understood using the usual technique of analysis into individual parts. For these integrated systems, the style of thinking that music teaches—chords, harmonies, progressions, patterns—may be essential. Indeed, music itself may provide the best technique for modeling such systems, an idea I've been toying with in my own work on hormonal physiology. What would the "symphony" of normal reproductive physiology sound like if we turned all our measurements into tones? How do the key and the tempo vary to create the harmonious rhythms of daily cycles? Could one identify the dissonances that indicate the presence of particular types of pathology? Intriguing possibilities.

Music has further advantages over many other forms of communication and analysis. "Music is unique in combining quality and quantity precisely and spontaneously so that sense impression can be measured and proportion can be experienced," writes Siegmund Levarie.[57] "The human sense of hearing has remarkable powers of pattern recognition," says the chemist Robert Morrison, "but hearing has been largely ignored as a means of searching for patterns in numerical data."[58] "We have really great computers between our ears," agrees Joseph Mezrich of AT&T Bell Labs.[59] In consequence, these and other researchers at Exxon Research and Engineering Company, AT&T Bell Labs, and Xerox Corporation have been investigating the ways in which the transformation of data into sound can facilitate the analysis of complex data.[60] Recent uses of musical analyses of data have included presentation of quantum states, infrared spectra of chemicals, DNA sequences, problems of taxonomy, economic patterns, and even urinalysis.[61] These are surely only the beginning.

Well, I see I'll never finish. Too bad, because I haven't touched on puzzles and games as sources of ideas, or developed the whole subject of play as thoroughly as I ought. Does physical activity—sports, games, and crafts—develop kinesthetic reasoning, for example? How can writing poetry, literature, and plays improve conceptual ability? Much more should be said about the idea that clarity of expression represents clarity of thought and thus the inventor needs unusually well-developed language skills. The uses of metaphorical thinking. Transformations between words, images, numbers, and music. And most interesting: the subject of handicaps. How are you limited if you lack one or more of these skills? Fascinating subject. Perhaps I'll write a book about these things one day. Well! I could go on for hours, but I'll try to sum up instead.

First, let me try to clarify one point. Despite all that I've said, I'm not advancing the hypothesis that one must be artistically or otherwise creative to become a successful scientist. Rather, I'm suggesting a more circumscribed hypothesis: that those people who utilize curiosity and imagination in their everyday lives will manifest these qualities in their scientific style as well. They will produce qualitatively different sorts of science than more conventionally trained scientists. That there are many distinct styles of scientific research is hardly a matter for question. Van't Hoff contrasted the detailed, factual investigations of Vauquelin and Kolbe to the imaginative, poetical style of Davy and himself.[62] In his *Grosse Männer*, Ostwald made much of the distinction between the classicist, a scientist interested primarily in the perfection of what exists, and the romanticist, for whom the imaginative exploration of the unknown is foremost.[63] Garland Allen has used the distinction to highlight the contrasting biological styles of Thomas Hunt Morgan (a nonartistic classicist) and E. B. Wilson (a romanticist, who not incidentally was considered the best amateur cellist in New York).[64] Brooke Hindle pointed out stylistic differences between the artistic visual thinker who designed the Brooklyn Bridge, John Roebling, and his contemporary nonvisual inventor, John Etzler.[65] Rosalind Yalow, who describes herself as unathletic, tone-deaf, and incapable of drawing, was the mathematical heavy in a collaboration with the "virtuoso chess and violin player" Solomon Berson, who some colleagues say contributed the main ideas and the brilliance.[66] Freeman Dyson has contrasted the imaginative, abstract, unifying ("Athenian") style of Albert Einstein with the factual, concrete, diversifying ("Manchester") style of Ernest Rutherford, whose only brush with the arts seems to have been an adolescent passion for photography.[67] "For Einstein, the electron must ultimately be understood as a clumping of waves in a non-linear field theory. For Rutherford, the electron remained a particle, a little beggar that he could see in front of him as plainly as a spoon."[68] Dyson shows that Rutherford and Einstein had as much trouble comprehending one another as van't Hoff and Kolbe had before them.[69]

The crux of the matter is that we bring to scientific research as much from our personal experience as from theory and experiment. It is this combination of personal knowledge and public knowledge that makes science. How we perceive determines what we perceive. The "eye of the mind" controls the "eyes of the forehead." Therefore the tools of thought that we are able to manipulate set the boundaries of what we can understand and create.

Please don't misunderstand me. I believe that we need both Vauquelins and Davys, Kolbes and van't Hoffs, Rutherfords and Einsteins. I am not arguing that one style of science is better than another; look at the Nobel laureates on the list of nonartistic scientists I assembled. What I am arguing is that a single mode of working, a single mode of training, a single mode of selection will not yield the diversity of scientific types we need. Fact-oriented, unimaginative classicists are too easy to train: They can be intensively drilled in the codified or normal aspects of science and rewarded for their ability to acquire information rather than for their ability to invent. They can profitably be narrow if what we are after is confirmatory or developmental science. But we must encourage educational diversification if we are to produce explorers. They will perceive the universe differently from other scientists only if they experience it differently.

Yet, our current attempts to improve science education seem to be focused solely upon the inculcation of more and more facts in the brains of ever-younger individuals. Introductory textbooks in every field get bigger and bigger, not because there are more principles to learn, but because there are more data. This is nonsense. Not only is it impossible to understand facts divorced from theory, it is impossible to understand theory without imagination, and it is impossible to develop imagination without the broad experience of the world developed through play. "Knowing that" is not "understanding how." Being able to solve mathematical equations is useless if you don't understand what the equations represent in real life.

Thus, if we are to train pioneers and explorers such as those we have studied here, we must accept that specialization does not suffice. Facts can be acquired at any time. What must be taught are the tools of thought that will give meaning to the facts. And so, if the scientific curriculum cannot accommodate the tools of thought I have outlined here, then those who would train themselves to be frontiersmen and frontierswomen will have to turn to other disciplines that already embody and teach those tools: to the fine arts, crafts, and literature, and to play—physical play, play with games, puzzles, and ideas. They must learn to understand the world their way by learning to create and recreate within it.

Do my ideas sound outrageous? Perhaps. Yet I keep good company. Listen to Jerome Wiesner, a former presidential science advisor and president of MIT, on how to develop creative scientists and engineers:

> *There must be encouragement and stimulation of imaginative and unconventional interpretations of experience in general; this is particularly true of problem-solving activities. It is important, especially in childhood and early youth, that novel ideas and unconventional patterns of action should be more widely tolerated, not criticized too*

soon and too often . . . We must explicitly encourage the development of habits and skills in looking for, and using, analogies, similes, and metaphors to juxtapose, readily, facts and ideas that might not at first appear to be interrelated. Early in this development we should foster a clear understanding of the special character and usefulness of the private, informal process of conducting a search for new ideas and insights; and of the distinction between it and the equally necessary but more elaborate and rigorous machinery needed for verification of results, and their systematic development for incorporation into the accepted body of knowledge.

We should help the maturing individual develop a personal "style," suited to his personality, abilities, and complement of knowledge, for approaching problems, research or other, in his special fields of interest. This "style" should reflect the particular mix of mental work habits he has developed to define a problem and to go from it to a spectrum of possible solutions. In a more fundamental sense it would also reflect the habitual pattern of resolving the apparent contradiction between the individual's private, imaginative, playful approach to research and the more formal and rigid reasoning process he would employ to establish and extend the result of that research so that it became "productive" as well as "creative."[70]

Listen also to Cyril Stanley Smith, emeritus professor of metallurgy and history of science at MIT:

I have slowly come to realize that the analytic, quantitative approach I had been taught to regard as the only respectable one for a scientist is insufficient. Analytical atomism is beyond doubt an essential requisite for the understanding of things, and the achievements of the sciences during the last four centuries must rank with the greatest achievements of man at any time: yet, granting this, one must still acknowledge that the richest aspects of any large and complicated system arise from factors that cannot be measured easily, if at all. For these, the artist's approach, uncertain though it inevitably is, seems to find and convey more meaning.[71]

Mitchell Feigenbaum concurs. He's a mathematician attempting to understand the structures of the observed universe using new forms of mathematics such as catastrophe theory:

It's abundantly obvious that one doesn't know the world about us in detail. What artists have accomplished is realizing there's only a small amount of this stuff that's important, and then seeing what it was. So they can do some of my research for me . . . I truly do want to know how to describe clouds. But to say there's a piece over there with that much density and next to it a piece with this much density—to accumulate that much detailed information, I think is wrong. It is certainly not how a human being perceives those things, and it is not how an artist perceives.[72]

The fine arts and crafts, then, are useful to scientists and technologists. Not merely sources of values or modes of personal expression, the fine arts and crafts also embody nonlogical but quite rational tools of thought that help us learn how to perceive, order, understand, and portray our universe. Thus, any educational system that fails to teach students how to play with and integrate the tools of thought embodied in the fine arts and crafts is producing intellectually handicapped students. I say this as a challenge: a general challenge to our Western mode of education that divides culture into arts and humanities on the one hand and sciences and technology on the other; and a particular challenge to objectivist-reductionists who seem to think that that which cannot be written as an equation is not knowledge. Personal knowledge is as important as public knowledge to the inventive scientist and technologist. The best of these tell us through their lives and work that there are not two cultures but one, and that the best science and technology can be produced only by individuals whose minds are accustomed to playing with and integrating every aspect of human activity. But until we understand how people form and recognize patterns, use metaphors, analogize, abstract, develop aesthetic sensibility, playact, play, manipulate, model, think kinesthetically—and how they correlate these tools of thought through transformational thinking (synscientia)—we will never succeed in understanding how inventive people invent, or how to train others to emulate them.

There is only one solution. We must heed the advice of C. H. Waddington: "The acute problems of the world can be solved only by *whole* men, not by people who refuse to be, publicly, anything more than a technologist, or a pure scientist, or an artist. In the world of today, you have got to be everything or you are going to be nothing."[73]

◀ TRANSCRIPT: Discussion

▷ **Imp:** A most provocative way to start the day! Comments, anyone? Jenny.

▷ **Jenny:** I just wanted to add a couple things to what you said about discoverers' being childlike. Don't try to defend yourself, Imp! One of the things I've noticed is that children tend to see through pomp, cant and meaningless civilities with remarkable clarity of vision. It's not without reason that in the ancient fairy tale, a child, and not an adult, perceives that the emperor is wearing no clothes. Maybe that's why it takes a childlike personality in a revolutionary scientist to look at the fairy tales spun by eminent colleagues and say: But there's nothing there! Such statements require the directness and disingenuousness that we associate with childish observations like: "Mommy! Look at that fat lady over there! Have you ever *seen* anyone so fat!" We grown-ups cringe, tell the child to shush, and turn away—even if it is true. The fact that we have learned not to see, and certainly not to comment, is what proves our "maturity." Since all of us are or were academics, I need hardly point out that we use the same test of maturity in our "ivory towers": How well does a colleague play the game; how comfortable does he make us feel? Yet if the first step in discovering is perceiving what no one else perceives or paying attention to the warts that no one wants to admit having, then perhaps social immaturity is a necessary attribute for an inventor.

That leads to my second point, Ariana. You claimed that scientists are humble. I beg to disagree. Looking around this table, I get no sense of humility at all. More like lots of big egos and an overabundance of self-assurance.

▷ **Ariana:** Sorry. I must not have been clear. When I said that many scientists are humble, I meant with respect to nature. A scientist who doesn't respect nature is no scientist. But respect for nature does not require an equal respect for the opinions of one's fellow scientists. Nature is always right, scientists rarely are. It's a matter of choosing which to defer to.

As to your first point, it's certainly true that the socialization we call professionalization teaches us our position and role within the scientific community and that "authority figures" don't like people who fail to respect their position.

▷ **Imp:** Richter? Nothing scathing to say?

▷ **Richter:** Only that I have doubts about the future of this "shotgun wedding" of science with art that Ariana is trying to foist on us.[1] You can mix oil and water as thoroughly as you want, but eventually they separate out again.

▷ **Ariana:** Unless you have an emulsifier, which, I believe I am correct in stating, is a chemical with affinities for both oil and water. You don't have to participate in the mixture, Richter, but don't assert that it shouldn't be done just because you don't want to do it.

▷ **Imp:** Touché. Constance?

▷ **Constance:** I just wanted to correct a misimpression that Ariana may have given unintentionally. What she's said about visual thinking, manipulative skill, aesthetics, and so forth among scientists applies just as well to engineers and inventors—even the imagining oneself as the object. Remember "Boss" Kettering and being a piston in a diesel engine—not the usual analytical approach.[2] Yet Thomas Hughes, a historian of technology, has noted that a common feature of what he calls "heroic inventors"—men such as Thomas Edison, Nikola Tesla, Ambrose Sperry, and Lee De Forest—is the use of such personally internalized metaphor and analogy. He says "the invention of machines, devices and processes by metaphorical thinking is similar to verbal creation, but the fascinating possibilities have not been much discussed, probably because persons interested in language are rarely interested in technology."[3]

In the same vein, Alexander Humboldt once prefaced one of his geological treatises with a picture of the god of poetry drawing back the veil to reveal nature. And Sir Alexander Kennedy, a British engineer, wrote: "Does the modern man object that all this is poetry and not science? Yes, truly it is poetry—the mere words stir one like a Beethoven symphony—but who among us is entitled to say where science ends and poetry begins, in matters about which we are so supremely ignorant?

May not the poetic vision be sometimes as far in advance of the scientific as the scientific is in advance of that of the ordinary commonplace mortal?"[4]

▷ **Ariana:** Van't Hoff would have agreed wholeheartedly.[5] And so do I!

▷ **Imp:** Anybody else? Okay, then you're up, Constance!

▲ CONSTANCE'S REPORT: History and Philosophy of Science in Science

> *I am not ashamed to agree with Herodotos and Machiavelli and Montaigne and Leibniz in believing that the great use of history is for the lessons it teaches us for our conduct today.*[1]
>
> *—Clifford Truesdell, mathematical physicist and historian of science*

Scientists are increasingly trained in a cultural vacuum. Few scientists have any knowledge of their rich humanistic birthright of history, philosophy, literature, and art, and in their ignorance they dismiss it as "something less in value than a mess of pottage."[2] It is not surprising, then, to find that science is usually done without regard to its history or its philosophy. Because history and philosophy of science have developed into independent disciplines during the past century, these aspects of science are left to experts who are rarely scientists. They therefore have little to say to working scientists, if for no other reason than that they don't know what working scientists do.[3]

Scientists generally respond by ignoring the historians and philosophers. Attempts by scientists to work in a totally objective present sometimes reach extremes: for example, recent studies show that 95 percent of citations in scientific papers are to publications less than five years old.[4] Citations of work older than ten years are extremely rare, and I have actually had scientists and inventors question whether any result or process more than ten years old can really be trusted. (The fact that the Leblanc soda process still works almost two hundred years after it was invented comes as a great surprise even to chemical engineers who should know better.) Since I spend a good deal of my time pointing out that their "novel" results are often covered by quite old patents, I have come to the conclusion that many scientists don't want to know what was done by their predecessors for fear that there will be nothing left for them to do.

References to philosophical problems of interpretation and to general questions of scientific methodology—as opposed to matters of specific mathematical or experimental technique—are similarly rare. I know of no studies of philosophical training among scientists (which does not, of course, mean that none exists) and I have met only one scientist who actually took a philosophy of science course. Training in any aspect of history of science seems to be equally rare. On the other hand, I have come across numerous statements concerning the irrelevance of philosophy, or even of general considerations of methodology, to science. These statements generally take the form of the following examples.

> *In their professional work the laboratory scientist and the scientific theoretician use the methods of experimental observation, mathematical deduction, and experimental verification in the grounding and building up of the structure of scientific theory. Arguments pro and con a scientific theory on the basis of common sense, abstract methodological philosophy, the general history of science, or the history of the particular theory in question, as such, carry no scientific weight.*[5]

Similarly, the discussion of the state of a field such as physical chemistry

> *will not include much history, particularly old history, because the development of the science of chemistry was marked by an almost continuous series of errors, such as phlogiston . . . The struggles of the earlier pioneers are a good deal easier to understand from the perspective of modern quantum mechanics, which has clarified so many perplexing problems.*[6]

Even the early master of history of science, George Sarton, went on record as saying:

> *Looking backward would hardly have helped the Stephensons, the Edisons, the Marconis, to solve their particular problems and to solve them as brilliantly as they did . . . the reading of history could not recommend itself to them except as a diversion . . . When a tough technician tells us that he does not care for history, that it is all "bunk"—there is really nothing that we can answer him.*[7]

Is that really true? Is there so little recognition that one cannot observe, induce, deduce, experimentally test, or otherwise operate as a scientist without encountering daily all of the problems of ontology and epistemology? Is heuristics so out of fashion that we actually believe that past successes (and failures) have nothing to teach us? Is there no recognition (outside of the pages of the *Journal of Chemical Education*, anyway) that every laboratory exercise practiced by students of science is a historical event stripped of its history? Do those most eager to discard the past not recognize the tenuous nature of their own historical positions vis-à-vis the future—that is to say, as perpetrators of "mistakes" that will be ignored as conscientiously in a hundred years as phlogiston is today?[8] Apparently so, for instead of learning from the methodological successes and failures of their predecessors, modern scientists ignore them. Perhaps they believe with Thomas Kuhn that science progresses by destroying its past.[9]

I was therefore surprised to find repeated references to the scientific use of what might broadly be considered history and philosophy of science throughout our sessions. Szent-Györgyi advocated the method of looking to the great results of the past as a way of beginning a new line of research. Peter Debye seems to have used the same approach at times. We saw the success of that approach in the revival of Berthollet's ideas by Guldberg and Waage, Pfaundler, Horstmann, and van't Hoff. Lothar Meyer and many other textbook writers and teachers (including Ostwald's teacher, Karl Schmidt) used the history of their field to illuminate outstanding contemporary problems. Van't Hoff turned to the history of science to validate his imaginative approach to scientific research, and was greatly influenced in his training and goals by Comtian positivism. Hunter suggested that one reason Pasteur discovered the asymmetry of the racemates and the subsequent phenomenon of asymmetric fermentation, rather than his patron Biot, was that Biot's vitalistic philosophy precluded the sorts of experiments carried out by Pasteur. Mitscherlich, as I pointed out, had different theoretical preconceptions that prevented him from observing molecular asymmetry. We often turned to Claude Bernard's philosophical insights to understand how experiments may best be invented and interpreted. Planck and Ostwald, too, took stands on important scientific issues based upon their neopositivistic philosophies, as did contemporaries such as Ernst Mach and Pierre Duhem. Additionally, Planck's biographer, Hans Kangro, suggested that it was typical of theoreticians such as these to take a historical approach to developing and defining their subject. Even Imp, Hunter, and Ariana were able to piece together apparently novel insights from fragments of ideas that were fifty or a hundred years old, which suggests to me that one way to find new problems is to read fifty- or one-hundred-year-old textbooks in search of questions that were never answered and data that have been lost. Imp suggests that the basis for his own research is not a problem of data but a methodological controversy with Francis Crick over what constitutes an acceptable theory. Dogma, accident, and untestable ideas, he maintains, have no place in science. This is a matter of philosophy.

So I began to wonder: Can history and philosophy of science in fact be useful or even essential to doing science? T. H. Huxley thought so: "Next to undue precipitation in anticipating the results of pending investigations, the intellectual sin which is commonest and most hurtful to those who devote themselves to the increase of knowledge is the omission to profit by the experience of their predecessors recorded in the history of science and philosophy."[10] Nor is he alone. In 1858 Pasteur recommended to the government official in charge of science education in France that the historical method be an integral part of all science teaching, since it alone "fashions the mind in the manner of inventors and thereby becomes an excellent guide to the intelligence . . . showing that nothing durable is made without great effort . . . giving the mind the habits of modesty, inviting the youth to respect authority and tradition, inspiring him to the cult of great men without making these great men demi-gods possessing supernatural and inaccessible talents in his eyes, but showing them to be men of diligence and devotion, virtues of which we are all capable."[11] Goethe wrote even more succintly that "the history of science is science itself," and was echoed by Ostwald: "The history of a science is . . . merely a *means of research*. It furnishes a *method* for the development of scientific conquests, but is not to be cultivated for itself without regard to its applications."[12]

In short, the study of history as a means of learning how to do science was hardly rare in the nineteenth century. Ostwald, for example, put his belief into action in several ways. He wrote a number of histories of chemistry and electrochemistry, founded his *Klassiker* series to preserve the greatest contributions to chemistry and engaged the best minds of his age to edit and introduce them (van't Hoff, Einstein, Ramsay, and so on). He taught his students historical and philosophical methodology as part and parcel of their education. Among his students, Alwin Mittasch wrote a treatise demonstrating the usefulness of knowing the history of chemistry in raising and solving important problems in catalysis; F. G. Donnan recommended the study of philosophy of science as essential to the progress of chemistry; and Wilder Bancroft examined the history of science for its methodological lessons.[13]

Goethe and Ostwald were not alone in their belief in history of science as science. Henri Poincaré wrote that "To foresee the future of mathematics, the true method is to study its history and its present state." Pierre Duhem believed that the historical method in physics was equivalent to logical analysis in providing an epistemological basis for theory, and Arrhenius likewise felt that the more firmly a new theory could be demonstrated to have deep historical roots in older science, the less likely it was to be discarded as a temporary aberration in thought. In other words, these men took a position quite the contrary to Kuhn and I. Bernard Cohen, who maintain that science progresses by a series of revolutions that obliterate the past. The ideas of Poincaré, Duhem, Arrhenius, Planck, and their colleagues were so novel that they preferred to emphasize the evolutionary character of their theories and the manner in which they integrated successfully into already established science.[14]

Perhaps most indicative of the importance these men placed on history of science is that when the first journal of the history of science, *Isis*, was founded in 1913, Arrhenius, Ostwald, and Poincaré were among the thirty-four-member "committee of patronage." They were joined by other scientific notables such as Moritz Cantor, Thomas Heath, Jacques Loeb, Gino Loria, William Ramsay, and Karl Sudhoff. Among their colleagues and predecessors who shared their interest was a who's who of chemistry and physics: Berthelot, Dumas, Duhem, Willstätter, and so on.

Many other famous scientists have drawn inspiration and knowledge from history of science. For Leibniz, history of science served "not only to give each man his own and to incite others to seek like glories, but also to prosper the art of invention by disclosing method through illustrious example." According to Hadamard, the French mathematicians Jules Drach and Evariste Galois viewed the history of mathematics in a similar fashion. They preferred to read mathematical results in the original form in which they were published, as difficult as the antiquated notations might be, because the textbook form concealed the "characteristic traits of the inventors. They desired to know as many different ways of inventing as possible. On the contrary, most investigators come to know only one inventor—themselves." The English mathematician Augustus de Morgan made the same point: "If he *is* to have his own researches guided in the way which will best lead him to success, he must have seen the curious ways in which the lower proposition has constantly been evolved from the higher," he wrote.[15] This process was accessible only through history.

Was the success of Leibniz, Drach, Galois, and de Morgan due to their unusual understanding of a variety of methods of invention? I cannot say for certain. But it is interesting that, for these men, it was not enough to know that a result could be achieved; they wanted to understand how results were invented, how style and result went hand in hand. The more methods of invention they could master, the greater their ability would be to raise and solve new problems. Or, as Imp pointed out on several occasions, the more ways a person has of perceiving problems and data, the greater one's chance of making a discovery or invention. Learning to think like the great masters of the past by recreating their discoveries may be an excellent form of self-training. And there is no doubt of the importance of autodidacticism in the training of eminent scientists, as our sessions have made clear.

The use of history of science isn't limited to historical figures. A number of modern researchers continue to draw problems and data from their knowledge of the history of science and encourage others to do so as well. Kuznetsov advocates this approach in an essay on the history of his field, catalysis. Clifford Truesdell, who has won international recognition for his work in both mechanics and the history of science, has constantly used what he calls "study of the masters" to learn the methods of research. He points to the work of James F. Bell in experimental solid mechanics and of Coleman and Noll in the 1960s as examples of successful application of old results to new problems. Coleman and Noll, for example, reinterpreted some of J. Willard Gibbs's classic thermodynamics work to invent the new field of thermomechanics.[16]

Other investigators have also championed the historical method in research. Tuzo Wilson arrived at one of his important contributions to geology by realizing that an early nineteenth-century study of the Hawaiian Islands contradicted a modern report. Rather than dismissing the old study as inaccurate, as most of his colleagues probably would have done, Wilson accepted the contradiction as real and of considerable consequence.[17] This led to his notion of oceanic hot spots. Chandler Brooks and Alfred Mirsky both second Szent-Györgyi in urging the renewal and rethinking of old studies in biology as a major source of new ideas and data for the future. Brooks goes so far as to suggest that "this is the moment to recommend that biologists study history—the history of the development of ideas."[18] And anyone familiar with the books and articles of Ernst Mayr, Michael Ghiselin, and Stephen Jay Gould is surely aware of the use to which these scientists put their knowledge of the history of evolutionary theory.

Some scientists have advocated the use of history of science for a different reason: the training of what we have been calling "global thinkers." As Ariana mentioned, there is a general agreement among Nobel Prize winners that to make important advances in science requires broad training in several fields of science and an ability to integrate concepts and data from them. The difficulty is that science seems to be growing so quickly that no scientist can keep up with his or her own specialty, let alone developments in several disciplines. How then are global thinkers to be trained?

In the first place, those who look only at the growth of scientific literature fail to perceive the concomitant process of codification and integration of scientific theories. Once a general theory or an acceptable model is advanced (such as the DNA double helix), a lot of specific studies no longer need be read or cited by most investigators.[19] The complexities that took dozens of important scientists many years to unravel can now be mastered by an introductory student in a very short time. In some sense, the history of a science is the history of how such insights are arrived at. Thus, Ernst Mayr suggests that history of science may provide an excellent means of training scientists more broadly:

> *Like Conant, I feel that the study of the history of a field is the best way of acquiring an understanding of its concepts. Only by going over the hard way by which these concepts were worked out—by learning all of the earlier wrong assumptions that had to be refuted one by one, in other words, by learning all the past mistakes—can one hope to acquire a really thorough and sound understanding. In science, one learns not only by one's own mistakes, but by the history of the mistakes of others.*[20]

The chemist Anna Harrison has made much the same point, saying that the standard way of teaching codified science—facts, figures, and results—may not be the best:

> *I worry about this, because I think it's not so much the* difficulty, *but the element of* adventure *that doesn't come over. I look at a course in physical chemistry, or something of that sort, and I think we don't really play fair with the younger people . . . In class you have all of these nice derivations and they follow one right after another. Some way or another the idea comes across that this is the way science developed. It didn't. It developed in a very erratic manner. After one has all the bits and pieces, someone who is good at writing a textbook can line it all up and make it look like it just proceeds in this fashion. It doesn't. The fun of it is the chase.*[21]

So, in Harrison's opinion, to train reasonable scientists we must convey not only facts but a knowledge of the process of science in the making.

History, then, is essential to understanding science. As Poincaré wrote, science is like a building. As the structure rises, we support its walls and arches with scaffolding. When at last the building is complete, the scaffolding is removed and the building must stand on its own. This is what gives science its objectivity and its timelessness. How the building was erected is irrelevant as long as the resulting structure is stable and enduring. And yet, Poincaré warns, if we do not include the scaffolding in our descriptions of how the building was erected, how shall the student understand the process by which it arose?[22]

Unless some scientists are willing to reveal the subjective elements in the process of building science—the admittedly capricious and personal scaffolds with which the edifice is initially buttressed—the knowledge of how to build the structures of science will become a mystery as deep as the engineering of the pyramids. Unfortunately, mystery-mongers abound. It is acceptable to take refuge in irrational deities such as chance to "explain" the insights of a Pasteur, a Fleming, or a Darwin. It is acceptable to present results as timeless and objective monuments to rationality without reference to their origins. The scaffolds are torn down, the workmen forgotten. We have failed to heed T. H. Huxley's admonition against the sin of ignoring the past and we will pay for it in the future.

But what of Huxley's other admonition, to attend to the philosophy of science? That advice has been followed by at least some eminent scientists. Despite the general disrepute in which most scientists hold philosophy of science, one finds Einstein writing:

> *How does a normally talented research scientist come to concern himself with the theory of knowledge? Is there not more valuable work to be done in this field? I hear this from many of my professional colleagues; or rather, I sense in the case of many more of them that this is what they feel.*
>
> *I cannot share this opinion. When I think of the ablest students whom I have encountered in teaching—i.e., those who distinguished themselves by their independence of judgement, and not only by mere agility—I find that they had a lively concern for the theory of knowledge. They liked to start discussions concerning the aims and methods of the sciences, and showed unequivocally by the obstinacy with which they defended their views that this subject seemed important to them . . .*
>
> *Concepts which have proved useful for ordering things easily assume so great an authority over us, that we forget their terrestrial origin and accept them as unalterable facts. They then become labelled as "conceptual necessities," "*a priori *situations," etc. The road of scientific progress is frequently blocked for long periods by such errors. It is therefore not just an idle game to exercise our ability to analyze familiar concepts, and to demonstrate the conditions on which their justification and usefulness depend, and the way in which these developed, little by little . . .*[23]

When one considers that practically every major physicist and mathematician of Einstein's generation—Bohr, Heisenberg, Planck, von Laue, Eddington, Millikan, Schrödinger, Russell, Whitehead, Jeans, Perrin, Poincaré, Rignano, Smuts, Minkowski, Weyl, Bridgeman, to name just a few—wrote about philosophical issues, then one sees the justice in Einstein's comments.

Please don't misunderstand me. Philosophy is not just an extracurricular interest of these physicists—though that would be interesting in and of itself given the correlations Ariana has presented; it affects how they construct and evaluate theories. I believe I mentioned at one of our earlier sessions Paul Forman's work on the genesis and reception of quantum mechanics in Germany and Great Britain, in which he shows quite clearly how understanding was largely determined by prior philosophical commitments. Mary Jo Nye has demonstrated much the same connection between physics and philosophy in France at the turn of the century. And certainly we have seen the role played by philosophy in the reception of van't Hoff's tetrahedral carbon atom. Those who considered atoms a metaphysical concept could not accept (and obviously did not invent) theories of space-filling atoms.[24] Philipp Frank, Alexandre Koyre, Edward Boring, and Robert Cohen have provided general discussions of many other examples.[25]

I also don't want to give the impression that philosophy of science is the preserve of physical scientists, though it is true that virtually all philosophy of science has been written by physicists or by philosophers studying physics. Biologists have also found philosophy important to their scientific research. Ramon y Cajal, for instance:

> *Philosophical studies above all constitute good preparation and excellent mental gymnastics for the laboratory man. It certainly does not escape our attention that many renowned investigators have come to science from the field of philosophy. It is idle to note that the investigator will occupy himself less with doctrine or philosophical creed—a creed which unfortunately changes every fifteen or twenty years—than with the criteria of truth and standards of critical judgement, with the exercise of which he will acquire flexibility and wisdom and will learn to question the apparent certainty of the strictest scientific systems, thus properly bridling the flight of his own imagination.*[26]

Ramon y Cajal's opinion is shared by Claude Bernard, of course, and Peter Medawar, Jacques Monod, and John Eccles.[27] The latter three men have each written about the profound influence of Karl Popper's writings on the way they conducted scientific experiments. Eccles, for example, describes how important it was to understand that, as depressing as it may be personally, it is beneficial to science to falsify one's own favorite theory. He could have found Poincaré saying the same thing, long before Popper: "The physicist who has just renounced one of his hypotheses ought . . . [to] be full of joy; for he had found an unexpected opportunity for discovery . . . Has the discarded hypothesis been barren? Far from that, it may be said it has rendered more service than a true hypothesis. Not only has it been the occasion of the decisive experiment, but without having made the hypothesis, the experiment would have been made by chance, so that nothing would have been derived from it."[28]

J. S. Haldane and Christian Bohr (Niels Bohr's father) fought over not only the interpretation of their common study of respiration but also the philosophical differences—especially the role of teleological thinking in biology—that led to their different interpretations.[29] (Despite its bad name, teleology has had, and in some scientists' opinions continues to have, a major role in the development of biological science.)[30] Haldane's son, J. B. S., in turn became one of the foremost proponents of dialectical materialism in England, along with other biologists, including Joseph Needham. The influence of philosophy on their choices of problems and techniques is clear.[31]

Indeed, it has always struck me as significant that the first nonvitalist theories of the origin of life, involving a chemical "soup," were invented by J. B. S. Haldane and A. I. Oparin, a Soviet scientist also working within dialectical materialism.[32] Could a vitalist even have considered such a possibility? Moreover, Francis Crick has stated that an important reason for his switch from physics to biology was "that he was an atheist and impatient to throw light into the remaining shadowy sanctuaries of vitalistic illusion." As with a great many molecular biologists, the catalyst for his career change was Schrödinger's little book *What Is Life?*, which is nothing if not philosophical. Crick, Monod, and François Jacob—the founders of molecular biology—have added their own philosophical tracts to this tradition, as have, more recently, evolutionists such as Ernst Mayr, Richard Levins, and Richard Lewontin.[33]

What strikes me most forcefully is that philosophical discussions seem to occur when a science is first emerging. Galileo wrote of methodology when he set out to found the "two new sciences" of mechanics and motion; Claude Bernard wrote as the founder of experimental physiology; Huxley wrote his philosophical and methodological treatises as a proponent of the new theory of natural selection, as did Herbert Spencer; the emergence of physical chemistry gave rise to the methodological tracts of Duhem and the energeticist movement of Ostwald and others; philosophical issues were as important as scientific ones to the theoretical physicists of the first two decades of this century as they laid new foundations for understanding the universe (just think of the Bohr-Einstein debates); philosophical issues demanded the attention of the molecular biologists as they founded their new science; and so forth.

Is there, then, something about the emergence of new ways of viewing science that also generates new ways of conceiving knowledge? Are the pioneers of science simultaneously the pioneers of philosophy? Leon Rosenfeld thinks so:

> *A tradition common to many pioneers in science has been the combination of achievement in actual discovery of natural laws with*

> *philosophical reflection on the nature of scientific thinking and the foundations of scientific truth. This combination is essential to such scientists in the sense that epistemological considerations played a decisive part in the success of their investigations and that, conversely, the results of the latter led them to deeper understanding of the theory of knowledge.*[34]

Pioneers explore the boundaries of nature and knowledge simultaneously. Therefore, it would seem that each "revolution" in science, if we are willing to accept the term—or perhaps I should say, each major shift in the pattern by which we organize knowledge—should be accompanied by philosophical and methodological upheaval. Indeed, the project to form a new "science of science" may add its share of insight to the theory of knowledge. I find that an intriguing possibility. To realize it, we need to delve much more deeply into how new sciences and new modes of thinking emerge hand in hand.

Consider, for example, the list of unsolved problems in biomedical sciences that Richter made at our first session—what cancer is, how drugs work, what thinking is, how healing occurs, and so forth. We have to ask whether these problems are unsolved because they are too difficult to solve, or because we have not yet developed the right philosophy of science for dealing with them. I'm thinking here of an old article in which Iago Galdston suggested that there is an ideological basis to all discovery. He cites, for example, Sigerist's statement that "I personally have the feeling that the problem of cancer is not merely a biological and laboratory problem, but it belongs to a certain extent to the realm of philosophy . . . All experiments require certain philosophical preparation. And I have the feeling that in the case of cancer many experiments were undertaken without the necessary philosophical background, and therefore proved useless."[35] Now, these sentences were written in 1932. Cancer is still largely an unsolved problem. Is this because we are stuck with a mode of explanation—say, one causative agent for one disease—that is not useful in this context?

A. C. Crombie has argued that philosophies delimit the range of answers acceptable in any period of history: "Dominant intellectual commitments have made certain kinds of questions appear cogent and have given certain kinds of explanation their power to convince, and have excluded others, because they have established, antecedent to any particular research, the kind of work that was supposed to exist and the appropriate methods of inquiry. They established in advance the kind of explanation that would give satisfaction when the supposedly discoverable had been discovered."[36] Evelyn Fox Keller has made the same point about the difficulties Barbara McClintock had in overcoming "master molecule" theories, such as the Central Dogma, that dominated (and still dominate) genetics and molelcular biology, and which precluded others' sharing McClintock's insights for many years.[37] Historically speaking, then, it appears that sometimes problems are unsolvable because they are posed within the wrong methodological and epistemological framework. To quote Werner Heisenberg: "What we observe is not nature in itself but nature exposed to our method of questioning."[38] It is not enough, then, to question nature; we must also question the nature of our questioning.

I conclude that the philosophy of science is a viable and useful subject of study for scientists, or could be were it perceived as an integral part of scientific investigation rather than a separate discipline practiced by nonscientifically trained practitioners. As Truesdell says, "The philosophy of science, I believe, should not be the preserve of senile scientists and of teachers of philosophy who have themselves never so much as understood the contents of a textbook of theoretical physics, let alone done a bit of mathematical research or even enjoyed the confidence of a creating scientist."[39] Rather, as Brillouin says, "The philosophical background of science is a very serious problem, still worth discussing, and most important for a better understanding of science."[40] I'll give Einstein the last word: "Science without epistemology is—insofar as it is thinkable at all—primitive and muddled."[41]

So much for the body of my talk. Imp encouraged us to stick our necks out and make some statements about the important problems and possibilities from our individual points of view, too. I am not very comfortable with this sort of thinking, but I'll give it a try.

I've already suggested that the most important lesson of the history and philosophy of science is that our knowledge of the past and our method of questioning determine what we find. It follows that scientists must steadfastly maintain a diversity of philosophies, paradigms, and traditions to avoid becoming trapped within a closed box of the sort Jenny has illustrated so well for us. Different traditions and philosophies will be needed to define and address the important questions of the future. Uniformity of training and opinion, along with deference to authority, may therefore be the greatest banes science must face. The only antidote—or perhaps vaccine would be a more appropriate metaphor—will be the training of scientists who care more about the origins and development of their field than they do about where their next grant will come from; who are willing to learn from the past rather than treat each present situation as unique; and who are philosopher-historian-scientists of the caliber of Poincaré and Duhem, Bernal and Needham, Holton and Mayr. Without such people, I just can't see how new themata will originate or how the global issues of science will be addressable.

I also wish to propose a complete shift in the way we analyze science. Typically, historians, philosophers, and scientists consider the most important aspect of science its codified solutions. Thus, we find myriads of books with titles such as *The Edge of Objectivity*, *The Art of the Soluble*, or *The Search for Solutions*. I believe that this view of science is backward. As George Wald says, "it isn't answers that make a scientist, it's questions . . . Science is a way of asking more and more meaningful questions. The answers are important mainly in leading us to new questions."[42] The most intriguing and important aspects of science thus concern not its ability to solve problems but its unique ability to pose them. This, it seems to me, is what Richter was trying to get us to see at our first session, and why Einstein, Bohr, Mark Kac, Stan Ulam, and so many others have emphasized that asking the right question is more than half the work of science. When we understand how scientists do that, then we will understand how to do science. Let me then, if I may be so bold, suggest that the proper title for a book about science should be *The Quest for Questions* and that rather than *Conjectures and Refutations* we need an analysis more along the lines of *Riddles and Surprises*.

There. That's as provocative as I can be.

◀ TRANSCRIPT: Discussion

▷ **Imp:** Bravo! Your final point is right on target. But we've got to turn the educational system upside down, too. We spend so much time trying to teach students how to answer pre-solved problems in the pre-approved manner that we fail to teach them how to pose new problems for themselves. Disgraceful!

▷ **Ariana:** And not very conducive to curiosity, either.

Can I bother you for a list of your scientist-historians and scientist-philosophers, Constance? After your cogent arguments for the importance of history and philosophy of science for scientists, I see that I shall have to consider adding a few more "correlative talents" to my current tables. I am particularly interested to see that many of your historical-philosophical scientists are on my nonartistic list. Perhaps this is another style that needs to be elucidated.

Truth to tell, I admit to being somewhat surprised by your talk. My impression from looking around at my colleagues is that a passion for the humanities (unlike the arts) is virtually incompatible with developed skill in science. I think Kuhn even stated that as a general principle.[1] So once again these unusual "gentlemen of science" may represent interesting anomalies.

▷ **Constance:** Quite possibly. I'm afraid I can't add much at the moment, except to say that Otto Loewi almost became an art historian instead of a scientist; Louis de Broglie was intent upon becoming a medieval historian until World War I introduced him to the wonders of radio technology; and J. B. S. Haldane had only a single degree and that was in classics. So at least the two passions are not incompatible.[2]

▷ **Hunter:** A point worth bearing in mind with regard to the recruitment of future scientists. As Jenny has gently reminded me from time to time throughout these sessions, why should we assume that the skills needed to be a successful scientist differ significantly from those needed to be a successful historian? The ability to excel at any endeavor may be an indicator of future excellence in science. Moreover, the examples you just mentioned, Constance, seem to indicate that early training in science seems not to be a prerequisite to later eminence. An observation that most definitely runs counter to the current trend in science education.

▷ **Constance:** Considering that most of the great German and Scandinavian scientists were trained in the *Gymnasium* tradition, which was almost all languages, classics, and history, one would think that the point would be obvious.

▷ **Richter:** But in the *Gymnasien* one had to master *something*, unlike the half-cooked mish-mash we serve up to students today. One at least learned how to learn.

▷ **Imp:** And to appreciate historical and philosophical perspectives other than one's own. Which brings us to your report, Richter—a new perspective on everything we've done so far, if I've understood your hints.

▲ RICHTER'S REPORT: The Evolution of Science

The working scientists of today are the product of the system of selection and education. It is not surprising that, with such a different social and economic environment, they should be different from those who were occupied in laying the foundations of modern science. In the early days the decision to occupy oneself with the study of science was a personal choice which very few people took, and then in spite of the very grave disabilities that attended the choice of such a useless vocation. It restricted the practice of science to wealthy men and those who could acquire patronage. Nowadays science is a definite profession. The process of selection that goes on inside scientific education is one that favors, on the one hand, technical proficiency and industry and, on the other, general social conformity.

—J. D. Bernal, crystallographer and historian of science

Nowhere . . . is the penalty on even the mildest case of nonconformity higher than in the United States.

—Erwin Chargaff, biochemist

I say that the greatest threat to the future of science is conformity—conformity of opinion, of action, of direction. Science progresses only where there is dissent. It is therefore fitting that my report follow Ariana's and Constance's. As you will see, I agree completely with Constance that a new philosophical approach is necessary to solving outstanding problems. Old data and theories can be profitably renewed. My own report is a case study in these prescriptions. Indeed, let me state at the outset my debt to Constance. Although the framework for my report is wholly my own, she supplied most of the examples. Our styles are complementary, even if our opinions are not.

On the other hand, I find Ariana's approach unhelpful. The cognitive, artsy-fartsy perspective gets us nowhere. We cannot know what happens inside our brains. The act of introspection necessarily disturbs the mind doing it so that at best we know the act of introspection and nothing more. The correlation between an unknowable internal process such as nonverbal thinking and an external manifestation such as artistic talent seems foolhardy, at best. People do perfectly good science without all this humanistic twaddle. And I believe that the process by which they do it can likewise be analyzed without recourse to unknowable mental mechanisms.

Nonetheless, I find that Ariana has made me pay attention to things I had previously ignored. Thus, while I disagree with her, I find her data illuminating and compatible with an entirely different and more profound interpretation. I am fascinated by the observation that the most successful scientists are often trained in unusual ways, or in odd combinations of subjects. Never mind Ariana's fanciful "tools of thought." The eclecticism is itself noteworthy. Creative scientists are *different*. Furthermore, I find myself marveling at the diversity of images invented by scientists for any given problem: the various models of DNA that Imp brought to our attention, the dozens of models of benzene rings, Constance's selection of periodic tables, and Imp's rearrangements of the genetic code.

But whereas Ariana highlights the diversity of visual patterns, I stress the diversity of logical solutions. Haüy, Biot, Laurent, Pasteur, and Mitscherlich all maintained different views of the connection between crystal form and chemical properties. Each theory was useful for solving one particular problem—atomic form, isomorphism, asymmetry, fermentation—and each was detrimental in preventing the solution of another. The "Arrhenius equation" describing the effect of temperature on reaction rates has passed through a dozen forms, each of which has strengths for particular applications and weaknesses for others. B. J. T. Dobbs has documented that Newton had access to multiple theories in trying to explain gravitation.[1] The wave-particle duality of light has never been resolved in favor of either theory. Ever since Einstein invented relativity theory, physicists have been inventing alternatives. I won't bore you with other examples, but simply state that the phenomenon of multiple solutions—complementary solutions—is not rare. It is simply ignored. Historians, bamboozled by scientific rhetoric about objectivity and demonstrable truth, have failed to perceive the diversity of opinion that is always present in science.[2] Scientists—I plead guilty myself—are trained to believe that each problem has only one correct answer and that this answer should have a mathematically provable formulation. We are just as blind. Solutions to scientific problems, far from being end points, are only beginnings. There is not *one* answer, but sets of answers. As Pauling stated, the object of research is not to find *the* answer, but to eliminate impossible answers. Consensus is not a prerequisite for scientific advance; dissent, I will argue, is.

Needless to say, when I began to work in science, I expected neither diversity nor dissent. Indeed, the paradigmatic view of science which I

Summary of Temperature-Dependence Equations

Differential form	Integrated form	Expression for *k*	Supported by
$\frac{d \ln k}{dT} = \frac{B + CT + DT^2}{T^2}$	$\ln k = A' - \frac{B}{T} + C \ln T + DT$	$k = AT^C e^{-(B-DT^2)/T}$	Van't Hoff, 1898; Bodenstein, 1899
$\frac{d \ln k}{dT} = \frac{B + CT}{T^2}$	$\ln k = A' - \frac{B}{T} + C \ln T$	$k = AT^C e^{-B/T}$	Kooij, 1893; Trautz, 1909
$\frac{d \ln k}{dT} = \frac{B + DT^2}{T^2}$	$\ln k = A' - \frac{B}{T} + \mathrm{DT}$	$k = Ae^{-(B-DT^2)T}$	Schwab, 1883; Van't Hoff, 1884; Spohr, 1888; Van't Hoff and Reicher, 1889; Buchböck, 1897; Wegscheider, 1899
$\frac{d \ln k}{dT} = \frac{B}{T^2}$	$\ln k = A' - \frac{B}{T}$	$k = Ae^{-B/T}$	Van't Hoff, 1884; Arrhenius, 1889; Kooij, 1893
$\frac{d \ln k}{dT} = \frac{C}{T}$	$\ln k = A' + C \ln T$	$k = AT^C$	Harcourt and Esson, 1895; Veley, 1908; Harcourt and Esson, 1912
$\frac{d \ln k}{dt} = D$	$\ln k = A' + DT$	$k = Ae^{DT}$	Berthelot, 1862; Hood, 1885; Spring, 1887; Veley, 1889; Hecht and Conrad, 1889; Pendelbury and Seward, 1889; Tammann, 1897; Remsen and Reid, 1899; Bugarszky, 1904; Perman and Greaves, 1908

Reproduced by permission of K. J. Laidler and the American Chemical Society.

held precludes both. I expected, following Kuhn and other philosophers of science, that scientists within any given discipline would share a common set of problems, techniques, solutions, and criteria. Furthermore, I expected that scientists trained within the currently accepted paradigm would be most likely to spot anomalies and correct them. Historically this seems not to be true. It is the peripheral types and oddly trained men who invent the new models of DNA, benzene, the periodic table, relativity theory, quantum mechanics, and so forth. The major proponents of the new DNA models, for example, are a pair of otherwise unpublished Germans; a small group in India, only one of whom worked previously with a recognized expert on DNA; and some Australians with no direct experience in the field at all—not, you'll notice, anyone in the U.S. or U.K., where the double helix was originally developed.[3] Ideas have social territories just as animals do.

I have also had to discard other preconceptions. I believed that once a theory is discarded, it is discarded forever. Kuhn, Lakatos, and other observers of scientific growth tell us that scientists must, and in fact do, choose the best model of the best theory and relegate the rest to the dustheap of history.[4] Practitioners who maintain the old theory are, in Kuhn's view anyway, banished from the field. For Popper and Lakatos, logical analysis should determine absolutely the best theory or the most promising research program.[5] I, as you are aware, also found this view appealing. It keeps science simple and uncluttered. It gives the field intellectual rigor. It prevents waste. Yet it does not account for the survival and utility of the old theories and supposedly moribund research programs such as Berthollet's. Was this an exception, or are many alternate theories usually kept in circulation? The answer is crucial, for if old theories do not die, then much of what we have believed about how science develops must be discarded.

Apparently old theories do not die—or at least not all do. As Constance points out, science does not discard its past. Old knowledge is renewed, reworked. Example: An article just appeared in *Nature* demonstrating that Kekulé's 1865 resonance model of benzene may be more accurate than the molecular orbital model we were all taught in college.[6] Example: Linus Pauling's banana-shaped sp2 + p hybridized double bonds of thirty years ago are apparently going to replace the Hückel sigma-pi orbitals we have been using in the interim.[7] Example: Heubner's hundred-year-old theory of suppression of puberty by the pineal has just been resurrected to explain the latest experimental results.[8] No answer seems ever to be sufficient or unquestionable. What we consider to be the best solution depends on what our criteria are. And historically, these keep changing.

Because I had not expected them, these phenomena of dissent, diversity, and survival worried me. So did the problem of why a demonstrably falsifiable theory such as the Bohr atom should be accepted in science. Why embrace it, if we know that it is inadequate? These anomalies make my "science is truth" position shaky. If science were truth, then once an answer was obtained it would be *the* answer. One would not accept an obviously incorrect theory. So why the elaboration of many answers when a sufficient one has already been generated? Why the constant reevaluation of old answers? What function does this diversity play in science? Should it and can it be eliminated so that only the correct solution is obtained? I began to seek alternative ways of perceiving these problems.

Then it struck me: a perceptual shift, a repatterning along the lines Jenny suggested to us, that answered all my questions at once. Science develops by an evolutionary process. Not by some revolutionary gestalt switch between the wrong answer and the right, but by a trial-and-error method that results in the class of all possible answers being generated. Each of these is then evaluated within a particular context of outstanding problems and unexplained data, and the best solution for that context is selected. This is the lesson of the Haüy-Biot-Laurent-Pasteur-Mitscherlich example. No single theory was sufficient to explain all aspects of chemical crystallography. Jenny's model for illumination also typifies the process. One first generates all of the possible five-line answers before moving on to an answer with four, three, one, or no lines. One then selects the answer that best fits one's needs.

I will not fuss now about whether the selection criteria are logical, empirical, aesthetic, historical, or sociological—all five types have been invoked in our discussions, to the dismay of some of us.[9] Suffice it to say that selection criteria exist and they differ from individual to individual. Perhaps the scientist must juggle all five. In the event, different selection criteria—or sets of selection criteria—cause different individuals to favor different solutions. More than one solution can therefore be viable at any given time. For the same reason, a demonstrably incorrect theory such as Bohr's planetary model of the atom could nonetheless flourish as long as no better-adapted theory was advanced to compete in the territory Bohr had mapped out. It need not even compete with van't Hoff's tetrahedral carbon atom, as long as the problem areas being addressed remained distinct—as long as organic chemists did not talk to physicists. Discovery is therefore, as Imp impressed upon us, context dependent; a conclusion you'll remember I resisted, but which I now believe to be a necessary consequence of an evolutionary model of scientific development.

So, right at the outset we have three of the most important prerequisites to an evolutionary model. Evolutionary processes require, among other things, a fund of diversity upon which a selective process can act and a range of selection factors that will lead to differential survival of ideas and practitioners. Science fits such a model. And before you accuse me of propounding some neo-Spencerian philosophy, permit me to state my case.

Can evolutionary theory be applied to cultural constructs such as science? Theoretically, yes. Herbert Spencer advocated the idea, as did Ostwald, Poincaré, Planck, and others. I am not saying that I advocate the writings of any of these men, but they did foresee the possibility. More recent investigators have also proposed the extension of evolutionary theory to culture. These include biologists and sociobiologists such as Julian Huxley, L. A. White, Jacques Monod, Ralph Gerard, L. L. Cavalli-Sforza, David Rindos, Jonas Salk, William Durham, Charles Lumsden, and E. O. Wilson. Not that these men agree with one another as to how to do it; nonetheless, there is general agreement that it can be done.

Here is the basic rationale. "The biologist," Julian Huxley has argued, "knows how fruitful has been the study of mechanisms of genetic transmission for understanding the process of biological evolution. He can properly suggest to the humanist that a study of the mechanisms of cultural transmission will be equally fruitful for understanding the process of human history."[10] And where better to start such a study than with the history of science? Wilhelm Ostwald correctly states that "the history of science furnishes the best material, the most sure, for the study of the laws that regulate the development of humanity. For this study of humanity does not escape the general rule of the natural and physical sciences that the phenomena to be scrutinized must be perceived first under the most simple conditions. It is in the history of science that one will find most easily these conditions."[11] Karl Popper agrees: "The growth of knowledge can be studied best by studying the growth of scientific knowledge."[12] Moreover, Popper's colleague Donald Campbell has interpreted Popper's falsification theory as the first step toward a general evolutionary model for epistemology in which there is selective elimination or retention of ideas through a form of what Campbell calls "natural selection."[13] I prefer to think of it as artificial selection, since it is done by thinking human beings.

Even Thomas Kuhn, who is best known for his revolutionary view of science, ends his *Structure of Scientific Revolutions* with the almost paradoxical statement that revolution is the mechanism by which the evolution of science occurs (shades of Gould and Lewontin):

> *The analogy that relates the evolution of organisms to the evolution of scientific ideas can easily be pushed too far. But with respect to the issues of this closing section it is very nearly perfect. The process described . . . as the resolution of revolutions is the selection by conflict within the scientific community of the fittest way to practice future science. The net result of a sequence of such revolutionary selections, separated by periods of normal research, is the wonderfully adapted set of instruments we call modern scientific knowledge. Successive stages in that developmental process are marked by an increase in articulation and specialization. And the entire process may have occurred, as we now suppose biological evolution did, without benefit of a set goal, a permanent fixed scientific truth, of which each stage in the development of scientific knowledge is a better exemplar.*[14]

What Kuhn does not describe is where the diversity of modes of doing science comes from, who generates them, or how. He also misses the point that evolution does not necessarily lead to a single best solution. What Hunter's histories have shown us, and Ariana's study has emphasized, is that the diversity has always been there—a very important point. I shall take this point and push it, as Kuhn warns us not to do, to its limit: I shall assert that evolutionary theory provides the basis for building Imp's "science of science."

To begin with, we need to consider the general characteristics of evolutionary processes. First, there must be a general and essentially random mechanism (or mechanisms) for generating diversity among the members of a population. Second, differences between individuals must be inheritable. Third, there must be a nonrandom selection process ensuring that all individuals do not survive to reproduce, or do not reproduce equally effectively. These three criteria increase the adaptivity of a population for its environment. Geographical isolation of part of the population results in speciation as it interbreeds and adapts under different selection pressures. Competition among species for common niches inevitably leads to the extinction of the less well adapted. Thus, some populations are able to reproduce to the exclusion of others. One prerequisite for evolution, then, is a changing environment in which new niches are constantly being created and subpopulations isolated. A homogeneous population in a static environment does not evolve. A bit simplistic, but a fairly accurate picture. Oh, yes—one other thing. Looking back at the historical record of anything that evolves should yield a treelike image such as Darwin introduced for describing phylogenies. This is a consequence of diversification and speciation. Many of the branches will, of course, terminate as a result of extinction.

Next we need cultural analogues to biological evolutionary mechanisms. I have summarized these in the accompanying table, based upon earlier comparisons by Ralph Gerard and L. L. Cavalli-Sforza.[15] In culture, what evolves are ideas, beliefs, languages, tools and instruments, habits of thought and action, and organizational patterns of work, of procreation, of invention, and so forth. Variation is introduced not by random mutation but along the lines of sexual recombination. People become heirs to odd mixtures of talents, ideas, habits, and tools. Another possible source of variation is "copy error," which may occur during information transfer (think of what happens in the game of "telephone"). Ideas may undergo alterations in the process of translation (ask any translator what happens when a passage in English is translated into another language and then back again). The slight changes in meaning that accompany such translations of an idea from one context to another may, in themselves, result in a new line of thought. Similarly, ideas, habits, organizational preferences, and so forth may undergo change due to hybridization when members of different cultures cooperate or conflict. Alternately, individuals may inherit from several cultures. And Lamarck-

Comparison of Biological and Cultural Evolution

	Biological	**Cultural**
What evolves	Genotypes of organisms, genetic profiles of populations	Ideas, beliefs, tools, organizational patterns, activities
Variation	Genetic mutation, hybridization	Genetic mutation and hybridization? Cultural hybridization, transmission error, translation
Transmission	Genetic	Genetic? Teaching/learning (active), mimicking (passive)
Selection	Natural and artificial	Environmental, social, cultural, institutional, individual
Isolation	Geographical	Geographical, social, institutional
Niche creation	Environmental change	Social and institutional environment

Based on Gerard, 1956, and Cavalli-Sforza, 1981.

ian inheritance is fully possible in cultural evolution: An individual or group of individuals may consciously attempt to alter their cultural patterns to become more adaptive (the reverse is also possible: consider the Shakers).

Transmission of culture is in some respects more complex than transmission of biological traits. Genetics may play a role in cultural transmission, as when proclivities are inherited, but as Ariana points out correctly—you see, we can agree!—genetics is not sufficient for explaining the evolution of culture. In addition we must consider teaching and learning (active culture transfer), and mimicking (a passive process, as the person being mimicked does not intend to transfer knowledge). Within teaching, Cavalli-Sforza outlines three major mechanisms: parent-child (one-to-one), teacher-learner (one-to-many), and social pressure (many-to-one). He shows that in practice it is much more complicated.[16]

We must also consider selection pressures. In both biological and cultural evolution we can assert the validity of what is popularly called "survival of the fittest." Determinants of fitness differ. In biology fitness is measured by the ability to survive and reproduce. In culture, fitness is more difficult to characterize. Biological survival to reproductive age and ability to reproduce are fairly well ensured in most human populations that support science, but other traits, such as acquired knowledge, skills, personality, and so forth, may be more important. The range of selective factors is certainly broad: In addition to personal competition, economics, social conventions, institutional hierarchies, and personal conflict are all active determinants of cultural adaptiveness. And here I want to emphasize that I am speaking not of the biological survival of the individual, but of his or her ability to pass on his or her cultural inheritance—in the case of scientists, the ability to train successful scientists and influence the course of future scientific research.

Finally, we need to consider analogues for genetic drift due to geographical isolation, "gene flow," migration, and other determinants of population profiles. Gerard and Cavalli-Sforza have again outlined various cultural analogues. For example, isolation may have other than geographical determinants in culture: Individuals or groups may be socially or institutionally isolated or ostracized. Furthermore, just as genes may be dominant or recessive, individuals conveying culture may also be more or less dominant or effective. Dominance need not imply elimination of ineffectual ideas, however. Charles Lyell's extended attack on Lamarck's evolutionary theory pointed out its manifest failings, yet effectively transmitted Lamarck's ideas to Darwin, who reworked them. Thus even incorrect or abandoned ideas may, like recessive traits, be passed from one generation to the next, creating new and interesting combinations.

One final point: Because it is not necessarily tied to genetic considerations, cultural evolution may occur much more quickly than biological evolution. A single individual may, through mass media or technology, convey new ideas, new modes of work, new tools, and so on to millions of progeny within a lifetime, changing the way they live. Just think of Edison's light bulb; the Wright brothers' airplane; Pasteur's vaccines; Wallace Carother's nylon; or the results of the Manhattan Project. On the other hand, a conservative culture may deny nonconforming individuals access to the means of cultural procreation (for instance, research, publishing, or teaching) and so prevent further evolution. This is a very real possibility in a society, such as ours, that emphasizes conformity. If the limits of conformity are too narrow, the "sports" that carry the promise of new evolutionary directions may be eliminated too soon, or denied access to cultural reproduction.

But enough. You get the idea. The question is whether science has the proper characteristics to represent a case of cultural evolution. I think it does.

Begin with what evolves: Theories and explanations. Tools and apparatus. Organizations for training and encouraging the use and invention of these things. Patterns of work (think of Kuhn's divisions of science into normal and revolutionary or our own dichotomy between exploratory and confirmatory research). Individual research programs. The scientific languages: verbal, mathematical, and, if we believe Ariana, visual and so forth. In short, what we generally call disciplines and the things that make up disciplines. These are the equivalents of distinct populations in nature.

Do we have evidence of the evolution of disciplines or their various components? Absolutely. All of the sciences have diversified from two roots—natural philosophy and natural history—into literally thousands of specialized fields, each with distinct practitioners utilizing distinct tools, publishing in distinct journals, and attending distinct meetings. Moreover, this diversification can be (and has been) illustrated as an evolutionary tree.[17] Within disciplines, particular developments also can be modeled by evolutionary trees, as Gerald Holton has done (without, perhaps, realizing it) in his essay on "Models for Understanding the Growth of Science." He presents two examples of the development of research programs over many years and including many investigators: the study of acoustic shock waves and the development of magnetic resonance. Both of these examples display the branching structure expected of evolutionary developments.[18]

Indeed, the growth of science overall and of its subdisciplines appears to obey very much the same laws of sigmoidal development as do biological populations.[19] Even individual research projects develop according to an evolutionary tree model. One idea or experiment leads to another. Some of these ideas can be developed, and branch out in unexpected directions. Others are abandoned (become extinct) as interest, insights, or funding dry up—or perhaps as competition makes the research unviable, or as the entire enterprise is shown to be invalid. Competition for practitioners also exists between techniques and apparatus within any given field. One observes changing profiles of use, with some techniques becoming dominant at one period and then giving way to others at a later date.[20] New disciplines often emerge hand-in-hand with new experimental and mathematical techniques and are subsequently limited by them.[21] The point is that science does not develop in anything like a straight line. The unexpected detours that we have discussed in our sessions are part and parcel of the process, and can be explained effectively by an evolutionary mechanism.

Indeed, this is Stephen Toulmin's point: that novelty in science often has the same unexpected and unintentional origins as novelty in nature.[22] Compare two examples. Feathers are thought to have originated as modified reptile scales, conferring on the organism the benefit of increased insulation. Reptiles that inhabited trees then found these proto-feathers adaptively advantageous for gliding and breaking falls by increasing air resistance. They could jump farther and more safely. They survived more often than their unfeathered brethren. Eventually, sufficiently large feathers made it possible to fly. Similarly, atomic numbers were originally assigned to elements simply to designate their order. No one suspected that the "oneness" of hydrogen or the "eightness" of oxygen had any meaning whatsoever. Yet Mendeleev's periodic table and quantum

POLYNEUROPATHIES	DEPOSITION DISORDERS			AGING OF THE BRAIN
	SULFATED MUCOPOLYSACCHARIDES	LIPIDS	POLYGLUCOSANS	

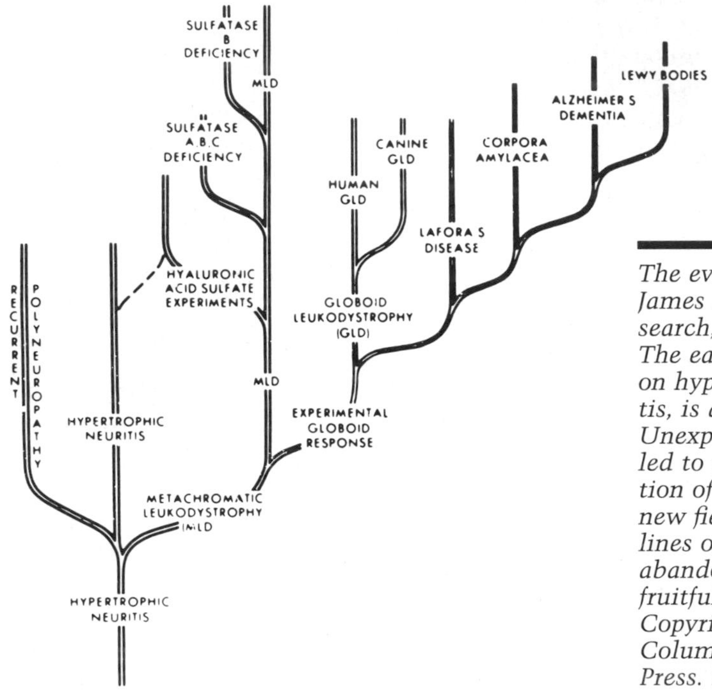

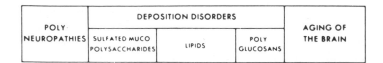

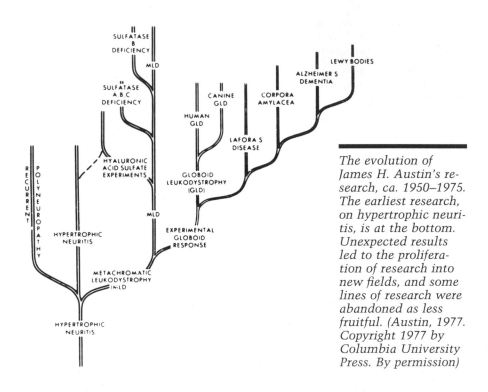

The evolution of James H. Austin's research, ca. 1950–1975. The earliest research, on hypertrophic neuritis, is at the bottom. Unexpected results led to the proliferation of research into new fields, and some lines of research were abandoned as less fruitful. (Austin, 1977. Copyright 1977 by Columbia University Press. By permission)

mechanics both make much of these numbers; they predict chemical activity. This is typical of what we've witnessed in our discussions. Ideas—Pasteur's cosmic asymmetric force, Fleming's bacteriophage hypothesis—are invented for one reason, and survive in some other modified and unforeseen form.

This is not to say that biological and cultural evolution are identical. One difference at least must be mentioned. Hybridization, which rarely arises in nature, occurs in the formation of many new disciplines and research programs. Physics and biology can fruitfully merge in the inception of molecular biology, for example; or a physical concept such as nuclear magnetic resonance can be transformed into a technique for studying the properties of chemical compounds and end up as a medical imaging device. These nodes of reintegration of previously split branches of the scientific tree are deemed by scientists and historians alike the most important events in the history of science.

What of modes of transmission? Think of the "scientific family trees" that Hunter and Constance presented. Whether scientific ability is passed from generation to generation by genetics, by apprenticeship, or by some combination of the two, modes of transmitting successful scientific traits exist. Having reviewed the relevant literature, I believe that the apprenticeship model accounts better for the data than does the genetic model—although genetics must set limits upon who can be apprenticed successfully. The transmission of attitudes, problems, and techniques from individual to individual is evident in the "tree" of physical chemists started by Berthollet and Laplace. We've also noted some cases compatible with a Lamarckian mechanism—Faraday, Darwin, and Einstein acquiring adaptive traits through self-training, which they bequeathed to their successors. So we have another basic element of an evolutionary system: a means of transmitting adaptive (and non-adaptive) traits—namely, teaching and learning.

Now, a mode of transmission is obviously useless unless there is a mechanism for creating diversity as well. If every scientist held the same theories and worked according to the same methods, no one would ever discover anything (or, alternatively, everyone would discover the same thing, which is no better). As Hunter, Imp, and Constance have made abundantly obvious, Berthollet, Pasteur, Fleming, van't Hoff, Arrhenius, and Ostwald were all trained differently, often much more broadly, than their colleagues. Furthermore, as Hunter pointed out, they mixed old and new theories, problems, and techniques in ways that most of their colleagues did not. If we view these mixtures of knowledge and skill as analogous to mixtures of genetic traits—in this instance, inherited through the traditions of what we've been calling codified science and science in the making—then each individual is literally a cultural hybrid. In terms of what each carries around in his head, he is identifiably different from his colleagues—as identifiably different as a mule is from a donkey or a horse, or a leopon from a leopard and a lion. Ariana's study indicates that such differences may be typical of successful scientists. Whether they also carry around a different mix of mental tools of thought, I leave to you to decide.

Other sources of diversity also exist; let me mention only two. First, experiment. All too often, the data that make a theory untenable result from an experiment designed to validate that theory. Thus, testing can lead to unexpected results, and unexpected results diversify the data base of science, making new theories possible. We've looked at plenty of cases, so I won't elaborate. Let me just say that without experiment to yield serendipitous results, theorizing would soon cease. Second, "copy error." In biology this refers of course to mistakes made in replicating DNA; it results in mutation. In cultural terms, it might mean a mistranslation or a misunderstanding of an idea or machine or social convention. Again, we've discussed a number of examples: Laënnec's mistaken analogy between the manner in which a piece of wood transmits sound and the way his stethescope worked, Thakray and Holton on Dalton's misunderstandings and misuses of Newtonian physics, the caloric theory, and so forth. If I recall properly, all of you seemed to think my position on these examples untenable. Nonetheless, the fact is that errors in comprehension can, and do, lead to novel ideas in science, as do misapplications of ideas from one field to another. If we are mainly interested in creating a diverse body of ideas and inventions to undergo selection at some later point in time, then we should welcome such errors. Whether one would want to go so far as to foster them, I cannot say. I don't imagine that being purposefully obtuse would be either sensible or productive.

Right. So now we have the equivalent of genetic diversity created by cultural hybridization, unexpected observations, miscontructions of existing knowledge. This is, in and of itself, insufficient. If the hybrid or mutant form is to be fecund, it must transmit its mix of knowledge to others. Some scientists fail at this point: Mendel, Herapath, Waterston, who knows how many others. Other scientists have been maladapted to the ecosystem that produced them. Think of the difficulties experienced by Einstein, van't Hoff, Arrhenius, and Ostwald in obtaining positions as researchers and teachers. They were denied, for a time, access to means of reproduction: students, colleagues, institutions, and in some instances, publications. When they finally did gain access to these means of reproduction, it was only at the peripheries of the scientific community.

This last fact strikes me as highly significant. Being a marginal member of the scientific community may be maladaptive in terms of most institutional contexts—one cannot obtain grants or get published—but being forced to the periphery may benefit the prototypes of new scientific species. It may protect those producing new ideas, inventions, or patterns of work from the fierce competition that characterizes all major scientific centers. Exclusion may give the maverick scientist a chance to mature before challenging the pack leaders for primacy. Just as geographical isolation is essential for the emergence of new biological species, isolation may be essential for speciation in cultural evolution as well.

Indeed, more than fifty years ago, the Australian physical chemist T. B. Robertson suggested that the geography of science might be an interesting topic.[24] No one seems to have taken his suggestion seriously, but they should have. As Hunter pointed out at our first session, while most Nobel laureates end up at major centers of science (or create them), most are the products of peripheral institutions and do their initial work on the peripheries of science. Think of Pasteur, pushed from Paris to the provincial towns of Strasbourg and Lille; or Fleming at St. Mary's Hospital. This phenomenon is particularly obvious with regard to the men who founded physical chemistry. If we take the most influential theoreticians and experimentalists of the new science to be Guldberg and Waage, Pfaundler. Horstmann, van't Hoff, Arrhenius, Ostwald, and J. Willard Gibbs, what do we find? Guldberg and Waage at the Royal Military Academy and the University of Christiania in Oslo; Pfaundler at the University of Innsbruck. Van't Hoff was trained at the Delft Polytechnich School, the universities of Leyden and Utrecht, and employed at the Veterinary College of Utrecht. Arrhenius was a product of Uppsala University and the Stockholm Högskola. Ostwald studied at the University of Dorpat in Tartu where he became a privatdocent, and then taught at the Realschule, finally graduating to a professorship at the Riga Polytechnic Institute after three years. J. Willard Gibbs was trained in mechanics at Yale University. Constance assures me that none of these schools had an international reputation in science at the time. Only Horstmann, trained and employed at Heidelberg, was at a recognized center of science, and he never rose above the rank of assistant professor, possibly because he was the only theoretician on the science faculty. One might therefore consider him a peripheral member of the institution.

Contrast the list I have just given with the contemporary centers of science. Using the *Dictionary of Scientific Biography* and J. R. Partington's *History of Chemistry* as guides, I determined that the centers of physical (especially thermodynamic) studies and of chemical (especially inorganic and electrolytic) studies during the period from 1860 to 1880 were clearly Heidelberg, Würzburg, Leipzig, Bonn, and Berlin in Germany; Zurich, Switzerland; Paris, France; and St. Petersburg, Russia. These cities had universities with faculty members of the caliber of Helmholtz, Clausius, Berthelot, and Mendeleev. Yet physicochemical theory did not originate in these centers of physical and chemical studies.

Indeed, if one looks at the development of physical chemistry within individual countries such as France, one finds that virtually all the major developments occur in the provinces rather than in the acknowledged scientific center of Paris. Raoult, Duhem, and Sabatier were all, like Pasteur, forced out of the centralized, bureaucratized Parisian world of prima donnas into peripheral institutions where they had less prestige but more freedom.[25] Perrin, the first physical chemist to break the Parisian barrier, did so only in 1910—twenty years after physical chemistry had been founded—when a position at the Sorbonne was created especially for him.

Now, speciation requires not only geographical isolation but the availability of new niches for the new species. Both J. D. Bernal and Ralph Gerard have suggested that the rate at which new knowledge enters universities is limited by the rate at which new appointments can be made or old institutional structures modified.[26] Gerard hypothesizes that innovation in science is closely related to institution building. We have seen examples in our sessions. It is clear that the reorganization of the French Academy of Sciences into the Institute, and the subsequent founding of the Institut d'Egypte by Napoleon, were crucial in allowing

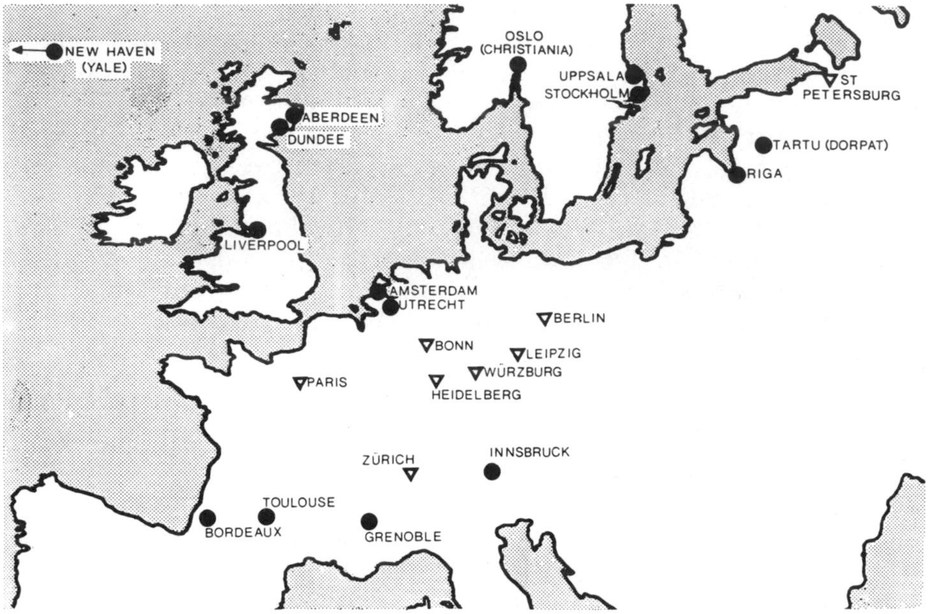

Map of Europe showing the major centers of physics and chemistry (triangles) *and the places physical chemistry originated* (circles) *during the period 1875–1890.*

new blood into the system. In just two years in Egypt (1798–1800), Berthollet invents mass action; the young Joseph Fourier begins his analytical theory of heat; Malus begins his physical experiments on light; J. B. Say invents Say's law of economics (that a new product will create its own market); the Rosetta Stone is discovered. In Paris, meanwhile, Laplace is writing some of his most important treatises, Proust discovers the chemical law of multiple proportions, and Lamarck invents his version of evolutionary theory. They are some of the most volatile years in the history of science. Can this be chance, or is it a result of the changes in the institutional structure of French science?

The same proliferation of niches made possible the physical chemistry of the latter part of the century. Initially, physical chemistry did not become institutionalized at any of the major centers of science. Virtually all of the first-generation positions were at new universities, or in universities that were growing sufficiently to establish a new department. Van't Hoff's position was created by the founding of the University of Amsterdam. Arrhenius became the first director of the Nobel Physical Chemistry Institute. Gibbs was appointed to a chair of mathematical physics created for him at Yale. In Britain, men like Thomas Carnelly and F. G. Donnan set up the first chemistry departments that included physical chemistry at Dundee, Aberdeen, and Liverpool—new universities founded between 1880 and 1890 as part of the university reform movement.[27] The same thing occurred in the United States. Physical chemistry established itself either in new universities such as the California Institute of Technology, Stanford University, and the University of Chicago, or at older institutions undergoing reformation and expansion of their chemistry programs, such as Harvard, Columbia, and Cornell.[28] In fact, Robert Bruce, studying the emergence of American science into world leadership at the beginning of this century, points to this institutional expansion following the Morill Land Grant Act of 1862, and the concomitant decentralization and pluralism, as the crucial factors permitting innovation.[29]

Furthermore, what is true of physical chemistry appears to be true for science in general. The introduction of engineering as an academic discipline in the United States occurred first in new institutions such as the Rensselaer Polytechnic Institute (founded 1824) and the Ohio Mechanics Institute (founded 1828), not at existing universities such as Harvard, Yale, Princeton, and Penn. The introduction of the new science of biochemistry occurred around the turn of the century in universities undergoing reform of their medical curricula or at new institutions: at the Carlburg Laboratory in Copenhagen (founded 1875); at the Pasteur Institute (1888); at what became the Lister Institute (1891); at Rockefeller Institute (1901); at the University of Liverpool (1902); at the Danish Serum Institute (1903); at Jacques Loeb's physiological laboratory, built in 1904 at the new University of California (1872); at the Kaiser Wilhelm Institute (1911); and so forth.[30]

American and German physiology and medicine during the nineteenth century demonstrate the same phenomenon. American physiology found its niches in virtually the same set of institutions as physical chemistry—a notable correlation, to my mind. In virtually every case, physiology was integrated into medical programs undergoing extensive reorganization or into new programs.[31] German physiology followed the same pattern, largely developing in newly established, independent facilities founded between 1868 and 1890 in Leipzig, Munich, Heidelberg, Berlin, Bonn, and so forth.[32] According to Ben-David, decentralization, brought about by rapid expansion of universities and facilities, increased intranational competition; fluid exchange of students and professors between universities was crucial to the success of the science.[33] When the system ceased to expand, became bureaucratized and centralized, German contributions to physiology and medicine ceased to attract worldwide admiration. This is not to say that the number of physiologists in

Centers of American Physical Chemistry (1890–1910) and Physiology (1890–1920)

Institution	Founded	New chemistry program	New biology or medicine program
Physical Chemistry and Physiology			
Harvard University	1636	1912[a]	1871, 1923
Yale University	1701	1878[b]	—
Columbia University	1754	1898[a]	1891
University of Michigan	1817	—	1881, 1923
Toronto University	1827	1887[c]	1887
University of Wisconsin	1848	1905[a]	1907[d]
Cornell University	1865	1890, 1898[a]	1898[d]
Johns Hopkins University	1876	—	—
Case Western (School of Applied Science)	1880	—	—
Stanford University	1885	—	—
University of Chicago	1890	—	—
Physical Chemistry			
Massachusetts Institute of Technology	1861	1903[a]	
University of California, Berkeley	1872	—	
California Institute of Technology (Throop)	1891	—	
Physiology			
University of Pennsylvania	1740		1875
St. Louis University	1832		1903[d]
Washington University (St. Louis)	1853		1899[d], 1914[a]
Rockefeller Institute	1901		—
Mayo Clinic	1915		—
University of California, Los Angeles	1919		—

Sources: Dolby, 1977; Geison, 1987; Servos, 1979, pp. 68–70, 149.
a. Laboratory built or organized.
b. Position created for J. Willard Gibbs.
c. University Federation Act passed by Province of Ontario.
d. Medical school founded.

Germany declined, nor the number of papers published. Rather, the limitation on senior positions meant that more and more scientists stayed at the *privatdocent* (postdoc) level longer and had less freedom in their research and opinions—exactly the situation we have today. There was no outlet for new ideas or new strategies of research. As Zloczower suggested, the growth of discoveries may be limited by the capacity for expansion or change inherent in the institutions of science.[34] An evolutionary model with its emphasis on an ever-shifting environment is fully compatible with such a conclusion.

Nye has drawn similar conclusions regarding the perceived general decline in the world status of French science, and the eventual shift of growth from Paris to the provinces at the turn of this century:

> *France's domination by its capital city had long frustrated reformers, and in science there is no doubt that the entrenchment of prominent scientists in the Parisian mecca resulted in inflexibility, overspecialization, and a lack of competition among teaching and research institutions. Since the country did not have a network of dynamic university centers, her national creativity could be weakened by a predominance of toiling mediocrities or intolerant* prima donnas *in the capital city. Indeed the conclusion by modern social psychologists that status differences within a group seem to increase conformity suggests that France's centralized, rigid academic hierarchy might have substantially inhibited impulses for originality within the scientific community.*[35]

As I mentioned before, large populations in an unchanging environment evolve extremely slowly. There are no new opportunities for "sports" to exploit, nor new selection pressures to alter the profile of the existing population.

Whatever the rate of overall change, individuals within a population compete for resources. Those who have access to resources protect their territory; those who lack resources invent strategies to acquire what they need. It is fairly obvious that resistance to new ideas—and indeed the entire peer review system—are means of protecting or controlling access to resources. The territorial imperative. Of more interest is the equivalent of interspecific competition. When a new species enters an existing ecosystem, interspecies competition results—the territorial imperative writ large. The cultural analogue would be competition between practitioners of neighboring disciplines for territory or resources lying between disciplines, or resistance to the incursion of a new hybrid. Both phenomena characterize the emergence of physical chemistry following the work of van't Hoff, Arrhenius, and Ostwald. The first manifestations of this competition occurred within the new science itself. Was physical chemistry to be the embodiment of Gunning, van't Hoff, and Ostwald's "allgemeine Chemie"—a coordinating discipline for the rest of science—or just a new specialty sandwiched between physics and chemistry? Both in Germany and in the United States, practitioners of the new field set forth alternative disciplinary programs, which competed for niche space within existing institutions. In both countries, those advocating physical chemistry as a new specialty succeeded in ousting the generalists after a couple of decades. Apparently it was simply too difficult to train and place large numbers of generalists in an already specialized scientific system—something Ariana might take note of.[36]

There was also competition with neighboring sciences, for example, with inorganic chemistry. Alfred Werner, the founder of coordination chemistry, actually resigned from the editorial board of the *Zeitschrift für anorganischer Chemie* in 1904 because it was being overrun by physicochemical articles and no longer addressed the needs of the inorganic chemist.[37] The new physical chemistry also met resistance from immunologists and physiologists. In 1901, Thorvald Madsen enlisted Arrhenius's aid in examining immunological reactions. Von Behring, Paul Ehrlich, and other immunologists were puzzled by the antibody-antigen reaction. Did the antibody destroy the antigen by cleaving, as an enzyme might do; did it neutralize it by a chemical reaction; or did it combine reversibly with it? Arrhenius and Madsen demonstrated with physicochemical methods that the reaction was reversible. Ehrlich refused to accept this conclusion or the techniques, claiming that physical chemistry had nothing to offer immunology.[38]

Similarly, in both England and France the physical chemists began encroaching on physiology, only to meet stiff resistance. In England, Frederick Gowland Hopkins warned physiologists in 1913 that the physicochemical approach to life was becoming too influential and that, unless something was done quickly, organic chemistry, which in his view ought to be the basis of physiology, would soon be ignored altogether. He went on to lead a very successful campaign to reinstate organic chemistry as the basis of physiological chemistry.[39] French physiologists trained in physicochemical techniques, such as Victor Henri and Emile Terroine at the Sorbonne, met similar resistance. Unsuccessful attempts were made by classically trained physiologists at the turn of the century to deny these physical chemists access to physiological problems.[40] Tuzo Wilson reported the same sort of territoriality when he first mixed geology and physics during the 1920s.[41]

In short, what happens is this. Hybridizers and generalists change the nature of scientific work, redefine disciplinary boundaries, and alter training requirements, thus compromising existing adaptations. They provoke resistance, which can be viewed as a natural consequence of competition for intellectual niche space. If Constance is correct that the emergence of new disciplines, and the rethinking of old ones, is always accompanied by philosophical innovation as well, then one would expect every major development in science to cause such a redefinition of niche space, and hence to encounter resistance. Let me assert that as a prediction.

What makes up a niche? Here is a sticky question. I have been maintaining, in accordance with most philosophical accounts, that a niche is defined by purely intellectual criteria: niches are unsolved problems and are filled by solutions. I am no longer sure that this representation is adequate. Clearly niches are also defined by personal and sociological factors. The evaluation of an idea is always carried out simultaneously with the evaluation of its originator. Access to the tools of scientific problem solving—and I am referring not to Ariana'a tools, but to the more conventional ones of equipment, funds, libraries, students and colleagues, and so forth—access to these tools is entwined with success at problem solving. Problem solving, conversely, is dependent on access to the tools. In short, I perceive a niche as a complex overlap, on the model of a Venn diagram, of a large variety of factors, including (though not necessarily limited to) unsolved problems, existing techniques for problem solving, resources available for problem solving, obstacles and inducements for solving particular problems, sources of competition, and reward systems. So, the scientific niche is analogous to a niche in an ecosystem and is defined by just as many factors. Among these must be considered logical, empirical, sociological, historical, and, in deference to Ariana, even aesthetic factors.

Enough; by now you either accept the plausibility of my position or do not. Clearly it will take a great deal of work to flesh it out. Consider this essay as a cultural analogue to Darwin's 1844 sketch of his theory of the origin of species. Give me fifteen more years of uninterrupted time to work out the consequences, as Darwin had, and we'll see whether an "Origin of Cultures" is possible.

Failing that, let me at least try to ward off unwarranted criticisms. One of the principles we have been aspiring to is to accept all data as valid, and therefore to invent theories that explain apparent contradictions. Let me, then, raise the most important apparent contradiction: scientific revolutions.

Kuhn maintains that science progresses by revolution, sudden changes of perspective that invalidate past theories.[42] I. B. Cohen claims to have documented the many forms that such scientific revolutions have taken.[43] Let us accept that there is an appearance of sudden change, a break with the past. How accurate is that perception, and from what perspective? That is the crux of the matter. For me, the most notable aspect of Cohen's book is his documentation that virtually every revolutionary from Copernicus to Einstein denied that he had created a revolution.[44] (I ignore charlatans such as Mesmer and Freud, who go to great lengths to claim revolutionary status they don't deserve.) Why this difference between personal and public perceptions of revolutions in science?

The key is the notion of stylistic fads in science. We characterize the science of any particular time, including our own, by what is most fashionable. We ignore that which is passé or in bad taste. So consider a thought experiment: You are Arrhenius or Einstein or someone of that nature. You are working outside the mainstream of science, focusing on problems that other scientists deem unsolvable or uninteresting. Your sources are, say, Berthollet's seventy-five-year-old mass action concept, or the Faraday-Maxwell fields-of-force ideas that have not yet penetrated academic physics. You utilize unusual techniques such as Edlund's method of measuring conductivities or non-Euclidean geometries. In short, you place yourself at the nexus of a set of largely unknown or ignored traditions of research. Now, you utilize these old, unusual, or unpopular sources to address some fundamental question everyone else thinks is old hat or unimportant. You find an anomaly or paradox. You resolve it. But to do so, you must repattern the rest of your field. You publish. What does this look like to the rest of the scientific community? I say it looks like revolution. The new pattern seems to come out of nowhere, and bears no resemblance to its predecessor. To those unfamiliar with the sources you have drawn upon and the problems you have addressed, your ideas appear revolutionary. They relate to nothing in their experience. They must suddenly rethink everything they thought they knew. Hence Kuhn's gestalt-shift analogy—or better, Jenny's puzzle model. For almost everyone but the inventor of such novel ideas, the genesis is completely hidden. And yet, as you have forced me to admit, these perceptual shifts have quite logical origins.

The consequences are obvious. The historian trying to reconstruct an internal history of what happened—as Kuhn has done with the so-called quantum revolution—finds that from the individual scientist's perspective (in this case, Planck's) there has been no revolution.[45] New ideas evolve slowly as a result of something like Constance's recursive process, one problem leading to a solution, leading to another problem, and so forth; only a few of these bear fruit. Novelty results from an elaboration or hybridization of existing ideas, innovation from tradition.[46] That's what the scientist sees himself doing.

But the historian reconstructing an external or social history sees just the opposite: a sudden emergence of unexpected novelty into an unprepared and unsuspecting scientific community. A huge surprise on a sociological scale. This is what Kuhn demonstrated in his *Structure of Scientific Revolutions*. So, in the history of science, as in evolutionary biology, as we have the problem of explaining the phenomenon that Gould and Lewontin call "punctuated equilibrium"—a historical record that appears to evolve in alternating stable states and sudden leaps. Once we understand the geography of speciation, we can see that these leaps are only apparent. The gradual evolution of a new species of organism or idea has taken place in obscure isolation while the major population centers of science have reached the sort of dynamic equilibrium Kuhn calls "normal science." Unless we can locate the isolated process of rapid change in time and place, the only records will be of the sudden emergence of the new species as it breaks out of its isolation to compete on the broad plain of science for existing niches with previously adapted species. Thus the problem of punctuation versus gradualism, of evolution versus revolution, is a false one resulting from a failure to consider perspective.

Now, make a leap of faith. Let us suppose my evolutionary model is correct. Several interesting consequences follow. First, we must totally rethink the so-called scientific method, abjuring the concept of a single truth. Science does not progress by finding one answer to one problem and trying to prove or disprove it, nor by inventing huge, monolithic paradigms and research programs that direct huge masses of scientists in their work. Science is still, and always will be, an individual process of discovery which advances only when scientists invent as many problem formulations as possible, elaborate as diverse a set of solutions as imaginable, and then select those problem-solution pairs that most adequately meet the logical, empirical, sociological, historical, and aesthetic criteria that the individual scientist applies in his daily work. We must accept the fact that no single problem-solution pair will satisfy the needs of every scientist, nor adequately describe every facet of nature, for the simple reason that every scientist is heir to a different mix of codified and uncodified science. Therefore it is not consensus for which we must strive, but the elaboration of as many adequate descriptions of nature as we can imagine—in short, the sort of complementarist view espoused by Bohr. Otherwise, we may miss the idea that has the greatest evolutionary value for the future, dismiss an idea such as Kekulé's benzene ring that can only be validated a hundred years hence or, worse, adopt and maintain a dogmatic belief in a set of ideas that quickly play out their potential and leave the field ever smaller and less interesting questions to address.

Second, it follows from my model that those who are trained and work entirely within an existing niche cannot make a discovery. They are Imp's stenokates. They may very well be adaptive within a narrowly defined region of knowledge and work, and they may elaborate or refine the existing pattern of research and teaching, but they will not be capable of evolving beyond the niche. Thus the importance of my earlier statement that a homogeneous population in a static environment is in equilibrium. Strict adherence to tradition will stifle development. We must therefore avoid adopting uniform curricula for the sciences; we must avoid like the plague prerequisites of courses or tests or accomplishments to become a scientist; we must abjure the systematization of the educational process. If we do not, every new scientist will be trained just like all the rest. We will lose the diversity necessary to the further evolution of science. Robertson warned us of this when he wrote that the "centripetal tendency" of investigation to congregate in large centers of science is "a serious handicap to the accelerated development" of science that leads to "a certain uniformity of thought which not infrequently becomes indistinguishable from prejudice."[47] Leopold Infeld and Ludwig Fleck sounded the same warning.[48] The point is this: If we clone scientists, we will produce cloned science. Clones are economical to produce, but infuriatingly uniform in their patterns of work and thought, as the Soviet Union and Japan have found out. These nations produce many scientists, but few scientific breakthroughs. Science can never be organized like an assembly line. Unlike industry, it must strive for maximum diversity, not maximum uniformity. We must, as Imp recommends, strive to produce eurokates—broadly adaptive individuals who can take advantage of rapidly changing sciences—not the stenokates who populate our scientific centers today.

Third, we absolutely must eradicate our present dependence upon a single source of scientific and technological funding, namely, government. As long as the sources of money are centralized, and the decisions concerning funding bureaucratized, there is no hope for the maverick. He will be selected out. On the contrary, if we are to foster mavericks, we must expand the base of financial support for science and technology as widely as possible—state, local, private, endowed, venture—and free its use as far as possible from all constraints. We must fund people rather than projects. We must fund scientists working at peripheral institutions of science equally with those working at established centers. We must fund unpopular topics as well as the latest fads. In other words, we must foster pluralism and competition—the mainstays of evolution—and allow many differently trained individuals to try their hands at many diverse problems, rather than try to build in success by planning. As Clifford Truesdell said, there is a corollary to the Peter Principle: A system that is perfectly organized is on the verge of collapse. It can tolerate no change. But science *is* change. It cannot therefore be organized; it can only be allowed to develop, or be prevented from doing so.

Fourth, we must seriously reconsider our selection criteria for scientists. It makes perfect sense to select members of a population on the basis of their ability to gather the most food, produce the most gadgets, and rock the boat as little as possible—as long as the purpose of that population is to maintain itself as efficiently as possible without change. But I reiterate: Science *is* change, "effective surprise," so it makes no sense whatsoever to choose scientists by these criteria. Pioneers are never efficient agriculturalists or manufacturers—they are hunters and gatherers. They make what they need or they find it. They cannot therefore be evaluated as we evaluate businesses. They must be evaluated by the blank spaces on the map they fill in or the paths they blaze through the wildernesses between disciplines. They must be evaluated in terms of the potential resources they discover, not the actual value of what they can carry home on their backs. More directly, we must stop evaluating scientists in terms of the grant dollars they accumulate, the number of publications they produce, and how prestigious their positions and titles are. The most important criterion must be the extent to which a scientist's work threatens to unsettle our most cherished beliefs.

Fifth, existing institutions must become more fluid. I have proposed that the great advances in science in the past have coincided with expansion and reorganization of the institutions of science. But because exponential growth of the economy, scientific funding, and the number of scientists cannot be maintained indefinitely, science can no longer depend on the expansion of existing institutions or the founding of new ones to stimulate its growth. Science will stagnate waiting upon the old to yield their places to the young, unless we purposely alter our existing institutions actively to promote change. I say this recognizing that I am asking the impossible; yet it must come to pass. We must emulate the plant and animal breeders who have produced outputs unbelievable to anyone unfamiliar with the use of hybridization, growth hormones, fertilizers, pest controls, and so forth. We must purposefully attempt to create new scientific hybrids instead of tracking people into majors at the age of twelve or eighteen. We must purposefully invent niches for these hybrids—"queer ducks," I've heard Jonas Salk call them—in existing institutions so that they can invigorate the existing stock. We must nourish their heresies and create the conditions in which they can flourish, providing artificial isolation if necessary, encouraging them to find or make the niches in which they will best grow, protecting them from the attacks of unimaginative colleagues and self-satisfied bureaucrats alike. And we must accept that only a few of them will succeed.

In short, we must study the conditions under which scientists and inventors flourish and then artificially recreate those conditions. We must invent a new science of science that studies scientists in exactly the same way that scientists study ecological determinants of the evolution of species. Should we fail to invent such a science, the alternative we face is the situation faced by agriculture a century ago: the over-exploitation of existing soil, crops, and breeds, with resulting periodic failures and famines. If the data gathered by Imp and Constance concerning the recent downturn in numbers of discoveries and inventions are accurate, we may already be witnessing just such a failure.

In conclusion, let me make some predictions. Every theory should be testable, and mine is no exception. One could, of course, test many of my assertions by historical study of other emerging sciences. Is the evolutionary picture I've drawn of physical chemistry typical? Are peripheral hybrids the originators of other sciences? Is the institutional picture the same? Philosophers call this sort of historical testing "retrodiction."[49] But I prefer to look forward. So let me make some predictions about the future.

First, the great inventions and discoveries will take place in countries having expanding economies that support the institutional growth or reconfiguration necessary to accommodate mavericks. Second, these mavericks can be expected to be oddly trained hybrids combining old and new theories, data, and techniques, and crossing established disciplinary boundaries. Third, they will most likely be trained at new and peripheral institutions or they will work in newly emerging fields rather than at recognized centers of science in established disciplines. Fourth, cultures (and on a more limited scale, institutions) that promote educational and ideological pluralism, intellectual freedom, and idiosyncratic behavior—the equivalent to tolerating genetic "sports"—will have a higher proportion of inventors and discoverers than will highly conformist and ideologically intolerant cultures or institutions. Fifth, the period preceding any great breakthrough in science will be characterized by the elaboration of a tremendous diversity of possible solutions from which the fittest will be selected. Sixth, consensus will be found only in the codified fields of science in which little or no research is presently being done. These six predictions would not, I believe, follow logically from anything Kuhn, Lakatos, Popper, Hanson, or most other analysts of science have said, and so they provide a reasonable test of the differences between our theories.

◀ TRANSCRIPT: Discussion

▷ **Imp:** The recalcitrant skeptic outdoes us all at the eleventh hour!

▷ **Ariana:** And should be reprimanded severely for not engaging in this sort of speculation more often.

▷ **Richter:** But I could not have reached this position save for my skepticism. What authority could have shown me this path?

▷ **Ariana:** Many. And warned you from it! But I *do* have one problem. How can you tell when someone is peripheral?

▷ **Richter:** Ah, a problem of definition for which I have no pat answer.

▷ **Imp:** Let me make some suggestions from personal experience, then. You know you're peripheral when you receive the abstract form for the neurosciences meeting and of the some two hundred and fifty categories they divide the field into, you can't find a single one that comes close to describing your research. Or you send a paper into *Science* and they ask for the names of experts in the field to act as referees and there aren't any because you've just invented the field. Or your subject never appears in the solicitations for grant applications through which NIH and NSF try to control the direction of science.

▷ **Hunter:** Or you can't find a job description that suits your multidisciplinary background and talents. The point is that peripherality, if we can call it that, can't be defined except with regard to the current state of science.

▷ **Richter:** Yes. Unfortunately there are some ambiguities in my theory as it stands. As you see, some are definitional: what is peripherality, or stasis, for example. And I frankly admit that I have not established firm boundary conditions for my ideas, partly because the data base is woefully tenuous. Constance?

▷ **Constance:** But not as tenuous as you imply. Edge and Mulkay's study of the emergence of radio astronomy in Britain certainly fits your evolutionary model.[1] The people involved were virtually all hybrids and worked on the institutional peripheries of science. I also understand that Ronald Giere and David Hull are preparing books employing the evolutionary analogy to explain various social and epistemological aspects of science.[2]

▷ **Jenny:** There are a couple of historians you ought to read as well, Richter. Paul Colinvaux has written a book called *The Fates of Nations: A Bio-*

logical Theory of History attempting an evolutionary interpretation of warfare from prehistory to the present.[3] He really utilizes only the concept of niche, but there are some similarities in what you both see happening. And then Roderick Seidenberg argues in his *Post-Historic Man* that as society increases in complexity, the drive to organize pushes us toward just the sort of consensual stasis you warn us against.[4] He predicts that the need to preserve large organizations will eventually make it impossible for the movers-and-shakers to move and shake. Hence the end of innovation.

▷ **Constance:** Or even extinction in the competition between ideas or cultural systems. And that's something I missed in your presentation of evolutionary theory, Richter: extinction. Don't you need an equivalent here?

▷ **Richter:** Certainly. You have a suggestion?

▷ **Constance:** Well, why not invoke "dead" theories such as the phlogistic and caloric theories of heat; Aristotle's and Driesche's entelechy (the driving force for embryological development); medieval concepts of the eye emitting rays of light; Berthelot's theory that all chemical reactions go in the direction that will create the most work; the belief that proteins carry genetic information—the dominant theory from 1850 till nearly 1950—that sort of thing. Essential branches of the historical development of these sciences that we leave out of textbooks because they turned out to be dead ends.

▷ **Imp:** Dead ends, but nonetheless the stock from which the new "species" arose. Along the same lines, you could also explain why ideas (like favorable mutations or new combinations of genes) probably have to arise several times—multiple discoveries, if you like—before they survive. Didn't you tell us, Constance, that the idea of the tetrahedral carbon atom was, what, about seventy years old and had been reinvented six or eight times by the time van't Hoff produced a viable form? That Lister, Pasteur, Burdon-Sanderson and others had observed the same sort of microbial antagonisms that led to the development of modern antibiotics by Fleming, Waksman, and their successors? The rediscovery of Mendel and Waterston. These examples suggest that variation isn't sufficient: The individual in whom the recombination or mutation occurs must be adapted to a niche appropriate to the expression and survival of the new trait. Therefore, the fact that a new idea or observation fails in one context is no assurance that it will not turn out to be the source of viable developments.

▷ **Hunter:** Okay, but let *me* play skeptic for a moment. Why do you want to do this sort of evolutionary analysis, Richter? Don't we already know all this? As you say, even Kuhn talks about evolution of science in his books, and we certainly use the idea as a metaphor when we teach science. So where's the originality? I mean, think of Martin Harwit's recent analysis of the history of astronomy. He's concluded there are seven traits common to discoveries:

> *1. The most important observational discoveries result from substantial technological innovation.*
> *2. Once a powerful new technique is applied . . . the most profound discoveries follow with little delay . . .*
> *3. A novel instrument soon exhausts its capacity for discovery . . .*
> *4. New . . . phenomena frequently are discovered by . . . researchers originally trained outside [the field] . . .*
> *5. Many of the discoveries of new phenomena involved the use of equipment originally designed for military use . . .*
> *6. The instruments used in the discovery of new phenomena often have been constructed by the observer and used exclusively by him . . .*
> *7. Observational discoveries of new phenomena frequently occur by chance—they combine a measure of luck with the will to pursue and understand an unexpected finding.*[5]

I think we can probably agree that most of the case studies we've looked at fit Harwit's conclusions, with a few modifications. Certainly most of our scientists were multidisciplinary and worked in fields outside their expertise. Most transferred techniques or theories from one field to another. Several, such as Pasteur and Fleming, invented their own techniques and equipment and utilized it themselves to make the crucial observations. No one found what he expected. So the question is whether an evolutionary system would foster these traits or not—and whether you even need an evolutionary formulation to make sense of it all.

▷ **Richter:** Allow me to tease your rather confused objections apart. First, why an evolutionary analysis? Because it is one thing to recognize analogies or to speak of science as evolving in a metaphoric sense and something totally different to assert that there are evolutionary *mechanisms* at work in science.

▷ **Constance:** Sure. The importance of Darwin wasn't his recognition that species evolve, but his invention of mechanisms by which evolution could occur.

▷ **Richter:** There is a second reason as well. The synthetic theory of evolution has not yet succeeded as a literal explanation for cultural evolution. Read the literature. Anthropologists and sociologists might as well be living pre-Darwin as far as their view of mankind goes. Virtually every definition of culture they produce sets mankind apart from the rest of nature and creates a gap unbridgeable save by miracle: man the exception. I refuse to believe it. Mankind evolved; culture evolved with it. Thus, the culture we have today resulted from evolution. It is necessary to understand it as an evolutionary process. The evolution of science is the place to start our studies of that process.

But on to your second point, that we already know everything I have said. Undoubtedly. Everything Darwin said had been said by someone before Darwin, but they had not discovered the organizing principle. It is not what we know, but how we know it and what it means that counts—the statement of principles and mechanisms that not only explain, but also predict and retrodict. You forget that theories, besides having an element of novelty—at least in the way they order data and in the way in which the basic principles are applied—must also explain the accumulated observations and principles that have preceded them. Your historical criterion of theory evaluation. So, of course what I say sounds familiar. It must. The question is whether anyone has ever produced a theory as encompassing as mine and as unified.

Which leads to your third point: Harwit's principles. Most of Harwit's traits come down to saying that discoveries occur when unusually trained people transfer information and techniques from one field to another and hybridize them. Within the field, the initial innovations come quickly, and then the most adaptive mode of working and thinking takes over. Thereafter, new innovations have little chance to develop except on the peripheries. All true. All subsumed in my theory.

But that point brings up some quibbles I do have with a few of Harwit's conclusions. From what I've read, he has a pronounced dislike of theory—a particular sore point with me. Apparently he found that theory predicts astronomical discoveries only about half the time. I don't think that's such a bad record, myself. Moreover, he ignores the fact that many of his "chance" discoveries occurred as a result of testing an incorrect theory. Therefore, add theory testing to his list of common traits—even when the theory turns out to be wrong.

Furthermore, to military development of new technologies must be added industrial developments. Many profound insights—Berthollet's, Pasteur's, most of thermodynamics—have occurred because scientists had knowledge of industrial techniques that presented insoluble problems or lacked theory. Also, while a new instrument soon exhausts its capability of discovery within a field, one of the most important spurs to discovery is the transfer of instrumentation from one field to another—hybridization. Think about Arrhenius's attempt to apply electrolytic equipment to the problem of molelcular weights, or the introduction of spectrophotometry from chemistry into astronomy. NMR techniques are yielding few surprises in chemistry anymore, but are being applied with the most exciting results in physiology and medicine. Thus, equipment doesn't lose its power of discovery; people do. They reify past practice rather than apply it to new situations and to the study of new phenomena. Also, in my experience (admittedly limited), directors of large, expensive technical installations tend to covet their equipment and prevent "outsiders" from using it. One must prove insider status to gain access to the equipment, although insiders have little probability of turning a field on its head. So it's a sociological problem rather than a technical problem. What needs to be done, it seems to me, is to allow big, expensive facilities and equipment to be used by the developers for the first decade, and then to turn them into national facilities for the use of any qualified scientist, regardless of field, on a first-come-first-served basis. Sure, a lot of crazy experiments would be tried in consequence. But the probability for surprise will increase.

▷ **Imp:** Which provides us with as good an introduction to Hunter's report on obstacles and inducements to exploratory science as we're likely to get. But first, I just want to say that there may be an opportunity to test at least one of your predictions. At the same time that the Soviet Union is trying to foster innovation by decentralizing control of science, the University Grants Committee in the United Kingdom is considering adopting a proposal to classify science departments into three levels: The top level would produce all of the Ph.D.s and do the primary research; the second level would grant master's and undergraduate degrees and do minor, inexpensive research; and the third level would be limited to introductory teaching and no research. The rationale is that they are currently "wasting" money with grants to second- and third-rate investigators at poorly equipped institutions.[6]

▷ **Hunter:** Such as van't Hoff at the Utrecht Veterinary College, Arrhenius in Stockholm, Ostwald at Tartu, or Fleming at St. Mary's.

▷ **Imp:** Well, as I pointed out at the outset of the Discovering Project, we put the very process of science in jeopardy whenever we make policy without a firm historical and philosophical perspective. So tell us, oh, Hunter: If Richter's right about the evolutionary nature of the enterprise, how may the unusual hybrids and "queer ducks" of science best be served that science itself may prosper?

▲ HUNTER'S REPORT: Obstacles and Inducements to Exploratory Research

As scientists we must accept that the world has limited resources. In all fields we must be alert to cost-effectiveness and maximization of efficiency. The major disadvantage in the present system is that it diverts scientists from investigation to feats of grantsmanship. Leo Szilard recognized the problem a quarter-century ago when he wrote . . . that progress in research could be brought to a halt by the total commitment of the time of the research community to writing, reviewing, and supervising a peer review grant system very much like the one currently in force. We are approaching that day. If we are to continue to hope for revolutions in science, the time has come to consider revolutionizing the mechanisms for funding science and scientists.

—Rosalyn S. Yalow, medical physicist

Writing projects was always an agony for me . . . I always tried to live up to Leo Szilard's commandment, "don't lie if you don't have to do." I had to. I filled up pages with words and plans I knew I would not follow. When I go home from my laboratory in the afternoon, I often do not know what I am going to do the next day. I expect to think that up during the night. How could I tell then, what I would do a year hence?

—Albert Szent-Györgyi, physiologist

Much of the freedom of science is now being legislated away, and we are approaching the Russian system of directed research—protestations to the contrary.

—Irvine H. Page, M.D.

I've found all three of the previous reports most interesting. I am particularly surprised and gratified that virtually all of my own observations and suggestions are consistent with Richter's evolutionary model. You'll see what I mean as I proceed.

My topic is obstacles and inducements to pioneering or exploratory research. How do age, education, employment, funding, equipment, organization, planning, and peer review affect the process of discovering? None of these things, I believe, is nearly as important as the cultivation of discoverers—rare, creative people who like to explore the frontiers of science and who are not afraid of their own ignorance. People—not money, equipment, facilities, granting agencies, or peer review boards—make discoveries and inventions. I believe that until we understand this fundamental point, science policy will only obstruct exploratory research. If science is to progress as rapidly as all of us desire, the obstacles must be removed and replaced with appropriate inducements. This report is an attempt to sketch what form these inducements might take.

Let me begin by painting a portrait of the pioneering scientists we've encountered in our sessions. I hope that you'll agree with me at the outset that there are different types of scientists: pioneers and explorers, developmental types, confirmers, axiomatists, technicians—classicists and romantics, Athenians and Manchestrians, Apollonians and Dionysians—call them what you will.[1] Certainly van't Hoff, cutting his rough block and leaving it for others to polish, is a different sort from Wilder Bancroft, one of Ostwald's American students, who wrote, "There is nothing I delight in more than a problem in which ninety percent of the work has already been done";[2] and both of these men differ again from Planck, a codifier whose goal was the complete mathematization and axiomatization of the second law of thermodynamics. As Ariana says, we need all these types and more if science is to progress. So, point one: A single mode of science education, of organization, of hiring and funding, will not generate and sustain the diversity of scientific types we need. The question I want to focus on in particular, however, is how to foster the pioneers and mavericks.

Education first. What kind of education makes a pioneering scientist? A diverse one. Ariana says it, Constance's report suggests it, Richter's theory depends on it. And from my own observations, I think there is no doubt of it. Three years ago, the National Academy of Sciences convened to honor the United States' 1983 Nobel laureates. The previous laureates Vassily Leontief, an economist; Herb Simon, an economist turned psychologist and computer programmer; David Baltimore, a virologist; and Murray Gell-Mann, a physicist, were asked to speak. All of them, without prior consultation, delivered the same message, distilled by Lewis Branscomb as: "The most dramatic progress was to be made by those who mastered the tools and ideas of a range of previously weakly related disciplines, and based their work on the emerging new mathematics."[3] Save for the fact that the experimentalist needs somewhat different tools than the mathematical theorist, I can hardly imagine a better summary of the training of Berthollet, Pasteur, van't Hoff, Arrhenius, and Fleming, not to mention the founders of fields such as molecular biology and geophysics. None of these scientists was trained within the existing patterns of professionalized education; arguably, they altered the face of science not by their expertise but by their highly trained eclecticism.

The lesson of these scientists is that scientific progress waits for no one. The map of science is constantly changing. The discoverer and inventor cannot know beforehand what she or he will discover or invent, and therefore needs not intensive and narrow training or expertise but a sure grasp of as wide a range of physical and mental tools as possible, and the ability and desire for self-motivated learning. As Sheldon Glashow has said, "Most successful people in physics made it by going off by themselves and learning what they wanted to."[4] What they learned from their mentors, if they had them, was not some body of pre-established knowledge but how to deal with the ambiguities and confusion of scientific exploration. Thus, as science changed, they knew how to retrain themselves to change with it.[5]

Mavericks are also unusually aware of the diversity of schools of thought concerning any particular subject. This awareness may stem from their autodidacticism, an innate preference for global thinking, or, as in the case of van't Hoff and Arrhenius, direct experience of different approaches during their important *Wanderjahren*. The earlier this acquaintance with the diversity of approaches to a subject occurs, the greater the potential young investigators have of recognizing some essential contradiction, paradox, or anomaly at the heart of their field. Thus, two changes in our educational system are needed to promote broad, independent thinkers: Students, both undergraduate and graduate, should be encouraged to spend one or more years studying at institutions other than their own so as to develop alternative views of their science; and they should be required to prepare a summary of theoretical and technical approaches to the general problems in their field and a list of the outstanding problems. Pauling has written about requiring students to defend a list of propositions or theses during examinations for degrees. I would add that some of these propositions should be P.B.I.s—partly baked ideas—as is common in some European countries. Examination of such ideas can be a semihumorous excursion into the unknown, a way of encouraging independent thinking and a test of students' imagination and courage.[6] The point is to prevent the dogmatic approach to knowledge so prevalent in our present educational system.

Assuming that potential mavericks can be identified and encouraged, what should we do with them? Historically, virtually every successful explorer had an early opportunity to do independent research. By "independent" I emphatically do not mean the sort of advisor-directed "independent" research that doctoral students and postdocs do these days, in which the problem, the techniques, and the evaluation are the advisor's. I mean that the young researcher identifies the problem, learns or invents the techniques for carrying out the research, and evaluates the results. By "early," I mean either by the age of twenty-two (preferably considerably earlier) or, if the researcher entered science late, within a year of beginning doctoral studies. In the case of our explorers, the early opportunity for independent research occurs by the age of twenty and is followed by institutionalized freedom of some sort immediately after graduate school. Examples are legion: Darwin, Maxwell, Joule, Planck, the Ionists, Einstein, J. J. Thomson, E. O. Lawrence, most members of the Manhattan Project, Dobzhansky, J. B. S. Haldane, H. J. Muller, James Watson—all were in their early twenties when they began their great works and either had some measure of independence (Darwin, Einstein, and Haldane) or had become full professors by age thirty.[7] Thomson, for example, replaced Rutherford as head of the Cavendish Lab at age twenty-eight, while Lawrence was the youngest full professor at the University of California. Similarly, when the Kaiser Wilhelm Institutes were set up in Germany at the beginning of this century, the heads of the institutes were, with one exception, under the age of forty, and one, Otto Warburg, was in his twenties.

Why are early opportunities so important? Do young scientists do more innovative work than older scientists? Certainly that is a widely held opinion. When the Institute of Advanced Studies was set up in Princeton, Warburg and another young head of one of the Kaiser Wilhelm Institutes, Fritz Haber, were asked for their advice about whom to appoint. Warburg's reply, approved by Haber, was, "It depends on whether you want scientists or oldsters beyond their prime."[8] Although many scientists would argue that the wisdom of "oldsters" more than makes up for their lack of creativity, most would concur that the creative years in science are the young years.[9] Hence the importance of giving young scientists as much time and freedom as possible, as early as possible in their careers.

"Assumptions!" I hear Imp grumble. "You've got to check your assumptions!" Is age per se the real determinant of discovery and invention? Certainly the mythology, especially in the physical sciences, is that if you haven't made your discovery by the age of thirty, you never will.[10] There are two possible reasons for this. One is that older scientists tend to become overwhelmed by administrative duties, peer review obligations, and fund raising. As James Watson once commented, "It's necessary to be slightly underemployed if you are to do something significant."[11] More disturbing is a widespread belief that scientists are simply burned out by the age of forty or so.[12] I find J. Z. Young, for example, writing that "There seems to be a limit beyond which new patterns and new connections are no longer easily formed. As we grow older the randomness of the brain becomes gradually used up. The brain ceases to be able to profit from experiment, it becomes set in patterns of laws. The well-established laws of a well-trained person may continue to be usefully applied to situations already experienced, though they fail to meet new ones."[13] In terms of Jenny's model of science, you codify your way of seeing the universe as a young scientist and then the pattern traps you. But does aging necessarily lead to mental ossification?

As with any hypothesis, the key to testing it is to seek exceptions. Two types of exceptions do exist: scientists who enter science at an advanced age, and those who continue to make important contributions into old age. Among those who entered science at an advanced age are William Herschel, who wrote his first scientific paper at the age of forty; Louis de Broglie, who began studying physics at the age of twenty-six; Robert Millikan, who, according to his own account, took up physics seriously only at the age of twenty-eight; and Leopold Infeld, who first undertook independent research at the age of thirty. There is also Francis Crick, whose training as a physicist was interrupted by World War II. He made his first major contribution to science at the age of thirty-five, when he wrote an important paper on the theory of helical diffraction. A year or so later, he solved the DNA structure with Watson. He received his Ph.D. afterward.[14] According to the myth, these men should have been too old to make a contribution. I suggest, therefore, that it is not age per se that is the controlling factor here, but the number of years one has been in a field. These men were novices at the age of thirty or forty, when most successful scientists are already full professors.

My second set of exceptions to the age rule confirms my hunch. A significant number of scientists make important discoveries or inventions well into their fifties and sixties. Berthollet, Pasteur, Fleming, Planck, Ostwald, Max Delbrück, Seymour Benzer, Leo Szilard, Linus Pauling, Donald Cram—all made major contributions to science after age forty. Because it usually takes about ten years to make a discovery into an acknowledged insight, I think we can agree that all of these men were doing important research into their sixties.

Now, what do these men have in common? In every case that I have been able to examine, researchers who continued to be productive past middle age changed fields regularly. In effect, they periodically returned to the state of a novice by taking up a new subject. They broke out of the patterns of work and thought to which they had become accustomed. As Benzer remarked, "The best way to have fun in science is to do something you are not trained for."[15] Such scientists need to explore.

Subrahmanyan Chandrasekhar is typical. He started his research career studying black holes in 1933; took up the problem of the distribution of stars in 1937; studied why the sky is blue during the forties; the behavior of hot fluids in magnetic fields in the fifties; the stability of rotating objects in the sixties; the general theory of relativity in the seventies; and black holes again in the eighties. And these are only his major research topics. Says one observer: "Just talking about 'Chandra's style' makes other astronomers tired. They can't understand how he regularly forces himself to abandon a subject and start over—how, in a discipline where a forty-year-old theoretician is considered way past his peak, a man of sixty-three could profitably *begin* analyzing what happens when things disappear down a black hole."[16] But Szent-Györgyi, another habitual field-changer, understood: "If one works for ten or twenty years on something, one needs a change of atmosphere. One gets stale; one doesn't see things."[17] "Once a man has missed the solution to a problem when he passes it by," said Szilard, "it is less likely he will find it next time."[18] Getting locked into one of Jenny's boxes for thirty or forty years simply condemns one to perceiving an ever narrower set of possibilities and repeating the same oversights and errors within an ever more limited framework. Thus, it's not physiological decay or mental ossification, or even being overwhelmed with other responsibilities, that causes creative decline in older scientists, but a failure of courage—courage to forsake the field of expertise in which they are respected or even revered for a field to which they have no more claim than anyone else.[19]

Many scientists apparently can't risk their prestige. Some are afraid to fail having once succeeded. One Nobel laureate who left research during middle age commented on this in a confidential interview. Rather than speak about himself, he said, suppose you're James Watson. Once you do something important, your colleagues say, "'He ought to do another thing like this.' [But] he knows that in a lifetime, his chance of doing this again—the equivalent of this—is almost zero—so he's going to be a little depressed by this, isn't he?"[20] Neither Watson nor the man speaking about him did any research after the age of forty. Setting impossible standards for oneself can interfere with trying again. Many scientists succumb to this syndrome without admitting it either to themselves or to others.

Chandrasekhar has proposed another obstacle to continued productivity by older scientists: arrogance.

> *How many young men after being successful and famous have survived for long periods of time? Not many. Even the very great men of the 1920s who made quantum mechanics—I mean Dirac, Heisenberg, Fowler—they never equalled themselves. Look at Maxwell. Look at Einstein . . . For lack of a better word, there seems to be a certain arrogance toward nature which people develop. These people have had great insights and made profound discoveries. They imagine afterwards that the fact they succeeded so triumphantly in one area means they have a special way of looking at science which must therefore be right. But science does not permit that.*[21]

Past performance indicates only that a scientist has the ability to do great work, not that he has the insight, energy, or humility to do so again.

The long-term discoverers and inventors are those who are constantly on the move, dissatisfied with past successes and eager to explore uncharted realms. Berthollet chucked his position in France at the age of forty-seven to adventure with Napoleon. Pasteur moved from crystallography to fermentation, spontaneous generation, germ theory, and finally vaccinology, totally ignorant of each new field he entered. When Dumas requested that he look into pebrine, a disease killing off the silkworms of Europe, Pasteur demurred, "But I know nothing of silkworms." Dumas replied, "So much the better! For ideas, you will have only those which shall come to you as a result of your own observations."[22] Peter Debye told a similar story. He received a Nobel Prize in 1936 for his research on dipole moments, the diffraction of electrons, and X-rays in gases, and became involved in polymer research shortly afterward: "At the beginning of the Second World War, R. R. Williams of Bell Labs came to Cornell to try to interest me in the polymer field. I said to him, 'I don't know anything about polymers, I never thought about them.'" Williams's answer, like Dumas's to Pasteur, was, "'That is why we want you.'"[23] Debye went on to say that this became his philosophy as well: "You should not ask for people who have already done in the university the things which they are going to apply in industry—this is the most nonsensical way of doing it. You should ask for people who have enough brain power that they at least have a feeling of how to handle a new problem. The specific nature of the problem is not important."[24] He has been echoed by Ostwald, J. B. S. Haldane, Fritz Haber, and MacFarlane Burnet.[25]

Besides their willingness to be novices, their work habits also set lifelong discoverers apart from their colleagues. A recent study shows that scientists contributing many important ideas to science over several decades of research ("long-term, high-impact scientists") usually investigate three or four problems simultaneously, explore others as they crop up, and constantly change the focus of their research. Most scientists investigate only one or two major problems during their careers, and they focus on these for long periods of time, often measured in decades. Not that an intensive preoccupation with a single problem is necessarily a counterproductive strategy in research; many eminent scientists have been fortunate enough to find an important problem early in their careers that required several decades to solve—Max Perutz's work on hemoglobin comes to mind. But these scientists contribute a single important insight to science, usually by middle age. More often than not, they spend the remainder of their careers elaborating on that insight, or administering other people's research. Thus, there is actually a negative correlation between staying in a field after making a major breakthrough and the likelihood of making another breakthrough. One insight per field is all we're generally allowed.[26]

All of the scientists we've discussed at our sessions fit the long-term, high-impact pattern. They changed research focus frequently and

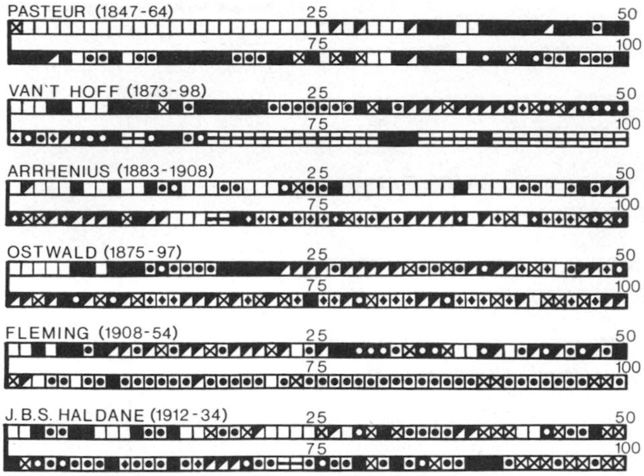

Graphic representations of the publication records of various scientists. Each box records one publication, in the order of the person's standard bibliography. Symbols indicate different fields of research ("X" represents a non-research publication). This group represents publications of scientists known for important contributions to several fields over several decades.

worked on many topics simultaneously, as you can see from these charts. You can compare these to the publication patterns of other eminent scientists who produced one grand discovery as young men, which they were never able to follow up. I've also included the patterns of some "average" scientists.

These results suggest that scientists who are continuously creative over the majority of their careers are those whose style is exploratory and eclectic. Indeed, studies of both biomedical researchers and chemists have shown that those producing the most patents or "discoveries" (admittedly a poorly defined term) were those who regularly combined several specialties or technical functions and often consulted for other research groups.[27]

I want to mention one other interesting difference between long-term, high-impact scientists, important short-term contributors, and low-impact scientists. Long-term, high-impact scientists rarely take on administrative duties and when they do, generally find the experience unsatisfactory. Scientists who make a single important contribution and most low-impact scientists spend a considerable part of their careers as administrators. It is not clear whether administrative obligations leave the short-term and low-impact scientists too little time for creative scientific work (many complain of this) or whether these scientists, having run out of ideas, turn to administrative work to maintain a powerful presence in their community. It's interesting to note, however, that in a study of industrial chemists, those producing many patents rarely expressed the desire to become administrators. *All* the chemists producing no patents wanted to become administrators.[28] I'll leave the interpretation to you.

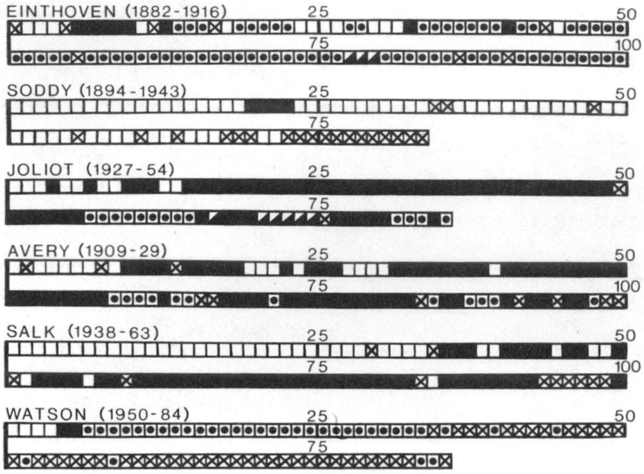

Publication records of scientists known for one major contribution.

What excites me about these results is the possibility that scientists having the greatest probability of producing high-impact research into their sixties or seventies may actually be identifiable by the time they are thirty-five or forty (assuming that they receive their doctorates during their mid-twenties). Most long-term, high-impact scientists have published at least five papers by the time they are forty years old, each of which has received ten citations in a given year, or one hundred citations over ten to fifteen years, and these scientists have the variegated publication pattern I've illustrated. These long-term explorers rarely have any interest in administrative duties (think of Chandrasekhar or Feynman) and when they do take them on, as did Pasteur for a brief period, they quickly abandon them. Rarely, long-term, high-impact scientists may become editors of journals, like van't Hoff and Ostwald, but usually this is a personal expedient taken to promote novel ideas. In general, scientists who remain productive are too busy with their own ideas to care about the direction of the rest of their field. Thus, past performance (as measured by impact on the field) coupled with publication pattern (exploratory or narrowly focused) and administrative interests (or lack thereof) appears to be a very good differentiator between successful, long-term explorers who will contribute to several fields and scientists with a single major contribution to make to one field.[29]

Now, assuming one can identify scientists having a high probability for carrying out a long series of successful explorations, how does one support them? Let's begin with the issues of time and planning. As we've witnessed repeatedly, discoveries are surprises; they cannot be planned. However, I believe that conditions fostering surprises can be planned. I suppose this belief comes from being a scientific heir, once removed, of Irving Langmuir, who said: "You can't plan to make discoveries. But you can plan work that will lead to discoveries . . . You can organize a laboratory so as to increase the probabilities that useful things will happen there. And, in so doing, keep the flexibility, keep the freedom . . . We know from our own experience that in true freedom we can do things that could never be done through planning."[30]

One reason that freedom is essential to the scientific explorer concerns the way insights are reached. Think about Pasteur's and Fleming's exploratory styles—Lorenz's "madness" and Delbruck's "sloppiness." These men are typical in working on many things and making many detours. Several studies have shown that chemists, physicists, and mathematicians rarely solve the problem they are working on by direct attack (less than 10 percent of the time). In 1912 Fehr reported that 75 percent of the scientists he interviewed stated that their most important discoveries occurred while they were working on some other unrelated problem.[31] Ninety percent found it necessary to abandon a problem, sometimes several times, before they were able to resolve it. (Many problems were never resolved.) A subsequent study of chemists found a variety of tactics used by successful problem solvers to increase their probability of success: 60 percent abandoned their problem and turned to other work; 45 percent found total relaxation (for example, going on vacation) effective; 47 percent reported that they would go over the problem repeatedly every night before going to bed; 15 percent turned to exercise or other physical activities; 14 percent reported that sitting and smoking helped them (although this might be classified as relaxation); and a few individuals reported that coffee or alcohol were effective.[32] (Obviously, most of the scientists interviewed utilized more than one means of promoting insights.) The point is that the best problem solvers seem to recognize that insight comes only by getting out of logical ruts and allowing intuition to suggest other possibilities. They also recognize that wanting to solve a problem does not guarantee that the problem is solvable. As Poincaré, Hadamard, and others report, one may have to wait years before

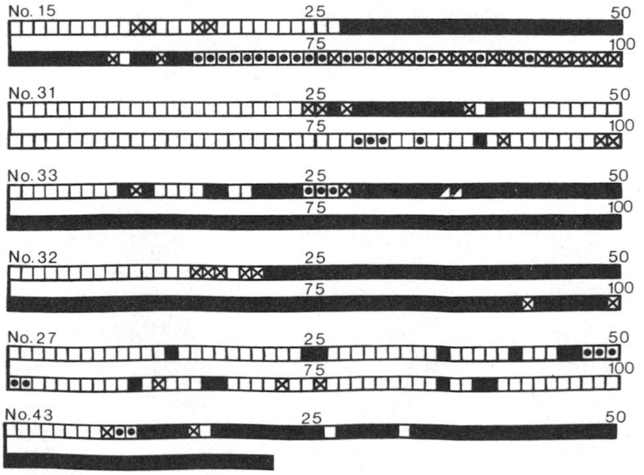

Publication records of scientists whose contributions were short-term (less than a decade) and high-impact (at least 100 citations to 1 or 2 papers over 15 years). This group includes a Nobel laureate and two persons multiply nominated for that prize.

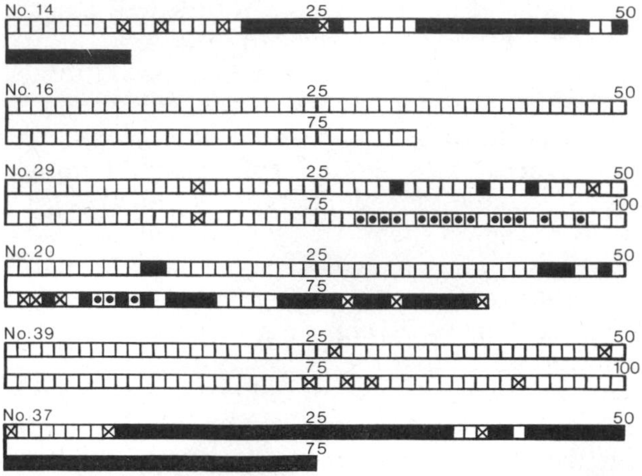

Publication records of low-impact ("average") scientists. The number of topics addressed and the number of topic changes made by long-term, high-impact scientists are significantly different ($p > 0.05$) from the corresponding numbers for short-term, high-impact scientists and for low-impact scientists.

the right information, insight, or reformulation of the problem occurs. This fits the models of illumination we developed in our sessions.

Now, this tells us that you can't assign a scientist any worthwhile problem in exploratory science and expect a solution to appear by brute force. On the contrary, trying to plan such insights, or to force breakthroughs, is almost sure to fail. (Note the dismal results of the "wars" on cancer and heart disease.) It's a matter of understanding the psychology of research, as J. J. Thomson once tried to explain:

> *If you pay a man a salary for doing research, he and you will want to have something to point to at the end of the year to show that the money has not been wasted. In promising work of the highest class, however, results do not come in this regular fashion, in fact years may pass without any tangible results being obtained, and the position of the paid worker would be very embarrassing and he would naturally take to work on a lower, or at any rate, different plane where he could be sure of getting year by year tangible results which would justify his salary. The position is this: You want this kind of research, but if you pay a man to do it, it will drive him to research of a different kind. The only thing to do is to pay him for doing something else and give him enough leisure to do research for the love of it.*[33]

Einstein said virtually the same thing.[34] I'm sure you've all heard his suggestion that the best place for a theoretician to work, after a patent office, would be a lighthouse, where the duties are minimal and the time for contemplation abundant. Darwin, Jacob Henle, Theodor Schwann, Emil du Bois-Reymond, Ramon y Cajal, Robert Koch, J. S. Haldane and many other scientists would probably agree, having done their best research in spare hours in their homes or private laboratories.[35] Unfortunately, we have effectively made such conditions impossible to achieve today by our regulations on who may buy chemicals, do animal experiments, have a laboratory, and so forth.

Time is of the essence, clearly. The question is how much time and how is it to be used. I've heard many of my colleagues assert that to be successful in research, one must spend fourteen hours a day, seven days a week in the lab—the Protestant work ethic. But we've already heard from Ariana of Békésy's response to this: eight hours of research, eight hours of leisure, eight hours of sleep. He's not atypical of the most creative scientists. Workers in the Cavendish Lab, including Rutherford and J. J. Thomson, rarely began before 10 A.M. or quit later than 6 P.M. A. V. Hill advised that a motto be put over every laboratory door reading: "It is better to work too little than to work too much."[36] And Arthur Hadley, a former president of Yale, observed that in his opinion the best work was "done by men who were a little lazy and had good consciences. Such men would not work for the love of being employed, often in time-consuming endeavors upon subjects which were not worthwhile; but if after due meditation they saw a thing was worth doing, their conscience would drive them to make the effort and to accomplish the task. These did the greatest work."[37] Thus, it is not necessarily time in the laboratory or office that is important, but "free" time to explore, idle, think, laze.

Let me be clear about this concept of laziness. The creative scientist is not so much lazy as discerning (though I've heard that Darwin and Crick, of all people, called themselves lazy!).[38] He or she recognizes when a problem is worth attacking and whether it's ready for solution or not. Moreover, he or she knows how much effort it's worth. Crick could not get down to his dissertation on hemoglobin structure because it was not, to his mind, one of the crucial questions facing science. The answer could only be a particular, and he wanted general, algorithmic answers. He wanted to go after something big like DNA structure. But even forty years ago, grad students weren't supposed to think big or independently. So Crick found it difficult to concentrate on what he was supposed to be doing. Lucky for us!

Similarly, I. I. Rabi, who coinvented the concept of nuclear magnetic resonance, took as a dissertation topic the extremely difficult problem of measuring the magnetic susceptibility of various inorganic crystals. Using techniques then available, the project could have taken him many years. Being another self-described "lazy" person, he gave himself a few weeks to think of a way to do the experiments more easily, and vowed to drop it in favor of something else if he could not come up with one. The problem was only worth so much effort. So he thought about it. He treated it as we treated the tennis match problem, searching for an insightful restatement. Sure enough, a couple of weeks later, without ever entering the lab, he figured out how to do his experiments. His solution was so simple that within six weeks, he had measured the magnetic susceptibilities of more crystals than everyone else had managed to do in the preceding decade.[39]

Other scientists limit themselves to one afternoon a week for work on the "big problem." They know they can't force insights. Hours in the lab may produce experimental results, but they don't bring ideas; freedom to relax or switch problems when and as needed, to follow one's nose, is the essential ingredient for insights. Thus, we must plan not for results but for the freedom to dispense one's time in accordance with the state of one's research and the opportunities available in the field.

What does the exploratory scientist need in the way of funding and equipment? Most of us, I suspect, came here thinking that as science progresses, research necessarily becomes more complicated and expensive. We must therefore pour ever larger sums of money into ever larger institutes crammed with ever more expensive equipment and run by ever larger staffs. Yet the scientists we remember as the great explorers of the last century never accepted the claim that money buys insight. Ramon y Cajal wrote in 1923, "At the present time there exist in Spain laboratories so richly endowed that they are the envy of the greatest scholars of other countries. Yet, in them little or nothing is produced. The fact is that our statesmen and educational institutions have forgotten two important things: that to declare oneself an investigator is not sufficient to be one; and that discoveries are made by men and not by scientific apparatus and overflowing libraries." We must, he says, recognize this essential truth: *"For scientific work the means are almost nothing, the man very nearly everything."*[40]

What is at issue here is whether one can have the freedom of the shack and the luxury of the palace at the same time.[41] For university vice presidents, of course, the bottom line is funding, overhead, and physical plant. New institutes of science with their fancy architectural facades look good on campus, encourage parents to pay exorbitant tuitions, and convince alumni to donate generously. The overhead pays for plush offices, large staffs, and frequent business trips. But the investigator is put in the unenviable position of prostrating himself to the idol of money whenever there's a decision to be made between a risky, exciting project that's unlikely to be funded and a sure bet that holds no promise of surprise. When tenure and even laboratory space depend on yearly grant renewals, self-preservation overrides scientific sense. While the universities prosper, science despairs.[42] But I digress.

Most pioneering or breakthrough scientists would not have been able to curry favor under our present system. As far as I can tell, almost every major discovery has resulted from crackpot ideas or attempts to solve "impossible" problems that no one took seriously at the time. We saw in our sessions how difficult it was for many of the greatest scientists to convince their colleagues even when they had results. Imagine the difficulties they would have had convincing them to fund the research that yielded those results! As it turns out, few major discoveries were made with prior peer approval. In consequence, the scientists had to rely upon existing resources or even their own money, and therefore almost no major discovery or invention was adequately funded or carried out with anything but the most primitive facilities. Let's take these issues separately: money, then facilities, and finally equipment.

Up until about World War II, most scientists had some financial independence. Lavoisier, Darwin, J. Willard Gibbs, Hans Driesch and his collaborator Curt Herbst, J. A. Le Bel, Rayleigh, the Haldanes, and many others were wealthy "amateurs." Norman Pirie believes that at least seven out of every ten of the men who worked with him at Cambridge had some inherited money, which "gave them a feeling of security and independence."[43] Most of the great biologists and immunologists—and here I include Fleming, Koch, Ramon y Cajal—supported themselves, at least for a time, as doctors with a private practice. Others, including J. J. Thomson, Almroth Wright, Arrhenius, and virtually every major biochemist during the first fifty years of this century, benefited from long-term, no-strings-attached fellowships such as the Grocer's Fellowships, Trinity Fellowships (which were unconditionally granted for six years and could be used for travel), and Rockefeller Foundation Fellowships. No one told them what to study, when to study it, or where.

Even institutional funding, when it was available, worked differently. J. J. Thomson has described the scientist's best friend as "a good balance at the bank," but emphasized that his bank account was under his own control, not under the control of an institution (in whose hands the money would have disappeared at the end of each fiscal year).[44] It could accumulate from year to year to be used as needed. Richard Goldschmidt, too, described a researcher's heaven at one of the Berlin Kaiser Wilhelm Institutes, where he had "the best laboratory in the world with unlimited space, technical help, assistants, multilingual secretaries, the best instruments and equipment, and complete freedom to use all this without bureaucratic interference, without requisition slips or purchasing departments. My time [was] completely my own, and I was even free to move anywhere and take my funds and salary with me if I thought my research required it."[45] Then he moved to U.C. Berkeley, and found out how the American bureaucrats watched every penny, monitored every requisition, and confiscated every unspent cent.

The point is that while these men valued money as an aid to research, they had the freedom to use it when and as they wished. When the money came with strings and monitoring, they found it a burden and an impediment. Indeed, Rustum Roy reports a survey of twenty of our most successful colleagues, all of whom agreed that they would take a 25 percent cut in their current funding if the system of competitive grants were replaced with long-term funding computed by a complex function of number of papers published, graduate students supported, and so forth.[46] Thus, it's not money per se that scientists want, but a secure base from which to launch their explorations and the freedom to use money as research possibilities change. Much of the waste in the current system is due to the innate insecurities created by constant site visits, reviews, renewals (and lack thereof), fluctuations in available funds, and the targeting and retargeting of resources by bureaucrats responding to political and popular wishful thinking.

I therefore find that many scientists have worked with little money, either by necessity or to circumvent bureaucracies. They teach us that ingenuity is far more powerful than any number of dollars. The facilities available to Pasteur, Wurtz, Berthelot, and Claude Bernard during the 1840s and 1850s have been described as utterly inadequate.[47] Van't Hoff, Arrhenius, and Ostwald all worked at fourth-rate institutions that had virtually no experimental facilities. Fleming's laboratory was a dingy room so cramped that when it was later converted into a rest area for doctors, it was barely large enough for two small daybeds. Two and sometimes three scientists worked in this room simultaneously.[48] Schally's isolations of brain peptides were carried out in an overcrowded, poorly equipped Veterans Administration Hospital facility.[49]

None of our breakthrough scientists worked with expensive equipment, either. Pasteur made his own apparatus, as did Arrhenius, Ostwald, and Fleming. Van't Hoff is reported to have used equipment for measuring transition points that "a well-trained physicist would only have used for preliminary experiments."[50] Howard Florey's initial grant in 1939 to investigate the possibility of purifying and characterizing penicillin was a mere twenty-five pounds.[51] The physiological research carried out by Starling used equipment so primitive that knowledgeable observers such as Otto Loewi were surprised that he was able to obtain results at all.[52] When Banting and Best isolated insulin each had no more than two hundred dollars in the bank, no salary, and no position; they were both living with friends. Meanwhile, better funded researchers made little headway on the problem.[53] The research done in Ernest Rutherford's Cavendish lab was usually carried out on an odd assortment of deal tables shoved together to create a surface for equipment of the chewing-gum and string variety, and C. T. R. Wilson, one of the most inventive members of the lab, supposedly never spent more than five pounds on an apparatus.[54] I believe Constance mentioned before the inexpensiveness and provisional nature of the equipment with which Baeyer, Rayleigh, and Roentgen produced their astounding results.[55] The original subatomic particle accelerator (cyclotron) was built for just a few dollars by E. O. Lawrence in 1930 and the second version for less than a hundred.[56] The atom was split by Otto Hahn in 1938 using apparatus that fit on a desktop and was so simple it could be assembled from widely available parts.[57] Donald Glaser built the first bubble chamber on a measly university grant of $750 after having been denied access to existing research facilities and funds from the major nuclear research agencies.[58]

So pervasive is this phenomenon that G. P. Thomson, Burnet, and Ramon y Cajal have each concluded that one of the "truths" of scientific progress is that the most important developments are carried out in ill-equipped laboratories with inadequate funds and amateurish techniques.[59] Perhaps it's worth noting that every single man I have just mentioned won a Nobel Prize. Many echoed Fleming's warning against working in a scientific "palace": "I have known research workers reduced to impotence by apparatus so fine and elaborate that they spend all their time playing with it." For the successful investigator, "the palace becomes . . . an ordinary laboratory. If the palace wins, he is lost."[60]

Breakthrough experiments are almost always simple experiments. And simple experiments are inexpensive. Any fool can do research with fifty thousand dollars' worth of equipment; it takes a clever person to think up an experiment using only some test tubes and pipettes. Take molecular modeling as an example. Pasteur carved his crystals from corks. Van't Hoff constructed his from cardboard. Pauling's original alpha helix was drawn on a piece of paper and folded to shape.[61] Subsequently he made more "complex" working models out of corks stuck into rubber hose. W. L. Bragg got Kendrew and Perutz going by tacking nails in a helix around a broom handle. Watson and Crick used cardboard models of the DNA bases for their initial work.[62] Today, if you try to model something, you'll immediately be criticized for not using the best models available—CPK kits that may run up to several thousand dollars per molecule, custom-cut rod and ball models that require access to a good machine shop, or even more expensive computer modeling systems. Yet I doubt these would have helped Pasteur and the rest invent their ideas.

Indeed, state-of-the-art equipment might have hindered them as it has hindered others. Remember, Le Bel erred about the tetrahedral carbon atom by sticking too close to the data. Bragg, Kendrew, and Perutz were misled about basic protein structures by paying too close attention to X-ray data, and Pauling was misled by the data concerning the structure of collagen.[63] Complexity and precision may actually forestall insight. If we accept Ariana's definition of science as the elaboration of all possible universes, successively constrained by an ever increasing body of observations and tests, then we must recognize that the role of precision is constraint. Before we constrain, we must first elaborate.

What makes for exploratory discoveries is knowing what to juxtapose and how to do it, which requires depth of insight coupled with the search for simplicity and beauty. The object in exploratory research is to cut the great rough block of the idea, as van't Hoff was wont to do, and allow other people to polish it—show that the phenomenon exists, as Roentgen, Hahn, and Glaser did. Precision can be achieved later.[64]

To summarize, the conditions that seem to foster the great explorers of science—the ecology of invention, as Richter might say—are as follows.[65] At a personal level, the pioneer needs:

A broad acquaintance with several emerging fields of science and mathematics, or technical developments relating to experimental technique.

An opportunity to train oneself, so that one's tools—mental or physical—differ from those of the run of the mill.

Early opportunities to begin independent research.

Underemployment. Duties must be sufficiently light as to leave the scientist plenty of leisure time for playing and thinking.
Financial independence. This applies both to earning a living (few great scientists did good work when they had to worry about where their next meal was coming from), and to funding research (exploratory research has rarely been approved by committees for the simple reason that such research yields surprises). In part, this requirement could be met by early institutional support in the form of long-term fellowships or early tenure.
Maximum diversity of research, achieved by working in several fields on several problems concurrently.
Frequent changes of field (every five to ten years).
Application of techniques and theories from one field to another.
The goal of simplicity and economy in apparatus, experimental procedure, and interpretation.

Some of these recommendations require institutional support or cooperation. Thus, our scientific institutions must learn to practice what George Burch has called "venture research,"[66] which follows these guidelines:

Fund people, not projects. Explorers never know what they are going to find or how they are going to find it. An experiment that can be planned in advance and whose outcome is unlikely to change our view of nature is not worth doing. The only way to encourage unexpected results and detours is to encourage stochastic blundering about; and the best way to encourage that is to find people who are good at aiming themselves into the unknown and letting them go. From my experience, it's pretty clear who these people are by the end of graduate school: They want to work on their own projects, not their professors' projects; they're independent and antidogmatic, self-motivated, self-evaluating, energetic, and unusually broadly trained.
Keep funding for exploratory research small. As we've seen, such research is inexpensive, simple, and quick (when it works), and usually involves individuals rather than groups. The investigator (I use the singular on purpose) either finds something or not. If not, he or she moves on. The object is to allow individual scientists to try as many things as possible, not to encourage empire building. Scientists with many ideas just want a chance to try them out and can think of many simple ways to do so.
Give explorers plenty of time. Exploratory funding must last for a sufficient period of time that the researcher can find or invent something novel and develop it into something that can stand on its own. Since no one can plan when an insight will occur, exploratory funding cannot be limited to six-month or yearlong periods. Anything less than five years is simply unreasonable and ten years is probably

about right. Too many exploratory projects die because they must justify themselves too soon and cannot.

Give scientists plenty of freedom. Exploratory funding must not have disciplinary or institutional strings attached. Successful exploratory scientists change fields often and retrain themselves frequently, and therefore need to be able to change both disciplinary and institutional affiliations as needed. This gets back to funding people rather than projects or institutions.

Reward successful explorers by encouraging further exploration. As I've said, I believe it's possible to identify many successful exploratory scientists by the age of forty to forty-five by their unique pattern of research and their previous successes. Ten years of minimal funding would get these scientists started. The successful ones should be given lifetime grants as early in their careers as possible to facilitate their exploratory, eclectic, field-changing style of research. Inducements might be added for changing fields, for training small numbers of protégés, and for abjuring administrative duties. Again, these grants need not be large—the object is simply to free the scientist from the necessity of applying for funding every year or two. Should he or she become involved with research that requires significant funding, then of course the normal channels could be addressed. In short, we must learn to reward trying as much as we reward success.

Free up some of the money that comes with every grant. If we insist on keeping the present system of peer-reviewed funding of large projects, then at least 15 percent of every grant should be uncommitted funds to be used at the discretion of the primary investigator. After all, if a scientist is worthy of a specific grant, surely he or she is worthy of receiving seed money for exploratory research as well. Better this than the lying to which Szent-Györgyi, Szilard, Debye, and others had to resort.

Subject exploratory research to a different kind of peer review than we use for confirmatory or developmental research. Exploratory research, as we've seen, has different rules. In consequence, its results and practitioners need to be evaluated in different ways. Journals should have special sections devoted to reporting exploratory results. Unsolved problems, negative results, contradictions, and antidogmatic hypotheses should have a prominent place in their pages. Peer review boards should evaluate candidates in terms of how iconoclastic and independent they are, how many new problems they've invented and phenomena they've discovered. The criteria should not be "Will it work?" but "Is it the sort of thing that will yield surprises whether it works or not?" or "If it turns out to be true, will we have to rethink the field?" Colleagues must learn to judge explorers not by how well they fit existing niches but by how uncomfortable they make them feel about what they think they know.

Create a small fund of money specifically for exploratory research. Such research need not compete with Big Science for money, manpower, or facilities. Its needs are very modest and also very different. Let those who are willing to work on small projects with little or no assistance do so. In the end, they will come up with results just as important as those of any large lab, and at far less expense.

Finally, we must encourage the development of exploratory research at those institutions best suited to do it: small colleges, backwater universities, and independent institutes where breadth of background, ability to work with few resources, and relative independence from mainstream science may be seen as spurs to innovation rather than insuperable obstacles to success.

What it comes down to is this: If we wish to encourage discovery and invention, we must gamble; and in order to gamble, we must learn how to play the odds. We must be willing to lose small amounts of money to remain in the running for the jackpot.

Now for the big question: To what extent does the current system of science foster the conditions most likely to yield exploratory discoveries, and to what extent does it impede them? Here we find a dismal state of affairs. Think about funding, for example. Who gets the big grants these days? Those most likely to make an outstanding breakthrough? Scientists under the age of thirty? Successful problem solvers changing fields? Hah! Try to find scientists under thirty who have independent funding! What you find instead are thousands upon thousands of doctoral candidates, seven or eight years into their theses, or postdocs facing their third or fourth or fifth nonrenewable position. Most of these, in my view, are overeducated, lopsided, narrow technicians. James Watson says outright: "If education is too long it'll probably kill you. As a scientist. That's why it's nice if in some way we can put people into a position where they can begin to do science not much after twenty."[67] Yet look around. Especially in fields that are growing slowly or have ceased to grow, such as space science and chemistry, there's a startling absence of young scientists under forty who have their own experimental groups.[68] Stasis—just as Richter's model predicts.

So when and where is the need for freedom and independence fulfilled in our system? Peter Debye said: "It's not filled at all. The real sense of independence comes only with great wealth, or with a man who says, 'It does not matter to me'"—people who are willing to forgo the usual rewards of the profession in return for personal freedom.[69] "I did what I did in spite of the educational system," reports Julius Axelrod, who, like Crick, earned his Ph.D. (in this case at age forty-four) only after doing some of his best work.[70] He is joined by several other eminent scientists who also escaped the "Ph.D. octopus," as William James called it: J. B. S. Haldane, whose only degree was a B.A. in Classics, and Nobel laureates Allan Cormack and Godfrey Hounsfield, who invented computer tomography. None of them had an advanced degree in science. These men made their own opportunities. How many other potential inventors are being lost because they look at the system and can't see how to circumvent it or find the rare opportunities for freedom within it?

And what about the "novice effect," which implies that if you don't have your great insight within ten years of entering a field you never will? In the present system, you're very fortunate to achieve tenure within ten years of receiving your Ph.D.; that means a minimum of fifteen years in the field already. Worse, several of my colleagues tell me that in their departments a young scientist who hasn't "settled down" by focusing on a single research topic is usually considered unfit to hire or doesn't receive tenure if he's lucky enough to be hired. We are so afraid of producing dilettantes and generalists that we are—to use Richter's language—selecting out those individuals most likely to do long-term, high-impact research as well. And that is a point worth stressing: We can foster change or we can prevent it. We must beware leveling the pyramid of talent and narrowing the distribution of eccentricity, for the result will surely be homogenization of thought. Science cannot survive that.

Young scientists aren't the only ones at risk. I've found that the situation is no better these days for established eclectic, high-impact scientists who wish to change fields. Gone are the days of just a few decades ago when a physicist could blithely move into biology. Even changes from, say, genetics to neurobiology are frowned upon today unless one has already studied something relevant, like the genetic expression of a neural protein. Robert Holley, for example, who has successfully changed fields several times in the past, tells me that he's found it more difficult each time.[71] Peer review boards are becoming more and more insistent on expertise in an area before they're willing to fund projects. Since there are no experts in the unknown, the demand is unreasonable. It's also possible, of course, that older scientists are encountering age discrimination, since everyone knows that scientists past the age of forty are over the hill. These shortsighted review boards should realize that they are, in effect, condemning potentially highly creative scientists to remain in fields in which they've already done their best work. They will not allow scientists to remedy their situations because we no longer trust the individual scientist, no matter how eminent or successful. Thus, field changes such as those that characterized the careers of Pasteur, Arrhenius, van't Hoff, and Debye are virtually unheard of today. Science is the loser. "Minerva," as Goldschmidt says, "covers her face and sends the owls away to catch mice."[72]

Another insanity of the present system is that the funding of research programs comes only after they have already been demonstrated to be viable—that is, after the key discoveries have been made. No doubt such a funding pattern is safe; we know there's something there to be developed. But such a policy is not often productive of unexpected inventions or surprising discoveries—the only ones that count for anything in science. On the contrary, it creates a Catch-22 situation: One must first discover or invent something to be funded, but one cannot discover or invent anything without first being funded. We all know how good scientists get out of this: They divert grant money from their funded projects to what they would really like to be doing, and use the unpublished results to apply for next year's grant, having already completed the research.[73] If this is dishonest, it's only because the system requires dishonesty.

The funding of science is backward—backward because it is responsive and not initiatory. We've put the horse behind the cart, and the carrot behind the horse. In this position the carrot acts as a prod, not an incentive. We can hardly wonder if the horse tries to turn itself around in the traces.

Unfortunately, the results of this policy are largely hidden, since they can be measured only by what does *not* happen. No one keeps track of whether research programs denied funding by NSF or NIH are less productive of discoveries or inventions than those that are funded. No one analyzes the efficiency of the system. We assume it works because many people publish many papers and some people manage to make some discoveries. But we could just as well assume, as Imp suggests, that the system doesn't work.

Imagine that Roentgen discovers X-rays today, or that Lise Meitner, Otto Hahn, and Fritz Strassmann have just reported the existence of atomic fission. What happens? First of all, in our system, all of them risk losing their funding if they follow up their results, since these are detours from their preplanned research programs. At best there will be a delay of several months while they apply to their funding agency for permission to detour. On the other hand, if they complete their current grant commitments before following up their odd results, they may be "scooped" or they may lose interest. Second, to continue their research, they have to apply for further funding, which may take several years. They might not be funded at all. Roentgen's results were met initially with almost universal disbelief, and the Meitner-Hahn-Strassmann work flew in the face of dogma concerning the inviolability of atomic structure. Third, other researchers wishing to reproduce or follow up their results would also have to forgo commitments on current grants and apply for new funding. Only the simplest and most technologically useful phenomena requiring very little equipment are likely to be reproduced quickly and to be followed up effectively. Given existing constraints, it might take months or years. (Just think about how many years it has taken to get geared up for AIDS research!) Compare this with the historical time frames. In both cases, the results were replicated and confirmed within days of their announcement; within weeks, dozens of the very best physicists in the world were publishing experimental and theoretical reports. As the old joke says, what used to take one scientist an hour to do, now takes a committee of twenty a month. Yet we call our system efficient!

The irony is that when we write our textbooks and when we speak of the ideas that shaped the history of science, we focus on the work of scientists such as Roentgen, Meitner, Hahn, and Strassmann. Why, then, have we created a system that would make it virtually impossible for such individuals to carry out their work or for their results to have the immediate impact they deserve?

This is not my opinion alone. Warren Weaver, Richard Goldschmidt, and Leo Szilard warned us thirty years ago that unless we changed our funding patterns we would begin to lose the revolutionaries. Rosalind Yalow, Linus Pauling, Howard Temin, Lewis Thomas, and Clifford Truesdell are warning us today.[74] Sigma Xi, the scientific research society, has reported in its "New Agenda for Science" that the majority of scientists in this country doubt that truly innovative research, interdisciplinary research, and nonmainstream research are or can be funded under the present system.[75] Dozens of articles in journals such as *Minerva* document the opinions of science watchers such as Mellanby, Shimshoni, Schulz, and others that large institutions, bureaucratic organization, preplanned research, the harnessing of research to the fiscal practices of business economics, and other common stimuli for research are not only ineffective but downright destructive of the ability of the individual scientist to function as a scientist.[76] Science, these people agree, must be disorganized to be effective. Even those who support Big Science, such as Lewis Branscomb and Alvin Weinberg, have wondered publicly whether the cross-disciplinary, individualistic mode of science that characterizes all great novelties in science can survive professionalization, departmentalization, and bureaucratization.[77] In short, there's widespread agreement that by trying to prevent the waste of dollars, we've created a system that is unbelievably wasteful of the most precious and important resource of science: creative human beings.

That's what worries me. Big Science has become so powerful, so money- and manpower-hungry, so dominant in setting educational policy, so bureaucratized, that the lone scientist may not exist any longer. The pioneers of the last generation see this already. Desmond Morris writes in his autobiography of his own struggles to remain free of the entrenched specialist attitudes of recognized disciplines, ruing the "self-censorship that stifles rebellious ideas at birth." "Rebel spirits like Louis Leakey and Alister Hardy," he maintains, "are sadly becoming a rare and vanishing species in the scientific world, and it is the poorer for the change."[78] Shortly before her death, Barbara McClintock felt she'd become a member of that vanishing species: "I no longer know whether I can be classified as a modern scientist or as an example of a beast on its way to extinction."[79] Julius Axelrod says of his fight against the bureaucratization of NIH that he doubts he could have survived there today.[80]

Luis Alvarez expressed the opinion that if he were a young man today, he wouldn't become a physicist because his forte is to tinker, and there's no longer room for lone tinkerers.[81] Leo Szilard moved from physics to biology for that reason.[82] Where would he move now? Indeed, consider what would happen to a Darwin or an Einstein today. Could Darwin carry out his twenty-year program to restructure biology within the present academic system? If he were a wealthy amateur, would anyone listen? Would anyone read Einstein, publishing from a patent office? Would a patent office even put up with such an employee?

I say that unless we start making provisions for such rare and valuable scientists now—for learning, thinking, and doing individually and eccentrically; for the disorganization of science—we'll have no great scientists in the future, and no great scientific breakthroughs, either. We cannot rely on mavericks to fight the system, for as the system gets bigger and more powerful, as competition for ever scarcer resources increases selection pressures and imposes greater conformity, more and more mavericks will follow the paths of Mendel, Herapath, and Waterston into near or perhaps real oblivion.[83]

The danger is manifest. If we do not constantly restock our basic supply of fundamentally new concepts and new phenomena, we will quickly be reduced to elaborating smaller and smaller implications and possibilities of existing knowledge. As we discussed at our first session, this may already be happening. More people doing more of the same. When I look at the articles published in *Science* and *Nature,* I cringe. Very few are surprising. Many are not about basic science but are technological applications of existing knowledge. Genetic engineering is, as its name says, engineering, not science; yet most of molecular biology and biochemistry seem to be devoted to it. In computer science, I'm told, almost every architecture and technique currently being developed was already in existence in 1956! If not for the transistor and the integrated circuit, we might still be where we were three decades ago. And it's the same in virtually every field. Where are the shocking and unpalatable new insights that threaten to repattern the world? Have the powers that be quietly and efficiently prevented them from being published, and thereby destroyed the careers of their proponents?

This is what I dread: that science will die from its own emphasis on uniformity, conformity, and relevancy—uniformity imposed by techniques of mass education that insist everyone learn to produce the same answer by the same path of reasoning; conformity imposed by the need to work in large groups and to appease and appeal to the ignorant masses of bureaucrats and legislators who've come to dominate the monetary aspects of the enterprise; and relevancy imposed by the need to demonstrate immediate and applicable results. We forget that the greatest breakthroughs in science have always come from individuals satisfying their own curiosity in the search for understanding, not from government-mandated programs. We ignore the studies that show that applied results in nearly every field come as often from so-called undirected basic research as from direct attempts to create the desired object or treatment.

In conclusion, I recommend to you the words of Garrett Hardin, written to commemorate the work of Darwin: "We cannot deliberately produce heretics, but we can make the conditions favorable for their spontaneous generation. We can see to it that a substantial minority have available to them that indispensable ingredient of heretical and creative thought, irresponsibility."[84] May we, as Richard Feynman exhorted his students to do, all achieve the freedom from responsibility that is necessary to think and do for ourselves.[85] May we create a new system of science that fosters rather than hinders the mavericks and explorers.

◀ TRANSCRIPT: Discussion

▷ **Imp:** Excellent! To summarize, we need a new slogan: Freedom, not "greedom"!

▷ **Jenny:** A meritocracy, such as the perpetrators of the French Revolution desired, instead of a timocracy or kakistocracy such as Hunter describes. (More fun words!) A timocracy is a state in which possession of wealth, property, and power is the prerequisite for office. In a kakistocracy, the least competent govern. From what I've heard, those with the fewest ideas and those whose creativity has been played out end up running science because those who actually make the discoveries and inventions have neither the time nor the desire to administer. Those who do administer favor those who add to their ostentatious trappings of power by supplying their coffers with money and their offices with underlings. Unable to do science—substitute history, medicine, law—they maintain their social status by telling others how to do it.

▷ **Hunter:** The Peter Principle. We assume that good scientists make good science administrators. Perhaps some do. But don't discount the Gunnings, Edlunds, and their peers who, though not terribly creative themselves, understand the needs of the truly creative quite well. It is them we must emulate.

▷ **Richter:** Agreed. But the world is no longer so simple. I question whether simple insights, simple experiments such as the ones you admire, are now possible. We have accumulated too much knowledge and too many techniques. Nothing obvious remains to be discovered. Thus, as I argued at our first session, the rate of discovery must slow and the problems become less tractable.

▷ **Hunter:** I emphatically disagree. In fact, I know of a number of cute experiments and theories that have appeared over the last couple years. Several doctors recently reported that packing the heart in plain old sugar after open-heart surgery prevents or helps to cure postsurgical infections, perhaps for much the same reasons that sugar prevents jelly from spoiling.[1] That's the sort of experiment that could have been tried by any surgeon anytime in the last two centuries. Why wasn't it? And here's another. A grad student at Caltech wanted to understand why sand moves the way it does. Complicated question, until you model the sand using dice and find out that cubic materials can have only three motions: slide, bounce and roll. Limits the possibilities, making the problem addressable. Again, why didn't someone think of that earlier?

In the same vein, a chemistry professor I know has his students—undergrads, mind you—busy inventing a periodic table of *compounds*. You don't even need fancy apparatus. Just a library and your brain. The only complication is that you have to work in four or more dimensions to chart the possibilities.[2] Anyone could have done this in the past hundred years. Then—Imp told me about this one—there are some guys who got tired of complicated assay procedures for bioactive compounds and invented a whole new series of tests based on naturally occurring enzymes: a banana slice for assaying catecholamines, and so forth. And in astronomy, a Frenchman named Antoine Labeyrie hit on a way to overcome the "resolution limit" that supposedly prevents earthbound telescopes from seeing stars as clearly as might be done from space. His technique, called "speckle interferometry," actually takes advantage of the failure of telescopes to focus shimmering starlight into a single point by photographing the separate "speckles" and then combining them using computer processing.[3] May not be the be-all and end-all of astronomical devices but it sure beats trying to figure out how to get the darn telescopes into space!

What it comes down to is this, Richter. We so overtrain scientists these days that technical proficiency with abstruse mathematics and experimental techniques replaces thinking. Students are drilled in the most sophisticated solutions we've invented and come to expect such sophistication of every aspect of science. Then they begin arguing as you do, and that's where the problem begins. Simple solutions aren't any rarer today than a hundred years ago. What prevents them from being invented and recognized is this unsound notion that everything simple has already been discovered; that what's left must therefore be complicated; and that complicated problems must have complicated solutions. Not true. Not true at all!

▷ **Imp:** Let me add that we also mistrain our students, as Constance has pointed out, by giving them the false notion that science develops in small steps, brick placed on brick, without scaffolding, without error, without leaps of wild imagination. In consequence, they learn to expect that research will proceed surely and incrementally if only it can be broken down into the small brick-by-brick steps they've been taught to think in.

Example: Biosphere II, an attempt to create a totally self-sustained, enclosed environment—a must if we're ever truly to explore space. I recently read a critique of the project by members of a government-sponsored agency. The members of the agency believed that the Biosphere group was jumping too far ahead and was bound to run into "quite a few problems . . . I'm amazed they'd take such a large step when we're going at it one meter at a time."[4] The government agency is asserting two things: first, that only problems whose solutions are already apparent should be addressed; and second, that the sole purpose of research is to solve problems. I disagree entirely. I believe that the most important aim of research is to *raise* problems, and that any research worth its name should address problems whose solutions are anything but obvious. That's how you generate the surprises.

▷ **Hunter:** Unfortunately, the solution seekers and the problem posers rarely operate on the same priorities. Warren Weaver reports that he was present at a Rockefeller Foundation board of trustees meeting at which was proposed "a rather flexible long-term grant in support of the general program of a first-rate scientist. After the necessarily broad presentation of the program, one of the trustees, an outstanding business leader, said: 'I am afraid I don't know exactly what this man proposes to do if he gets the grant.' The question was answered by another trustee, a scientist: 'Why, if he knew just what he was going to do, he wouldn't need to do it!'"[5]

▷ **Imp:** Right. So, in this spirit of doing it to find out if it is worth doing, I now present my own report!

▲ IMP'S REPORT: A Manual of Strategies for Discovering

> *I, myself, being fairly ignorant of scientific literature, could find more knowledge new to me in an hour's time spent in the library than I could find at my workbench in a month or a year. It is not truth I am searching for, it is* new *truth . . . A scientific researcher has to be attracted by these [blank] spots on the map of human knowledge, and if need be, be willing to give his life for filling them in.*
>
> *—Albert Szent-Györgyi, biochemist and physiologist*

> *The only interesting fields of science are the ones where you still don't know what you're talking about.*
>
> *—I. I. Rabi, physicist*

> *It was certainly better to imagine myself becoming famous than maturing into a stifled academic who had never risked a thought.*
>
> *—James D. Watson, molecular biologist*

What do breakthrough scientists do that average scientists don't? What makes them different? I'm convinced that famous scientists aren't any more intelligent than those who aren't famous. I'm convinced that successful ones aren't right any more often than their colleagues, either. I believe that the architects of science are simply more curious, more iconoclastic, more persistent, readier to make detours, and more willing to tackle bigger and more fundamental problems. Most important, they possess intellectual courage, daring. They work at the edge of their competence; their reach exceeds their grasp. As they stretch themselves, they stretch science. Thus, they not only succeed more often and out of all proportion; they also fail more often and on the same scale. Even their failures, however, better define the limits of science than the successes of more conventional and safe scientists, and thus the pioneers better serve science. The problem I have therefore posed is: How can one best survive on the edge of ignorance? What strategies and tactics must the cosmic explorer master to survive in the unknown?

Here, for what it's worth, is my stab at answering these questions: propositions and principles. Games! Dedijer's "rules of thumb." Coining Bacon's currency of invention to purchase knowledge. But updated, of course, presented as a dialectical system, each bit of advice balanced by its counterproposition. Contradictions and implications, interwoven inseparably—complementarity incarnate.

But first some caveats. These principles are strategies, not infallible algorithms for making discoveries or inventions. They are no more useful than strategies in chess or in warfare. The person applying them must know how to apply them and when to apply them—something that I hope our earlier sessions have illuminated. Merely knowing the following list of strategies and tactics will not confer the understanding necessary to explore nature successfully, any more than reading about how to play a violin will teach one how to play. Indeed, as in the old joke about the violinist who asks how to get to Carnegie Hall, the scientist might similarly ask how to get to Stockholm. The answer is the same: Practice, practice, practice!

I also want to stress that these principles are not for every scientist. They may not work for Big Science, and many certainly don't apply to developmental or confirmatory research. Moreover, these are not strategies for quick professional success. Recall the lonely and frustrating path that men such as Pasteur, van't Hoff, Arrhenius, Szent-Györgyi and Einstein traveled during their first dozen years as scientists. No one can challenge existing belief systems, habits of work, and power structures and expect to be welcomed by the scientific community with open arms. Yet one cannot successfully break from the herd professionally—become one of Hunter's mavericks—and remain personally and socially a member of it. What one accomplishes depends on how much one is willing to sacrifice.

For the complacent, for those who are satisfied with their present position in science, and for those whose fondest desire is to become the editor of several journals, to sit on all the relevant NSF committees, to become head of the department at thirty-five and president of the university at fifty—in short, to do the things that Pasteurs, Mendels, Darwins, McClintocks, Curies, Maxwells, Einsteins, Feynmans never do—the body of my report is not for you. To you, the scientists who refused to attend these sessions, I say:

Train narrowly. Think small. Proceed incrementally.
Nothing ventured, nothing lost. Discover accidentally.
Be objective, not subjective. Never make an err'r.
Follow fashion, not your passion. No criticism dare.
Papers published measure progress (profession'ly, that is)
Grantsmanship, collegiality replace that mental fizz . . .

Well, there's more, but you get the idea. And a hint of things to come! But since no one here needs this sort of advice, I shall turn to more joyous thoughts: strategies for pioneering research.

Young's principle: One cannot possess a useless talent or skill. To begin at the beginning, one cannot train oneself too widely for a career in science. As Thomas Young wrote, "it is impossible to possess any qualification which one may not want, and capabilities are but light burdens."[1] As long as one learns well and deeply, breadth of learning, like experience of life itself, can only confer insight. Chance favors the prepared mind; so be a eurokate, not a stenokate.

Kettering's principle: Action creates results. There are two kinds of experience: direct and vicarious. The former is always to be preferred. As Kettering once said, "I have never heard of anyone stumbling on something sitting down."[2] Once you begin to act, you make things happen. The more ways you "twist the lion's tail," to use Francis Bacon's term, the more you'll make it roar. Explorers are trained by exploring.

Disraeli's principle: Court serendipity by being eccentric.[3] To act different, be different. (But not so different that no one takes you seriously!) Your probability of discovering or inventing something different increases as your experiences, hobbies, skills, knowledge, philosophy, and goals become increasingly unusual.

Truesdell's advice: Learn from the masters.[4] One of the most useful ways to be unusual is to emulate the masters of exploration. Learn from their experiences. The more different styles of research you can master, the greater your chances of raising and solving an important problem. Since Constance has already given us an impressive disquisition on this subject, I shall say no more.

Pauling's principle: Try many things. Like training, the search for scientific problems must also be broad. Those who look in one place find one thing. Thus, Pauling believes that the key to success is trial and error: "Just have lots of ideas and throw away the bad ones."[5] Although we may quibble over what distinguishes good ideas from bad ones, the advice has many advocates. Richet, for example:

> *A fisherman, in order to catch fish in a district of which he is rather ignorant, casts his line in various parts of the river. In the same way, it is necessary to try right and left all manner of experiments, one of which may perhaps be fruitful. But in order to do much, one must not delay too long over any one of these experiments. A first rough hasty one should be attempted to see if it gives indication of immediate results . . . If the result is nil, one does not continue, and will*

> *not have lost much time. But if one is successful, this first success is insufficient. Then ensues a long period of work, demanding a technique more and more perfect . . .*[6]

In other words, survey the field until you find an object worthy of your undivided attention.

Salk's advice: "Do what makes your heart leap!"[7] Listening to some scientists, one gets the impression that finding a research problem is like falling in love or going on an adventure. Jonas Salk's advice to young scientists stresses not only the emotional involvement the scientist must have with his subject, but the commitment that goes along with such involvement. Discovering and inventing are not part-time occupations; one must live with one's research if it is to bear fruit. As Szent-Györgyi says, "If I look at myself objectively, the first thing I notice is that I find myself running every morning, at an early hour, very impatiently, to my laboratory. My work does not finish when I return from my workbench in the afternoon. I go on thinking about my problems all the time . . . Why do I do this? . . . I have to. I'm miserable if I can't."[8] So, like Amundsen or Peary searching for the Poles, you must find a goal that drives you despite the dangers.

The principle of problem choice: Think big. No one ever became famous for circumnavigating a pond. The same principle applies in science. One's program of research must have sufficient facets and ramifications to yield many unexpected surprises. Important discoverers and inventors usually attempt to solve the largest, most important *system* of problems they can imagine. As Peter Medawar says, "any scientist of any age who *wants to make important discoveries must study important problems*. Dull or piffling problems yield dull or piffling answers. It is not enough that a problem be 'interesting'—almost any problem is interesting . . ."[9] So find a problem that is interesting *and* promises to change the maps of science.

G. P. Thomson's principle: Importance does not correlate with difficulty. Ah, you might respond, but addressing big problems is like attempting to scale Mount Everest. Such adventures are not to be undertaken by just anybody. Don't fool yourself. Thomson points out that "In choosing a line of work many men are too modest. They are inclined to shrink from the really exciting things as too difficult. This is a fallacy. The difficulty of research has no relation to its interest or importance."[10] As we've seen repeatedly, mountains can often be reduced to molehills by asking the right question. Charles Richet agrees: "Truth to tell, this distinction, between easy unfruitful problems and hard problems rich in consequences, is theoretical rather than real. For if the experimenter has original ideas, he will always find some means of bringing them to light; he will be able to transform a subordinate question into a fundamental one. If, on the contrary, he is of ordinary intelligence, he will reduce a fundamental question to a trivial one."[11] So find the easy, subordinate problem that provides the key to a system of difficult ones, and place your problem statement in the broadest possible context. The successful pioneer is often simply the first to try.

The principle of problem formulation: The formulation of a problem is often more essential than its solution. According to Einstein and Infeld, a solution "may be merely a matter of mathematical or experimental skill. To raise new questions, new possibilities, to regard old problems from a new angle, requires creative imagination and works real advance in science."[12] We have discussed various strategies for guiding this search for questions at our previous sessions, so I will remind you of only the most important here: search for problem statements promising algorithmic rather than unique solutions; divide your problem area into a logically connected "tree" of subproblems until an addressable subproblem is formulated; choose techniques of investigation appropriate to each type of subproblem; always begin by assuming that your problem can be solved by analogy to a question which has already been answered (alternatively: no problem is unique); reason from what is known to what is not.[13]

The lost key conundrum: Dare to explore where there is no light. Where does one find problems? Not where answers already exist. There is an old story about a drunk who loses his key in a dark alley. A policeman wandering by later that night finds the drunk on his hands and knees under the street lamp at the corner. "Hey! What are you doing there?" "Looking for my key." "Where'd ya lose it?" "In the alley." "Then why are you looking under the lamp?" "It's too dark to see in the alley." Like the drunk, too many scientists choose their research projects within the sphere of existing light. They are scared to be ignorant, scared to flounder. They are what Peter Medawar calls "philagnoists"—lovers of their own ignorance. Not so the best scientists, who seek out the unknown. Peter Carruthers, head of theoretical physics at Los Alamos, speaks for many when he says: "There's a special tension to people who are constantly in the position of making new knowledge. You're always out of equilibrium. When I was young, I was deeply troubled by this. Finally, I realized that if I understood too clearly what I was doing, where I was going, then I probably wasn't working on anything very interesting."[14] Don't be paranoid of the void.

The pioneering urge: The frontiers of science are the richest sources of novelty. A corollary to the "Lost Key Conundrum" is that there will be a crowd searching for the key under the light. If you assume that keys to understanding nature are fairly randomly spread about, your chances of finding one are much better out in the dark because you're likely to be the only one searching there. "I'm afraid I'm like Daniel Boone," says John Wheeler. "Whenever anyone moved within a mile of him, he felt crowded and moved on."[15] Robert Holley has also said that he likes to pioneer new areas and leaves them as soon as it becomes clear that other scientists will develop them.[16] Exploratory scientists "quit the beaten track," to use De la Rive's description of Faraday's method.[17] So move on, look for wildernesses or for fields undergoing rapid change, where there are no experts. Lack of competition means increased time to search and a wider area to search in.[18]

Szent-Györgyi's advice: Renew old knowledge. [19]Another source of rich ideas is previously plowed fields now abandoned. Each new theory requires the rethinking and reformulation of everything that has been discovered or invented before. Therefore, as Constance has documented so well, looking back through the history of a field can yield the greatest surprises.

Medawar's advice: Challenge expectation. Peter Medawar has suggested a criterion for evaluating the surprise potential of a particular bit of research: "If an experiment does not hold out the possibility of causing one to revise one's views, it is hard to see why it should be done at all."[20] One might say the same of a new theory as well: If it isn't crazy, it probably isn't important.

Pasteur's method: Find a contradiction between theory and data. As Hunter so clearly showed us, a particularly useful indicator of potential discoveries is a contradiction between theory and data. One of the two must be wrong, and therefore something is sure to be learned by investigating the discrepancy.

Bacon's principle: Truth comes out of error more rapidly than out of confusion. Another way that Szent-Györgyi recommends to create surprises is to invent a huge new theory and then try to disprove it. This advice clearly presupposes that the scientist is willing—even eager—to err. Few of us are. One of the greatest obstacles to discovering and inventing is fear of failure or of making a mistake. The myth of objectivity has led to the myth of personal infallibility. The scientist who always does what's right (and thereby publishes error-free research) is often commended for being unusually sound. "Sound?" says a character in a C. P. Snow novel. "They call Pritt sound because he has never done anything wrong. I should like to know when he has ever done anything right."[21] Scientists who never err contribute nothing to science. Listen to these comments concerning the very best scientists: "Everyone, including Einstein and Newton, could be wrong as often as he was right."[22] "As happens to many imaginative researchers, Baeyer erred often, not only in theory."[23] "During the first two years after he took his medical degree [Claude Bernard's] scientific achievements can very nearly be reduced to a catalogue of errors."[24] "Many of Bernal's ideas were right, but many were also wrong."[25] "Pauling [is bold] to the point of rashness . . . a lot of his ideas *were* wrong."[26] As Goethe wrote, "Man errs so long as he is striving." So be bold: Dare to err.

Delbrück's principle: Be sloppy enough that something unexpected happens, but not so sloppy that you can't tell what happened.[27] Correlated with fear of error is fear of sloppiness. Many scientists are hamstrung by trying to control every variable and make every measurement exact. Although this may be a virtue in developmental and confirmatory research, it is a sin for the explorer. As Fleming commented while walking through one of the spotless stainless-steel factories that produced penicillin during the 1950s, "But I could never have *discovered* it here!"

Lord Dainton's advice: *"Cherchez le paradoxe."* Another common error among researchers is to ignore things that don't make sense. Too many young scientists of my acquaintance think that if something is incomprehensible or creates paradoxical results (or if nonsense is spouted by an authority figure), they must be too stupid to understand what's going on. Trust yourself. When comprehension fails, science begins. Lord Dainton says, "The best advice to the young, imaginative scientist is that valuable novelty is most likely to be found in observations that do not conform with current orthodoxy. '*Cherchez le paradoxe*' is not a bad motto . . ."[28] More succinctly, Bohr is supposed to have said, "No paradox, no progress."[29]

Bernard's principle: All data are valid. One way to find contradictions and paradoxes is to remember Claude Bernard's dictum that all data are valid, even when they appear to contradict one another. It is precisely in such instances that the most interesting aspects of experiments become manifest; if each result is reproducible, then the conditions under which they are produced are not sufficiently defined. It is just such questions of defining conditions of validity that often yield surprises, for they often reveal the boundaries beyond which theories fail. Therefore, embrace contradiction.

Imp's first principle: Play contradictions. Nature is not, unfortunately, always kind enough to produce contradictions and paradoxes on command. One way to circumvent this difficulty is to invent the contradictions and paradoxes yourself. Nothing is so obviously true or unquestionably correct that you can't learn something by asserting the opposite.

Imp's second principle: Play implications. Another strategy for locating problems is to push every idea to its limits. At the least, one finds the conditions bounding the validity of the idea. At best, one finds that the theory or technique fails. Then the fun begins!

Macfarlane's law: "When a number of conflicting theories coexist, any point on which they all agree is the one most likely to be wrong."[30] Macfarlane's law provides a tactic for locating the specific problem underlying any given contradiction or paradox. That which is so obvious as to be taken for granted is usually the source of the difficulty. Therefore, never ignore the obvious (for it may not be obvious at all), and always check for hidden assumptions (they wouldn't need to hide if they were correct!).

Kuhn's principle: Revolutions follow the recognition of anomalies. All of the preceding principles regarding problems can be subsumed under Kuhn's principle that revolutions in science follow the recognition of anomalies. One must therefore train oneself to know what the scientific community expects, and then search instead for the unexpected pieces that don't fit. As Alexander Fleming suggests: "Never neglect any appearance or any happening which seems to be out of the ordinary: more often than not it is a false alarm, but it *may* be an important truth."[31]

Robert Good, former director of the Sloan-Kettering Cancer Institute, says similarly: "For years I have been trying to teach my students that opportunism is the real way of science. If you pay attention to things that don't fit, you are more likely to make discoveries than if you try to find things that do fit."[32] So discard your blinders and dare to detour.

Fermi's admonition: Never try to solve a problem until you can guess the answer. Detours presuppose a prior path. As simplistic as this may sound, there is no point in beginning research without a goal in mind. This is the basis of Pauling's stochastic approach to research and Polya's heuristics. You must follow your guess into the unknown, keeping your eyes open for surprises. Fermi, however, warned us to follow only those guesses which define the answer to a problem—not a specific answer, mind you, but the magnitude or shape the answer can be expected to take. Will a measurement be in the range of a thousand or a million? Will it be a quadratic function or a hyperbolic? Unless you know these things beforehand, you can wander in the unknown forever and achieve nothing, for the paths there are infinite.

Darwin's opinion: "Without speculation there is no good and original observation."[33] Expectations concerning the unknown can be formed only by speculation. Therefore, follow Darwin's and Richet's advice: "Be as bold in the conception of hypotheses as rigorous in their demonstration."[34]

Grimm's law: Any rough-and-ready rule will be aggrandized automatically into a law of ideal and perfect completeness. One way to speculate boldly is to employ Grimm's law, the logical precursor to Imp's second principle ("push it till it fails"). Grimm's law is usually invoked in a negative sense. In keeping with my first principle, however, I shall turn it on its head. I assert that every observation and every idea has the potential to illuminate some vast area of nature if only the investigator will magnify the minute into the universal. In this sense, I am using Grimm's law as a corollary to the principle of problem choice: Think big; think globally; imagine every phenomenon to be an example of the sublimity of the mundane.[54] Emulate Szent-Györgyi: "I make the wildest theories, connecting up the test tube reaction with the broadest possible philosophical ideas . . ."[55] In every grain of sand a universe . . .

Huxley's principle: Every day some beautiful theory is destroyed by cruel, harsh reality. The necessary palliative for unrestrained Grimmness and for Bacon's principle of the divinity of error is the recognition that every theory has a flaw and every observation has its limitations. A handful of results and theories will last a lifetime or two, some just a few years, and most but a few days. Great scientists not only blow their own bubbles, they also pop most of them themselves. Faraday's exuberant inflation of the smallest fact to the grandest theory was accompanied by an equally rigid and uncompromising self-criticism: he "always cross-examined an assertion."[56] Self-criticism is a scientist's best friend. Better to slay your own brain-children before they leave home than to watch someone else do it after you've become fond of them.

Arrhenius's principle: Things that are said to be impossible are the most important for the progress of science. Another strategy for locating important problems is to try the things that the scientific community believes to be impossible but can't prove to be so. As Edwin Land, the inventor of the Polaroid process and the Land camera, says: "Don't do anything that someone else can do. Don't undertake a project unless it is manifestly important and nearly impossible. If it is manifestly important, then you don't have to worry about its significance. Since it's nearly impossible, you know that no one else is likely to be doing it, so if you do succeed, you will have created a whole domain for yourself."[57] Frédéric Joliot-Curie was a bit more succinct: "The further an experiment is from theory, the closer it is to a Nobel Prize."[58] In short, dare to question the dogmas and preconceptions of your field; challenge the existing boundaries. Nature, not human opinion, must be the arbiter of what is possible.

Bohr's principle: "We all agree your theory is crazy; what divides us is whether it is crazy enough to be correct."[59] Related to the impossibility of solving certain problems is the craziness of their answers. Niels Bohr once criticized Wolfgang Pauli for being too reasonable. Freeman Dyson has elevated this criticism into a principle:

> *The objection that [new theories] are not crazy enough applies to all the attempts which have so far been launched at a radically new theory of elementary particles. It applies especially to crackpots. Most of the crackpot papers which are submitted to* The Physical Review *are rejected, not because it is impossible to understand them, but because it is possible. Those which are impossible to understand are usually published. When the great innovation appears, it will almost certainly be in a muddled, incomplete form. To the discoverer himself it will be only half-understood; to everyone else it will be a mystery. For any speculation which does not at first glance look crazy, there is no hope.*[60]

No doubt this is why Lewis Thomas reports that the surest sign of great events in science is the word "'impossible' spoken as an expletive, followed by laughter."[61] If a thing isn't unthinkable to somebody, then it isn't worth thinking at all.

Monod's principle: Precision encourages imagination. Your craziness cannot, however, be haziness. Jacques Monod stated that too many people equate creativity in science with sloppy thinking and with rule breaking. On the contrary, he asserts that "The severest scientific exactitude, rather than forbidding, actually authorizes and encourages enthusiasm for the boldest speculations."[62] Therefore, the wilder the ideas you wish to propose, the better they must be anchored by the accepted techniques of science.

G. P. Thomson's advice: Demonstrate, don't measure.[35] The object in exploratory research, as opposed to developmental or confirmatory research, is to confirm or expose the existence of a new phenomenon, not to characterize it in detail. Accuracy is unimportant. Hypotheses should be tested qualitatively. Take Szent-Györgyi's isolation of vitamin C. The standard assay procedures for anti-oxidant activity took two or three hours to perform. He was looking for an unknown compound and estimated that to isolate such a compound would take several thousand isolations and tests. We're talking several man-years—or should I say "person-years"?—simply to test the compounds he isolates, not counting the time required for the isolations themselves. So the first thing he did was to look for the simplest oxidation reaction: the peroxidase reaction we all do in intro biology courses. Takes a couple of minutes to set up and a few seconds to read. Nothing to measure: just a color reaction to observe. But that was sufficient for Szent-Györgyi's purpose. He found his anti-oxidant.[36] Roentgen's demonstration of the existence of X-rays is another example. The phenomenon was as clear as his photograph of the bones in his wife's hand, yet he never made a measurement and couldn't identify the cause. Thus, when exploring nature, strive to elucidate the "what" and leave the "how," "how much," and "how often" for a later stage of research.

George's strategy: Vary the conditions over the widest possible range.[37] Once Nature has revealed a phenomenon, one must then cajole her into speaking further by creating the most diverse conditions. Why? Because the nuclear structure of the atom was discovered when Rutherford set a graduate student to make sure no alpha particles were reflected at large angles from a sheet of gold foil. Unexpectedly there were.[38] Because pulsars were discovered during a set of control scans of the heavens where nothing was expected to be found at all.[39] Because Ringer's solution could have been invented by anyone who had taken the time to ask, "Why sodium chloride and no other salt? Why distilled water and not tap?" You can't be surprised if you won't look, and the history of science indicates all too clearly that those who are surprised are those who are willing to look to the side and detour. So don't stop exploring just because you've found a replicable phenomenon—this is when the real exploratory research should *begin*.

Langmuir's principle: Turn it on its head.[40] Sometimes neither serendipity nor planning cooperates with the desires of the scientists. Turn nature's recalcitrance to your advantage. When a desired effect cannot be achieved because undesirable factors interfere, invert your research: deliberately increase the undesirable factors so as to exaggerate their bad effects. Hans Geiger and Walther Müller were measuring cosmic rays but kept picking up "spurious" or "stray" discharges. To track down the source of this "noise" they decided to maximize it. This was the origin of the Geiger counter.[41] Two of the developers of fault-tolerant computer hardware and software—Algirdas Avižienis and Antonin Svoboda—solved many of the problems only when they recognized that everyone else was trying to design perfect systems that were error-proof. They asserted the opposite: Every system will necessarily be flawed, so how can a system be designed to tolerate mistakes?[42] Turning your research on its head often yields useful surprises.

Taton's principle: Synthetic discoveries are made by those whose research is diversified.[43] Hunter has demonstrated through his examples of Berthollet, Pasteur and van't Hoff that many scientists succeed in solving important problems by exploring a variety of fields simultaneously and finding their intersections. This is certainly an adaptive strategy in Richter's evolutionary sense, since research seems to branch in unexpected directions and to hybridize with unforeseen results. Indeed, some problems can be solved only in the light of others. Manfred Eigen provides an example of how this strategy works in practice. Typically, he set out to solve an "impossible" problem: how to measure the rates of "infinitely fast" chemical reactions—meaning reactions that reach equilibrium in less than a thousandth of a second. Most of these reactions, as Hunter mentioned a few weeks ago, involve electrolytes. Eigen is a global thinker who kept track of seemingly unrelated topics such as sound absorption by seawater during sonar experiments. Seawater is composed almost completely of electrolytes (ionized salts) and, at boundaries between temperature layers, absorbs a lot of sound. How? Suddenly, the two problems coalesced. Perhaps the frequency of sound absorbed could be used as a measure of the rate of association and dissociation of the complexes. This was the origin of his relaxation techniques for measuring ultrafast chemical reactions.[44] Thus, Eigen succeeded only because he "looked to the side" to survey surrounding fields. Having noted how frequently this was true in his own work, H. J. Muller once asked his colleagues, "Must we geneticists become bacteriologists, physiological chemists and physicists, simultaneously with being zoologists and botanists? Let us hope so."[45]

The novice effect: Ignorance is bliss. Another way to facilitate interdisciplinary perspectives is to welcome your own ignorance as a challenge to be overcome. Hunter has already documented for us the surprising fact that many discoveries are made by young scientists just moving into a field and by older scientists with little or no formal training in that particular science. Pasteur and his invention of the germ theory of disease is a prime example. Physicist Luis Alvarez and his meteoritic-impact theory of mass extinctions is another: "I will," he writes, "probably be remembered longest for work done . . . in a field about which I knew absolutely nothing until I was sixty-six years old. The field is geology."[46] Alvarez goes on to say that his ignorance of the field was crucial to his discovery because the technique he used (measurement of iridium) had already been tried and found unsatisfactory by previous investigators. "Fortunately, I hadn't heard of their work [or] I'm sure we wouldn't have bothered."[47] Too often results and theories become facts and dogmas, causing alternatives to be discarded prematurely. Outside perspectives are needed. Thus, Peter Carruthers recommends roving: "You'd be surprised how many times just posing a problem to a colleague in a different field will give an important lead. It's because each field has become so developed and so dug in that you can't see into the next trough. It's at the intersections you get the nice overviews."[48] So change fields: Make the effort to climb some hills and see what's in the neighboring valleys!

Burnet's advice: "Do as large a proportion as possible of your experiments with your own hands."[49] When you climb those neighboring hills make sure you do your own observing. Many scientists assign all experimental work to lab techs and postdocs. But as Hunter and Ariana have made clear, only the prepared mind will note and attach significance to an anomaly. Each individual possesses a specific blend of personality, codified science, science in the making, and cultural biases that will match particular observations. If you don't do your own observing, the discovery won't be made. Never delegate research.

Planck's principle: "Only when I have convinced myself."[50] A corollary of doing your own experimentation is to do your own thinking as well. Accept nothing until you have satisfied yourself that there is no other way to think about it. This means, in effect, you must train yourself—be an autodidact, learn your subject your way. Whatever you do, don't accept an idea just because everyone else does. "I was never able to make a fact my own without seeing it," wrote Faraday.[51] "I never pay any attention to anything by 'experts,'" wrote Feynman. "I calculate everything myself."[52] Einstein did the same, eschewing even published tables for the values of constants.[53] These scientists warn us that you don't understand anything till you can do it yourself. Understanding is personal and cannot be taught.

Occam's razor: Seek simplicity. Like the explorers of old, the pioneering scientist must travel light. Einstein is supposed to have said: "Make it as simple as it can be, and no simpler." Enough said.

Szent-Györgyi's principle: Nature is parsimonious. Related to the law of simplicity is the law of parsimony. Do not multiply objects or ideas unnecessarily, but rather explore their combinations; as Szent-Györgyi admonishes us, apples, bananas, and men all utilize the same chemicals according to the same chemical principles simply by emphasizing different aspects of the system.[63] All of nature is woven in this parsimonious way.

Bates's principle: Simple solutions come only from detailed understanding of the complications.[64] Skill at recognizing and using simplicity comes no more easily than skill at rock climbing. What looks to a novice like an impossibly sheer wall requiring crampons, pitons, and belays may appear as easy as a ladder to an experienced rock climber. It is only when one has enough skill to appreciate the difficulties that the ascent becomes simple. This is the lesson of the tennis-match problem, and of van't Hoff's recognition that what his colleagues had thought to be many problems requiring diverse solutions—molecular asymmetry, valency, isomorphism, and so forth—were in fact facets of one problem having one simple solution: the shapes of atoms. In each case, simplicity was arrived at not by simplification, but by the most thorough understanding of the principles involved. Do not, therefore, confuse simplicity with simplification.

Dirac's principle: Seek beauty.[65] Scientists, as Ariana made clear in her report, seek beauty as avidly as any artist. Often this means overlooking or ignoring the ugliness of reality, or seeing through it; and that means obeying Maier's law.

Maier's law: If the data don't fit the theory, ignore the data.[66] Heisenberg once commented to Einstein that he had abandoned several theories because "a good theory must be based on directly observable magnitudes." Einstein surprised Heisenberg by replying, "Nonsense . . . Perhaps I could put it more diplomatically by saying that it may be heuristically useful to keep in mind what one has actually observed. But on principle, it is quite wrong to try founding a theory on observable magnitudes alone. In reality the very opposite happens. It is the theory which decides what we can observe."[67] Therefore, neither limit yourself to what is known nor rely upon its accuracy. As Crick says of his DNA work with Watson: "There's a perfectly sound reason—it isn't just a matter of aesthetics or because we thought it was a nice game—why you should use the *minimum* of experimental data. The fact is, you remember, that we knew that Bragg and Kendrew and Perutz had been *misled* by the experimental data. And therefore every bit of experimental evidence *we* had got at any one time we were prepared to throw *away*, because we said it may be misleading."[68]

Agassi's law: Not all data supporting a theory are to be believed.[69] The complement to Maier's law is Agassi's. In science, as in life, we tend to believe what fits our preconceptions, even twisting data to match our desires. We thereby put false constructions on evidence that would better bear a different interpretation. Unfortunately, the most difficult data to undermine are those which we have no reason to doubt. Anyone who has paid attention to the recent spate of scientific frauds will have noticed that all of them were perpetrated by foisting off false data matching the expectations of the scientific community. The lesson is clear: Doubt most that which you would most believe—Ariana's "faith of a heretic."

Richter's rule: A new theory must account for all existing data as fact or artifact. Maier's and Agassi's laws require yet another complementary rule, or else we could pick and choose our data to fit any theory. Under this rule a theory must set explicit boundary conditions establishing which data may be ignored as artifactual and why.

Ah, I feel Jenny poking me in the ribs. Right. Enough. Let me just say that I'm also compiling a list of specific tactics for employing these strategies, but they're too detailed for our present purposes.[70] The present list must suffice as a start.

And as a conclusion? Well, you know me well enough by now to know that, as much as I'd like to consider what we've accomplished here to be of some significance, I don't want us to take anything we've said too seriously. Science, after all, is no more than a very controlled form of play, and when the fun goes out of it, it's time to find another game. In that spirit, I therefore dedicate to you, my Discovering Project companions, the following rhyme, which I've entitled "The Trick to Being a Maverick."

Train broadly. Think big. Proceed by leap and fill.
Nothing ventured, nothing gained. Intuition: logic kill.
Embrace ignorance. Change fields. Become an autodidact.
Dare to err. Dumb questions air. Eschew the true and tried act.
Act childlike. Think simple. Turn it head o'er tails.
Always try it. Don't deny it. Push it till it fails.
Speculate. Correlate. Curiosity appease.
Ya gotta have themata—aesthetically they please.
Be a maverick. Challenge dogma. Try impossible things.
Dream well. Imagine better. Build a theory that sings.
Predict events. Experiment, but never deputize.
Court precision and skepticism. It's good to criticize.
Play act. Abstract. Form patterns and recognize.
Model it. Feel it. Metaphor and analogize.
Heuristics, stochastics—these guide you in the void.
Insistence. Persistence. Don't be paranoid.
Haziness. Craziness. Goal: implausibility.
Laughter first. Misunderstandings. Last respectability.
Joy in work. Research, fun. Covet independency.
Freedom seek. Self satisfy. Court irresponsibility.
Do these things and you shall find surprises unexpected;
Detours left which turn out right, old dogmas now corrected.
So do us all a favor: Start thinking good thoughts now.
Discovering and inventing: There's no better life, I vow!

◀ TRANSCRIPT: Discussion

▷ **Imp:** Ha! There you have it all rolled up in a nutshell! Comments?

▷ **Richter:** A criticism in the devil's advocate mode. I find many of your principles at odds with one another. Could you not be more consistent?

▷ **Imp:** Absolutely! And lose everything that's important. Look, I'm not trying to produce some fancy, idealistic methodology of science. I'm trying to help real scientists deal with the mess of ambiguities that characterize real research. Thus, I give you not a series of steps to follow, but rather a homeostatic *system* of strategies and tactics that produce a self-regulating push into the unknown. I'm sorry if regulatory systems always seem to involve a balancing of opposites, but they do.

▷ **Richter:** Fine. But you have failed to heed your own advice: Turn it on its head. Balance it. How do you avoid unnecessary detours, for example? Here I find Ernst Mayr's list of unsuccessful strategies more useful. Do not, advises Mayr, tackle problems lying in two domains, neither of which has a well-formulated theory. Many nineteenth-century geneticists who failed to develop workable theories did so because they attempted to formulate a theory of inheritance and of embryological development simultaneously. The complexity of the combined problem hid the simplicity of the inheritance problem. Successful combinations of fields occur only where one of the fields provides a model for investigating the other. Thus, one might be warned against using memory in the immune system as a model for memory in the mind since we do not yet understand basic functions of immunological memory such as self-nonself recognition.

Mayr also identifies other poor research strategies: Failure to partition general problem areas into addressable subproblems. The repeated confirmation of already accepted principles by the elaboration of an infinity of examples. The recording of observations without any attempt to explain the results or to utilize them to test a theory. Failure to carry through a research program to its logical conclusion in the face of counterintuitive results. Failure to abandon a research program that is played out. Failure to consider alternative explanations. Premature attempts to formulate laws of nature. (Here I disagree—Mayr misses the function of simplification and error in the progress of science.) Failure to acknowledge that even incorrect theories are usually based upon correct observations.[1]

I could add a few more myself: Becoming paralyzed by the existing literature. Disdain for "old observations." Premature adherence to a theory before alternatives have been investigated (or, stated another way, failure to keep your options open as long as possible). Undue reliance upon a single method or technique of demonstration. Disdain, especially among experimenters, for theory. Failure to define theoretically or to identify experimentally boundary conditions for the validity of hypotheses and results. Inability to admit error or contradiction in one's work. Belief that if it (whatever "it" is) existed, we would have found it already. Never going beyond one's data, and therefore failing to generalize or test implications.

Enough. The point is we must pay just as much attention to what does *not* work and why it does not as we do to what does work and why it does.

▷ **Constance:** Which I think we've done to some extent, Richter. Remember, Hunter and I spent a fair amount of time trying to figure out why one person—Pasteur or Fleming—makes a discovery when his colleagues—Mitscherlich, Biot, Allison, or whoever—don't. Surely that counts for something.

▷ **Imp:** And I think I hinted at most of your conclusions in my report, and certainly with my first bit of rhyme. But if I ever write another version of this manual, I'll heed your advice. In the meantime, and without actually checking to make sure, I'll wager that simply adopting the opposite strategy of each of the ones I've listed will yield a nonadaptive strategy.

▷ **Jenny:** Unless, of course, you're onto deep truths whose opposites also contain a grain of truth. That's a joke, Imp! But seriously, may I suggest we adjourn? This has been a very long day, and since I made dinner reservations at Gullivers for all of us at seven, we really ought to be on our way.

Voilà! Allons-y!

Postscript

The present author has been able to watch a great many attempts at building new physical theories. He is in a position to remember how many of these inventions failed definitely, how many succeeded, and how they all had to be later modified and corrected.

—Leon Brillouin, physicist (1964)

So, we're always walking the tightrope between alternatives. We're scared of the big project that might take a lifetime and lead nowhere . . . the short piece of work essentially without meaning.

—Eleazar Lipsky, physician and novelist (1959)

A balance that does not tremble cannot weigh. A man who does not tremble cannot live.

—Erwin Chargaff, biochemist (1978)

▶ IMP'S JOURNAL: Ambiguities

Tactics and strategies all very well. But there's psychological side to discovering, too. Dealing with ambiguities, inconsistencies, contradictions. Emotionally.

Hardest part of doing science is knowing you'll never pin down nature. Not really. Snow has character compare discovering to making love. "Suddenly your unconscious takes control. And nothing can stop you. You know that you're making old Mother Nature sit up and beg. And you say to her, 'I've got you, you old bitch.' You've got her just where you want her. Then to show there's no ill-feeling, you give her an affectionate pinch on the bottom."[1] Right. What he never wrote about was how old Mother Nature then withholds herself and makes *you* beg! Very frustrating!

Case in point: amino acid pairing. Finally found two possible instances in physiological systems: fibrinogen aggregation and reproductive hormone control. Doolittle found small peptide that interferes with fibrinogen aggregation. [Editor's note: fibrins are blood proteins involved in forming blood clots.] Binding site unknown. Predict according to amino acid pairing hypothesis. One site on another small peptide—fibrinopeptide A. Test for binding, and it works.[2] Unfortunately, fibrinogen too big to look for binding sites on it using same techniques. Too bad, 'cause approach might work with fibronectin, etc. Stymied.

Then, reading about LHRH [luteinizing hormone–releasing hormone] antagonists, find ref. to Orts, Bruot and Sartin's work on BPART [bovine pineal antireproductive tripeptide].[3] BPART interferes with physiological activity of LHRH. Both found in pineal gland. Both act on testes. And exact amino acid pairs. Do they bind? Yes![4] Very specifically. Much more specific than "molecular sandwiches." Incredible. Suggests every peptide hormone may have chemically complementary peptide antagonist—antihormones; antitransmitters. Could be isolated by mutual binding. May have complementary receptor systems, too. Perfect homeostatic control system. Very exciting.

Moreover, LHRH-BPART pair isn't predicted by Mekler-Blalock-Smith hypothesis, whereas several of my negative controls are. Making progress, right?

Wrong! Just as I finish my research, Bost and Blalock announce synthesis of sequence complementary to ACTH [adrenocorticotropic hormone, the "stress" hormone] *according to his hypothesis* and damned if they don't claim to get binding, too![5] Of course, they didn't bother testing my hypothesis or cite my results. In fact, have only one control. Worse: Complementary sequence has 70 percent homology with sequence predicted by my hypothesis, but backwards. Does that matter? How good is their test? On other hand, claim antibody to the ACTH receptor recognizes ACTH complementary sequence! That means opposite strand of DNA may code for receptor sequences! Moreover, if two *linear* sequences of proteins have tertiary [folded, three-dimensional] complementarity as well, as Blalock's results suggest, then unsolved problem of protein folding within grasp. Complementary amino acid sequences would confer complementary folded structures.

Can all this be true? Is old Mother Nature tantalizing me just to frustrate me in the end? Or is she playing fast and loose with Blalock? Or both of us? Who's begging now?

Well, one thing's certain. No matter how crazy I was to challenge Central Dogma, it's paid off in unexpected results. I'm where I wanted to be: the frontier! Now we'll see just how useful my "Manual of Discovering" is in practice. And discover its limitations, too.

▼ JENNY'S NOTEBOOK: Conclusions Are Questions

Needless to say, it's taken months to get the various sessions, journals, and notebooks transcribed, edited, and in order. Not to mention the usual problems of getting our friends to return the material with their corrections. Fortunately, they allowed most of their comments to stand—even a few that in retrospect sound a little embarrassing—in order to give an accurate picture of the process we went through. So today, *finally*, I dumped the whole thing on Imp's desk so he could take one more critical look at it before we turn it over to the publisher. "I've finished my read-through," I announced.

Imp was putting on his running shoes. He's back to his usual cheerful, energetic self. "What do you think?" he asked.

"A typically Impish effort," I said. "Joking aside, I think there's some important stuff here. The reinterpretation of chance discoveries as unexpected surprises resulting from a logic of research was an eye-opener for me—certainly made your search for strategies of research more plausible. So did the idea that the context of validation could be identical to the context of discovery. I mean, that isn't what I was taught about scientific method and the role of crucial experiments: One hypothesized and then tested, right? So much for proof and disproof. Instead, we argued that multiple hypotheses are generated, all of them flawed. These are evaluated according to diverse personal criteria that come down to ascertaining the extent to which each promises to alter scientific practice.

And where do these multiple hypotheses come from? Induction and deduction? Inference? Again we disagreed with accepted dogma. Hypotheses, we concluded, are generated by problems—mismatches between data and theory—and are defined by the criteria implicit in the mismatch. Since induction and deduction are methods of proof and disproof, not methods for problem generation, new methods must be invented to describe scientific thinking. The focus of explaining scientific discovery, we agreed, must become the role of the individual scientist as a nexus for unique sets of data, theories, problems, and personal proclivities.

"These repatternings of how we understand science provided the context for all the innovations that followed. Ariana's definition of science as the elaboration of all possible explanations of phenomena successively constrained by an ever increasing body of tests and observations; her various tools of thought, transformational thinking, and "synscientia"; Constance's model of the invention process, her quintessential criteria for evaluating novelty, and her description of themata as guides to research; Hunter's emphasis on apprenticeship and autodidacticism as complementary components of scientific training, his demonstration of the fecundity of mixing odd combinations of disciplines and traditions in forming new sciences, his inducements to innovation, and his concept of venture research; Richter's problem types—I'd never thought about how many different problems there are or how each needs a different method for its solution—and, of course, his evolutionary theory of scientific development; your insistence that dissent rather than adherence to a paradigmatic dogma typifies exploratory science, and your idea of fault tolerance (though I must say, I consider myself quite adept at the latter already!). More generally, I was intrigued by the recurring themes of global thinking, pattern forming, guessing, incorrect hypotheses, planning for detours, development of intuition and a feeling for one's subject. How's that for a summary?"

Imp stood up and stretched. "Don't forget your model for illumination." He was massaging my ego, and we both knew it—which didn't make it any less pleasant.

"I don't know. Looking back on it, I'm surprised we—perhaps Richter deserves most credit—managed to make it all fit together reasonably coherently," I continued. "What I wonder is how different our conclusions would have been had we convened a different group of people. Say Ariana had been too busy, or Richter."

"Yes, I've thought about that, too," replied Imp as he began a few warm-up exercises. "The creative syntheses would've been different, of course. Or perhaps the syntheses wouldn't have occurred. We might have been like the blind men and the elephant, each giving a separate, noncomplementary picture of our subject.[1]

"But I'd like to think that any group of scientists would've sketched out the same area of overlaps we did—with different emphases. Can't expect a philosopher to focus on the same questions as a historian, or expect a theoretical physicist to view science exactly as I do. If we'd had a sociologist or anthropologist, we might have gotten some different views. But where the philosopher, historian, psychologist, theoretician, and experimentalist meet—well, I think we've pretty well mapped out the common ground. Given the data, we've at least delimited the types of solutions that are possible. And some of the more interesting ways in which the data can be organized meaningfully. Not bad for a start."

"A start? You aren't satisfied?"

"Never." Imp grinned. "The most important stuff is yet to come."

"What stuff?"

Imp thought for a moment and then broke into a big smile. "All the problems we didn't solve, the connections we didn't make, and the questions we didn't ask!" He meant it, too. "It doesn't matter what we've accomplished here—or whether we've accomplished anything. The only important thing is what effect it has. On what Ariana and Constance do that they wouldn't have done otherwise. On Richter's creativity. On the way Hunter teaches and does research. On how our students do science. Or, in your case, perhaps history. On funding, on freedom to do research, on whom we cultivate to do science. Have we made people think about the subject in a different way? That's what counts. And we can't control that.

"If you want to know my one, deep hope," he went on, "it's that this manuscript reaches one aspiring Darwin or Curie, laboring privately for decades to transform science; one van't Hoff or McClintock working in a tenuous position and dreaming of something better; one Arrhenius whose department doesn't consider him fit to continue as a scientist; one Einstein struggling alone in his patent office; one Ramanujan training himself in a little town in India. If what we've said and done here gives such a person the courage to do science on his own and in his own way—you know I mean her, too—then we have succeeded in producing something important. If we can provide such a person a single strategy for research, a single tactic for survival, that he hadn't considered before, that might be all that's needed. If we can reassure him that he's not the first and won't be the last to have to struggle so hard, perhaps that will bolster his courage to keep up the fight."

"Or maybe it could help one Gunning or Edlund or Schmidt to be able to recognize and foster such a person," I suggested.

"Yes, or—think big!—even some enlightened billionaire philanthropist! After all, both the Rockefeller Institute and the MacArthur Fellows program had their inspiration in books or essays by scientists writing about the unmet needs of the truly creative scientist." Imp paused. "But I'd be satisfied with helping a single individual. Who knows? Even so small a thing might change the world." And with that, he took off out the door.

We can hardly expect a committee to acquiesce in the dethronement of tradition. Only an individual can do that, an individual who is not responsible to the mob. Now that the truly independent man of wealth has disappeared, now that the independence of the academic man is fast disappearing, where are we to find the conditions of partial alienation and irresponsibility needed for the highest creativity?

—Garrett Hardin, biologist and historian of science (1959)

Whatever system we adopt, it must be kind to rebels, and there must be no good-conduct prizes . . . If we do not train our students to think for themselves—perhaps it would be fairer to say, if we do not allow them to think for themselves—no opportunities that we provide in later life will be of much avail.

—W. W. C. Topley, physician (1940)

The danger comes when scientists allow themselves *to be organized, when they begin to* respect and obey *pronouncements on science by academies, universities, societies, and, finally, governments. May that day never come!*

—Clifford Truesdell, mathematical physicist and historian of science (1984)

Notes

PROLOGUE: ON FACT AND FICTION

1. Bernal, 1939, xv.
2. See Beveridge, 1950/1957, 219.
3. Quoted in Bernal, 1954, 5.
4. Snow, 1934/1958, 258–260.
5. Tolstoy, 1899/1984, 99.
6. Quoted in Goldsmith, 1980, 230.
7. Quoted in Bernard, 1927/1957, v.
8. Weisskopf, 1977, 410.
9. Chesterton, 1924/1956, 3.

PREPARATIONS: TOWARD A SCIENCE OF SCIENCE

Jenny's Notebook: Impish Ideals

1. Bronowski, 1958, 60.
2. Snow, 1934/1958, 13.
3. See Arthus, 1943, 373.
4. See Snow, 1934/1958, 258–260.

Imp's Journal: Dissatisfactions

1. See Poincaré, 1905, 14.
2. Truesdell, 1984, 91.
3. See Huxley, 1956, 33–69.
4. Szent-Györgyi, 1961, 49.

Jenny's Notebook: The Problem Area Defined

1. "The Advancement of Learning," book II, quoted in Moulton and Schifferes, 1960, 133.
2. Quoted in McConnell, 1969, 251–252, and Weber, 1973, xv–xvi.
3. Quoted in Judson, 1980, 3.
4. Quoted in Winkler, 1985.
5. Pauling, 1961, 46.
6. Harwit, 1981, 9.
7. Ritterbush, 1968, iv.

Imp's Journal: Who Cares?

1. See Klemm, 1977, x.

Transcript: Courting Novelty

1. See Richet, 1927, 129; Alvarez, 1932; G. P. Thomson, 1957, 11; Feldman and Knorr, 1960, 12; Hilts, 1984, 142; Lightman, 1984; Lehman, 1953.

2. Sadoun-Goupil, 1977, 6, 48, 103–108.
3. Herold, 1962; Crosland, 1967; Sadoun-Goupil, 1977, 43–47.
4. *La Décade Egyptienne*, 1799–1801, I: 10–14, 78, 80, 83, 129, 221, 293, 295, 296; II: 5, 9, 32, 59, 99, 100, 128, 166, 167, 232, 264; III: 290, 292, 298.
5. Guerlac, 1959; Edelstein, 1971.
6. *Science Digest*, Dec. 1984.
7. Berthollet, 1791.
8. Berthollet, 1789; Lemay, 1932; Haynes, 1938, 13–17; Musson and Robinson, 1969, chap. 8.
9. Gillispie, 1957.
10. Buonaparte, 1859, V, no. 3952.
11. Quoted in Lowinger, 1941, 1, n. 3.
12. See Coleby, 1938, 39–47; Duncan, 1962, 189–194.
13. Bartlett, 1976, 000.
14. Holmes, 1962, 108.
15. Kuhn, 1962/1970, 52–61.
16. Gillispie, 1957, 170.
17. Vandermonde et al., 1786; C. S. Smith, 1968, 275–348.
18. Lavoisier, 1789, 54–55.
19. Berthollet, 1791, 12.
20. Ibid., 1–10.
21. Baumé, 1773, 22; Laplace and Lavoisier, 1783; Guerlac, 1976.
22. Lavoisier, 1789, 54.
23. Laplace, 1784, xii–xiii.
24. Laplace, 1796, 196–197.
25. See Duncan, 1962; Smeaton, 1963, 60–61.
26. Berthollet, 1795.
27. Crosland, 1967, 103, 235; Court, 1972.
28. *La Décade Egyptienne*, 1799, II, 99–100; Berthollet, 1800.
29. Berthollet, 1801a, 4–5; Berthollet, 1801b.
30. Berthollet, 1801a; Berthollet, 1803, 1.
31. Kuhn, 1962/1970.

Jenny's Notebook: The Arts of Scientists

1. Tuchman et al., 1986.
2. Besant, 1908.
3. Duchamp, "Disks Bearing Spirals Made for Anemic Cinema," in Tuchman et al., 1986, 264.
4. See Sekuler and Levinson, 1977, 61.
5. Vitz and Glimcher, 1984, 27–29.
6. See M. Gardner, 1986; M. Gardner, 1959; M. Gardner, 1961; M. Gardner, 1977; Golomb, 1954.

DAY ONE: THE PROBLEM OF PROBLEMS

Imp's Journal: Discovering a Problem

1. Crick, 1958; Crick, 1970.
2. Feynman, 1985, 166; Wald, 1966, 27; Cannon, 1932, 61.

Transcript: How Does Science Grow?

1. Cannon, 1932, 61.
2. Kuhn, 1977, 340–352; Popper, in Schilpp, 1974, 1174–1180; Gombrich, in Schilpp, 1974, 925–957, in Miller, 1983, 231, and in McConnell, 1983, 145–173.

3. Bronowski, 1978; C. S. Smith, 1981; J. Miller, 1983, 214.
4. Hoffmann, 1987; 1988; Weisskopf, 1980; Chandrasekhar, 1988.
5. Ossowska and Ossowski, 1935/1964–65. See also Walentynowicz, 1982; Price, 1963; Goldsmith and Mackay, 1964; Weinberg, 1967; Shils, 1968.
6. See Poincaré, 1913/1946, 210; Thomson, 1937, 62; Needham, 1929, 251.
7. Zuckerman, 1977, 96–143.
8. See Szent-Györgyi, 1961, 47.
9. Krebs, 1967, 1441.
10. Dedijer, 1966, 275.
11. See Beveridge, 1950.
12. See Feldman and Knorr, 1960, 4.
13. See Bush, 1960, 23; Holton, 1975, 216–217; Vijh, 1987, 9.
14. Brinkman et al., 1986. See also Thomsen, 1986, 245.
15. Wilde, 1931, 43.
16. Quoted in Thomsen, 1986, 26–27.
17. Price, 1963, 1–32. See also Ben-David, 1960; Harwit, 1981, 13–54; Russell et al., 1977, 330–331.
18. Parkinson, 1963, 193.
19. I. Cohen, 1985.
20. American Chemical Society, 1976; *Modern Photography*, 1987; *Popular Photography*, 1987.
21. Ramo, 1987; Kanigel, 1987, 50.
22. Zuckerman, 1977, 45, 84, 90, 171.
23. Beaver, 1976.
24. Mellanby, 1974.
25. Szilard, 1961.
26. Quoted in Weber, 1973, 3n.
27. Dyson, 1958.
28. M. Wilson, 1972, 29–30.
29. See Bernard, 1927/1957, 41.
30. See Millikan, 1950, 29; M. Wilson, 1954, 407; Brillouin, 1964, 46; Kirchner, 1984.
31. See Schachman, 1979, 364.
32. See Burnet, 1968, 65.
33. See Mellanby, 1967; Mellanby, 1974; Weiss, 1971; Ziman, 1969.
34. Gray, 1962; Braun et al., 1988; Nolting and Feshbach, 1980.
35. Selye, 1977, 287.
36. Szent-Györgyi, 1966, 68.
37. Glass quoted in Kneller, 1978, 39; Burnet, 1968, 11–12. See also Crick, 1967; Badash, 1972.
38. Carmichael, 1930, 14–15; Peacock, 1979, 54.
39. Meadows, 1972, 283.
40. Quoted in Millikan, 1950, xiii.
41. Ibid.
42. Snow, 1934/1958, 168.
43. Benét, 1942.
44. See Wilson, 1961, 101.
45. See Russell et al., 1977; Meadows, 1972; Armytage, 1957; Higgs, 1985.

Jenny's Notebook: Implications, Contradictions

1. Diderot, 1796/1966, 202.
2. See Schaffner, 1980; Burnet, 1963; Jerne, 1976.

Transcript: What's Worth Investigating?

1. Weissmann, 1985, xviii; Thomas, 1983, 151.
2. J. Maxwell, 1875, 357.
3. Krebs and Shelley, 1974, 17.
4. Arthur Yuwiler, personal communication.
5. Willstätter, 1958, 56.
6. Crick, 1974, 766.
7. Hull, 1974; Hempel, 1966; Poincaré, 1913/1946; Duhem, in Lowinger, 1941; Carmichael, 1930.
8. Valéry, 1929, 26.
9. See Ramon y Cajal, 1947, 75; Bancroft, 1928, 170.
10. J. J. Thomson, 1937, 16; G. P. Thomson, 1957, 11. See also Crowther, 1968, 250; Mach, 1943, 367.
11. Richet, 1927, 44; Burnet, 1968, 30; Medawar, 1979, 17. See also Osborne, 1927, 309; Ramon y Cajal, 1947, 75–76.
12. Szent-Györgyi, 1966. See also Brooks, 1966, 11–12.
13. See Carmichael, 1930, 182–183; Poincaré, 1913/1946; Lowinger, 1941.
14. Bernard, 1927/1957, 177.
15. Krebs and Shelley, 1975, 95.
16. Heisenberg, 1958, 35.
17. Quoted in Livingston, 1973, 286. See also Thomson, 1957, 9; Beveridge, 1980; Koestler, 1976, 95; Lowinger, 1941, 126–127.
18. Adapted from Halmos, 1968.
19. See Truesdell, 1984, 594–639.
20. Quoted in Judson, 1979, 20–21.
21. See Root-Bernstein, 1982c.
22. Planck, 1958.
23. Danielli, 1966.
24. Polya, 1962.
25. See Szent-Györgyi, 1966; Ramon y Cajal, 1947, 120; Medawar, 1979, 13.
26. See Carlson, 1981, 246; Szent-Györgyi, 1966, 116–117; Rayleigh, 1942, 99; Feynman, 1985; M. Wilson, 1972, 360; Koestler, 1976, 111ff.
27. Beveridge, 1950, 40.
28. Quoted in Hardin, 1959, 84.
29. Holton, 1973, 355.
30. Judson, 1984, 35.
31. I. Cohen, 1982, 248.
32. See Price, 1963; Weinberg, 1967.

Imp's Journal: Dogma Denied

1. Watson, 1965, 297–298.
2. Lehninger, 1970, 632–633; Temin, 1981; Fraenkel-Conrat, 1979, 47.
3. Crick, 1958.
4. Quoted in Judson, 1979, 337.
5. Foster, 1899, 231.
6. Quoted in Judson, 1979, 337.
7. Crick, 1970, 562.
8. See Crick, 1967.
9. See Keller, 1983, 5–8, 172–179; Keller, 1985, 170–172.
10. Chargaff, 1974, 778. See also Chargaff, 1978, 106–107; Chargaff, 1963.

DAY TWO: PLANNING OR CHANCE?

Imp's Journal: Alternative Hypotheses

1. See Mekler, 1969; Mekler, 1980; Ildis, 1980; Blalock and Smith, 1984.
2. Root-Bernstein, 1982a.
3. Root-Bernstein, 1982b.

Transcript: Planning

1. See Hull, 1974; Reichenbach, 1938; Braithwaite, 1955; Popper, 1959; Hempel, 1966; Feyerabend, 1981; Kneller, 1978, 68–95.
2. Feyerabend, 1961.
3. Schiller, 1917.
4. Hanson, 1961.
5. See Cannon, 1945; Nicolle, 1932; Beveridge, 1950; Dale, 1948; Burnet, 1968; Medawar, 1967.
6. See Koestler, 1976.
7. Polya, 1962, 1.
8. Poincaré, 1913/1946, 438.
9. Bernard, 1927/1957, 166.
10. Quoted in Perkin, 1923, 71.
11. Quoted in Halacy, 1967, 12–14.
12. Maxwell, 1974a, 1974b. See also Kneller, 1978, 80–87.
13. Polanyi, 1958.
14. Hanson, 1958; 1961; Caws, 1967; Maxwell, 1974a, 1974b.
15. Harris, 1970; Gutting, 1973. See also Kneller, 1978, 89–91.
16. Nickles, 1980.
17. See Mullin, 1962.
18. Quoted in R. Vallery-Radot, 1901/1919, 86.
19. Ibid., 83.
20. Pasteur, 1922, I, 19–30.
21. Bernal, 1953, 182.
22. Biographical material and background from Dagognet, 1967; Dubos, 1950; Dubos, 1960; Duclaux, 1920; Geison, 1974; R. Vallery-Radot, 1884; R. Vallery-Radot, 1901/1919; P. Vallery-Radot, 1954; P. Vallery-Radot, 1968.
23. See Pasteur, 1860; Bernal, 1953; Mauskopf, 1976; Kottler, 1978; Roll-Hansen, 1972.
24. Mauskopf, 1976.
25. See Melhado, 1980, 114.
26. Ibid., 86.
27. See Lowinger, 1941, 107.
28. Ritterbush, 1968, 38.
29. See George, 1936, 84.
30. Root-Bernstein, 1985.
31. Goldsmith, 1980, 225–226.
32. Dubos, 1950, 26, 96; P. Vallery-Radot, 1954; Wrotnowska, n.d.; Perreux, 1962.
33. See Partington, 1964, 376–377.
34. Root-Bernstein, 1985.
35. Nye, 1980.
36. Franks, 1981.
37. Langmuir, 1953/1986.
38. Pasteur, 1860.
39. R. Vallery-Radot, 1901/1919, 39.

40. Watson, 1968, 126.
41. Quoted in R. Vallery-Radot, 1884, 24.

Jenny's Notebook: Personal Knowledge

1. Debye, 1966, 81.
2. See Rae, 1972, 316.
3. Keller, 1983, 197–207; Keller, 1986; Goodfield, 1981, 63.
4. Knudtson, 1985, 66–72; Salk, 1983, 7.
5. Polanyi, 1967, 16.
6. Quoted in Judson, 1980, 6.
7. Hadamard, 1945, 142–143.
8. Ulam, 1976, 17.
9. A. Roe, 1951; A. Roe, 1953, 141ff.

Transcript: Chance

1. R. Vallery-Radot, 1901/1919, 70ff; Duclaux, 1920, 40ff; Bernal, 1953, 200ff; Kottler, 1978, 90; Geison, 1974, 355ff.
2. See Gassman et al., 1985.
3. Dubos, 1950, 106–107. See also Dubos, 1960, 34–35.
4. Bernal, 1953, 207; Kottler, 1978, 90; Duclaux, 1920, 44–46; R. Vallery-Radot, 1900, 73–74.
5. Pasteur, 1922, II, 18, 21, 129–130; Pasteur, 1940, 345–348.
6. Pasteur, 1922, I, 314–344.
7. See Land, 1973, 163.
8. Pasteur, 1922, I, 329–344, 360–363, 369–380; Kottler, 1978.
9. Pasteur, 1922, I, 157.
10. P. Vallery-Radot, 1968, 10; R. Vallery-Radot, 1901/1919, 71.
11. Papiers Pasteur (hereinafter PP), Corresp., 11: 432–433, 6: 254–258; Pasteur, 1922, I, 413–465; Pasteur, 1940, 326n; P. Vallery-Radot, 1968, 10–13 and notes; R. Vallery-Radot, 1901/1919, 73.
12. PP, Corresp., 6: 257v, 258.
13. Prévost, 1977, 101; Pasteur, 1922, I, 391–405; R. Vallery-Radot, 1901/1919, 81; PP, I, 70.
14. Pasteur, 1922, I, 329–344, 360–363, 369–380; PP, Registres de Laboratoire, 6: 17–55, 153.
15. Pasteur, 1940, 326.
16. Pasteur, 1940, 325n, 335.
17. R. Vallery-Radot, 1884, 30–31.
18. PP, Cours de Chimie, Faculté des Sciences de Lille, 19v, 20.
19. PP, Corresp., 11: 437, 437v.
20. PP, Registres de Laboratoire, 1857, cahier 2: 44, 70.
21. Taton, 1957, 138–141.
22. Root-Bernstein, 1983e, 387.
23. Chevreul, 1858; PP, III, 403.
24. Pasteur, 1922, II, 21.
25. Ibid., I, 314–344.
26. PP, III, 406.
27. Pasteur, 1922, II, 28; I, 314–344.
28. Dubos, 1950, 106–107; Dubos, 1960, 34–35; Bernal, 1953, 207; Kottler, 1978, 90; Geison, 1974, 350; R. Vallery-Radot, 1884, 73–74; P. Vallery-Radot, 1968, 52.
29. PP, III, 396, 405v, 406.

30. Pasteur, 1922, II, 129–130.
31. Ibid., I, 369–380.
32. Elstein et al., 1978.
33. See Nickles, 1980.
34. See Nagel and Newman, 1960, 12.
35. Thomson, 1937, 28.
36. Huxley, 1882, 173–174. See also Bernard, 1927/1957, 40.
37. Medawar, 1969, 33.
38. Hardy, 1874, chap. 5.
39. See Bliss, 1982.
40. Quoted in Judson, 1980, 69. See also Ramon y Cajal, 1947, 83.
41. See Goldschmidt, 1956, 150–175; Willstätter, 1965, 65.
42. Bernard, 1927/1957, 31.

Jenny's Notebook: The Eye of the Mind

1. Quoted in R. Jones, 1978/1979, 41.
2. Pearl, 1923, 85–89; Root-Bernstein, 1983c.
3. See McCain and Segal, 1978, 25–26, fig. 1; Kirchner, 1984.
4. Shaler, 1909; Rapport and Wright, 1964, 43–47.
5. Bibby, 1960, 111.
6. Quoted in R. Jones, 1978/1979, 41.
7. Quoted in Margenau and Bergamini, 1964, 112.
8. Frisch, 1979, 72.
9. Quoted in Gregory, 1916, 86.
10. J. Thomson, 1937, 226.
11. See Root-Bernstein, 1985; Root-Bernstein, 1987c; Goldschmidt, 1956, 18; Ramon y Cajal, 1947, 134–135; Zigrosser, 1955–1976, 14–15.

Imp's Journal: Surprises

1. Root-Bernstein, 1983b; Root-Bernstein, 1984b; Root-Bernstein and Westall, 1984c; Root-Bernstein, 1987a.

DAY THREE: LOGIC OF RESEARCH, SURPRISE OF DISCOVERY

Imp's Journal: Competition

1. Root-Bernstein, 1982a, 1982b.
2. Mekler, 1969; Mekler, 1980; Cook, 1977; Kauffman, 1986; Blalock and Smith, 1984.
3. Grafstein, 1983.

Jenny's Notebook: The Body Is Part of the Mind

1. See Lipsky, 1959, 248; Taton, 1957, 44.
2. See Olmsted and Olmsted, 1961, 53; Olby, 1966, 106–107; Gregory, 1916, 6; Clark, 1968, 84–91, 161–162; Bell, 1937, 421ff.
3. Michelson in Livingston, 1973, 111–115; Muller in Carlson, 1981, 174–175; Szent-Györgyi in Wilson, 1972, 10. Other examples: Metchnikoff in DeKruif, 1926, 208ff; Banting in Bliss, 1982, 106ff; Bohr, Born, and Pauli in Heilbron, 1985, 391; Archibald Couper, Ignaz Semmelweis, Ludwig Boltzmann, Wallace Carothers in *Dictionary of Scientific Biography.*
4. Forman, 1981, 24.

Transcript: The Probability of Discovering

1. Quoted in Cannon, 1945, 68; and in Austin, 1978, appendix A.
2. Austin, 1978.
3. Quoted in Judson, 1980, 69.
4. Karl Popper, in Krebs and Shelley, 1975, 22.
5. Manfred Eigen, in Ibid.
6. M. Wilson, 1972, 14.
7. Pauling, 1961, 46. See also Smith, 1910, 142.
8. See Cannon, 1945; Dale, 1948; Beveridge, 1950, 37–55; Shapiro, 1987; Hannan et al., 1987.
9. Cannon, 1945, 71.
10. Olmsted and Olmsted, 1952/1961, 83ff; Holmes, 1974.
11. Richet, 1927.
12. See Duclaux, 1896/1920, 280–285; R. Vallery-Radot, 1901/1919, 392; Cannon, 1945, 72; Dubos, 1950, 327.
13. Cadeddu, 1985.
14. Pasteur, 1880. See also Ramon y Cajal, 1893/1951, 139.
15. Cannon, 1945, 72; Bliss, 1982, 26n; Beveridge, 1957, 38; Dale, 1948, 453.
16. Houssay, 1952, 112–116; Bliss, 1982, 26.
17. Houssay, 1952.
18. See Eiseley, 1965.
19. See Snow, 1934/1958; Lipsky, 1959.
20. Maurois, 1959; Hare, 1970; Hughes, 1974; Macfarlane, 1984.
21. Quoted in Maurois, 1959, 109–110.
22. Hughes, 1974, 41–42.
23. See Feynman, 1985.
24. Macfarlane, 1984, 74–76.
25. Ibid., 146.
26. Ibid., 263.
27. Bustinza, 1961, 181.
28. Quoted in Macfarlane, 1984, 246.
29. Quoted in Hardin, 1959, 134.
30. Thomsen, 1986a, 27.
31. Moulton and Schifferes, 1960, 244; Judson, 1980, 4; Weber, 1973, v, 203; Debye, 1966, 82–84; Feynman, 1985; Read, 1947; Escarpit, 1969, 258.
32. See Hill, 1960, 200; A. Taylor, 1966, 55.
33. Quoted in Maurois, 1959, 109.
34. Quoted in Morris, 1979, 60.
35. Ibid.
36. Ibid., 60–62.
37. Colebrook, 1954, 70–103.
38. Macfarlane, 1984, 73.
39. Colebrook, 1954, 74.
40. Quoted in Maurois, 1959, 114.
41. Fleming, 1922, 306.
42. Macfarlane, 1984, 98.
43. See Hughes, 1974, 41; Macfarlane, 1984, 15–16.
44. See Judson, 1979, 48–49.
45. Carlson, 1981, 128–129; J. Thomson, 1937, 341.
46. Quoted in Judson, 1980, 69.
47. Quoted in Fowles, 1969, 183.
48. Davy, 1840, 352.

49. Moulton and Schifferes, 1960, 300.
50. Macfarlane, 1979, 98–100.
51. Cannon, 1945.
52. Einstein and Infeld, 1938, 36.

Jenny's Notebook: Patterns in Mind Space

1. Doyle, n.d., 407 ("The Reigate Puzzle"). See also ibid., 467–468 ("The Naval Treaty").
2. M. Wilson, 1961, 233–234.
3. R. Lewis, 1944, 22–24; R. Lewis, 1945.
4. See Root-Bernstein, 1980, 369–374.
5. Lipscomb, 1982, 7.
6. See Young, 1987.
7. Brillouin, 1964.
8. Szent-Györgyi, 1955, 64; Weiss, 1970; Weiss, 1971.
9. See Patterson, 1988.

Transcript: The Fun of Discovering

1. Fleming, 1929a, 226.
2. Hare, 1974.
3. Macfarlane, 1984, 246.
4. Colquhoun, 1975.
5. Macfarlane, 1984, 118.
6. Hare, 1974; Hughes, 1974, 52.
7. Macfarlane, 1984, 119–120.
8. Ibid., 120, plate 4a.
9. Quoted in Maurois, 1959, 125.
10. Ibid., fig. facing p. 192.
11. Fleming, 1929a, fig. 1.
12. Ibid., 226.
13. Maurois, 1959, fig. facing p. 192.
14. Macfarlane, 1984, 141.
15. Hughes, 1974, 50.
16. Taton, 1957, fig. facing p. 96.
17. Fleming, 1944.
18. Macfarlane, 1984, 253.
19. See Bernard, 1927/1957, 39.
20. Maurois, 1959, 125.
21. Quoted in ibid., 129–130.
22. Ibid., 162.
23. See Medawar, 1979, 90.
24. Macfarlane, 1984, 117.
25. Ibid., 108; Ludovici, 1952, 100–101.
26. Maugh, 1987.
27. Fleming, 1929a, 226.
28. Ibid., table 2.
29. Macfarlane, 1984, 121, plate 4.
30. Quoted in Maurois, 1959, 131.
31. Scott, 1947.
32. See Goldsmith, 1980, 144; Hodgkin, 1977, 1; Medawar, 1964, 7–12; Holton, 1986, vii.
33. Szent-Györgyi, 1966, 111–128.

34. Quoted in Maurois, 1959, 131.
35. Quoted in ibid., 109.
36. See Taton, 1957, 113; Macfarlane, 1984, 31, 136.
37. Weisskopf, 1977, 409.
38. See Taton, 1959.
39. See ibid., 62n; Brown, 1977, 723; Price, 1975, 144–148.
40. Barber and Fox, 1958.
41. Quoted in Judson, 1980, 81–85.
42. Burnet, 1968, 53–54.
43. Ibid., 126.
44. Taton, 1957; Hadamard, 1941.
45. Merton, 1957.
46. Watson, 1968; Wade, 1980. See also Kuhn, 1977, 66–104.
47. See Harris, 1970, 323–324.
48. Huxley, 1900, II, 464.
49. Szent-Györgyi, 1957.
50. Quoted in Hardin, 1959, 84.
51. See Eiseley, 1979.
52. Thackray, 1965/1966; Thackray, 1966. See also Greenaway, 1966.
53. Holton, 1973, 385–386.
54. See Caws, 1969; Weiss, 1969.
55. See Lewis Thomas, in Judson, 1980, 69.

Imp's Journal: Unexpected Connections

1. Prusiner, 1982. See also Root-Bernstein, 1983a.
2. See Taubes, 1986.
3. Root-Bernstein and Westall, 1984c; Root-Bernstein, 1987a.

DAY FOUR: CREATING UNITY FROM DIVERSITY

Imp's Journal: Thematum

1. J. S. Haldane, 1931.
2. Dubos, 1965.
3. See C. Smith, 1987.

Transcript: Modeling the Process

1. See Carmichael, 1930.
2. See Hanson, 1967, 334–336; Harwit, 1980.
3. Kuhn, 1977, 171–172.
4. See Schilling, 1958.
5. Bates et al., 1977; Rodley et al., 1976; Cyriax and Gäth, 1978; Sasisekharan and Pattabiraman, 1976; Stokes, 1982, 1986a, 1986b.
6. Watson and Crick, 1953, 128–129.
7. Crick et al., 1979.
8. Brillouin, 1964, 37–41; Torrance, 1965, 663–665; Beveridge, 1980, 6, 55–56; Bernard, 1927/1957, 24; Duhem, in Lowinger, 1941, 35–37; Fleck, 1979, 94; Michael, 1977, 156–165.
9. Wallas, 1926.
10. Hardin, 1959, 122; Krebs and Shelley, 1975, 90.
11. Kuhn, 1979.
12. Clark, 1984, 293–302.

13. Bernard, 1927/1957, 25.
14. Fermi, 1954, 47.
15. Singer, 1981, 42–43; Davis and Hersh, 1981, 220.
16. Thomsen, 1980, 11.
17. Mendelssohn, 1973, 140–141.
18. See Hardin, 1959, vi, 292.
19. Bruner, 1962.
20. Brush, 1976.
21. See Root-Bernstein, 1980.
22. See Rabkin, 1987.
23. See Molland, 1985, 224–225.
24. See Szent-Györgyi, 1961, 48–49.
25. See Molland, 1985.
26. See Russell, 1930; Robertson, 1931, 32–33; Osler, 1951, 1; Truesdell, 1984, 91.
27. Van't Hoff, 1878/1967.
28. W. Ostwald, 1909.
29. See E. Cohen, 1912; E. Cohen, 1961; W. Ostwald, 1899; Walker, 1914; Jorissen and Reicher, 1912.
30. E. Cohen, 1912, 1–20; E. Cohen, 1961, 949–950.
31. Van't Hoff to Ostwald, 27 July 1901, in Korber, 1969.
32. Comte, 1869, III, 31–36.
33. Quoted in E. Cohen, 1912, 27.
34. W. C. Williams, 1984. See also Whitaker, 1969; Root-Bernstein, 1987c.
35. Hoffmann, 1987; Hoffmann, 1988a.
36. Read, 1947, 212.
37. Holmes, 1962; Lindauer, 1962.
38. Quoted in E. Cohen, 1961, 951.
39. Van't Hoff, 1930b, 4.
40. Kaufmann, 1961.
41. See Kangro, 1975, 7.
42. Quoted in E. Cohen, 1912, 55.
43. Quoted in E. Cohen, 1912, 63–66.
44. Quoted in E. Cohen, 1912, 54.
45. Quoted in Walker, 1914, 265.
46. Wurtz, 1869, 182–183.
47. Pasteur, 1860.
48. Miller, 1975, 1.
49. See E. Cohen, 1912, 79; Snelders, 1973.
50. See Sementsov, 1955; Snelders, 1973; Snelders, 1974.
51. Van't Hoff, 1874; van't Hoff, 1904.
52. Snelders, 1973, 266; Larder, 1967.
53. See Miller, 1975.
54. See Snelders, 1973, 271; Kekulé, 1867.
55. See Sementsov, 1955; Snelders, 1974.
56. Shropshire, 1981, 130.
57. Van't Hoff, 1874.
58. See Snelders, 1973, 272–275.
59. See Elstein et al., 1978.
60. Gruber, 1980; Lauden, 1980. See also Bancroft, 1928, 172.
61. Kuhn, 1977, 70.
62. Sementsov, 1955, 99–100.
63. See Judson, 1980, 581–582.

Jenny's Notebook: Falsification

1. Popper, 1958; Popper, 1962. See also Kneller, 1978, 48–67.
2. See Lowinger, 1941, 143–146.
3. E. Cohen, 1912, 127ff; van't Hoff, 1877/1967.
4. Lakatos, 1963; Harvey, 1978, 739; Kuhn, 1962. See also Kneller, 1978, 68ff.
5. See Gamow, 1966.
6. Dedijer, 1966, 275–276.
7. Kirchner, 1984; Feynman, 1985, 253–254.
8. Van't Hoff, 1876a.
9. Van't Hoff, 1876b, 1876c.
10. Snelders, 1974, 4; E. Cohen, 1912, 118–119.
11. See Lowinger, 1941, 125.
12. Medawar, 1969. See also Buck, 1975; Beveridge, 1980, 56–57.
13. Matthias, 1966.
14. A. M. Taylor, 1966, 32.
15. See Quine, 1951; Harré, 1981, 18–19.
16. See Lowinger, 1941, 134–135.
17. J. J. Thomson, 1937, 396.
18. W. Ostwald and Nernst, 1889; Travers, 1956, 91; Dolby, 1976.
19. See Root-Bernstein, 1984a.
20. Harré, 1981, 19.
21. Kirchner, 1984. See also Feynman, 1985, 342–343.

Transcript: Global Thinking

1. Ladenburg, 1905; Harrow, 1927, 122; Root-Bernstein, 1980, 4–6.
2. E. Cohen, 1912; Eugster, 1971.
3. Reisenfeld, 1931; Walker, 1933; Snelders, 1970.
4. W. Ostwald, 1926; Walden, 1904; Körber, 1974.
5. Planck, 1949; Kangro, 1975.
6. Fox, 1974; Holmes, 1962; Lindauer, 1962.
7. See W. Ostwald, 1919; Walden, 1904; Körber, 1974.
8. See Judson, 1979, 356–357.
9. Quoted in Meissner, 1951, 113.
10. See Root-Bernstein, 1980.
11. Nachmansohn, 1972, 1.
12. Quoted in G. Ostwald, 1953, 35.
13. Quoted in Garrett, 1963, 186.
14. Szent-Györgyi, 1966a, 120; 1966b.
15. Perkin, 1923, 70.
16. Crowther, 1968, 28.
17. Ibid., 63–64.
18. Walden, 1904, 30–40.
19. See Mellanby, 1974, 75.
20. Waldrop, 1982.
21. A. Avizienis, personal communication.
22. G. Ostwald, 1953, 53.
23. Quoted in Walden, 1904, 28.
24. Van't Hoff, 1894, 7.
25. Van't Hoff, 1878–1881, 13.
26. E. Cohen, 1912, 112–125; Jorissen and Reicher, 1912, 25–26.
27. See Lodge, 1885, 723.
28. W. Ostwald, 1887, 1–2; H. Jones, 1913, 203.

29. Willstätter, 1965, 44.
30. Mach, 1926/1976, 129.
31. See Willstätter, 1965, 37.
32. Poincaré, 1913/1946, 385.
33. Quoted in Kangro, 1975, 8.
34. See Livingston, 1973; Millikan, 1927; Fermi, 1954.
35. Quoted in Jorissen and Reicher, 1912, 34. See also Walker, 1913, 264–265.
36. W. Ostwald, 1919, I, 117. See also Walden, 1904.
37. Planck, 1949, 14.
38. Holton, 1973, 13.
39. See Wittgenstein, 1969; N. Maxwell, 1974a, 1974b.
40. Holton, 1973, 191, 192.
41. Quoted in Shropshire, 1981, 129.
42. Crowther, 1968, 249.
43. Papanek, 1971, vii.
44. T. P. Hughes, 1985, 21.
45. See J. Bernstein, 1978, 55; Rabi, 1970, 92; Weisskopf, 1972, 144–145. See also Gregory, 1916, 36.
46. Quoted in Overbye, 1984, 185.
47. Goldschmidt, 1956, 132.
48. Reisenfeld, 1931; Arrhenius, 1912.
49. Meissner, 1951.
50. Brush, 1976, 186.
51. Reisenfeld, 1931; Walker, 1933; Root-Bernstein, 1981.
52. See W. Ostwald, 1919; Walden, 1904.
53. Planck, 1949, 21.
54. Jorissen and Reicher, 1912, 30–35.
55. See Forman, 1974; Goldschmidt, 1956; Sachse, 1928; Mendelssohn, 1973, 40, 132.
56. Van't Hoff, 1878–1881; van't Hoff, 1894, 5–7; Walker, 1913, 263.
57. Horstmann, 1903.
58. Van't Hoff, 1894, 6–7.
59. Arrhenius, 1912b, 86.
60. See Carlson, 1981.
61. Arrhenius, 1912b, 85.
62. Krikorian, 1975; Bünning, 1975.
63. Krikorian, 1975; Root-Bernstein, 1980, 295–297.
64. Arrhenius, 1912b, 86.
65. Krikorian, 1975, 49.
66. Traube, 1864.
67. Traube, 1867.
68. Rudolph, 1976; Bünning, 1975, 1–38.
69. Krikorian, 1975, 50–52.
70. E. Cohen, 1912; Robinson, 1974, 557.
71. Daub, 1971, 308–310.
72. Clausius to Arrhenius, 23 June 1884.
73. Van't Hoff, 1894, 8.
74. Hadamard, 1945, 142–143.
75. See George, 1936, 235, 253–264.
76. Dyson, 1979, 75–76.
77. See Bronowski, 1978; Root-Bernstein, 1984d.
78. Quoted in Lowinger, 1941, 109–110.
79. Van't Hoff, 1894, 8.

80. W. Ostwald, 1891, 116.
81. Van't Hoff, 1884.
82. W. Ostwald, 1891, 116.
83. Van't Hoff, 1884; 1886.
84. Planck, 1887a, 1887b.
85. See Root-Bernstein, 1980; Drennan, 1961; H. Jones, 1913.

Jenny's Notebook: Renewing Old Knowledge

1. De Kruif, 1936.
2. Davidson, 1924.
3. Root-Bernstein, 1982d.
4. McClure and Lam, 1940; McClure et al., 1944.
5. Szent-Györgyi, 1966, 65.
6. Litwack and Kritchevsky, 1964, 330.
7. Szent-Györgyi, 1966, 66–67.
8. See Judson, 1980.

DAY FIVE: INSIGHT AND OVERSIGHT

Imp's Journal: Illumination

1. Westall and Root-Bernstein, 1983; Root-Bernstein and Westall, 1986b; Westall and Root-Bernstein, 1986.
2. Krueger et al., 1982.
3. Root-Bernstein and Westall, 1983.
4. Silverman et al., 1986; Root-Bernstein and Westall, 1986b; Root-Bernstein and Westall, in press.

Jenny's Notebook: Patterns on Paper

1. Mazurs, 1957/1974.
2. See Kuhn, 1977, 410.
3. See Root-Bernstein, 1984f.

Transcript: Insight and Oversight

1. Eccles, 1948; Eccles, 1970; Krebs and Shelley, 1975, 95.
2. Wallas, 1926.
3. Koestler, 1964, 101–120.
4. Taton, 1959, 74.
5. Platt and Barker, 1931.
6. See Hadamard, 1945, 8–20.
7. Mach, 1926/1976, 116; Crick and Mitchison, 1984.
8. Loewi, 1960; Loewi, 1965.
9. Judson, 1980, 7.
10. Matthias, 1966, 41.
11. See Darwin, 1958, 120–121; Wallace, 1905, I, 361–363; Ramon y Cajal, 1947, 61; Judson, 1980, 6–7; Mach, 1926/1976, 116; Axelrod, 1981, 30; Koestler, 1976, 117; Hadamard, 1945, 15–16; Ostwald, 1919, I, 117; Willstätter, 1965, 70; J. J. Thomson, 1937, 82; M. Wilson, 1972, 53; Mendelssohn, 1973, 133.
12. R. Wilson, 1966, 54–55.
13. See Hadamard, 1945, 10.
14. See Judson, 1980, 6.
15. Pauling, 1963, 47. See also Shropshire, 1981, 159.

16. Pauling, 1963, 47.
17. Ibid., 46.
18. Poincaré, 1913/1946, 390–392.
19. Ibid., 392–393. See also Carruthers, quoted in Broad, 1984, 60.
20. See Hadamard, 1945, 33; Ramon y Cajal, 1947, 42; Koestler, 1976, 169; J. J. Thomson, 1937, 82.
21. Poincaré, 1913/1946, 211.
22. Arrhenius, 1903/1966.
23. Snow, 1934/1958, 99.
24. M. Wilson, 1961.
25. Koestler, 1976; Rothenberg, 1979.
26. See Poincaré, 1913/1946.
27. J. B. S. Haldane, 1939; Levins and Lewontin, 1985.
28. Kuhn, 1962; Kuhn, 1970; Kuhn, 1977.
29. See Reisenfeld, 1931, 13; Arrhenius, 1912; Arrhenius, 1913.
30. Arrhenius to J. A. Bladin, 14 June 1881 and 22 June 1881.
31. See Cackowski, 1969; Wotiz and Rudofsky, 1984.
32. Root-Bernstein, 1980, 46–60, 153–160.
33. Arrhenius, 1912, 353.
34. Faraday, 1839–1855, par. 662–664.
35. Partington, 1964, 25.
36. Ibid., 1964, 663–668.
37. Clausius, 1857; Williamson, 1851. See also Partington, 1964, 668–670; Arrhenius, 1884a.
38. Kauffman, 1972; Kauffman, 1976; Lund, 1965.
39. Arrhenius, 1884b.
40. Arrhenius, 1913; Reisenfeld, 1931, 13.
41. See Root-Bernstein, 1980; Bernstein, 1978.

Jenny's Notebook: Modeling Illumination

1. Bost et al., 1985.
2. Weisburd, 1987, 299.
3. E. Cohen, 1912, 54.
4. Brillouin, 1964, 44.
5. George, 1936, 244.
6. Courtesy of Tom van Sant.
7. Bronowski, 1958, 65.
8. See Silverman, 1986.
9. Good, 1962, 120–132.
10. See Cairns-Smith, 1974, 54–55.
11. See ibid., 57.
12. Maxwell quoted in Rukeyser, 1942, 439; Gibbs, 1873.
13. See Poincaré, 1913/1946, 211–213; J. J. Thomson, 1930, 179; Snow, 1934/1958, 133–134, 260–261.
14. De Broglie in Moulton and Schifferes, 1960, 550; Eddington in Weber, 1973, 109. See also George, 1936, 29.
15. Quoted in Beveridge, 1980, 35.
16. See R. Vallery-Radot, 1900, 83.
17. Barber, 1960. See also Krebs, 1966; Frankel, 1979; Shinn, 1980.
18. Mahoney, 1976. See also Gordon, 1977; Peters and Ceci, 1982.
19. Forman, 1979, 11.

DAY SIX: COMPLEMENTARY PERSPECTIVES

Ariana's Report: Who Discovers, How?

1. M. Wilson, 1972, 11–12.
2. Quoted in Bernard, 1927, xix.
3. Duhem in Lowinger, 1941, 1, n. 3; Sarton, 1941, 128; Kubie, 1953/1954; Nachmansohn, 1972, 1.
4. Holton, 1973, 366–374.
5. Ramon y Cajal, 1937, 29.
6. Holton, 1975, 210.
7. Guthrie, 1921; Goertzels and Goertzels, 1962.
8. Williams-Ellis and Willis, 1954, 181.
9. See Crowther, 1968, Introduction.
10. See Forbes, 1987; Michael, 1977; Mitroff, 1974; McPherson, 1964; McClelland, 1962; Eiduson, 1962; Snow, 1959; Deutsch and Shea, 1957; Roe, 1953.
11. See Michael, 1977, 147; Thurstone, 1964, 15; Kock, 1978, 103–123.
12. Eiduson, 1962, 258.
13. Ibid., 260.
14. Ramon y Cajal, 1893.
15. Galton, 1874.
16. Möbius, 1900.
17. W. Ostwald, 1909.
18. E. Cohen, 1912; van't Hoff, 1878/1967.
19. Van't Hoff, 1878, 12.
20. Bernal, 1939, 85–86.
21. Quoted in Galton, 1892/1972, 20.
22. L. Huxley, 1900, I, 235.
23. Szent-Györgyi, 1961, 49.
24. Quoted in Dubos, 1959, 95.
25. Gregory, 1916, 56.
26. Hindle, 1981.
27. Ferguson, 1977.
28. A. Miller, 1984.
29. Emerton, 1984.
30. Gruber, 1978; S. Roe, 1981; Lapage, 1961; Ritterbush, 1968.
31. Ramon y Cajal, 1947, 36ff.
32. See Root-Bernstein, 1984.
33. See Baltzer, 1967; Zigrosser, 1955/1976; John-Steiner, 1985.
34. Taton, 1957, 42.
35. Cannon, 1945, 34–45; Hilts, 1975; Taylor, 1963.
36. See Fergusson, 1977, 832.
37. Holton, 1973, 370–376.
38. A. Roe, 1958, 141–142.
39. A. Roe, 1951.
40. Shepard, 1982; John-Steiner, 1985.
41. H. Gardner, 1983.
42. Ramon y Cajal, 1937, 36–47.
43. Quoted in Rayleigh, 1942, 99.
44. Lorenz, 1952, 12.
45. Poincaré, 1913/1946, 365–368.
46. Wechsler, 1978; Curtin, 1982; Waddington, 1969.
47. Ratliff, 1974, 15–16.

48. Quoted in Curtin, 1982, 26–27.
49. Davis and Hersh, 1981, 310–311.
50. Quoted in Lehto, 1980, 112.
51. Garrett, 1963, 13.
52. Garrison, 1948, 190.
53. Hofstadter, 1979.
54. Sylvester, 1864, 613n.
55. Kassler, 1982; Kassler, 1984.
56. See Root-Bernstein, 1987; L'Echevin, 1981; Marmelszadt, 1946; Gamow, 1966, 80–83.
57. Levarie, 1980, 237.
58. Quoted in Peterson, 1985, 348.
59. Quoted in ibid., 349.
60. See ibid., 348–350.
61. Ibid.; Brandmüller and Claus, 1982, 302; Boxer, 1987; Schwarz, 1988.
62. Van't Hoff, 1878–1967.
63. W. Ostwald, 1909.
64. Allen, 1978, 3, 21.
65. Hindle, 1984.
66. See Overbye, 1982.
67. See D. Wilson, 1983, 28.
68. Dyson, 1982, 49.
69. See also Rubin, 1980.
70. Wiesner, 1965, 531–532.
71. C. Smith, 1978, 9.
72. Quoted in Gleick, 1984, 33–34.
73. Waddington, 1972, 36.

Transcript: Discussion

1. See Weisburd, 1987, 300.
2. See Rae, 1972, 316.
3. Hughes, 1985, 22.
4. Hill, 1962, 43.
5. Van't Hoff, 1878/1967, 18.

Constance's Report: History and Philosophy of Science in Science

1. Truesdell, 1984.
2. Read, 1947, xx.
3. See Home, 1983.
4. See Carter, 1974, 25.
5. Lowinger, 1941, 164.
6. E. Wilson, 1986, 70.
7. Sarton, 1952, 7. See also George, 1936, Preface; Goldschmidt, 1960, 29; Rosenthal-Schneider, 1980, 27.
8. See John Ferguson, in Read, 1947, 298–299.
9. Kuhn, 1977, 345.
10. T. Huxley, 1887, 237.
11. L. Pasteur, 1858b, 1858c.
12. W. Ostwald, 1912, 3.
13. Mittasch, 1926; Donnan, 1930; Bancroft, 1928. See also Farber, 1966.
14. Poincaré, 1913/1946, 369; Arrhenius, 1908, v–vi. See also Lowinger, 1941, 9–10, 161–162.

15. Leibniz quoted in Truesdell, 1984, 586; Hadamard, 1945, 11; de Morgan quoted in Richards, 1987, 7.
16. Truesdell, 1984, 450–452. See also Kuznetsov, 1966.
17. T. Wilson, 1981, 129–130.
18. Brooks, 1966, 11–12. See also Brooks and Cranefield, 1959; Mirsky, 1966, 37.
19. See J. B. S. Haldane, 1969.
20. Mayr, 1982, 20.
21. Quoted in Shropshire, 1981, 55–56.
22. Poincaré, 1913/1946, 463.
23. Quoted in Holton, 1973, Preface.
24. See Forman, 1974; Nye, 1974; Forman et al., 1975.
25. Frank, Koyre, Boring, and Cohen in Frank, 1954.
26. Ramon y Cajal, 1947, 65.
27. Medawar, 1967; Medawar, 1969; Monod quoted in Beveridge, 1980, 56–57; Eccles, 1970. See also Buck, 1975.
28. Poincaré, 1913/1946, 134.
29. J. S. Haldane, 1931.
30. See Cannon, 1944, 108–115; Waife, 1960; Agass, 1984.
31. J. B. S. Haldane, 1939. See also Wersky, 1978.
32. Bernal, 1967, 199–251.
33. Crick, 1967; Monod, 1969; Monod, 1971; Levins and Lewontin, 1985.
34. Rosenfeld, 1970, 239.
35. Sigerist quoted in Galdston, 1939. See also Yonge, 1985/1986.
36. Crombie, 1984, 14.
37. Keller, 1983, 172–179; Keller, 1985, 170–172.
38. Heisenberg, 1958, 58.
39. Truesdell, 1984, ix.
40. Brillouin, 1964, ix.
41. Quoted in Rosenthal-Schneider, 1980, 27.
42. Wald, 1961.

Transcript: Discussion

1. Kuhn, 1977, 155–156.
2. See Loewi, 1960, 4–5; Gamow, 1966, 81; Clark, 1968, 57.

Richter's Report: The Evolution of Science

1. See Donovan, 1977, 75.
2. See L. Laudan, 1977; L. Laudan, 1982.
3. See Stokes, 1982; Stokes, 1983; Stokes, 1986.
4. Kuhn, 1959; Lakatos, 1976.
5. Popper, 1959; Popper, 1962; Lakatos, 1976.
6. Cooper et al., 1986.
7. Palke, 1986.
8. Kolata, 1984.
9. See Root-Bernstein, 1984.
10. J. Huxley, 1956, 3–4.
11. W. Ostwald, 1912, Preface.
12. Popper, 1959, 15.
13. Campbell, 1974.
14. Kuhn, 1959, 172–173.
15. Gerard, 1956; Cavalli-Sforza, 1971; Cavalli-Sforza, 1986.
16. Cavalli-Sforza, 1986; Cavalli-Sforza and Feldman, 1981.

17. Margenau and Bergamini, 1964, 86–99.
18. Holton, 1973, 416–420.
19. See Harwit, 1981; Holton, 1973, 42–421; Salk and Salk, 1981.
20. See Rabin, 1987.
21. See Borell, 1987.
22. Toulmin, 1961, 113–114.
23. Burnet, 1968, 36.
24. Robertson, 1931, 68–71.
25. Nye, 1986.
26. Bernal in Goldsmith, 1980, 179; Gerard, 1957.
27. Morton, 1969; Root-Bernstein, 1980.
28. Servos, 1979.
29. Bruce, 1987.
30. See Kohler, 1982; Fruton, 1972, 9–13.
31. Geison, 1987.
32. Frank, 1987, 32.
33. Ben-David, 1960.
34. Zloczower, 1960.
35. Nye, 1977, 375–376.
36. Hiebert, 1978; Servos, 1982.
37. Kaufmann, 1966, 678.
38. Ehrlich, 1899; Ehrlich, 1904.
39. Hopkins, 1913.
40. Terroine, 1959/1965, 500.
41. T. Wilson, 1981, 118.
42. Kuhn, 1959; Kuhn, 1977.
43. I. Cohen, 1985.
44. Ibid.
45. Kuhn, 1978.
46. See Schon, 1963; Kuhn, 1977.
47. Robertson, 1931, 68–69.
48. Infeld, 1941, 299; Fleck, 1979, 82.
49. Root-Bernstein, 1981; Root-Bernstein, 1984.

Transcript: Discussion

1. Edge and Mulkay, 1976.
2. Giere, 1988; Hull, 1988.
3. Colinvaux, 1980.
4. Seidenberg, 1950.
5. Harwit, 1981, 18–22.
6. Turney, 1987.

Hunter's Report: Obstacles and Inducements to Exploratory Research

1. See Ostwald, 1909b; Dyson, 1982; Szent-Györgyi, 1972.
2. Bancroft, 1928, 170.
3. Branscomb, 1986, 4.
4. Quoted in Fisher, 1984, 28.
5. See Broad, 1984.
6. See Pauling, 1950; Wise, 1962, 9.
7. See M. Wilson, 1972, 169.
8. Quoted in Goran, 1967, 144–115.

9. See Richet, 1927, 128–129; Alvarez, 1932, 24; Lehman, 1953; G. P. Thomson, 1957, 11; Feldman and Knorr, 1960, 12; Watson, in Judson, 1979, 40–45; Beveridge, 1980, 101; Davis and Hersch, 1981, 62; Hilts, 1984, 142; Lightman, 1984.
10. See Lightman, 1984.
11. Quoted in Judson, 1979, 20.
12. See ibid., 44–45.
13. Quoted in Beveridge, 1980, 101.
14. See Harwit, 1981, 65; Millikan, 1950, 123; Infeld, 1944; Judson, 1979.
15. Personal communication to Robert Langridge.
16. Tierney, 1984, 1.
17. Szent-Györgyi, 1966, 119.
18. Szilard, 1966, 28.
19. See Root-Bernstein, 1984.
20. Subject 15, in Eiduson, Scientist Project Interviews.
21. Quoted in Tierney, 1984, 5–6.
22. Pasteur, 1920–1939, VI, 447.
23. Debye, 1966, 80.
24. Ibid., 83.
25. W. Ostwald, 1909; J. B. S. Haldane, 1963, 32–39; Haber, in Goran, 1967, 108–109; Burnet, 1968, 35.
26. See Root-Bernstein et al., in press.
27. Pelz and Andres, 1966, 54–79; Andrews, 1979; Finkelstein et al., 1981.
28. McPherson, 1964, 417.
29. Root-Bernstein et al., in press.
30. Quoted in Halacy, 1967, 12–14.
31. Fehr, 1912.
32. Platt and Baker, 1931.
33. Quoted in Rayleigh, 1942, 199–200.
34. See Dukash and Hoffmann, 1979, 57.
35. See Rothschuh, 1973, 201, 223.
36. Quoted in Lusk, 1932, 48–49.
37. Quoted in ibid., 49.
38. See L. Huxley, 1927, 36; Judson, 1979, 41.
39. See Bernstein, 1978.
40. Ramon y Cajal, 1947, 108, 107. See also Osler, 1972, 3.
41. See Shropshire, 1981, 79–80.
42. See Weaver, 1968; Debye, 1966, 84; Truesdell, 1984, 457.
43. Quoted in Wersky, 1978, 22.
44. J. J. Thomson, 1937, 125.
45. Goldschmidt, 1960, 305.
46. Roy, 1979.
47. Richet, 1927, 154.
48. Ludovici, 1952, 129.
49. Wade, 1981.
50. Van Deventer, quoted in Walker, 1912, 265.
51. Macfarlane, 1984, 171.
52. Loewi, 1960, 10.
53. Bliss, 1982, 74.
54. Crowther, 1968, 28.
55. See ibid., 63–64; Nitske, 1971, 275–276.
56. Margenau and Bergamini, 1964, 132–133.
57. Clark, 1984, 298.
58. Weber, 1973, 48.

59. Thomson, 1957, 10; Burnet, 1968, 83; Ramon y Cajal, 1947, 107. See also Taylor, 1987.
60. Quoted in Macfarlane, 1984, 231.
61. Judson, 1980, 121.
62. Perutz, 1987.
63. See Judson, 1979, 113, 41.
64. See Thomson, 1957, 10.
65. Root-Bernstein, 1988; Vijh, 1987; Mellanby, 1974.
66. Burch, 1976.
67. Quoted in Judson, 1979, 45. See also Wilson, 1972, 103.
68. Eugene Levy, quoted in Greenberg, 1986, 88.
69. Debye, 1966, 84.
70. Quoted in Shropshire, 1981, 53.
71. Personal communication to author.
72. Goldschmidt, 1949, 226.
73. See Debye, 1966, 84; Szent-Györgyi, 1972, 955.
74. Weaver, 1968; Goldschmit, 1949; Szilard, 1961; Yalow, 1986; Pauling in Shropshire, 1981, 162; Temin in ibid., 62–63; Thomas, 1974, 134–140; Truesdell, 1984.
75. Sigma Xi, 1987.
76. See Goldschmidt, 1949; Weaver, 1958, 178; Conant, in Wilson, 1954, 423; Schultz, 1980, 645; Cottrell, 1962, 393; Shimshoni, 1965, 448; Hardin, 1959, 296; Mellanby, 1974, 82; Yalow, 1986.
77. Branscomb, 1986; Weinberg, 1967. See also Kadanoff, 1988.
78. Morris, 1981, 100, 286.
79. Quoted in Keller, 1983, 207.
80. Axelrod, 1981, 25–29, 47–57.
81. Alvarez, 1987. See also Oldendorff, 1972.
82. Szilard, 1966, 26.
83. See Truesdell, 1984, 402.
84. Hardin, 1959, 294.
85. Feynman, 1985, 346.

Transcript: Discussion

1. *Science News*, 24 August 1985, 125.
2. Thomsen, 1987, 87.
3. See Bates, 1984.
4. Weisberg, 1987.
5. Weaver, 1958, 173. See also Fleck, 1979, 86.

Imp's Report: A Manual of Strategies for Discovering

1. Quoted in Peacock, 1855, 97–98.
2. Quoted in Austin, 1978, 72.
3. Quoted in ibid., 78.
4. Truesdell, 1984, 440ff.
5. Pauling, 1977. See also Francis Crick, in Judson, 1979, 41.
6. Richet, 1927, 121–122.
7. Personal communication to author.
8. Szent-Györgyi, 1961, 48.
9. Medawar, 1979, 13.
10. G. P. Thomson, 1957, 9.
11. Richet, 1927, 132–133.

12. Einstein and Infeld, 1938.
13. See Polya, 1952; Polya, 1954; Danielli, 1966; Root-Bernstein, 1982.
14. Quoted in Broad, 1984, 54.
15. Quoted in Overbye, 1984, 177–178.
16. Personal communication to author.
17. Quoted in Agassi, 1971, 173.
18. See J. T. Wilson, 1981, 122.
19. Szent-Györgyi, 1963; Szent-Györgyi, 1966.
20. Medawar, 1979, 94.
21. Snow, 1934/1958, 213.
22. M. Wilson, 1961, 205.
23. Willstätter, 1965, 56.
24. Olmsted and Olmsted, 1961, 55.
25. Goldsmith, 1980, 227.
26. Judson, 1979, 41.
27. See Beveridge, 1930, 65; Merton, 1975, 335.
28. Dainton, 1987, 17.
29. Quoted in Thomsen, 1986, 27.
30. Macfarlane, 1984, 253.
31. Quoted in Maurois, 1959, 109.
32. Quoted in Beveridge, 1980, 30.
33. Quoted in Hardin, 1959, 102.
34. Richet, 1927, 123.
35. G. P. Thomson, 1961/1968, 129–131.
36. Szent-Györgyi, 1966a.
37. George, 1936, 289. See also G. P. Thomson, 1961/1968, 128–132.
38. Gamow, 1966.
39. Judson, 1980, 81–85.
40. See Jaffe, 1957, 207.
41. Trenn, 1986.
42. Avižienis, personal communication to author.
43. Taton, 1957. See also Eiseley, 1979, 74.
44. Judson, 1980, 78–80.
45. Quoted in Judson, 1979, 49. See also Poincaré, 1913/1946, 4.
46. W. Alvarez, 1987, 251.
47. Ibid., 253.
48. Quoted in Broad, 1984, 60.
49. Burnet, 1963, 86–87.
50. Quoted in Kangro, 1975, 8.
51. Quoted in Agassi, 1971, 23.
52. Feynman, 1985, 255.
53. Infeld, 1944.
54. See Burnet, 1963, 86–87.
55. Szent-Györgyi, 1963/1965, 467. See also Tyndall, 1872, 44, 108; Agassi, 1971, 31.
56. Quoted in Agassi, 1971, 13.
57. Quoted in Chakravarty, 1987, 83.
58. Quoted in Weber, 1973, 63.
59. Quoted in Dyson, 1958, 79–80.
60. Ibid., 80.
61. Thomas, 1974, 140.
62. Monod, 1969, 2.
63. Szent-Györgyi, 1963; Szent-Györgyi, 1966.

64. Bates, 1984.
65. Dirac, 1963. See also Curtin, 1982, 36.
66. Maier, 1960.
67. Heisenberg, 1971, 62–63.
68. Quoted in Judson, 1979, 113.
69. Agassi, 1975, 127–154.
70. Root-Bernstein, in press.

Transcript: Discussion

1. Mayr, 1982, 832–834, 843, 848.

POSTSCRIPT

Imp's Journal: Ambiguities

1. Snow, 1951, 320.
2. Root-Bernstein and Westfall, 1984e.
3. Orts et al., 1980a; Orts et al., 1980b.
4. Root-Bernstein and Westfall, 1986c.
5. Bost et al., 1985.

Jenny's Notebook: Conclusions Are Questions

1. See Atkinson, 1985.

References

UNPUBLISHED SOURCES

A book such as this should not, perhaps, need extensive documentation. The only real tests of its conclusions are the daily practice of science and the extent to which its conclusions resonate with the experience of the reader, providing an interpretative framework for his or her experience. Nonetheless, since modern scholars are rapidly entering a second Age of Scholasticism, in which nothing may be written without the justification of acknowledged authority, I have done my best to satisfy the demands of modern scholarship.

Although I have, wherever possible, found written corroboration for the opinions and positions espoused by the characters in this book, much of the dialogue has been drawn from personal conversations and collaborations with scientists and historians of science at Princeton University; the Salk Institute for Biological Studies; the Veterans Administration Hospital in Brentwood, Calif.; the University of California, Los Angeles; MacArthur Prize Fellows meetings; and various conventions. People who have provided particularly important or interesting points of view have been listed in the Acknowledgments, Notes, or both.

I have also benefited greatly from access to confidential interviews, psychological test results, bibliographies, and citation data compiled by the late Bernice Eiduson (Neuropsychiatric Institute, UCLA) and her assistants, Maurine Bernstein and Helen Watts Schlichting. Dr. Eiduson compiled information on forty scientists beginning in 1959 and retested and reinterviewed each subject again in 1969 and 1978. Although Dr. Eiduson published a number of articles based upon her studies, much remained to be done and she was kind enough to grant me permission to use the material just prior to her death. The material is particularly interesting since four of her subjects have received Nobel Prizes since 1959, two others have been multiply nominated for that award, and another became a Presidential Science Advisor. These extraordinary individuals are balanced by an equal number who never published more than twenty-five papers during their careers and failed to establish a reputation of any significance in science. Thus, Dr. Eiduson's research has provided an invaluable source of attitudes and anecdotes as well as a check upon my own ideas and experience.

I must thank James Watson and Jonas Salk for copies of their curricula vitae, which I used in my study of the publication records of eminent scientists. The other publication records are from standard sources such as biographies or biographical memoirs printed by the Royal Society (U.K.) or the National Academy of Sciences (U.S.).

I have used a variety of manuscript sources. Most of Louis Pasteur's correspondence and laboratory notebooks reside in the collection of the Bibliothèque Nationale in Paris. I have listed these under the general heading of *Papiers Pasteur* in the References, and have limited my citations to material not otherwise available in printed sources. Among the material examined for evidence of the development of Pasteur's cosmic asymmetric force hypothesis and its influence on his various discoveries are his unpublished *Correspondance*, 1840–1895; unpublished *Papiers*, 1848–1860 (vols. 1–3); *Registres de Laboratoire*, 1853–1865 (vols. 5–25); and lecture notes, *Cours de Chimie, Faculté des Sciences de Lille*, 1856.

The Alexander Fleming papers are mainly housed at the British Museum in London. An extensive file of photographs and related material is in the Audio Visual Communications Department of St. Mary's Hospital Medical School, London. Unfortunately, Fleming's wife died just as I began my research on his discoveries and her will has been in probate up to the present. Because permission of the trustee is necessary to quote from and reproduce his papers, and because the new trustee was not named before my book was finished, I was unable to cite anything other than the material published or mentioned in Gwyn Macfarlane's biography. I am nonetheless confident that when future scholars go to check my reconstructions against the manuscripts, they will find that I have accurately described what is there.

The J. H. van't Hoff Archive is at the Museum Boerhave, Leiden. Since correspondence with the archivist revealed that no laboratory notebooks or unknown letters were in the collection, I have not used anything unknown to his biographers. On the other hand, my account of Svante Arrhenius's research is informed by a number of manuscript sources, including his correspondence, which is housed in the archives of the Kungliga Vetenskapsakademiens Bibliotek, Stockholm, and a manuscript autobiography, *Levnadsröon*, in the possession of the Arrhenius family. I must thank Pamela van Atta for her help in translating some of the Swedish material, which would otherwise have been inaccessible to me.

PUBLISHED SOURCES

Abelson, P. H. 1965. "Relation of Group Activity to Creativity in Science," *Daedalus* 94: 603–614.

Adrian, Edgar Lord. 1961. "Creativity in Science," *Proceedings of the Third World Congress of Psychiatry*. Toronto: University of Toronto Press, McGill University Press. 1: 41–44.

Agassi, Joseph. 1975. *Science in Flux*. The Hague: Dordrecht.

—— 1979. "Art and Science," *Scientia* 114: 127–140.

—— 1984. "A Holomechanical Model for Research in the Life Sciences," *Journal of Social and Biological Structures* 7: 75–79.

Albers, Josef. 1975. *Interaction of Color*, rev. ed. New Haven, Conn.: Yale University Press.

Alverez, Luis W. 1987. *Adventures of a Physicist*. New York: Basic Books.

Alvarez, W. C. 1932. "The Influence of Dr. Cannon's Work upon Medical Thought and Progress," pp. 10–25 in *Walter Bradford Cannon*. Cambridge, Mass.: Harvard University Press.

American Chemical Society Editors. 1976. "Most Important Chemical Discoveries of the Past Century," *Chemical and Engineering News* 6 (April).

Ancker-Johnson, Betsy. 1973. "Physicist," *Annals of the New York Academy of Sciences* 208: 23–28.

Armytage, W. H. G. 1957. *Sir Richard Gregory: His Life and Work*. New York: St. Martin's.

Arrhenius, Svante. 1884a. "Über die Gültigkeit der Clausius-Williamsonschen Hypothese," *Berichte der deutschen chemische Gesellschaft* 15: 49–52.

—— 1884b. "Recherches sur la Conductibilité galvanique des Electrolytes," *Bihang till Kungliga Vetenskaps-Akademiens Handlingar*, nos. 13 and 14.

—— 1886. "On the Conductivity of Mixtures of Aqueous Acid Solutions," *Report of the British Association for the Advancement of Science* 56: 310–312.

—— 1887a. "Letter on Electrolytic Dissociation," *Sixth Circular of the British Association Committee for Electrolysis* (May).

—— 1887b. "Försök att beräkna dissociationen (aktivitetskoefficienten) hos i vatten lösta kroppar" [Investigation of the calculation of dissociation (activity coefficient) of dissolved substances in water], *Öfversigt öfver Kunglige Vetenskaps-Akademiens Förhandlingar*: 405.

—— 1887c. "Über die Dissoziation der in Wasser gelösten Stoffe," *Zeitschrift für physikalische Chemie* 1: 631–648.

—— 1887d. "Über die Einwirkung des Lichtes auf das elektrische Leitungs vermögen der Haloidsalze des Silbers," *Zeitschrift der Wiener Akademie der Wissenschaft* 96: 831.

—— 1887e. "Über das Leitungsvermögen der phosphoreszierenden Luft," *Weidemanns Annalen der Physik und Chemie* 32: 545.

—— 1888. "Erik Edlund: Nécrologue," *Lumière électrique* 29: 632–633.

—— 1892. "Gültigkeit der Beweises von Herrn Planck für die van't Hoff-sche Gesetz," *Zeitschrift für physikalische Chemie* 9: 5.

—— 1903. "Development of the Theory of Electrolytic Dissociation (Nobel Lecture, 1903)," pp. 45–58 in *Nobel Lectures: Chemistry, 1901–1921*. Amsterdam: Elsevier, 1966. Also in *Popular Science Monthly* 55 (1904): 385–396.

—— 1907. *Theories of Chemistry*. London: Longmans, Green.

—— 1912a. "Electrolytic Dissociation," *Journal of the American Chemical Society* 34: 353–364.

—— 1912b. *Theories of Solutions*. New Haven, Conn.: Yale University Press.

—— 1913. "Aus der Sturm- und Drangzeit der Lösungstheorien," *Chemisch Weekblad* 10: 584–599.

Arthus, Maurice. 1921/1943. "Maurice Arthus' Philosophy of Scientific Investigation" (preface to Arthus, *L'Anaphylaxie à l'Immunité*), trans. H. E. Sigerest, *Bulletin of the History of Medicine* 14: 365–390.

Ashton, S. V., and C. Oppenheim. 1978. "A Method of Predicting Nobel Prizewinners in Chemistry," *Social Studies of Science* 8: 341–348.

Asimov, Isaac, ed. 1985. *Great Science Fiction Stories by the World's Great Scientists*. New York: Donald Fine, Inc.

Atkinson, J. W. 1985. "Models and Myths of Science: Views of the Elephant," *American Zoologist* 25: 727–736.

Austin, J. H. 1978. *Chase, Chance, and Creativity: The Lucky Art of Novelty.* New York: Columbia University Press.
Axelrod, Julius. 1981. "Biochemical Pharmacology," pp. 25–37 in *The Joy of Research,* W. Shropshire, ed. Washington, D.C.: Smithsonian Institution Press.
Badash, Lawrence, 1972. "The Completeness of 19th Century Science," *Isis* 63: 48–58.
Bailey, Edward. 1962. *Charles Lyell.* London: Thomas Nelson.
Baker, Robert A., ed. 1969. *A Stress Analysis of a Strapless Evening Gown.* Garden City, N.Y.: Doubleday.
Bancroft, Wilder D. 1928. "The Methods of Research," *Rice Institute Pamphlet* 15: 167–285.
Barber, Bernard. 1961. "Resistance by Scientists to Scientific Discovery," *Science* 134: 596–602. Also in Barber and Hirsch, 1962, 539–556.
—— and Renée C. Fox. 1958. "The Case of the Floppy-Eared Rabbits: An Instance of Serendipity Gained and Serendipity Lost," *American Journal of Sociology* 64: 128–136. Also in Barber and Hirsch, 1962, 525–538.
—— and Walter Hirsch, eds. 1962. *The Sociology of Science.* New York: Free Press.
Barrett, J. T. 1986. *Contemporary Classics in Clinical Medicine.* Philadelphia: ISI Press.
Barringer, Herbert, George Blanksten, and Raymond Mack, eds. 1965. *Social Change in Developing Areas: A Reinterpretation of Evolutionary Theory.* Cambridge, Mass.: Schenken.
Barron, Frank. 1958. "The Psychology of Imagination," *Scientific American* 199 (September): 150–166.
Bates, R. H. T. 1984. "Notes for the Seminar 'Image Processing for Industry, Agriculture and Science.'" Unpublished ms.
Baum, Harold. 1982. *The Biochemists' Songbook.* New York: Pergamon Press.
Baynes, K., and F. Pugh. 1981. *The Art of the Engineer.* Woodstock, N.Y.: Overlook Press.
Beaver, Donald de B. 1976. "Reflections on the Natural History of Eponymy and Scientific Laws," *Social Studies of Science* 6: 89–98.
Bell, Charles. 1870. *Letters of Sir Charles Bell.* London: John Murray.
Bell, Eric T. 1937/1961. *Men of Mathematics.* New York: Simon and Schuster. Reprint, New York: Dover.
Bell, J. F. 1973. "The Experimental Foundations of Solid Mechanics," in Flügge's *Encyclopedia of Physics,* vol. VIa, ed. C. Truesdell. Berlin: Springer-Verlag.
Ben-David, Joseph. 1960. "Scientific Productivity and Academic Organization in Nineteenth-Century Medicine," *American Sociological Review* 25: 828–843. Also in Barber and Hirsch, 1962, 305–328.
Benét, Stephen Vincent. 1942. "Schooner Fairchild's Class," pp. 286–300 in *Selected Works of Stephen Vincent Benét.* New York: Holt, Rinehart and Winston.
Bernal, J. D. 1953. *Science and Industry in the Nineteenth Century.* London: Routledge and Kegan Paul.
—— 1939/1967. *The Social Function of Science.* Reprint, Cambridge, Mass.: MIT Press.
—— 1967. *The Origin of Life.* Cleveland, Ohio: World.
Bernard, Claude. 1927/1957. *An Introduction to the Study of Experimental Medicine,* trans. H. C. Greene. New York: Macmillan. Reprint, New York: Dover.
Bernstein, Jeremy. 1978. *Experiencing Science.* New York: Basic Books.
—— 1985. "Retarded Learner [John Archibald Wheeler]," *Princeton Alumni Weekly* (9 October): 28–31, 38–42.

Bernstein, R. S. 1979. "Svante Arrhenius and Electrolytic Dissociation—A Revaluation," pp. 201–212 in *Selected Topics in the History of Electrochemistry*, George Dubpernell and J. H. Westbrook, eds. Princeton: The Electrochemical Society.
Berthelot, Marcellin. 1866. *Comptes rendus* 63: 518.
—— 1876. *Comptes rendus* 82: 441.
Berthollet, C. L. 1789. *Annales de Chimie* 2: 163–173.
—— 1791. *Elémens de l'art de la teinture*. Paris. English ed., trans. William Hamilton. London: Stephan Couchman, 1791.
—— 1795. "Débats," *Séances des Ecoles Normales*, 4 vols. Paris: L. Reynier. I: 205, 311, 423; II: 188, 369; III: 107, 356; IV: 99, 320.
—— 1800."Observations sur le Natron," *Journal de Physique*, 51: 5–9.
—— 1801a. *Recherches sur les lois de l'affinité*. Paris. English ed., trans. M. Farrell. Baltimore: Philip Micklin, 1809.
—— 1801b. "Nouvelle Leçons (en continuant) Chemie," vol. IX, pp. 5–48 in *Séances des Ecoles Normale. Paris: L'Imprimerie du Cercle-Social.*
—— 1803. *Essai de statique chimique*. Paris: Demonville et Soeurs.
Besant, Annie, and C. W. Leadbeater. 1908. *Occult Chemistry*. London: Theosophical Publishing House.
Beveridge, W. I. B. 1950. *The Art of Scientific Investigation*. New York: Norton.
—— 1980. *Seeds of Discovery*. New York: Norton.
Bibby, Cyril. 1960. *T. H. Huxley: Scientist, Humanist, and Educator*. New York: Horizon Press.
Bingham, Roger. 1984a. "Outrageous Ardor: Carleton Gajdusek," in Hammond, 1984, 11–22.
—— 1984b. "Tuzo Wilson: Earthquakes, Volcanoes, and Visions," in Hammond, 1984, 189–196.
Biology and the Future of Man. 1976. Proceedings of the international conference held at the Sorbonne. Paris: Universities of Paris.
Bischoff, C. A. 1894. *Handbuch der Stereochemie*. Frankfurt: Bechhold.
Bishop, George. 1965. "My Life among the Axons," *Annual Review of Physiology* 27: 1–17. Reprinted in *The Excitement and Fascination of Science*. Palo Alto: Annual Reviews, Inc.
Blackwell, Richard J. 1969. *Discovery in the Physical Sciences*. Notre Dame, Ind., and London: University of Notre Dame Press, 1969.
Blalock, J. E., and E. M. Smith. 1984. "Hydropathic Anti-complementarity of Amino Acids Based on the Genetic Code," *Biochemical and Biophysical Research Communications* 121: 203–207.
Bliss, Michael. 1982. *The Discovery of Insulin*. Chicago: University of Chicago Press.
Bonner, James. 1959. "Creativity in Science," *Engineering and Science* 22: 13.
Bonner, John Tyler. 1964. "Analogies in Biology," pp. 251–255 in J. R. Gregg and F. T. C. Harris, eds., *Form and Strategy in Science*. Dordrecht: Reidel.
Bost, K. L., E. M. Smith, and J. E. Blalock. 1985. "Similarity between the Corticotropin (ACTH) Receptor and a Peptide Encoded by an RNA That Is Complementary to ACTH mRNA," *Proceedings of the National Academy of Sciences* 82: 1372–1375.
Boutelier, Glenn D., and A. H. Ullman. 1980. "Finding Your Chemical Roots—a Chemical Genealogy," *Educational Chemistry* 17: 108–109.
Bowman, William. 1891. "In Memoriam F. C. Donders," *Proceedings of the Royal Society* 49.
Bowser, Hal. 1987. "Maestros of Technology," *Invention and Technology* (Summer): 24–30.

Boxer, S., ed. 1987. "Play the Right Bases and You'll Hear Bach," *Discover* (March): 10–12.

Braithwaite, R. B. 1955. *Scientific Explanation*. Cambridge, England: Cambridge University Press.

Brandmüller, Josef, and Reinhart Claus. 1982. "Symmetry: Its Significance in Science and Art," *Interdisciplinary Science Reviews* 7: 296–308.

Brannigan, A. 1982. *The Social Basis of Scientific Discoveries*. Cambridge, England: Cambridge University Press.

Branscomb, Lewis M. 1986. "The Unity of Science," *American Scientist* 74: 4.

Braun, T., W. Glänzel, and A. Schubert. 1988. "The Newest Version of the Facts and Figures on Publication Output and Relative Citation Impact of 100 Countries, 1981–1985," *Scientometrics* 13: 181–188.

Brillouin, Leon. 1961. "Thermodynamics, Statistics, and Information," *American Journal of Physics* 29: 326–327.

—— 1964. *Scientific Uncertainty and Information*. New York: Academic Press.

Brinkman, W. F., et al. 1986. *Physics through the 1990s*. Washington, D.C.: National Academy of Sciences.

Broad, William J. 1984. "Tracing the Skeins of Matter [interview with Peter A. Carruthers]," *New York Times Magazine* (May 6): 54–62.

Brock, W. H., N. D. McMillan, and R. C. Mollan, eds. 1981. *John Tyndall: Essays on a Natural Philosopher*. Dublin: Royal Dublin Society.

Broda, Engelbert. 1983. *Ludwig Boltzmann*, trans. L. Gay. Woodbridge, Conn.: Oxbow Press.

Bronowski, Jacob. 1958. "The Creative Process," *Scientific American* 199: 59–65.

—— 1978. *Origins of Knowledge and Imagination*. New Haven, Conn.: Yale University Press.

Brooke, J. H. 1976. "Charles-Adolphe Wurtz," in *Dictionary of Scientific Biography*, C. C. Gillispie, ed. New York: Scribner's. Vol. 14, pp. 529–532.

Brooks, Chandler M. 1966. "Trends in Physiological Thought," pp. 9–13 in *The Future of Biology*. New York: New York University Press.

—— and Paul F. Cranefield, eds. 1959. *The Historical Development of Physiological Thought*. New York: Hafner.

Brooks, W. K. 1902. "The Lesson on the Life of Huxley," p. 710 in *Smithsonian Institution Report, 1900*. Washington, D.C.: Government Printing Office.

Brown, J. Douglas. 1965. "The Development of Creative Teacher-Scholars," *Daedalus* 94 (Summer): 615–631.

Brown, Ronald A. 1977. "Creativity, Discovery and Science," *Journal of Chemical Education* 54: 720–724.

Bruce, Robert V. 1987. *The Launching of Modern American Science, 1846–1876*. New York: Knopf.

Bruner, Jerome S. 1962. "The Conditions of Creativity," pp. 1–30 in *Contemporary Approaches to Creative Thinking*, H. E. Gruber, G. Terrell, and M. Wertheimer, eds. New York: Atherton Press.

Brush, Stephen G. 1974. "Should the History of Science Be Rated X?" *Science* 183: 1164–1172.

—— 1976. "John J. Waterston," in *Dictionary of Scientific Biography*, C. C. Gillispie, ed. New York: Scribner's. Vol. 14, pp. 184–186.

Buck, Carol. 1975. "Popper's Philosophy of Science for Epidemiologists," *International Journal of Epidemiology* 4: 159.

Bünning, Erwin. 1975. *Wilhelm Pfeffer*. Stuttgart: Wissenschaftliche Verlagsgesellschaft.

Buonaparte, Napoleon. 1859. *Correspondance*. Paris: Plon, 1859.

Burch, George E. 1976. "On Venture Research," *American Heart Journal* 92: 681–683.
Burnet, F. M. 1963. *The Integrity of the Body*. Cambridge, Mass.: Harvard University Press.
—— 1968. *Changing Patterns: An Atypical Autobiography*. Melbourne: Heinemann.
—— 1972. "Immunology as a Scholarly Discipline," *Perspectives in Biology and Medicine* 16: 1.
Burton, Alan C. 1975. "Variety—the Spice of Science as Well as Life: The Disadvantages of Specialization," *Annual Review of Physiology* 37: 1–12.
Bush, Vannevar. 1960. *Science, the Endless Frontier*. Washington, D.C.: National Science Foundation.
Cackowski, Z. 1969. "A Creative Problem Solving Process [Kekulé's invention of the Benzene ring]," *Journal of Creative Behavior* 3: 185–193.
Cadeddu, Antonio. 1985. "Pasteur et le choléra des poules: révision critique d'un récit historique," *History and Philosophy of the Life Sciences* 7: 87–104.
Cahan, David. Unpublished. "The Magician from Schwerin: August Kundt's Art of Experimental Physics."
Cairns-Smith, A. G. 1974. "The Methods of Science and the Origins of Life," pp. 53–58 in K. Dose et al., eds., *The Origin of Life and Evolutionary Biochemistry*. New York: Plenum.
Campbell, Donald T. 1974. "Evolutionary Epistemology," in *The Philosophy of Karl Popper*, ed P. A. Schilpp. La Salle, Ill.: Open Court. Vol. 1, pp. 411–463.
Cannon, Walter B. 1932. "Closing Remarks," pp. 59–65 in *Walter Bradford Cannon*. Cambridge, Mass.: Harvard University Press.
—— 1945/1965. *The Way of an Investigator*. New York: Hafner.
Carlson, Elof Axel. 1981. *Genes, Radiation, and Society: The Life and Work of H. J. Muller*. Ithaca, N.Y.: Cornell University Press.
Carmichael, R. D. 1930. *The Logic of Discovery*. Chicago: Open Court Publishing.
Carter, G. 1974. *Peer Review, Citations, and Biomedical Research Policy*. Santa Monica, Calif.: Rand Corporation.
Cavalli-Sforza, Luigi L. 1971. "Similarities and Dissimilarities of Sociocultural and Biological Evolution," pp. 535–541 in *Mathematics in the Archeological and Historical Sciences*, ed. F. R. Hodson, D. G. Kendall, and P. Tautu. Edinburgh: Edinburgh University Press.
—— 1973. "Models for Cultural Inheritance: I, Group Mean and Within-Group Variation," *Theoretical Population Biology* 4: 42–55.
—— 1986. "Cultural Evolution," *American Zoologist* 26: 845–855.
—— and M. W. Feldman. 1981. *Cultural Transmission and Evolution: A Quantitative Approach*. Princeton, N.J.: Princeton University Press.
Caws, Peter. 1965. *The Philosophy of Science: A Systematic Account*. Princeton, N.J.: Van Nostrand, 1965.
—— 1969. "The Structure of Discovery," *Science* 166: 1374–1380.
Chagnon, N. A., and W. Irons. 1979. *Evolutionary Biology and Human Social Behavior: An Anthropological Perspective*. North Scituate, Mass.: Duxbury Press.
Chakravarty, S. N. 1987. "The Vindication of Edwin Land," *Forbes* (May 4): 83.
Chance, Burton. 1940. "Richard Bright, Traveller and Artist—with Illustrations," *Bulletin of the History of Medicine* 8: 909–933.
Chandrasekhar, S. 1987. *Truth and Beauty: Aesthetics and Motivations in Science*. Chicago: University of Chicago Press.
Chargaff, Erwin. 1963. *Essays on Nucleic Acids*. New York: Elsevier.
—— 1968. Review of Watson's *Double Helix*, in *Science* (March 29).

—— 1971. "Preface to a Grammar of Biology," *Science* 172: 637–642.
—— 1974. "Building the Tower of Babble," *Nature* 248: 776–778.
—— 1978. *Heraclitean Fire: Sketches from a Life before Nature.* New York: Rockefeller University Press.
Chesterton, G. K. 1924/1956. *Tales of the Long Bow.* New York: Sheed and Ward.
Chevreul, M. E. 1858. Book review, *Journals des savants* (August): 507–527.
Church, Richard, and M. M. Buzman, eds. 1945. *Poems of Our Time, 1900–1942.* London: J. M. Dent.
Claparède and Flournoy. 1945. "Inquiry into the Working Methods of Mathematicians," trans. J. Hadamard, in Hadamard, 1945, 136–141. Original publication in *L'Enseignement Mathématique* 4 (1902); 6 (1904).
Clark, Ronald W. 1984. *Einstein: The Life and Times.* New York: Abrams.
—— 1985. *The Life of Ernst Chain: Penicillin and Beyond.* New York: St. Martin's.
Clausius, Rudolf. 1857. "Electricitätsleitung in Elektrolyten," *Annalen der Physik* 101: 338–360.
Cohen, Ernst. 1912. *Jacobus Henricus van't Hoff: Sein Leben und Wirken.* Leipzig: Akademische Verlagsgesellschaft.
—— 1933. "Kamerlingh Onnes Memorial Lecture," in *Memorial Lectures Delivered Before the Chemical Society, 1914–1932.* London: The Chemical Society. Vol. 3, pp. 91–107.
—— 1961. "J. H. van't Hoff," pp. 948–958 in *Great Chemists*, trans. R. E. Oesper. New York: Interscience.
Cohen, I. Bernard. 1982. "Uncovering Discovery [review of Brannigan, 1982]," *Nature* 297: 248.
—— 1985. *Revolution in Science.* Cambridge, Mass.: Harvard University Press.
Cole, K. C. 1983. "Victor Weisskopf: Living for Beethoven and Quantum Mechanics," *Discover* (June): 49–54.
Cole, Stephen. 1979. "Age and Scientific Performance," *American Journal of Sociology* 84: 958–977.
Colebrook, Leonard. 1954. *Almroth Wright: Provocative Doctor and Thinker.* London: Heniemann.
Coleby, L. J. M. 1938. *The Chemical Studies of P. J. Macquer.* London: George Allen and Unwin.
Colinvaux, Paul. 1980. *The Fates of Nations.* New York: Simon and Schuster.
—— 1982. "Towards a Theory of History: Fitness, Niche and Clutch of *Homo sapiens*," *Journal of Ecology* 70: 393–412.
Colquhoun, D. B. 1975. "Alexander Fleming," *World Medicine* (January 29): 41–43.
Comroe, J. H., and R. D. Dripps. 1976. "Scientific Basis for the Support of Medical Research," *Science* 192: 105–111.
Comte, Auguste, 1869. *Cours de philosophie positive,* 3d ed. Paris: Balliere. 6 vols.
Cook, Norman D. 1977. "The Case for Reverse Translation," *Journal of Theoretical Biology* 64: 113–135.
Cooper, D. L., J. Gerratt, and M. Raimondi. 1986. "The Electronic Structure of the Benzene Molecule," *Nature* 323: 699–701.
Court, S. 1972. "The *Annales de Chimie*, 1789–1815," *Ambix* 19: 113–128.
Cranefield, Paul. 1966. "The Philosophical and Cultural Interests of the Biophysics Movement of 1847," *Journal of the History of Medicine 21: 1–7.*
Crick, F. H. C. 1958. "On Protein Synthesis," Symposium of the Society for Experimental Biology, *The Biological Replication of Macromolecules* 12: 138.
—— 1967. *Of Molecules and Men.* Seattle: University of Washington Press.

—— 1968. "The Origin of the Genetic Code," *Journal of Molecular Biology* 38: 367–379.

—— 1970. "Central Dogma of Molecular Biology," *Nature* 227: 561–563.

—— 1974. "The Double Helix: A Personal View," *Nature* 248: 766–769.

—— 1981. *Life Itself: Its Origin and Nature*. New York: Simon and Schuster.

——, J. C. Wang, and W. R. Bauer, 1979. "Is DNA Really a Double Helix?" *Journal of Molecular Biology* 129: 449–461.

—— and Mitchison, G. 1983. "The Function of Dream Sleep," *Nature* 304: 111–114.

Crombie, A. C. 1984. "What Is the History of Science?" *Times Higher Education Supplement* (March 2): 14–15.

Crosland, Maurice. 1967. *The Society of Arcueil*. London: Heinemann.

Crowther, J. G. 1968. *Scientific Types*. London: Barrie and Rockliff.

Curie, Eve. 1937. *Madame Curie*, trans. V. Sheean. Garden City, N.Y.: Garden City Publishing Co.

Curtin, Deane W., ed. 1982. *The Aesthetic Dimension of Science: 1980 Nobel Conference*. New York: Philosophical Library.

Cyriax, B. and R. Gäth. 1978. "The Conformation of Double-Stranded DNA," *Naturwissenschaften* 65: 106–108.

Dagognet, Francois. 1967. *Methodes et doctrine dans l'oeuvre de Pasteur*. Paris: Presses Universitaire de France.

Dainton, Lord. 1987. "Hunt the Paradox and Fate May Smile," *The Scientist* (July 13): 17.

Dale, Henry. 1948. "Accident and Opportunism in Medical Research," *British Medical Journal* (September 4): 451–455.

—— 1954. *An Autumn Gleaning: Occasional Lectures and Addresses*. London: Pergamon Press.

Dalziel, K. 1982. "Axel Hugo Theodor Theorell," *Biographical Memoirs of Fellows of the Royal Society* 29: 585–621.

Danielli, James F. 1966. "What Special Units Should Be Developed for Dealing with the Life Sciences and What Specializations of Program Are Most Likely to be Needed in the Future?" pp. 90–98 in *The Future of Biology*. New York: SUNY Press.

Darlington, C. D. 1969. *The Evolution of Man and Society*. New York: Simon and Schuster.

Darwin, Charles. 1958. *The Autobiography of Charles Darwin, 1809–1882*. New York: Norton.

Daub, Edward E. 1971. "Rudolf Clausius," in *Dictionary of Scientific Biography*, C. C. Gillispie, ed. Vol. 3. New York: Scribner's.

Davidson, E. C. 1925. "Tannic Acid in the Treatment of Burns," *Surgery, Gynecology and Obstetrics* 41: 202–221.

Davis, Philip J. and Rueben Hersh. 1981. *The Mathematical Experience*. Boston: Birkhauser.

Davy, Sir Humphry. 1840. "Parallels between Art and Science," *The Collected Works of Sir Humphrey Davy*, John Davy, ed. London: Smith and Cornhill. Vol. 8, pp. 306–308.

De Beer, Gavin R. 1953. "Glimpses at Some Historical Figures of Modern Zoology," in *Science, Medicine and History*, E. A. Underwood, ed. London: Oxford University Press. Vol. 2, pp. 233–242.

Debye, Peter. 1966. Interview, pp. 77–86 in *The Way of the Scientist*. New York: Simon and Schuster.

La Décade Egyptienne, Journal Littéraire et d'Economie Politique. 1798–1801. 3 vols. Cairo: Imprimerie Nationale. Reprinted as the first 3 vols. of S. Bous-

tany, ed., *The Journals of Bonaparte in Egypt, 1798–1801,* 10 vols. Cairo: Al-Arab Bookshop, 1971.
Dedijer, Steven. 1966. Interview, pp. 266–277 in *The Way of the Scientist.* New York: Simon and Schuster.
De Kruif, Paul. 1926. *Microbe Hunters.* New York: Harcourt Brace.
—— 1936. *Why Keep Them Alive?* New York: Harcourt Brace.
Delaunay, Albert. 1959. *Journal d'un Biologiste.* Paris: Plon.
Dembart, Lee. 1985. "An Unsung Geometer Keeps to His Own Plane," *Los Angeles Times* (July 14), section 4, p. 3.
Deutsch and Shea, Inc. 1957. *A Profile of the Engineer: A Comprehensive Study of Research Relating to the Engineer.* New York: Industrial Relations Newsletter, Inc.
Dick, Auguste. 1981. *Emmy Noether, 1882–1935,* trans. H. I. Blocher. Boston: Birkhauser.
Dirac, P. A. M. 1963. "The Evolution of the Physicists' Picture of Nature," *Scientific American* (May): 45–53.
Dixon, Bernhard. 1973. *What Is Science For?* London: Collins.
"Dr. Crypton." 1986. "Fermilab: Where Science Is Art," *Science Digest* (February): 35–44, 74.
Dolby, R. G. A. 1976. "Debates over the Theory of Solution . . ." *Historical Studies in the Physical Sciences* 7: 297–404.
—— 1977. "The Transmission of Two New Scientific Disciplines from Europe to North America in the Late Nineteenth Century," *Annals of Science* 34: 287–310.
Dolman, Claude E. 1978. "Reflections on Fleming," *Chemistry* 51: 6–10.
Donnan, F. G. 1930. "Science and Philosophy: A Proposed International Conference," *Nature* 125: 857.
Donovan, Arthur. 1987. "Explaining Scientific Change: Theoretical Claims and Historical Cases," *Isis* 78: 75–76.
Doyle, Sir Arthur Conan. n.d. *The Complete Sherlock Holmes.* Garden City, N.Y.: Doubleday.
Drennan, O. J. 1961. "Electrolytic Solution Theory: Foundations of Modern Thermodynamic Considerations," Ph.D. dissertation, University of Wisconsin.
Dresselhaus, Mildred S. 1973. "Electrical Engineer," *Annals of the New York Academy of Sciences* 208: 17–22.
Dubos, René. 1950. *Louis Pasteur: Free Lance of Science.* Boston: Little, Brown.
—— 1960. *Pasteur and Modern Science.* Garden City, N.Y.: Doubleday.
—— 1965. *Man Adapting.* New Haven, Conn.: Yale University Press.
—— 1976. *The Professor, the Institute, and DNA.* New York: Rockefeller University Press.
Duclaux, Emil. 1896/1920. *Pasteur: Histoire d'un Esprit.* Paris: Sceaux. English ed. trans. E. F. Smith and F. Hedges, *Pasteur: The History of a Mind.* Philadelphia: W. B. Saunders.
Duhem, Pierre. 1899. "Une science nouvelle: la chimie physique," *Revue philomathique de Bordeaux et du Sud-Ouest*: 205–219; 260–280.
Duncan, A. M. "Some Theoretical Aspects of Eighteenth Century Tables of Affinity," *Annals of Science* 18 (1962): 177–194; 217–237.
Dyson, Freeman J. 1958. "Innovation in Physics," *Scientific American* 199: 74–82.
—— 1979. "The World of the Scientist—Part II," *The New Yorker* (August 13): 64–88.
—— 1982. "Manchester and Athens," pp. 41–62 in *The Aesthetic Dimension of Science,* E. W. Curtin, ed. New York: Philosophical Library.

Eccles, John C. 1958. "The Physiology of Imagination," *Scientific American* 199: 135–146.

—— 1970. *Facing Reality: Philosophical Adventures by a Brain Scientist*. New York: Springer-Verlag.

Edelstein, S. 1971. "The Role of Chemistry in the Development of Dyeing and Bleaching," pp. 288–289 in *The Journal of Chemical Education: Selected Readings in the History of Chemistry*, A. Ihde, ed. New York: Journal of Chemical Education.

Edge, David O., and Michael J. Mulkay. 1976. *Astronomy Transformed—The Emergence of Radio Astronomy in Britain*. New York: Wiley.

Egerton, Judy. 1986. *British Watercolors*. London: The Tate Gallery.

Ehrlich, Paul. 1957. "Physical Chemistry versus Biology in the Doctrines of Immunity [1899]," in *Gesammelte Arbeiten*, F. Himmelweit, ed. Berlin: Springer-Verlag. Vol. 1, p. 414.

Eiduson, Bernice. 1962. *Scientists: Their Psychological World*. New York: Basic Books.

—— and Linda Beckman, eds. 1973. *Science as a Career Choice: Theoretical and Empirical Studies*. New York: Russell Sage Foundation.

Einstein, Albert, and Leopold Infeld. 1938. *The Evolution of Physics*. New York: Simon and Schuster.

Eiseley, Loren. 1965. "Darwin, Coleridge, and the Theory of Unconscious Creation," *Daedalus* 94 (Summer): 588–602.

Elstein, A. S., L. S. Schulman, and S. A. Sprafka. 1978. *Medical Problem Solving: An Analysis of Clinical Reasoning*. Cambridge, Mass.: Harvard University Press.

Emerton, Norma E. 1984. *The Scientific Reinterpretation of Form*. Ithaca, N.Y.: Cornell University Press.

Escarpit, Robert. 1969. "Humorous Attitude and Scientific Inventivity," *Impact of Science on Society* 19: 253–258.

Eugster, Hans P. 1971. "The Beginnings of Experimental Petrology," *Science* 173: 481–489.

Ewald, P. P. 1972. "Carl Heinrich Hermann," in *Dictionary of Scientific Biography*, C. C. Gillispie, ed. New York: Scribner's.

Faraday, Michael. 1839–1855. *Experimental Researches in Electricity*. London: Quaritch. 3 vols.

Farber, Eduard. 1966. "From Chemistry to Philosophy: The Way of Alwin Mittasch (1869–1953)," *Chymia* 11: 156–178.

Feldman, Arnold S., and Klaus Knorr. 1960. *American Capability in Basic Science and Technology*. Princeton, N.J.: Center of International Studies.

Feleki, László. 1969. "Keeping Laughably up with Science," *Impact of Science on Society* 19: 259–268.

Ferguson, Eugene S. 1977. "The Mind's Eye: Nonverbal Thought in Technology," *Science* 197: 827–836.

Fermi, Laura. 1954. *Atoms in the Family*. Chicago: University of Chicago Press.

Feyerabend, Paul K. 1961. "Comments on Hanson's 'Is There a Logic of Scientific Discovery?'" pp. 35–37 in *Current Issues in the Philosophy of Science*. New York: Holt, Rinehart and Winston.

—— 1975. *Against Method: Outline of an Anarchist Theory of Knowledge*. Atlantic Highlands, N.J.: Humanities Press.

—— 1981. *Rationalism and Scientific Method: Problems of Empiricism*. Cambridge, England: Cambridge University Press. 2 vols.

Feynman, Richard P. 1985. *"Surely You're Joking, Mr. Feynman!"* New York: Norton.

Field, George B. 1981. "Theoretical Physics," pp. 65–78 in *the Joy of Research*, W. Shropshire, ed. Washington, D.C.: Smithsonian Institution Press.

Fisher, Arthur. 1984. "The Charm of Physics: Sheldon Glashow," pp. 24–35 in *A Passion to Know: Twenty Profiles in Science*, A. L. Hammond, ed. New York: Scribner's.

Fleck, Ludwig. 1979. *Genesis and Development of a Scientific Fact*. Chicago: University of Chicago Press.

Fleming, Alexander. 1922. "On a Remarkable Bacteriolytic Substance Found in Secretions and Tissues," *Proceedings of the Royal Society* 93: 306.

—— 1924. "On the Antibacterial Power of Egg-white," *Lancet* 1: 1303.

—— 1926. "A Simple Method of Removing Leucocytes from Blood," *British Journal of Experimental Pathology* 7: 231.

—— 1929a. "On the Antibacterial Action of Cultures of a Penicillium, with Special Reference to their Use in the Isolation of *B. influenzae*," *British Journal of Experimental Pathology* 10: 226–236.

—— 1929b. "Lysozyme—a Bacteriolytic Ferment Found Normally in Tissues and Secretions," *Lancet* 1: 217.

—— 1932. "Lysozyme," *Proceedings of the Royal Society of Medicine (Section of Pathology)* 26: 1.

—— 1944. "The Discovery of Penicillin," *British Medical Bulletin* 2: 4.

—— and V. D. Allison. 1925. "On the Specificity of the Proteins of Human Tears, *British Journal of Experimental Pathology* 6: 87.

Forbes, Peter. 1987. "Muse in a Test Tube," *The Scientist* (December 14): 24.

Forman, Paul. 1974. "The Financial Support and Political Alignment of Physicists in Weimar Germany," *Minerva* 12: 39–66.

—— 1979. "The Reception of an Acausal Quantum Mechanics in Germany and Britain," pp. 1–49 in *The Reception of Unconventional Science* (AAAS Selected Symposium No. 25), S. H. Mauskopf, ed. Boulder, Colo.: Westview Press.

—— 1981. "Einstein and Research," pp. 13–24 in *The Joy of Research*, W. Shropshire, ed. Washington, D.C.: Smithsonian Institution Press.

——, J. L. Heilbron, and S. Weart. 1975. "Physics circa 1900," *Historical Studies in the Physical Sciences* 5: 5–185.

Foster, Michael. 1899. *Claude Bernard*. London: T. Fisher Unwin.

Fowles, J. 1969. *The French Lieutenant's Woman*. New York: Signet.

Fox, Robert. 1974. "The Rise and Fall of Laplacian Physics," *Historical Studies in the Physical Sciences* 4: 89–136.

Fraenkel-Conrat, Heinz. 1979. "Comments," p. 47 in *The Origins of Modern Biochemistry: A Retrospect on Proteins*, P. R. Srinivasan, J. S. Fruton, and J.T. Edsall, eds. Annals of the New York Academy of Sciences, vol. 325.

Frank, Robert G. 1987. "American Physiologists in German Laboratories, 1865–1914," pp. 11–46 in *Physiology in the American Context, 1850–1940*, G. Geison, ed. Baltimore, Md.: William and Wilkins.

Frankel, Henry. 1979. "Continental Drift Theory," pp. 50–89 in *The Reception of Unconventional Science* (AAAS Selected Symposium No. 25), S. H. Mauskopf, ed. Boulder, Colo.: Westview Press.

Franklin, Kenneth J. 1953. *Joseph Barcroft, 1872–1947*. Oxford: Blackwell Scientific.

Franks, Felix. 1981. *Polywater*. Cambridge, Mass.: MIT Press.

French, A. P., ed. 1979. *Einstein: A Centenary Volume*. Cambridge, Mass.: Harvard University Press.

Friedman, Bruno. 1969. "The Editor Comments," *Impact of Science on Society* 19: 223–224.

Frisch, O. R. 1970. "Lise Meitner," *Biographical Memoirs of the Fellows of the Royal Society* 16: 405–420.

—— 1979. *What Little I Remember*. Cambridge, England: Cambridge University Press.
Fruton, Joseph F. 1972. *Molecules and Life: Historical Essays on the Interplay of Chemistry and Biology*. New York: Wiley.
The Future of Biology. 1965. A symposium sponsored by Rockefeller University and SUNY. New York: SUNY Press.
Gaffron, Hans. 1970. "Resistance to Knowledge," *The Salk Institute Occasional Papers* 2: 1–61.
Galdston, Iago. 1939. "The Ideological Basis of Discovery," *Bulletin of the History of Medicine* 7: 729–735.
Galton, Francis. 1892/1972. *Hereditary Genius: An Inquiry into Its Laws and Consequences*. Reprint, Gloucester, Mass.: Peter Smith, 1972.
—— 1874/1970. *English Men of Science: Their Nature and Nurture*. London: Macmillan. Reprint, Frank Cass.
Gamow, George. 1966. *Thirty Years That Shook Physics*. New York: Doubleday.
Garard, Ira D. 1969. *Invitation to Chemistry*. New York: Doubleday.
Gardner, Howard. 1983. *Frames of Mind: The Theory of Multiple Intelligences*. New York: Basic Books.
Gardner, Martin. 1959. *Mathematical Puzzles and Diversions*. New York: Simon and Schuster.
—— 1977. "Mathematical Games," *Scientific American* 236 (January): 110–121.
—— 1986. "Puzzles and Science," in *Puzzles Old and New*, brochure accompanying exhibit of puzzles organized by the Craft and Folk Art Museum of Los Angeles.
Garfield, Eugene. 1970. "Citation Indexing for Studying Science," *Nature* 227: 669–671.
Garrett, Alfred B. 1963. *The Flash of Genius*. Princeton, N.J.: Van Nostrand.
Gassmann, E., J. E. Kuo, and R. N. Zare. 1985. "Electrokinetic Separation of Chiral Compounds," *Science* 230: 813–814.
Geison, Gerald. 1974. "Louis Pasteur," in *Dictionary of Scientific Biography*, C. C. Gillispie, ed. New York: Scribner's.
——, ed. 1987. *Physiology in the American Context, 1850–1940*. Baltimore, Md.: William and Wilkins.
George, William H. 1936. *The Scientist in Action: A Scientific Study of His Methods*. London: Williams and Norgate.
Gerard, Ralph. 1957. "Problems in the Institutionalization of Higher Education: An Analysis Based on Historical Materials," *Behavioral Science* 2: 134–146.
——, E. Kluckhohn, and A. Rapoport. 1956. "Biological and Cultural Evolution: Some Analogies and Explorations," *Behavioral Science* 1: 6–34.
Gernand, H. W., and W. J. Reedy. 1986. "Planck, Kuhn, and Scientific Revolutions," *Journal of the History of Ideas* 47: 469–485.
Gibbs, J. W. 1928. "Graphical Methods in the Thermodynamics of Fluids," in *The Scientific Papers of J. Willard Gibbs*. New York: Longmans Green, 1928. Vol. 1, pp. 1–32.
Giere, R. N. 1988. *Explaining Science: A Cognitive Approach*. Chicago: University of Chicago Press.
Gilbert, G. K. 1886. "The Inculcation of Scientific Method by Example, with an Illustration Drawn from the Quarternary Geology of Utah," *American Journal of Science* 31: 286–299.
Gillispie, C. C. 1957. "The Discovery of the Leblanc Process," *Isis* 48: 152–170.
——, ed. 1970–1977. *Dictionary of Scientific Biography*. New York: Scribner's.
Gingerich, Owen, ed. 1975. *The Nature of Scientific Discovery*. Washington, D.C.: Smithsonian Institution Press.

Gleick, James. 1984. "Solving the Mathematical Riddle of Chaos [interview with Mitchell Feigenbaum]," *New York Times Magazine* (June 10): 31–71.
Goertzel, V., and M. G. Goertzel. 1962. *Cradles of Eminence*. Boston: Little, Brown.
Goldschmidt, Richard. 1949. "Research and Politics," *Science* 109: 219–227.
—— 1953. "Otto Bütschli, Pioneer of Cytology (1848–1920)," in *Science, Medicine and History*, E. A. Underwood, ed. London: Oxford University Press. Vol. 2, pp. 223–232.
—— 1956. *Portraits from Memory: Recollections of a Zoologist*. Seattle: University of Washington Press.
—— 1960. *In and Out of the Ivory Tower: The Autobiography of Richard B. Goldschmidt*. Seattle: University of Washington Press.
Goldsmith, Margaret. 1946. *The Road to Penicillin: A History of Chemotherapy*. London: Lindsay Drummond.
Goldsmith, Maurice. 1965. "Toward a Science of Science [interview with J. D. Bernal]," *Science Journal* (March): 88–92.
—— 1980. *Sage: A Life of J. D. Bernal*. London: Hutchinson.
—— and A. Mackay, eds. 1964. *The Science of Science*. London: Souvenir Press.
Golomb, S. W. 1954. "Checkerboards and Polyominoes," *American Mathematical Monthly* 61: 675–682.
Good, Irving, J. 1962. "Botryological Speculations," pp. 120–132 in *The Scientist Speculates*, I. J. Good, ed. New York: Basic Books.
Goran, Morris. 1967. *The Story of Fritz Haber*. Norman, Okla.: University of Oklahoma Press.
Gordon, Bonnie B., ed. 1985. *Songs from Unsung Worlds: Science in Poetry*. Boston: Birkhauser.
Gordon, Michael. 1977. "Evaluating the Evaluators," *New Scientist* (February 10): 342–343.
Grafstein, Daniel. 1983. "Stereochemical Origins of the Genetic Code," *Journal of Theoretical Biology* 105: 157–174.
Gray, George W. 1962. "Which Scientists Win Nobel Prizes," pp. 557–565 in *The Sociology of Science*, B. Barber and W. Hirsch, eds. New York: Free Press.
Greenaway, F. 1966. *John Dalton and the Atom*. London: Heinemann.
Gregg, J. R., and F. T. C. Harris, eds. 1964. *Form and Strategy in Science*. Dordrecht: Reidel.
Gregory, Richard. 1916. *Discovery, or the Spirit and the Service of Science*. London: Macmillan.
Grobel, Lawrence. 1986. "The Remarkable Dr. Feynman," *Los Angeles Times Magazine*, vol. 2, no. 16 (April 20): 14–19.
Grotthuss, C. J. D. von. 1806. "Memoire sur la décomposition de l'eau et des corps qu'elle tient en dissolution à l'aide de l'électricité galvanique," *Annales de chimie et de physique* 58: 54–74.
Gruber, Howard. 1978. "Darwin's 'Tree of Nature' and Other Images of Wide Scope," pp. 120–140 in *On Aesthetics in Science*, J. Wechsler, ed. Cambridge, Mass.: MIT Press.
—— 1980. "The Evolving Systems Approach to Creative Scientific Work: Charles Darwin's Early Thought," pp. 113–130 in *Scientific Discovery: Case Studies*, T. Nickles, ed. Dordrecht: Reidel.
Guerlac, Henry. 1959. "Some French Antecedents in the Development of the Chemical Revolution," *Chymia* 5: 77–81.
—— 1976. "Chemistry as a Branch of Physics: The Collaboration of Lavoisier and Laplace," *Historical Studies in the Physical Sciences* 7: 240–276.
Guldberg, C. M., and P. Waage. 1879. "Chemische Affinität," *Journal für praktische Chemie* 19: 1–46.

Guthrie, Leonard G. 1921. *Contributions to the Study of Precocity in Children.* London: Eric G. Millar.
Gutting, Gary. 1943. "Conceptual Structures and Scientific Change," *Studies in the History and Philosophy of Science* 4 (November): 212–216.
Guye, Philippe. 1903. "Editorial Introduction," *Journal de chimie physique* 1: 1–6.
Hadamard, Jacques. 1945. *The Psychology of Invention in the Mathematical Field.* Princeton, N.J.: Princeton University Press.
Haeckel, Ernst. 1904/1974. *Kunstformen der Natur.* Leipzig: Verlag des Bibliographischen Institutes. Reprint, New York: Dover.
Halacy, D. S., Jr. 1967. *Science and Serendipity: Great Discoveries by Accident.* Philadelphia.
Haldane, J. B. S. 1939. *The Marxist Philosophy and the Sciences.* New York: Random House.
—— 1976. *The Man with Two Memories.* London: Merlin Press.
Haldane, J. S. 1931. *The Philosophical Basis of Biology.* Garden City, N.Y.: Doubleday, Doran.
Hallpike, C. R. 1985. "Social and Biological Evolution, I: Darwinism and Social Evolution," *Journal of Social and Biological Structures* 8: 129–146.
Halmos, P. R. 1968. "Mathematics as a Creative Art," *American Scientist* 36: 375–389.
Hannan, P. J., R. Roy, and J. F. Christman. 1988. "Chance and Drug Discovery," *ChemTech* 18: 80–83.
Hanson, N. R. 1958. *Patterns of Discovery: An Inquiry into the Conceptual Foundations of Science.* Cambridge, England: Cambridge University Press.
—— 1961. "Is There a Logic of Discovery?" pp. 20–35 in *Current Issues in the Philosophy of Science.* New York: Holt, Rinehart and Winston.
—— 1967. "An Anatomy of Discovery," *Journal of Philosophy* 64: 321–352.
Hardin, Garrett. 1959. *Nature and Man's Fate.* New York: Holt, Rinehart and Winston.
Hardy, Thomas. 1874. *Far From the Madding Crowd.* London: Smith Elder.
Hare, Ronald. 1970. *The Birth of Penicillin.* London: George Allen and Unwin.
Harré, Rom. 1981. *Great Scientific Experiments.* Oxford: Phaidon.
Harris, E. E. 1970. *Hypothesis and Perception: The Roots of Scientific Method.* London: George Allen and Unwin.
Harrow, Benjamin. 1927. *Eminent Chemists of Our Time,* 2d ed. New York: Van Nostrand.
Hartley, Harold. 1933. "Theodore William Richards Memorial Lecture," in *Memorial Lectures Delivered before the Chemical Society, 1914–1932.* London: The Chemical Society. Vol. 3, pp. 131–163.
Harwit, Martin. 1981. *Cosmic Discovery: The Search, Scope and Heritage of Astronomy.* Brighton: Harvester Press.
Hawkins, D. T., W. E. Falconer, and N. Bartlett. 1978. *Noble Gas Compounds: A Bibliography, 1962–1976.* New York: IFI/Plenum.
Haynes, William. 1938. *Chemicals in the Industrial Revolution.* Princeton, N.J.: Princeton University Press.
Heath, A. E. 1947. "Analogy as a Scientific Tool," *Rationalist Annual*: 51–58.
Heidelberger, Michael. 1977. "A 'Pure' Organic Chemist's Downward Path," *Annual Review of Microbiology* 31: 1–12.
Heilbron, J. L. 1985. "Artes compilationis [review of J. Nehra and H. Rechenberg, *The Historical Development of Quantum Theory*]," *Isis* 76: 388–393.
Heisenberg, Warner. 1958. *Physics and Philosophy.* New York: Harper and Brothers.

—— 1970. *The Physicist's Conception of Nature*, trans. A. J. Pomerans. Westport, Conn.: Greenwood Press.

—— 1971. *Physics and Beyond: Encounters and Conversations*. New York: Harper and Row.

Hempel, C. G. 1966. *Philosophy of Natural Science*. New York: Prentice-Hall.

Henderson, Lawrence J. 1925. *The Order of Nature*. Cambridge, Mass.: Harvard University Press.

—— 1927/1957. Introduction to Claude Bernard, *An Introduction to the Study of Experimental Medicine*. New York: Macmillan. Reprint, New York: Dover.

Herival, John. 1975. *Joseph Fourier: The Man and the Physicist*. Oxford: Clarendon Press.

Herold, J. Christopher. 1962. *Buonaparte in Egypt*. New York: Harper and Row.

Hesse, Mary B. 1966. *Models and Analogies in Science*. Notre Dame, Ind.: University of Notre Dame Press.

Hiebert, Erwin N. 1971. "The Energetics Controversy and the New Thermodynamics," pp. 67–86 in *Perspectives in the History of Science and Technology*, D. H. D. Roller, ed. Norman, Okla.: University of Oklahoma Press.

—— 1978. "Nernst and Electrochemistry," pp. 180–200 in *Selected Topics in the History of Electrochemistry*, G. Dubpernell and J. H. Westbrook, eds. Princeton, N.J.: Electrochemical Society.

Higgs, Edward. 1985. "Counting Heads and Jobs: Science as an Occupation in the Victorian Census," *History of Science* 23: 335–349.

Hill, A. V. 1927. "Obituary: Professor W. Einthoven," *Nature* 120: 591.

—— 1960/1962. *The Ethical Dilemma of Science*. New York: Rockefeller Institute Press; London: Scientific Book Guild.

Hilts, Philip J. 1984. "Robert Wilson: Lord of the Rings," pp. 139–147 in *A Passion to Know: Twenty Profiles in Science*, A. L. Hammond, ed. New York: Scribner's.

Hindle, Brooke. 1981. *Emulation and Invention*. New York: New York University Press.

—— 1984. "Spatial Thinking in the Bridge Era: John Augustus Roebling versus John Adolphus Etzler," *Annals of the New York Academy of Sciences* 424: 131–148.

Hinshelwood, Cyril. 1965. "Science and Scientists," *Nature* 207: 1055–1061.

Hittorf, W. 1853–1859. "Über die Wanderungen der Ionen," *Annalen der Physik* 89: 177–211; 98: 1–34; 103: 1–56; 106: 337–411.

Hodges, Laurent. 1987. "Color It Kodachrome," *Invention and Technology* (Summer): 47–53.

Hodgkin, Alan L. 1977. *The Pursuit of Nature: Informal Essays on the History of Physiology*. Cambridge, England: Cambridge University Press.

Hoffmann, Roald. 1987. "Plainly Speaking," *American Scientist* 75 (July–August): 418–420.

—— 1988a. "How I Work as Poet and Scientist," *The Scientist* (March 21): 10.

—— 1988b. "Nearly Circular Reasoning," *American Scientist* 76 (March–April): 182–185.

Hofstadter, D. R. 1979. *Gödel, Escher, Bach*. New York: Basic Books.

Holmes, F. L. 1962. "From Elective Affinities to Chemical Equilibria: Berthollet's Law of Mass Action," *Chymia* 8: 105–145.

—— 1974. *Claude Bernard and Animal Chemistry*. Cambridge, Mass.: Harvard University Press.

Holt, Rackham. 1943. *George Washington Carver*. Garden City, N.Y.: Doubleday Doran.

Holton, Gerald. 1973. *Thematic Origins of Scientific Thought: Kepler to Einstein*. Cambridge, Mass.: Harvard University Press.

—— 1975. "Mainsprings of Scientific Discovery," pp. 199–217 in *The Nature of Scientific Discovery*, O. Gingerich, ed. Washington, D.C.: Smithsonian Institution Press.

—— 1978. *The Scientific Imagination: Case Studies*. Cambridge, England: Cambridge University Press.

—— 1986. "Foreword," pp. i–xii in *Contemporary Classics in the Physical, Chemical and Earth Sciences*. Philadelphia, Pa.: ISI Press.

Hopkins, Frederick G. 1913. "The Dynamic Side of Biochemistry," *Nature* 92: 213–223.

Horstmann, A. F. 1903. *Abhandlungen zur Thermodynamik chemischer Vorgänge*, J. H. van't Hoff, ed. Leipzig: Wilhelm Engelmann, 1903.

Houssay, B. A. 1952. "The Discovery of Pancreatic Diabetes: The Role of Oscar Minkowski," *Diabetes* 1 (March–April): 112–116.

Hoytink, G. J. 1970. "Physical Chemistry in the Netherlands after van't Hoff," *Annual Review of Physical Chemistry* 21: 1–16.

Hughes, Thomas P. 1985. "How Did the Heroic Inventors Do It?" *Ameican Heritage of Invention and Technology* (Fall): 18–25.

Hughes, W. Howard. 1974. *Alexander Fleming and Penicillin*. London: Priory Press.

Hull, David L. 1974. *Philosophy of Biological Science*. New York: Prentice-Hall.

—— 1988. *Science as a Process: An Evolutionary Account of the Social and Conceptual Development of Science*. Chicago: University of Chicago Press.

Huskey, V. R., and H. D. Huskey. 1980. "Lady Lovelace and Charles Babbage," *Annals of Computing* 2: 299–329.

Huxley, Aldous. 1956. *Tomorrow and Tomorrow and Tomorrow*. New York: Harper and Brothers.

Huxley, Julian. 1956. "Evolution, Cultural and Biological," p. 3–25 in *Current Anthropology*, W. L. Thomas, Jr., ed. Chicago: University of Chicago Press.

Huxley, Leonard. 1900. *Life and Letters of Thomas Henry Huxley*. New York: Appleton. 2 vols.

—— 1927. *Charles Darwin*. New York: Greenberg.

Huxley, T. H. 1899. "On Science and Art in Relation to Education," in *Collected Essays*. New York: Macmillan. Vol. 3, pp. 160–188.

—— 1935. *Diary of the Voyage of the H.M.S. Rattlesnake*, ed. Julian Huxley. London: Chatto and Windus.

Ildis, R. G. 1980. "The Principle of Cross-Stereo-Complementarity and the Symmetry of the Genetic Code," *Mendeleev Chemistry Journal* (English translation) 25: 431–434.

Jackson, A. Y. 1943. *Banting as an Artist*. Toronto: Ryerson Press; Boston: Bruce Humphries, Inc.

Jaffe, Bernard. 1957. *The Story of Chemistry*. New York: Premier Books.

James, William. 1987. "The Ph.D. Octopus [1903]," pp. 67–74 in *Essays, Comments, and Reviews*. Cambridge, Mass.: Harvard University Press.

Jerne, Neils K. 1976. "The Immune System: A Web of V-Domains," *The Harvey Lectures* 70: 93–110.

Jewkes, John, David Sawers, and Richard Stillerman. 1958. *The Sources of Invention*. London: Macmillan.

John-Steiner, Vera. 1985. *Notebooks of the Mind: Explorations of Thinking*. Albuquerque, N.M.: University of New Mexico Press.

Jolly, W. P. 1972. *Marconi*. London: Constable.

Jones, Harry C. 1913. *A New Era in Chemistry*. New York: Van Nostrand.

Jones, R. V. 1971. "Sir Harold Hartley, F.R.S.: An Appreciation . . ." *Notes and Records of the Royal Society of London* 26: 1–3.

—— 1978/1979. "Through Music to the Stars: William Herschel, 1738–1822," *Notes and Records of the Royal Society of London* 33: 37–56.
Jorissen, W. P. and L. T. Reicher. 1912. *J. H. van't Hoff's Amsterdamer Periode, 1877–1895*. Helder, Holland: C. der Boer.
Judson, Horace F. 1979. *The Eighth Day of Creation: Makers of the Revolution in Biology*. New York: Simon and Schuster.
—— 1980. *The Search for Solutions*. New York: Holt, Rinehart and Winston.
—— 1984. "Behind the Painted Mask: The Unexpected Legacy of Claude Lévi-Strauss," *The Sciences* (March–April): 26–35.
Kadanoff, Leo P. 1988. "The Big, the Bad, and the Beautiful," *Physics Today* (February): 9–10.
Kangro, Hans. "Max Karl Ernst Ludwig Planck," in *Dictionary of Scientific Biography*, C. C. Gillispie, ed. New York: Scribner's.
Kanigel, Robert. 1987. "One Man's Mousetraps," *New York Times Magazine* (May 17): 48–54.
Kassler, Jamie C. 1982. "Music as Model in Early Science," *History of Science* 20: 103–139.
—— 1984. "Man—A Musical Instrument: Models of the Brain and Mental Functioning before the Computer," *History of Science* 22: 59–92.
Kaufmann, Walter A. 1961. *The Faith of a Heretic*. Garden City, N.Y.: Doubleday.
Kekulé, August. 1861–1866. *Lehrbuch der organische Chemie*. Erlangen, Germany.
Keller, Evelyn Fox. 1983. *A Feeling for the Organism: The Life and Work of Barbara McClintock*. San Francisco: W. H. Freeman.
—— 1984. "Barbara McClintock: The Overlooked Genius of Genetics," pp. 121–126 in *A Passion to Know: Twenty Profiles in Science*, A. L. Hammond, ed. New York: Scribner's.
—— 1985. *Reflections on Gender in Science*. New Haven, Conn.: Yale University Press.
Kirchner, Helmut O. 1984. "Fashions in Physics," *Interdisciplinary Science Reviews* 9: 160–171.
Klemm, W. R., ed. 1977. *Discovery Processes in Modern Biology*. Huntington, N.Y.: R. E. Krieger.
Klieneberger-Nobel, Emmy. 1980. *Memoirs*. New York: Academic Press.
Kneller, George F. 1978. *Science as a Human Endeavor*. New York: Columbia University Press.
Knight, D. M. 1967. *Atoms and Elements: A Study of Theories of Matter in England in the Nineteenth Century*. London: Hutchinson.
Knudtson, Peter M. 1985. "S. Ramon y Cajal: Painter of Neurons," *Science 85* (September): 66–72.
Koblitz, Ann Hibner. 1983. *A Convergence of Lives: Sofia Kovalevskaia, Scientist, Writer, Revolutionary*. Boston: Birkhauser.
Kock, Winston E. 1978. *The Creative Engineer: The Art of Inventing*. New York: Plenum.
Koeppel, Tonja A. 1975. "Significance and Limitations of Stereochemical Benzene Models," pp. 97–113 in *Van't Hoff—Le Bel Centennial* (American Chemical Society Symposium Series, Vol 12), B. O. Ramsay, ed. Washington, D.C.: American Chemical Society.
Koestler, Arthur. 1976. *The Act of Creation*. London: Hutchinson.
Kohler, Robert. 1982. *From Medical Chemistry to Biochemistry: The Making of a Biomedical Discipline*. Cambridge, England: Cambridge University Press.
Kohlrausch, F. 1885. "Über die Lietvermögen einiger Elektrolyte in ausserst verdünnter Lösung," *Annalen der Physik* 26: 161–226.

—— 1888. "A Review of the Present Condition of the Theory of Electrolysis of Solutions," *The Electrician* 21: 466–467, 504–507.
Kohn, Alexander. 1969. "The Journal in which Scientists Laugh at Science," *Impact of Science on Society* 19: 259–268.
Kolata, Gina. 1984. "Puberty Mystery Solved," *Science* 223: 272.
Körber, H.-G. 1969. *Aus dem Wissenschaftlichen Briefwechsels Wilhelm Ostwalds*, Part 2. Berlin: Akademie-Verlag.
—— 1974. "C. W. W. Ostwald," in *Dictionary of Scientific Biography*, C. C. Gillispie, ed. New York: Scribner's.
Kottler, Dorian. 1978. "Louis Pasteur and Molecular Dissymmetry," *Studies in the History of Biology* 2: 57–98.
Kovalevskaya, Sofya. 1978. *A Russian Childhood*, trans. B. Stillman. New York: Springer-Verlag.
Krebs, Hans A. 1966. "Theoretical Concepts in Biological Sciences," pp. 83–95 in *Current Aspects of Biochemical Energetics*, N. O. Kaplan and E. P. Kennedy, eds. New York: Academic Press.
—— 1967. "The Making of a Scientist," *Nature* 215: 1441–1445.
—— and J. H. Shelley, eds. 1975. *The Creative Process in Science and Medicine*. Amsterdam: Excerpta Medica.
Krikorian, A. D. 1975. "Excerpts from the History of Plant Physiology and Development," pp. 9–97 in *Historical and Current Aspects of Plant Physiology: A Symposium Honoring F. C. Steward*. Ithaca, N.Y.: Cornell University Press.
Krueger, J. M., J. R. Pappenheimer, and M. L. Karnovsky. 1982. "Sleep-Promoting Effects of Muramyl Peptides," *Proceedings of the National Academy of Sciences (USA)* 79: 6102–6106.
Kubie, L. S. 1953/1954. "Some Unsolved Problems of the Scientific Career," *American Scientist* 41: 596; 42: 104.
Kuhn, T. S. 1962/1970. *The Structure of Scientific Revolutions*. Chicago: University of Chicago Press.
—— 1977. *The Essential Tension*. Chicago: University of Chicago Press.
—— 1978. *Black Body Theory and the Quantum Discontinuity, 1894–1912*. Oxford: Clarendon.
Kuznetsov, V. I. 1966. "The Development of Basic Ideas in the Field of Catalysis," *Chymia* 11: 179.
Ladenburg, Albert. 1905. *Lectures on the History of the Development of Chemistry since the Time of Lavoisier*, trans. L. Dobbin. Edinburgh: The Alembic Club.
Lagrange, Emile. 1938. *Robert Koch: Sa vie et son oeuvre*. Paris: Legrand.
Laidler, Keith J. 1985. "Chemical Kinetics and the Origins of Physical Chemistry," *Archive for the History of the Exact Sciences* 32: 43–75.
Lakatos, Imre. 1963. "Proofs and Refutations," *British Journal of the Philosophy of Science* 14: 1–25, 120–139; 221–245.
—— 1976. "Review of S. Toulmin's *Human Understanding*," *Minerva* 14 (Spring): 128–129.
Land, Barbara. 1973. *Evolution of a Scientist: The Two Worlds of Theodosius Dobzhansky*. New York: Thomas Y. Crowell.
Lapage, Geoffrey. 1961. *Art and the Scientist*. Bristol: John Wright and Sons.
Laplace, P. S. de. 1784. *Théorie du mouvement et de la figure élliptique des planetes*. Paris: P.-D. Pierres.
—— 1796. *Exposition du système du monde*. Paris: Imprimerie du Cercle-Social.
—— and A. Lavoisier. 1783/1920. *Memoire sur la chaleur*. Paris. Reprinted, Gauthier-Villars.

Larder, D. F. 1967. "Historical Aspects of the Tetrahedron in Chemistry," *Journal of Chemical Education* 44: 661–666.

Laudan, Larry. 1977. *Progress and Its Problems*. Berkeley: University of California Press.

—— 1982. "Two Puzzles about Science: Reflections on Some Crises in the Philosophy and Sociology of Science," *Minerva* 20: 253–268.

Laudan, Rachel. 1980. "The Method of Multiple Working Hypotheses and the Development of Plate Tectonic Theory," pp. 331–344 in *Scientific Discovery: Case Studies*, T. Nickles, ed. Dordrecht: Reidel.

Lavoisier, Antoine L. 1789. "Note of Mr. Lavoisier on Tables of Affinities," pp. 45–55 in Richard Kirwan, *Essay on Phlogiston and the Constitution of Acids*. London: Johnson.

Lehman, Harvey C. 1953. *Age and Achievement* Princeton, N.J.: Princeton University Press.

Lehninger, Albert L. 1970. *Biochemistry*. New York: Worth Publishers.

Lehto, Olli. 1980. "Rolf Nevanlinna," *Suomalainen Tiedeakatemia Academia Scientiarum Finnica Vuoskirja* (Yearbook): 108–112.

Lemay, Pierre. 1932. "Berthollet et l'emploi du chlore pour le blanchiment des toiles," *Revue d'Histoire de la Pharmacie* 78.

L'Engle, Madeleine. 1962. *A Wrinkle in Time*. New York: Farrar, Straus, and Giroux.

Levarie, Siegmund. 1980. "Music as a Structural Model," *Journal of Social and Biological Structures* 3: 237–245.

Levere, T. H. 1975. "Arrangement and Structure—A Distinction and a Difference," pp. 18–32 in *Van't Hoff—Le Bel Centennial* (American Chemical Society Symposium Series, vol. 12), B. O. Ramsay, ed. Washington, D.C.: American Chemical Society.

Levi-Montalcini, Rita. 1988. *In Praise of Imperfection: My Life and Work*, trans. Luigi Attardi. New York: Basic Books.

Levins, Richard, and Richard Lewontin. 1985. *The Dialectical Biologist*. Cambridge, Mass.: Harvard University Press.

Lewis, D. 1982. "Cyril Dean Darlington," *Biographical Memoirs of Fellows of the Royal Society* 29: 114–157.

Lewis, Ralph. 1944. "The Field Inoculation of Rye with Claviceps Purpurea (Fr.) Tul.," Ph.D. dissertation, Department of Botany, Michigan State College of Agriculture and Applied Science (now Michigan State University).

—— 1945. "The Field Inoculation of Rye with Claviceps Purpurea," *Phytopathology* 35: 353–360.

Lightman, Alan P. 1984. "Elapsed Expectations," *New York Times Magazine* (March 25): 68.

Lindauer, M. W. 1962. "The Evolution of the Concept of Chemical Equilibrium from 1775 to 1923," *Journal of Chemical Education* 39: 384.

Lipscomb, William N. 1982. "Aesthetic Aspects of Science," pp. 1–24 in *The Aesthetic Dimension of Science*, D. W. Curtin, ed. New York: Philosophical Library.

Lipsky, Eleazar. 1959. *The Scientists*. New York: Appleton-Century-Crofts.

Litwack, G., and D. Kritchevsky. 1964. *Actions of Hormones on Molecular Processes*. New York: Wiley.

Livingston, D. M. 1975. *The Master of Light: A Biography of Albert A. Michelson*. New York: Scribner's.

Lodge, Oliver. 1885. "On Electrolysis," *Report of the British Association for the Advancement of Science* 55: 723–772.

Loewi, Otto. 1958. "A Scientist's Tribute to Art," pp. 389–392 in *Essays in Honour of Hans Tietze*, Ernst Gombrich, ed. New York: Gazette des Beaux Arts.

—— 1960. "An Autobiographical Sketch," *Perspectives in Biology and Medicine* 4: 3–25.

—— 1965. *The Workshop of Discoveries*. Lawrence, Kans. University of Kansas Press.

Lorenz, Konrad. 1952. *King Solomon's Ring*. New York: Crowell.

—— 1971. "Knowledge and Freedom," pp. 231–261 in *Hierarchically Organized Systems in Theory and Practice*, ed. P. Weiss. New York: Hafner Press.

Lowinger, Armand. 1941. *The Methodology of Pierre Duhem*. New York: Columbia University Press.

Ludovici, L. J. 1952. *Fleming: Discoverer of Penicillin*. London: Andrew Dakers.

Lund, E. W. 1965. "Guldberg and Waage and the Law of Mass Action," *Journal of Chemical Education* 42: 548–550.

Luria, Salvadore E. 1984. *A Slot Machine, A Broken Test Tube: An Autobiography*. New York: Harper and Row.

Lusk, Graham. 1932. "The Life of a Professor," pp. 45–56 in *Walter Bradford Cannon*. Cambridge, Mass.: Harvard University Press.

Lyons, Albert S., and R. Joseph Petrucelli. 1978. *Medicine: An Illustrated History*. New York: Abrams.

Macfarlane, Gwyn. 1984. *Alexander Fleming: The Man and the Myth*. Cambridge, Mass.: Harvard University Press.

Mach, Ernst. 1926/1976. *Erkenntnis und Irrtum*, 5th ed. English edition, *Knowledge and Error: Sketches on the Psychology of Enquiry*, trans. T. J. McCormack and P. Foulkes. Dordrecht: Reidel.

—— 1943. *Popular Scientific Lectures*, trans. T. J. McCormack. 5th ed. La Salle, Ill. Open Court.

Mahoney, Michael J. 1976. *Scientist as Subject: The Psychological Imperative*. Cambridge, Mass.: Ballinger.

Maier, N. R. F. 1960. "Maier's Law," *American Psychologist* 15: 208–212.

Mansfield, Richard S., and Thomas V. Busse. 1981. *The Psychology of Creativity and Discovery: Scientists and Their Work*. Chicago: Nelson-Hall.

Margenau, Henry, and David Bergamini. 1964. *The Scientist*. New York: Time-Life Books.

Marmelszadt, Willard. 1946. *Musical Sons of Aesculapius*. New York: Froeben Press.

Maslow, Abraham H. 1966. *The Psychology of Science: A Reconnaissance*. New York: Harper and Row.

Matthias, Bernd. 1966. Interview, pp. 35–45 in *The Way of the Scientist*. New York: Simon and Schuster.

Maugh, Thomas H. 1987. "Frog Leads Researcher to Powerful New Antibiotics," *Los Angeles Times* (xxx), section 1, pp. 1 and 24.

Maurois, André. 1959. *The Life of Sir Alexander Fleming, Discoverer of Penicillin*, trans. G. Hopkins. New York: E. P. Dutton.

Mauskaupf, Seymour. 1976. "Crystals and Compounds: Molecular Structure and Composition in Nineteenth-Century French Science," *Transactions of the American Philosophical Society* 66: 55–80.

Maxwell, James Clerk. 1875. "On the Dynamical Evidence of the Molecular Constitution of Bodies," *Nature* 11 (March): 357–359; 375–377.

Maxwell, Nicholas. 1974. "The Rationality of Scientific Discovery: Part I, The Traditional Rationality Problem," *Philosophy of Science* 41: 123–153. "Part II, An Aim-Oriented Theory of Scientific Discovery," 247–295.

Mayr, Ernst. 1982. *The Growth of Biological Thought: Diversity, Evolution, and Inheritance*. Cambridge, Mass.: Harvard University Press.

Mazurs, E. G. 1957/1974. *Graphic Representations of the Periodic System during One Hundred Years*. University, Ala.: University of Alabama Press.

McCain, Garvin, and Erwin M. Segal. 1973. *The Game of Science*, 2d ed. Monterey, Calif.: Brooks/Cole.

McClelland, David C. 1962. "On the Psychodynamics of Creative Physical Scientists," pp. 141–174 in *Contemporary Approaches to Creative Thinking*, H. E. Gruber, G. Terrell, and M. Wertheimer, eds. New York: Atherton Press.

McClure, Roy D., and C. R. Lam. 1940. "Problems in the Treatment of Burns: Liver Necrosis as a Lethal Factor," *Southern Surgery* 9: 223.

—— and H. Romence. 1945. "Tannic Acid and the Treatment of Burns: An Obsequy," *Annals of Surgery* 121: 454–460.

McConnell, James V. 1969. "Confessions of a Scientific Humorist," *Impact of Science on Society* 19: 241–252.

—— 1985. "Learning Theory," pp. 250–263 in *Great Science Fiction Stories by the World's Great Scientists*, I. Asimov, ed. New York: Donald Fine.

McConnell, R. B., ed. 1983. *Art, Science and Human Progress*. London: John Murray.

McPherson, J. H. 1964. "Prospects for Future Creativity Research in Industry," pp. 414–423 in *Widening Horizons in Creativity*, C. W. Taylor, ed. New York: Wiley.

Medawar, Peter B. 1964. "Is the Scientific Paper a Fraud?" pp. 7–12 in *Experiment*, D. O. Edge, ed. London: British Broadcasting Corporation.

—— 1967. *The Art of the Soluble*. London: Methuen.

—— 1969. *Induction and Intuition in Scientific Thought. Philadelphia: American Philosophical Society.*

—— *1979. Advice to a Young Scientist*. New York: Harper and Row.

Meadows, A. J. 1972. *Science and Controversy: A Biography of Sir Norman Lockyer*. Cambridge, Mass.: MIT Press.

Meige, Henry. 1925. *Charcot Artiste*. Paris: Masson.

Meissner, Walter. 1951. "Max Planck, the Man and His Work," *Science* 113: 75–81.

Mekler, L. B. 1969. "On Specific Selective Interaction between Amino Acid Residues of Polypeptide Chain," *Biofizika* 14: 581–584. In Russian.

—— 1980. "A General Theory of Biological Evolution: A New Approach to an Old Problem," *Mendeleev Chemistry Journal* (English translation) 25: 333–360.

Melhado, Evan M. 1980. "Mitscherlich's Discovery of Isomorphism," *Historical Studies in the Physical Sciences* 11: 87–123.

Mellanby, Kenneth. 1967. "A Damp Squib," *New Scientist* 33: 626–627.

—— 1974. "The Disorganization of Scientific Research," *Minerva* 12: 67–82.

Mendelssohn, Kurt. 1973. *The World of Walther Nernst: The Rise and Fall of German Science, 1864–1941*. Pittsburgh: University of Pittsburgh Press.

Merton, Robert K. 1957. "Priorities in Scientific Discovery: A Chapter in the Sociology of Science," *American Sociological Review* 22: 635.

—— 1961. "Singletons and Multiples in Scientific Discovery: A Chapter in the Sociology of Science," *Proceedings of the American Philosophical Society* 105: 470–486.

—— 1975. "Thematic Analysis in Science: Holton's Concept," *Science* 188: 335–338.

Meyer, Lothar. 1883. *Die modernen Theorien der Chemie*. Breslau: Maruschke and Berendt.

Michael, William B. 1977. "Cognitive and Affective Components of Creativity in Mathematics and the Physical Sciences," pp. 141–172 in *The Gifted and the Creative: A Fifty-Year Perspective*, J. C. Stanley, W. C. George, and C. H. Solane, eds. Baltimore, Md.: Johns Hopkins University Press.
Miles, Ashley. 1982. "Reports by Louis Pasteur and Claude Bernard on the Organization of Scientific Teaching and Research," *Notes and Records of the Royal Society of London* 37: 101–118.
Miller, Arthur I. 1984. *Imagery in Scientific Thought: Creating Twentieth-Century Physics*. Boston: Birkhauser.
Miller, Jane A. 1975. "M. A. Gaudin and Early Nineteenth Century Stereochemistry," pp. 1–17 in *Van't Hoff—Le Bel Centennial*, ed. O. B. Ramsay. Washington, D.C.: American Chemical Society.
Miller, Jonathan. 1983. *States of Mind*. New York: Pantheon.
Millikan, Robert Andrews. 1927. *Evolution in Science and Religion*. New Haven, Conn.: Yale University Press.
—— 1950. *The Autobiography of Robert A. Millikan*. New York: Prentice-Hall.
Mitroff, Ian I. 1974. *The Subjecive Side of Science: A Philosophical Inquiry into the Psychology of the Apollo Moon Scientists*. New York: American Elsevier.
Mitchison, Dick, and Naomi Mitchison. 1978. *The Two Magicians*. London: Dennis Dobson.
Mitchison, Naomi. 1975. *Solution Three*. New York: Warner Books.
Mitscherlich, Eilhard. 1844–1847. *Lehrbuch der Chemie*, 4th ed. Berlin: E. S. Mittler. 2 vols.
Modern Photography. 1987. "150 Years of Photography" (September): 36–41.
Möbius, P. J. 1900. *Die Anlage zur Mathematik*. Leipzig: J. U. Barth.
Molland, A. George. 1985. "Discovering Western Science [review of D. Boorstin's *The Discoverers*]," *Isis* 76: 224–227.
Monod, Jacques. 1969. *From Biology to Ethics*. San Diego: Salk Institute.
—— 1971. *Chance and Necessity*. New York: Knopf.
Moore, Ruth. 1966. *Niels Bohr: The Man, His Science, and the World They Changed*. New York: Knopf.
Morris, Desmond. 1962. *The Biology of Art*. New York: Knopf.
Morton, R. A. 1969. *The Biochemical Society: Its History and Activities, 1911–1969*. London: The Biochemical Society.
—— 1979. *Animal Days*. New York: Bantam Books.
Moulton, F. R., and J. J. Schifferes, eds. 1960. *The Autobiography of Science*, 2d ed. Garden City, N.Y.: Doubleday.
Muller, H. J. 1943. "E. B. Wilson—An Appreciation," *American Naturalist* 77: 5–37, 142–172.
Mullin, A. A. 1962. "The Logic of Logic," p. 364 in *The Scientist Speculates*, I. J. Good, ed. New York: Basic Books.
Musson, A. E., and E. Robinson. 1969. *Science and Technology in the Industrial Revolution*. Manchester, England: Manchester University Press.
Nachmansohn, David. 1972. "Biochemistry as Part of My Life," *Annual Review of Biochemistry* 41: 1–28.
Nagel, Ernest, and James R. Newman. 1960. *Gödel's Proof*. New York: New York University Press.
Needham, Joseph. 1929. *The Skeptical Biologist*. London: Chatto and Windus.
Negrin, Howard. 1977. "Georges Cuvier: Administrator and Educator," Ph.D. dissertation, New York University.
Nemec, B. 1953. "Julius Sachs in Prague," in *Science, Medicine and History*, E. A. Underwood, ed. London: Oxford University Press. Vol. 2, pp. 211–216.
Neufield, Arthur N. 1986. "Reproducing Results," *Science* 234: 11.

Nickles, Thomas, ed. 1980. *Scientific Discovery, Logic, and Rationality*. Dordrecht: Reidel. Vols. 56 and 60 of Boston Studies in the Philosophy of Science.
Nicolle, Charles. 1932. *Biologie de l'Invention*. Paris: Alcan.
Nitske, W. Robert. 1971. *The Life of Wilhelm Conrad Roentgen, Discoverer of the X-Ray*. Tucson, Ariz.: The University of Arizona Press.
Nolting, L. E., and Feshback, M. 1980. "R and D Employment in the USSR," *Science* 207: 493–503.
North, J. D. 1976. "James Joseph Sylvester," in *Dictionary of Scientific Biography*, C. C. Gillispie, ed. New York: Scribner's.
Novak, B. J., and G. R. Barnett. 1956. "Scientists and Musicians," *Science Teacher* 23: 229–232.
Nye, Mary Jo. 1972. *Molecular Reality: A Perspective on the Scientific Work of Jean Perrin*. London: MacDonald.
—— 1974. "Gustav Le Bon's Black Light: A Study in Physics and Philosophy in France at the Turn of the Century," *Historical Studies in the Physical Sciences* 4: 163–196.
—— 1977. "Nonconformity and Creativity: A Study of Paul Sabatier, Chemical Theory, and the French Scientific Community," *Isis* 68: 375–391.
—— 1980. "N-Rays: An Episode in the History and Psychology of Science," *Historical Studies in the Physical Sciences* 11: 125–156.
—— 1986. *Science in the Provinces: Scientific Communities and Provincial Leadership in France, 1860–1930*. Berkeley: University of California Press.
Oblonsky, Jan G. 1980. "Eloge: Antonin Svoboda, 1907–1980," *Annals of Computing* 2: 284–298.
Øhrstrøm, Peter. 1985. "Guldberg and Waage on the Influence of Temperature on the Rates of Chemical Reactions," *Centaurus* 28: 277–287.
Ölander, A., et al. 1959. Arrhenius centennial volume, *Kungliga Svenska Vetenskapsakademiens Årsbok* 5.
Oldendorff, Wiliam H. 1972. "Science Education," *Science* 176: 966.
Olmsted, J. M. D., and E. H. Olmsted. 1961. *Claude Bernard and the Experimental Method in Science*. New York: Collier Books.
Oppenheimer, Jane M. 1967. *Essays in the History of Embryology and Biology*. Cambridge, Mass.: MIT Press.
Oppenheimer, Robert. 1956. "Analogy in Science," *American Psychologist 11: 127–135.*
Orts, R. J., B. C. Bruot, and J. L. Sartin. 1980. "Inhibitory Properties of a Bovine Pineal Tripeptide, Threonylseryllysine, on Serum Follicle-Stimulating Hormone," *Neuroendocrinology* 31: 92–95.
——, T.-H. Liao, J. L. Sartin, and B. C. Bruot. 1980. "Isolation, Purification and Amino Acid Sequence of a Tripeptide from Bovine Pineal Tissue Displaying Antigonadotropic Properties," *Biochemica Biophysica Acta* 628: 201–208.
Ossowska, Maria, and Sanislaw Ossowski. 1935/1964–65. "The Science of Science," *Minerva* 3: 72–82.
Ostwald, Grete. 1953. *Wilhelm Ostwald, mein Vater*. Stuttgart: Berliner Union.
Ostwald, Wilhelm. 1875. "Über die chemische Messenwirkung des Wassers," *Journal für praktische Chemie* 12: 264–270.
—— 1876. "Volumische Studien. I. Ueber das Bertholletsche Problem," *Poggendorf's Annalen* 8: 154–168.
—— 1884. "Notiz über das elektrische Leitungsvermögen der Säuren," *Journal für praktische Chemie* 30: 93–95.
—— 1887. "An die Leser," *Zeitschrift für Physikalische Chemie* 1: 1–2.
—— 1891. *Solutions*, trans. M. M. Pattison-Muir. London: Longmans, Green.
—— 1899. "Jacobus Henricus van't Hoff," *Zeitschrift für physikalische Chemie* 31: v–xviii.

—— 1905/1907. *Kunst und Wissenschaft*. Leipzig: Veit. English edition, *Letters to a Painter on the Theory and Practice of Painting*, trans. H. W. Morse. Boston: Ginn.

—— 1909. "Svante August Arrhenius," *Zeitschrift für physikalische Chemie* 69: v–xx.

—— 1909/1912. *Grosse Männer*. Leipzig: Akademische Verlag. French edition, *Les grands hommes*, trans. M. Dufour. Paris: Flammarion.

—— 1912. *L'Evolution de l'electrochimie*. Paris: Félix Alcan.

—— 1926–1927. *Lebenslinien: Eine Selbstbiographie*. Berlin: Klasing. 3 vols.

—— and W. Nernst. 1889. "Freie Ionen," *Zeitschrift für physikalische Chemie* 3: 11.

Outram, Dorinda. 1984. *Georges Cuvier*. Manchester, England: Manchester University Press.

Overbye, Dennis. 1982. "Rosalyn Yalow: Lady Laureate of the Bronx," *Discover* (June): 40–48.

—— 1984. "Messenger at the Gates of Time: John Wheeler," pp. 177–186 in *A Passion to Know: Twenty Profiles in Science*, A. L. Hammond, ed. New York: Scribner's.

Paget, James. 1901. *Memoirs and Letters of Sir James Paget*, Stephen Paget, ed. London: Longmans, Green.

Palmaer, Wilhelm. 1930/1961. "Svante Arrhenius," in *Buch der Grossen Chemiker*, ed. G. Bugge. Berlin: Verlag Chemie. Vol. 2, pp. 443–462. English edition, trans. and abbreviated by R. E. Oesper, pp. 1094–1109 in *Great Chemists*, E. Farber, ed. New York: Interscience.

Palmer, W. G. 1965. *A History of the Concept of Valency to 1930*. Cambridge, England: Cambridge University Press.

Papanek, Victor. 1971. *Design for the Real World: Human Ecology and Social Change*. New York: Pantheon.

Parergon. 1940, 1942, 1947. Evansville, Ind.: Mead Johnson.

Parkinson, C. N. 1962. "Parkinson's Law in Medical Research," *New Scientist* 13: 193–195. Also in Baker, 1963/1969, 189–195.

Partington, J. R. 1964. *A History of Chemistry*. London: Macmillan.

Pasteur, Louis. 1905/1922. "Recherches sur la dissymmétrie moléculaire des produits organiques naturels." English trans., *Researches on Molecular Asymmetry* (Alembic Club Reprints, no. 25). Edinburgh: Alembic Club. *Oeuvres*, 1922, vol. 1, pp. 314–344.

—— 1920–1939. *Oeuvres de Pasteur*, Pasteur Vallery-Radot, ed. Paris: Masson. 7 vols.

—— 1940–1957. *Correspondance, 1840–1895*, Pasteur Vallery-Radot, ed. Paris: Flammarion. 4 vols.

Patterson, John W. 1988. "Knotty Problems," *Science News* 133 (March 19): 179.

Pauling, Linus. 1952. "Use of Propositions in Examinations for the Doctor's Degree," *Science* 116: 667.

—— 1963. "The Genesis of Ideas," in *Proceedings of the Third World Congress of Psychiatry, 1961*. Toronto: University of Toronto Press, McGill University Press. Vol. 1, pp. 44–47.

—— 1977. "Linus Pauling: Crusading Scientist," transcript of broadcast of NOVA, no. 417, J. Angier, executive producer. Boston: WGBH-TV.

—— 1981. "Chemistry," pp. 132–146 in *The Joys of Research*, W. Shropshire, Jr., ed. Washington, D.C.: Smithsonian Institution Press.

Payne-Gaposchkin, Cecilia H. 1984. *Cecilia Payne-Gaposchkin: An Autobiography and Other Recollections*. Cambridge, England: Cambridge University Press.

Peacock, George. 1855. *Life of Thomas Young*. London: John Murray.

Pearl, Raymond. 1923. *Introduction to Medical Biometry and Statistics*. Philadelphia, Pa.: Saunders.

Peierls, Rudolf. 1981. "Otto Robert Frisch," *Biographical Memoirs of Fellows of the Royal Society* 28: 283–306.

Perkin, William Henry. 1933. "Baeyer Memorial Lecture," *Memorial Lectures Delivered before the Chemical Society, 1914–1932*. London: The Chemical Society. Vol. 3, pp. 131–163.

Perreux, Gabriel. 1962. *Pasteur au pays d'Arbois*. Dole: Presses Jurassiennes.

Perutz, Maurice F. 1987. "I Wish I'd Made You Angry Earlier," *The Scientist* (February 23): 19.

Peters, D. P. and S. J. Ceci. 1982. "Peer Review Practices of Psychological Journals: The Fate of Published Articles Submitted Again," *Behavioral and Brain Science* 5: 187–255.

Peterson, Ivars. 1985. "The Sound of Data," *Science News* 127: 348–350.

Pfaundler, Leopold. 1876. "Horstmann's Dissociationstheorie und die Dissociation fester Körper," *Berichte der deutschen chemische Gesellschaft* 9: 6.

Pfeffer, W. F. P. 1877. *Osmotische Untersuchungen: Studien zur Zellenmechanik*. Leipzig: Englemann.

Planck, Max. 1887a. "Über das Princip der Vermehrung der Entropie: Dritte Abhandlung—Gesetze des Eintritts beliebiger thermodynamischer und chemischer Reactionen," *Wiedemann's Annale der Physik* 32: 462–503. *Physikalische Abhandlungen und Vorträge*. Braunschweig: Vieweg, 1958. I, 232–273.

—— 1887b. "Über die molekulare Konstitution verdünnter Lösungen," *Zeitschrift für physikalische Chemie* 1: 577–582. (*Physikalische Abhandlungen*, I, 274–279).

—— 1890. "Über den osmotischen Druck," *Zeitschrift für physikalische Chemie* 6: 187–189. (*Physikalische Abhandlungen*, I, 327–329).

—— 1892. "Erwiderung auf einen von Herrn Arrhenius erhobenen Einwand," *Zeitschrift für physikalische Chemie* 9: 636–637. (*Physikalische Abhandlungen*, I, 433–434).

—— 1949. *Scientific Autobiography and Other Papers*, trans. Frank Gaynor. New York: Philosophical Library.

—— 1958. "Phantom Problems in Science," in *The Development of Modern Science*, G. Schwartz and P. W. Bishop, eds. New York: Basic Books. Vol. 2, pp. 956–965.

Platt, W., and R. A. Baker. 1931. "The Relationship of the Scicntific 'Hunch' to Research," *Journal of Chemical Education* 8: 1969.

Plauche, W. C., and J. C. Edwards. 1988. "Images and Emotion in Patient-Centered Clinical Teaching," *Perspectives in Biology and Medicine* 31 (Summer): 602–609.

Poincaré, Henri. 1913/1946. *The Foundations of Science: Science and Hypothesis; The Value of Science: Science and Method*, trans. G. B. Halsted. Lancaster, Pa.: Science Press.

Polanyi, Michael. 1958. *Personal Knowledge: Towards a Post-Critical Philosophy*. Chicago: University of Chicago Press.

Polya, George. 1962. *Mathematical Discovery: On Understanding, Learning, and Teaching Problem Solving*. New York: John Wiley. 2 vols.

Popper, Karl. 1958. *The Logic of Scientific Discovery*. London: Hutchinson.

—— 1962. *Conjectures and Refutations: The Growth of Scientific Knowledge*. New York: Basic Books.

Popular Photography. 1987. "Fifty Innovations That Changed the World" (January): 52–57.

Porter, J. R. 1972. "Louis Pasteur Sesquicentennial," *Science* 178: 1249–1254.

Prévost, Marie-Laure. 1977. "Manuscrits et correspondance de Pasteur à la Bibliothèque Nationale," *Bulletin de la Bibliothèque Nationale* (Paris) 2 (September): no. 3, 99–107.
Price, Derek J. 1956. "The Exponential Curve of Science," *Discovery* 17: 240–243.
—— 1963. *Little Science, Big Science*. New York: Columbia University Press.
—— 1975. *Science since Babylon*, enl. ed. New Haven, Conn.: Yale University Press.
Prusiner, Stanley B. 1982. "Novel Proteinaceous Infectious Particles Cause Scrapie," *Science* 216: 136–144.
Quine, W. V. O. 1951. "Two Dogmas of Empiricism," *The Philosophical Review* 60: 20–43.
Rabi, I. I. 1970. *Science: The Center of Culture*. New York: World Publishing.
Rabkin, Yakov M. 1987. "Technological Innovation in Science: The Adoption of Infrared Spectroscopy by Chemists," *Isis* 78: 31–54.
Rae, John B. 1973. "Charles Franklin Kettering," in *Dictionary of Scientific Biography*, C. C. Gillispie, ed. New York: Scribner's.
Ramo, Simon. 1987. "Why We're behind in Technology," *Los Angeles Times* (March 15), section 4, p. 5.
Ramon y Cajal, Santiago. 1937. *Recollections of My Life*, trans. E. H. Craigie and J. Cano. Cambridge, Mass.: MIT Press.
—— 1951. *Precepts and Counsels on Scientific Investigation: Stimulants of the Spirit*, trans. J. M. Sanchez-Perez. C. B. Courville, ed. Mountain View, Calif.: Pacific Press Publishing Association.
Rankine, W. J. M. 1874. *Songs and Fables*. London: Macmillan.
Rapport, Samuel, and Helen Wright. 1964. *Science: Method and Meaning*. New York: Washington Square Press.
Ratliff, Floyd. 1974. "Georg von Békésy: His Life, His Work, and His 'Friends,'" pp. 9–27 in *The George von Békésy Collection*, Jan Wirgin, ed. Malmö: Allhems Förlag.
Rayleigh, Lord. 1942. *The Life of Sir J. J. Thomson, O.M.*. Cambridge, England: Cambridge University Press.
Read, John. 1947. *Humour and Humanism in Chemistry*. London: G. Bell and Sons.
Reichenbach, Hans. 1938. *Experience and Prediction*. Chicago: University of Chicago Press.
Rensberger, Boyce. 1984. "Margaret Mead: An Indomitable Presence," pp. 37–46 in *A Passion to Know: Twenty Profiles in Science*, A. L. Hammond, ed. New York: Scribner's.
Rich, Alexander, and Norman Davidson, eds. *Structural Chemistry and Molecular Biology*. San Francisco: W. H. Freeman.
Richards, Joan L. 1987. "Augustus de Morgan, the History of Mathematics, and the Foundations of Algebra," *Isis* 78: 7–30.
Richet, Charles. 1927. *The Natural History of a Savant*, trans. Sir Oliver Lodge. London: J. M. Dent.
Riedman, Sarah R. 1960/1974. *The Story of Vaccination*. New York: Rand McNally, 1960. Folkestone, England: Bailey Bros. and Swinfen.
Rindos, David. 1985. "Darwinian Selection, Symbolic Variation, and the Evolution of Culture," *Current Anthropology* 26: 65–87.
Ritterbush, Philip C. 1968. *The Art of Organic Forms*. Washington, D.C.: Smithsonian Institution Press.
—— 1970. "The Shape of Things Seen: The Interpretation of Form in Biology," *Leonardo* 3: 305–317.

—— 1972. "Aesthetics and Objectivity in the Study of Form in the Life Sciences," pp. 25–60 in *Organic Form: The Life of an Idea*, G. S. Rousseau, ed. London: Routledge and Kegan Paul.

Robertson, T. Brailsford. 1931. *The Spirit of Research*, J. W. Robertson, ed. Adelaide: F. W. Preece and Sons.

Robinson, Gloria. 1974. "Wilhelm Friedrich Philipp Pfeffer," in *Dictionary of Scientific Biography*, C. C. Gillispie, ed. New York: Scribner's.

Rodley, G. A., R. S. Scobie, R. H. T. Bates, and R. M. Lewitt. 1976. "A Possible Conformation for Double-Stranded Polynucleotides," *Proceedings of the National Academy of Sciences (USA)* 73: 2959–2963.

Roe, Anne. 1951. "A Study of Imagery in Research Scientists," *Journal of Personality* 19: 459–470.

—— 1953. *The Making of a Scientist*. New York: Dodd, Mead.

Roe, Shirley Ann. 1981. *Matter, Life, and Generation: Eighteenth-Century Embryology and the Haller-Wolff Debate*. Cambridge, England: Cambridge University Press.

Roll-Hanson, Nils. 1972. "Louis Pasteur—A Case Against Reductionist Historiography," *British Journal of the Philosophy of Science* 23: 347–361.

Root-Bernstein, R. S. 1980/1981. "The Ionists: Founding Physical Chemistry, 1872–1890." Ph.D. dissertation, Princeton University. Ann Arbor, Mich.: University Microfilms.

—— 1982a. "Amino Acid Pairing," *Journal of Theoretical Biology* 94: 885–894.

—— 1982b. "On the Origin of the Genetic Code," *Journal of Theoretical Biology* 94: 895–904.

—— 1982c. "The Problem of Problems," *Journal of Theoretical Biology* 99: 193–201.

—— 1982d. "Tannic Acid, Semipermeable Membranes, and Burn Treatment," *Lancet* (November 20): 1168.

—— 1983a. "Protein Replication by Amino Acid Pairing," *Journal of Theoretical Biology* 100: 99–106.

—— 1983b. "The Structure of a Serotonin and LSD Binding Site of Myelin Basic Protein," *Journal of Theoretical Biology* 100: 373–378.

—— 1983c. "Mendel and Methodology," *History of Science* 21: 275–295.

—— and F. C. Westall. 1983d. "Sleep Factors: Do Muramyl Peptides Activate Serotonin Binding Sites?" *Lancet* (March 19): 653.

—— 1983e. "Galileo: Seeing and Perceiving," *Science News* 124: 387.

—— 1984a. "On Defining a Scientific Theory," pp. 64–94 in *Science and Creationism*, Ashley Montagu, ed. Oxford: Oxford University Press.

—— 1984b. "Molecular Sandwiches as a Basis for Structural and Functional Similarities of Interferons, MSH, ACTH, LHRH, Myelin Basic Protein, and Albumins," *FEBS Letters* 168: 208–212.

—— and F. C. Westall. 1984c. "Serotonin Binding Sites: Structures of Sites on Myelin Basic Protein, LHRH, MSH, and ACTH," *Brain Research Bulletin* 12: 425–436.

—— 1984d. "Creative Process as a Unifying Theme of Human Cultures," *Daedalus* 113 (Summer): 197–219.

—— and F. C. Westall. 1984e. "Fibrinopeptide A Binds Gly-Pro-Arg-Pro," *Proceedings of the National Academy of Sciences (USA)* 81: 4339–4342.

—— 1984f. "On Paradigms and Revolutions in Science and Art," *Art Journal* (Summer): 109–118.

—— 1985. "Visual Thinking: The Art of Imagining Reality," *Transactions of the American Philosophical Society* 75: 50–67.

——, F. Yurochko, and F. C. Westall. 1986a. "Clinical Suppression of Experimental Allergic Encephalomyelitis by Muramyl Dipeptide 'Adjuvant,'" *Brain Research Bulletin* 17: 473–476.

—— and F. C. Westall. 1986b. "Complementarity between Antigen and Adjuvant in the Induction of Autoimmune Diseases—A Dual Antigen Hypothesis," *Journal of Inferential and Deductive Biology* 2: 1–37.

—— and F. C. Westall. 1986c. "Bovine Pineal Antireproductive Tripeptide Binds to Luteinizing Hormone–Releasing Hormone: A Model for Peptide Modulation by Sequence Specific Peptide Interaction?" *Brain Research Bulletin* 17: 519–528.

—— 1987a. "Catecholamines Bind to Enkephalins, Morphiceptin, and Morphine," *Brain Research Bulletin* 18: 509–532.

—— 1987b. "Harmony and Beauty in Biomedical Research," *Journal of Molecular and Cellular Cardiology* 19: 1–9.

—— 1987c. "Tools of Thought: Designing an Integrated Curriculum for Lifelong Learners," *Roeper Review* 10: 17–21.

—— 1988. "Setting the Stage for Discovery," *The Sciences* (May–June): 26–35.

—— 1989a. "Who Discovers and Invents," *Research Technology Management* 32 (Jan.–Feb.): 43–50.

—— 1989b. "Strategies of Research," *Research Technology Management* 32 (May–June): 36–41.

—— and F. C. Westall. In press. "Serotonin Binding Sites, II: Muramyl Dipeptide Binds to Serotonin Binding Sites on Myelin Basic Protein, LH-RH, and MSH-ACTH," *Brain Research Bulletin*.

——, M. Bernstein, and H. W. Schlichting. Submitted. "Identification of Long-Term, High-Impact Scientists with Notes on Their Methods of Working," *Minerva*.

Roscoe, H. E. 1906. *The Life and Experiences of Sir Henry Enfield Roscoe*. New York: Macmillan.

Rosenthal-Schneider, Ilse. 1980. *Realty and Scientific Truth: Discussions with Einstein, von Laue, and Planck*. Detroit: Wayne State University Press.

Ross, Ronald. 1910. *Philosophies*. London: Murray.

—— 1928. *Poems*. London: E. Mathews and Marrot.

Rothenberg, Albert. 1979. *The Emerging Goddess: The Creative Process in Art, Science and Other Fields*. Chicago: University of Chicago Press.

Roy, Rustum. 1979. "Proposals, Peer Review, and Research Results," *Science* 204: 1154–1156.

Rubin, Lewis P. 1980. "Styles in Scientific Explanation: Paul Ehrlich and Svante Arrhenius on Immunochemistry," *Journal of the History of Medicine* 35: 397–425.

Rudolph, G. 1976. "Moritz Traube," in *Dictionary of Scientific Biography*, C. C. Gillispie, ed. New York: Scribner's.

Rukeyser, Muriel. 1942. *Willard Gibbs*. Garden City, N.Y.: Doubleday, Doran.

Russell, A. S. 1930. "The Necessity for Genius in Scientific Advance," *The Listener* (May 28): 949.

Russell, C. A. 1971. *A History of Valency*. New York: Humanities Press.

——, N. G. Coley, and G. K. Roberts. 1977. *Chemists by Profession*. Milton Keynes, England: Open University Press.

Sadoun-Goupil, Michèlle. 1977. *Le Chimiste Claude-Louis Berthollet, 1748–1822: Sa vie—son oeuvre*. Paris: J. Vrin.

Sahlins, Marshall, and Elman Service, eds. 1960. *Evolution and Culture*. Ann Arbor, Mich.: University of Michigan Press.

Salk, Jonas. 1983. *Anatomy of Reality*. New York: Columbia University Press.

Sanchez-Perez, J. M. 1947. "Some Reminiscences of S. R. Cajal: A Farewell Tribute," *Bulletin of the Los Angeles Neurological Society* 12 (March): 1.

Sarton, George. 1941. "The History of Medicine versus the History of Art," *Bulletin of the History of Medicine* 10: 122–135.

Sasisekharan, V., and N. Pattabiraman. 1976. "Double Stranded Polynucleotides: Two Typical Alternative Conformations for Nucleic Acids," *Current Science* 45: 779–783.

Schachman, Howard K. 1979. "Summary Remarks: A Retrospect on Proteins," pp. 363–373 in *The Origins of Modern Biochemistry: A Retrospect on Proteins*, P. R. Srinivasan et al., eds. Vol. 325 of Annals of the New York Academy of Sciences.

Schaffner, Kenneth F. 1980. "Discovery in the Biomedical Sciences: Logic or Irrational Intuition?" pp. 171–205 in *Scientific Discovery: Case Studies*, T. Nickles, ed. Dordrecht: Reidel. Vol. 60 of Boston Studies in the Philosophy of Science.

Scherr, George H., ed. 1983. *The Best of the Journal of Irreproducible Results*. New York: Workman.

Schilleer, F. C. S. 1917. "Scientific Discovery and Logical Proof," in *Studies in the History and Methods of the Sciences*, C. Singer, ed. Oxford: Clarendon Press.

Schilling, Harold K. 1928. "A Human Enterprise," *Science* 127: 1324–1327.

Schilpp, P. A., ed. 1974. *The Philosophy of Karl Popper*. La Salle, Ill.: Open Court Press. 2 vols.

Schoenfeld, A. H. 1979. "Explicit Heuristic Training as a Variable in Problem-Solving Performance," *Journal for Research in Mathematics Education* 10: 174–187.

Schon, Donald A. 1967. *Invention and the Evolution of Ideas*. London: Social Science Paperbacks. Formerly published as *Displacement of Concepts*. London: Tavistock, 1963.

Schultz, Theodore W. 1980. "The Productivity of Research: The Politics and Economics of Research," *Minerva* 18: 644–651.

Schwartz, Joel. 1988. "Musical Urinalysis," *Omni* (February): 33.

Science Digest. 1984. "America's Top 100 Young Scientists" (December): 40–71.

Scott, W. M. 1947. "Alexander Fleming," *Veterinary Record* 59: 680.

Sekular, Robert, and Eugene Levinson. 1977. "The Perception of Moving Targets," *Scientific American* 236 (January): 60–73.

Selye, Hans. 1977. "Biological Adaptations to Stress," pp. 266–288 in *Discovery Processes in Modern Biology*, W. R. Klemm, ed. Huntington, N.Y.: Krieger.

Sementsov, A. 1955. "The Eightieth Anniversary of the Asymmetrical Carbon Atom," *American Scientist* 43: 97–100.

Sergeant, Howard, ed. 1980. *Poems from the Medical World*. Lancaster, England: MTP Press, 1980.

Servos, John W. 1979. "Physical Chemistry in America, 1890–1933: Origins, Growth, and Definition." Ph.D. dissertation, Johns Hopkins University.

—— 1982. "A Disciplinary Program That Failed: Wilder D. Bancroft and the *Journal of Physical Chemistry*, 1896–1933," *Isis* 73: 207–232.

Shaler, Nathaniel. 1909. *The Autobiography of Nathaniel Southgate Shaler*. Boston: Houghton Mifflin.

Shankland, Robert S. 1973. "Karl Rudolph Koenig," in *Dictionary of Scientific Biography*, C. C. Gillispie, ed. New York: Scribner's.

Shapiro, Gilbert. 1987. *Skeleton in the Darkroom: Stories of Serendipity in Science*. New York: Harper and Row.

Shelley, Mary. 1818/1981. *Frankenstein*. London. Reprint, New York: Bantam Books.

Shepard, Roger N., ed. 1982. *Mental Images and Their Transformations.* Cambridge, Mass.: MIT Press.

Sherman, Paul D. 1981. *Colour Vision in the Nineteenth Century.* Bristol, England: Adam Hilgar.

Shimshoni, Daniel. 1965. "Israeli Scientific Policy," *Minerva* 3: 441–456.

Shinn, Terry. 1980. "Orthodoxy and Innovation in Science: The Atomist Controversy in French Chemistry," *Minerva* 18: 539–555.

Shropshire, Walter, ed. 1981. *The Joys of Research.* Washington, D.C.: Smithsonian Institution Press.

Sigma Xi. 1987. *A New Agenda for Science.* New Haven, Conn.: Sigma Xi.

Silverman, D. H. S., J. K. Krueger, and M. L. Karnovsky. 1986. "Specific Binding Sites for Muramyl Peptides on Murine Macrophages," *Journal of Immunology* 136: 2195–2201.

Silverman, William A. 1986. "Subversion as a Constructive Activity in Medicine," *Perspectives in Biology and Medicine* 29: 385–391.

Singer, I. M. 1981. "Mathematics," pp. 38–46 in *The Joy of Research,* W. Shropshire, ed. Washington, D.C.: Smithsonian Institution Press.

Smeaton, W. A. 1963. "Guyton de Morveau and Chemical Affinity," *Ambix* 11: 55–64.

Smith, Alexander. 1910. *Introduction to Inorganic Chemistry.* New York: Century.

Smith, Cyril Stanley, ed. 1968. *Sources for the History of the Science of Steel, 1532–1786.* Cambridge, Mass.: MIT Press.

—— 1981. A Search for Structure: Selected Essays on Science, Art, and History. Cambridge, Mass.: MIT Press.

—— 1987. "The Tiling Patterns of Sebastien Truchet and the Topology of Strucural Hierarchy," *Leonardo* 20: 373–385.

Smith, Homer William. 1935. *The End of Illusion.* New York: Harper and Brothers.

—— 1953. *From Fish to Philosopher.* Boston: Little, Brown.

Smith, John K., and David A. Hounshell. 1985. "Wallace H. Carothers and Fundamental Research at DuPont," *Science* 229: 436–442.

Smith, T. F., and H. J. Morowitz. 1982. "Between History and Physics," *Journal of Molecular Evolution* 18: 265–282.

Snelders, H. A. M. 1970. "Svante August Arrhenius," in *Dictionary of Scientific Biography,* C. C. Gillispie, ed. New York: Scribner's.

—— 1973. "The Birth of Stereochemistry: An Analysis of the 1874 Papers of J. H. van't Hoff and A. J. Le Bel," *Janus* 60: 261–278.

—— 1974. "The Reception of J. H. van't Hoff 's Theory of the Asymmetric Carbon Atom," *Journal of Chemical Education* 51: 2–7.

Snow, C. P. 1934/1958. *The Search.* New York: Scribner's.

—— 1966/1967. *Variety of Men.* New York: Scribner's.

Snyder, E. E. 1940. *Biology in the Making.* New York: McGraw-Hill.

Sommer, Jack. 1987. "A New Agenda for Science," *American Scientist* 75 (March–April): 223–224.

Speert, Harold. 1980. *Obstetrics and Gynecology in America: A History.* Chicago: American College of Obstetricians and Gynecologists.

Spence, R. 1970. "Otto Hahn," *Biographical Memoirs of the Fellows of the Royal Society* 16: 279–314.

Steinbeck, John, and Edward F. Ricketts. 1941/1971. *Sea of Cortez.* Mamaroneck, N.Y.: Paul P. Appel.

Stent, Gunther S. 1978. *Paradoxes of Progress.* San Francisco: W. H. Freeman.

Stern, Nancy. 1980. "John William Mauchly: 1907–1980," *Annals of the History of Computing* 2: 100–103.

Stewart, David M. 1984. "The Secret of Stradivari," *American Way* (October): 179–182.
Stokes, T. D. 1982. "The Double Helix and the Warped Zipper—An Exemplary Tale," *Social Studies of Science* 22: 207–240.
—— 1983. "The Side-By-Side Model of DNA: Logic in a Scientific Invention," Ph.D. dissertation, University of Melbourne.
—— 1986. "Reason in the *Zeitgeist*," *History of Science* 24: 111–123.
Storr, Anthony. 1972. *The Dynamics of Creation*. New York: Atheneum.
Sugar, H. S., and C. C. Foster. 1981. "Maximilian Salzmann: Ophthalmic Pioneer and Artist," *Survey of Ophthalmology* 26: 28–30.
Sweeley, C. C., J. F. Holland, D. S. Towson, and B. A. Chamberlin. 1987. "Interactive and Multi-Sensory Analysis of Complex Mixtures by an Automated Gas Chromatography System," *Journal of Chromatography* 399: 173–181.
Sylvester, Joseph. 1964. "Algebraical Researches Containing a Disquisition on Newton's Rule for the Discovery of Imaginary Roots," *Philosophical Transactions of the Royal Society of London* 154: 579–666.
Szent-Györgyi, Albert. 1957. *Bioenergetics*. New York: Academic Press.
—— 1960. *Introduction to a Submolecular Biology*. New York: Academic Press.
—— 1963. "On Scientific Creativity," *Proceedings of the Third World Congress of Psychiatry, 1961*. Toronto: University of Toronto Press, McGill University Press. Vol. 1, pp. 47–50.
—— 1966a. "In Search of Simplicity and Generalizations (50 Years of Poaching in Science)," pp. 63–76 in *Current Aspects of Biochemical Energetics*, N. O. Kaplan and E. P. Kennedy, eds. New York: Academic Press.
—— 1966b. Interview, pp. 111–128 in *The Way of the Scientist*. New York: Simon and Schuster.
—— 1972. "Dionysians and Apollonians," *Science* 176: 966.
Szilard, Leo. 1961. *The Voice of the Dolphins*. New York: Simon and Schuster.
—— 1966. Interview, pp. 23–34 in *The Way of the Scientist*. New York: Simon and Schuster.
—— 1978. *Leo Szilard: His Version of the Facts*. S. Weart and G. Szilard, eds. Cambridge, Mass.: MIT Press.
Taton, René. 1957. *Reason and Chance in Scientific Discovery*, trans. A. J. Pomerans. New York: Philosophical Library.
Taubes, Gary. 1986. "The Game of the Name is Fame: But Is It Science?" *Discovery* 7 (December): 28–52.
Taylor, Alfred M. 1966. *Imagination and the Growth of Science*. London: John Murray.
"[C. F.] Taylor Sculpture Is Unveiled." 1987. MIT *Tech Talk*, April 1: 7.
Taylor, Philip L. 1987. "Lessons from the Michelson-Morley Experiment," *The Scientist* (July 13): 11.
Temin, Howard M. 1981. "Oncology and Virology," pp. 58–64 in *The Joys of Research*, W. Shropshire, ed. Washington, D.C.: Smithsonian Institution Press.
Terroine, Emile F. 1959. "Fifty-five Years of Union between Biochemistry and Physiology," *Annual Review of Biochemistry* 28: 1–15.
Thackray, Arnold. 1965/1966. "Documents Relating to the Origins of Dalton's Chemical Atomic Theory," *Memoirs and Proceedings of the Manchester Literary and Philosophical Society* 108, no. 2.
—— 1966. "John Dalton—Accidental Atomist," *Discovery* (September): 28–34.
Thomas, Lewis. 1974. *Lives of a Cell*. New York: Bantam Books.
—— 1983. *Late Night Thoughts on Listening to Mahler's Ninth Symphony*. New York: Viking Press.

Thomsen, Dietrick E. 1980. "A Dozen Participants in Search of a History," *Science News* 118: 10–12.
—— 1986a. "Going Bohr's Way in Physics," *Science News* 129: 26–27.
—— 1986b. "Physics to the End of the Century," *Science News* 129: 245.
—— 1987. "A Periodic Table for Molecules," *Science News* 131: 87.
Thomson, George P. 1957. *The Strategy of Research.* Southampton, England: University of Southampton Press.
—— 1961/1968. *The Inspiration of Science.* Oxford: Oxford University Press. Reprint, New York: Anchor Books.
Thomson, J. J. 1930. "Tendencies of Recent Investigations in the Field of Physics," *The Listener* (January 29): 177–179, 210.
—— 1937. *Recollections and Reflections.* New York: Macmillan.
Thomson, Keith S. 1983. "The Sense of Discovery and Vice Versa," *American Scientist* 71: 522–524.
Tierney, John. 1984. "Quest for Order: Subramanyan Chandrasekhar," pp 1–8 in *A Passion to Know: Twenty Profiles in Science,* A. L. Hammond, ed. New York: Scribner's.
Todd, John. 1980. "John Hamilton Curtiss, 1909–1977," *Annals of the History of Computing* 2: 104–110.
Tolstoy, Leo. 1984. *Resurrection,* trans. Vera Traill. New York: New American Library.
Topley, W. W. C. 1940. *Authority, Observation and Experiment in Medicine: The Linacre Lecture, 1940.* Cambridge, England: Cambridge University Press.
Torrance, E. Paul. 1965. "Scientific Views of Creativity and Factors Affecting its Growth," *Daedalus* 94 (Summer): 663–681.
Toulmin, Stephen. 1961. *Foresight and Understanding: An Inquiry into the Aims of Science.* New York: Harper.
Traube, Moritz. 1864. "Experimente zur Theorie der Zellbildung," *Medicinische-Zentrallblatt* 39.
—— 1867. "Experiment zur Theorie der Zellbildung und Endosmose," *Archive für Anatomie und Physiologie*: 87–165.
Trautz, M. 1930. "August Friedrich Horstmann," *Berichte der deutschen chemische Gesellschaft* 63: 21a–86a.
Travers, Morris. W. *A Life of Sir William Ramsay.* 1956. London: Edward Arnold.
Truesdell, Clifford. 1984. *An Idiot's Fugitive Essays on Science.* New York: Springer-Verlag.
Tsilikis, J. D. 1959. "Simplicity and Elegance in Theoretical Physics," *American Scientist* 47: 87–96.
Tuchman, Maurice, et al. 1986. *The Spiritual in Art: Abstract Painting, 1890–1985.* New York: Abbeville Press.
Turney, Jon. 1987. "Research Tier Plan Splits U.K. Scientists," *The Scientist* 1 (June 15): 1–2.
Tyndall, John. 1872. *Faraday as Discoverer.* New York: Appleton.
—— 1897. *On the Study of Physics: Fragments of Science—A Series of Detached Essays, Addresses and Reviews.* 6th ed. New York: Appleton. 2 vols.
Ulam, S. M. 1976. *Adventures of a Mathematician.* New York: Scribner's.
Valéry, Paul. 1929. *Introduction to the Method of Leonardo da Vinci,* trans. T. McGreevy. London: J. Rodker.
Vallery-Radot, Pasteur. 1954. *Pasteur inconnu.* Paris: Flammarion.
—— 1968. *Pages illustrés de Pasteur.* Paris: Hachette.
Vallery-Radot, Réné. 1884. *M. Pasteur: Histoire d'un savant par un ignorant.* Paris: J. Hetzel.

—— 1901/1919. *La Vie de Pasteur.* Paris: Flammarion. English edition, *The Life of Pasteur,* trans. R. L. Devonshire. London: Constable.

—— 1912. *Pasteur dessinateur et pasteliste (1836–1842).* Paris: A. Marty, E. Paul.

Van't Hoff, Jacobus H. 1873. "Über eine neue Synthese der Propionsäure," *Berichte der deutschen chemische Gesellschaft* 6: 1107.

—— 1874a. "Beiträge zur Kenntniss der Cyanessigsäure," *Berichte der deutschen chemische Gesellschaft* 7: 1382; 1571.

—— 1874b. *Voorstel tot uitbreiding der tegenwoordig in de scheikunde gebruikte structuur-formules in de ruimte . . .* Utrecht: J. Greven. Other editions: *La Chimie dans l'espace,* Rotterdam: P. M. Bazendijk, 1875; *Lagerung der Atome in Raume,* Braunschweig: F. Vieweg, 1877; *Chemistry in Space,* Oxford: Clarendon Press, 1891.

—— 1876. "Die Identität von Styrol und Cinnamol, ein neuer Körper aus Styrax," *Berichte der deutschen chemische Gesellschaft* 9: 5.

—— 1878/1967. "Imagination in Science," trans. Georg F. Springer, *Molecular Biology, Biochemistry, and Biophysics* 1: 1–18.

—— 1878–1881. *Ansichten über die organische Chemie.* Braunschweig: Vieweg.

—— 1884. *Etudes de dynamique chimique.* Amsterdam: F. Muller.

—— 1886a. "L'equilibre chimique dans les systèmes gazeux ou dissous à l'état dilué," *Archives Néerlandaises* 20: 239–302.

—— 1886b. "Lois de l'equilibre chimique dans l'état dilué," *Kungliga Svenska Vetenskaps-Akademiens Handlingar* 21, no. 17.

—— 1887/1929. "Die Rolle des osmotischen Drucks in der Analogie zwischen Lösungen und Gasen," *Zeitschrift für physikalische Chemie* 1: 481–508. English version trans. James Walker, pp. 5–42 in *The Foundations of the Theory of Dilute Solutions.* London: Alembic Club.

—— 1888. "Dissociationstheorie der Elektrolyte," *Zeitschrift für physikalische Chemie* 2: 781–786.

—— 1894. "Wie die Theorie der Lösungen entstand," *Berichte der deutschen chemische Gesellschaft* 27: 6–19.

—— 1901/1966. "Osmotic Pressure and Chemical Equilibrium (Nobel Lecture, 1901)," pp. 5–10 in *Nobel Lectures: Chemistry, 1901–1921.* Amsterdam: Elsevier.

—— 1903a. "Lebensbericht [A. Horstmann]," in *Ostwald's Klassikern der Exakten Naturwissenschaften,* no. 137. Leipzig: Engelmann.

—— 1903b. *Physical Chemistry in the Service of the Sciences,* trans. A. Smith. Chicago: University of Chicago Press.

Vaughn, M. K., et al. 1980. "Effect of Synthetic Threonylseryllysine (TSL), a Proposed Pineal Peptide, on Reproductive Organ Weights and Plasma and Pituitary Levels of LH, FSH, and Prolactin in Intact and Castrated Immature and Adult Male Rodents," *Neuroendocrinology Letters* 2: 235–240.

Vijh, Ashok K. 1987. "Spectrum of Creative Output of Scientists: Some Psychosocial Factors," *Physics in Canada* 43: 9–13.

Viola, H. J., and C. J. Margolis, eds. 1986. *Magnificent Voyagers: The U.S. Exploring Expedition, 1838–1842.* Washington, D.C.: Smithsonian Institution Press.

Vitz, Paul C., and A. B. Glimcher. 1984. *Modern Art and Modern Science: The Parallel Analysis of Vision.* New York: Praeger.

Waddington, C. H., ed. 1972. *Biology and the History of the Future.* Edinburgh: Edinburgh University Press. Chicago: Aldine-Atherton.

—— 1975. *The Evolution of an Evolutionist.* Edinburgh: Edinburgh University Press.

Wade, Nicholas. 1981. *The Nobel Duel*. New York: Anchor Press.
Waife, S. O. 1960. "In Defense of Teleology," *Perspectives in Biology and Medicine* 4: 1–2.
Waksman, Selman A. 1954. *My Life With the Microbes*. New York: Simon and Schuster.
Wald, George. 1958. "Innovation in Biology," *Scientific American* 199 (September): 100–113.
—— 1961. "Foreword," in G. Ames and R. Wyler, *Biology: An Introduction to the Science of Life*. New York: Golden Press.
—— 1966. "On the Nature of Cellular Respiration," pp. 27–32 in *Current Aspects of Biochemical Energetics*, N. A. Kaplan and E. P. Kennedy, eds. New York: Academic Press.
Walden, Paul. 1904. *Wilhelm Ostwald*. Leipzig: Wilhelm Engelmann.
Waldrop, M. M. 1982. "NASA Looks for Thomas Edisons," *Science* 218: 870–871.
Walentynowicz, Bohdan, ed. 1982. *Polish Contributions to the Science of Science*. Dordrecht: Reidel.
Walker, James. 1914. "Van't Hoff Memorial Lecture," in *Chemical Society Memorial Lectures*. London: Gurney and Jackson. Vol. 2, pp. 255–271.
—— 1933. "Arrhenius Memorial Lecture," in *Chemical Society Memorial Lectures*. London: Chemical Society. Vol. 3, pp. 109–130.
Wallace, Alfred Russell. 1905. *My Life: A Record of Events and Opinions*. London: Chapman and Hall.
Wallas, Graham, 1926. *The Art of Thought*. New York: Harcourt, Brace.
Watson, J. D. 1965. *Molecular Biology of the Gene*. New York: W. A. Benjamin.
—— 1968. *The Double Helix*. New York: Atheneum.
—— and F. H. C. Crick. 1953. "The Structure of DNA," *Cold Spring Harbor Symposia on Quantitative Biology* 18: 123–131.
The Way of the Scientist: Interviews from the World of Science and Technology. 1966. New York: Simon and Schuster.
Weaver, Warren. 1968. "The Encouragement of Science," *Scientific American* 199: 170–178.
Weber, R. L., compiler, and E. Mendoza, ed. 1973. *A Random Walk in Science*. New York: Crane, Russak. London: Institute of Physics.
Weinberg, Alvin. 1967. *Reflections on Big Science*. Cambridge, Mass.: MIT Press.
Weisberg, Louis. 1987. "Unorthodox Science Fuels Biosphere Space Trial," *The Scientist* (May 18): 7.
Weiss, Paul A. 1964. "Life on Earth (by a Martian)," *The Rockefeller Institute Review* 2, no. 6, pp. 8–14.
—— 1971. "The Growth of Science: Knowledge Explosion?" pp. 134–140 in *Within the Gates of Science and Beyond: Science in Its Cultural Commitments*. New York: Hafner.
—— 1969. "'Panta' rhei'—And So Flow Our Nerves," *American Scientist* 57: 287–305.
——, ed. 1971. "The Basic Concept of Hierarchic Systems," pp. 1–44 in *Hierarchically Organized Systems in Theory and Practice*. New York: Hafner Press.
Weisskopf, Victor. 1972. "The Significance of Science," *Science* 176: 138–146.
—— 1977. "The Frontiers and Limits of Science," *American Scientist* 65: 405–411.
—— 1980. "L'art et la science," *CERN* (March): 31–52.
Weissmann, Gerald. 1985. *The Woods Hole Cantata: Essays on Science and Society*. New York: Raven Press.

Wells, D. B., H. D. Humphrey, and J. J. Coll. 1942. "The Relation of Tannic Acid to the Liver Necrosis Occurring in Burns," *New England Journal of Medicine* 26: 629–636.
Wersky, G. 1978. *The Visible College: The Collective Biography of British Scientific Socialists of the 1930s*. New York: Holt Rinehart and Winston.
Westall, F. C., and R. S. Root-Bernstein. 1983. "An Explanation of Prevention and Suppression of Experimental Allergic Encephalomyelitis," *Molecular Immunology* 20: 169–177.
—— 1986. "The Cause and Prevention of Post-Infectious and Post-Vaccinal Encephalopathies in Light of a New Theory of Autoimmunity," *Lancet* (August 2): 251–252.
Whitaker, Paul F. 1969. *More than Medicine*, R. N. Whitaker, ed. New York: Carlton Press.
White, Leslie A. 1959. *The Evolution of Culture*. New York: McGraw-Hill.
—— 1975. *The Concept of Cultural Systems*. New York: Columbia University Press.
Wiesburd, Stefi. 1987. "The Spark: Personal Testimonies of Creativity," *Science News* 132 (November 7): 298–300.
Wiesner, Jerome B. 1965. "Education for Creativity in the Sciences," *Daedalus* (Summer): 527–537.
Wilde, Oscar. 1931. *The Picture of Dorian Gray*. New York: Modern Library.
Wilkinson, Lise. 1971. "William Brockedon, F.R.S. (1787–1854)," *Notes and Records of the Royal Society of London* 26: 65–72.
Williams, L. Pearce. 1965. *Michael Faraday*. London: Chapman and Hall.
Williams, Trevor I., ed. 1976. *A Biographical Dictionary of Scientists*, 2d ed. New York: Wiley-Interscience.
—— 1984. *Howard Florey: Penicillin and After*. Oxford: Oxford University Press.
Williams, William Carlos. 1984. *The Doctor Stories*, compiled by Robert Coles. New York: New Directions.
Williams-Ellis, A., and E. C. Willis. 1954. *Laughing Gas and Safety Lamp*. New York: Abelard-Schuman.
Williamson, Alexander W. 1851. "Theory of Aetherification," *Chemical Gazette* 9: 294.
Willstätter, Richard. 1965. *From My Life: The Memoirs of Richard Willstätter*, trans. L. S. Hornig. New York: W. A. Benjamin.
Wilson, David. 1983. *Rutherford: Simple Genius*. London: Hodder and Stoughton.
Wilson, E. Bright. 1986. "One Hundred Years of Physical Chemistry," *American Scientist* 74: 70–77.
Wilson, Mitchell. 1954. *American Science and Invention*. New York: Bonanza Books.
—— 1961. *Meeting at a Far Meridian*. Garden City, N.Y.: Doubleday.
—— 1972. *Passion to Know*. Garden City, N.Y.: Doubleday.
Wilson, Robert R. 1966. Interview, pp. 46–55 in *The Way of the Scientist*. New York: Simon and Schuster.
Winkler, Karen J. 1985. "Historians Fail to Explain Science to Laymen, Scholar Says," *Chronicle of Higher Education* (August 7): 7.
Wise, Mervyn E. 1962. "Dutch pbis," pp. 9–11 in *The Scientist Speculates*, I. J. Good, ed. New York: Basic Books.
Wittgenstein, Ludwig. 1969. *On Certainty*, trans. D. Paul and G. E. M. Anscombe. Oxford: Blackwell.
Wotiz, J. H., and S. Rudofsky. 1984. "Kekulé's Dreams: Fact or Fiction?" *Chemistry in Britain* 20: 720–723.

Wrotnowska, Denise. n.d. "Pasteur's First Vocation," *Organorama* 2(6): 17–21.
—— 1981. "Pasteur: Première recherches sur la fermentations (1855–1857)," *Clio Medica* 15: 191–199.
Wurtz, Adolphe. 1869. *A History of Chemical Theory*, trans. Henry Watts. London: Macmillan.
Yalow, Rosalyn. 1986. "Peer Review and Scientific Revolutions," *Biological Psychiatry* 21: 1–2.
Yonge, Keith A. 1985/1986. "The Philosophical Basis of Medical Practice," *Humane Medicine* 1: 25–29; 2: 26–32.
Yost, Edna. 1959. *Women of Modern Science*. New York: Dodd, Mead.
Young, Donn C. 1987. "Viewing Stereo Drawings," *Science* 235: 623.
Zigrosser, Carl, ed. 1955–1976. *Ars Medica: A Collection of Medical Prints Presented to the Philadelphia Museum of Art by Smith-Kline Corporation*. Philadelphia, Pa.: Philadelphia Museum of Art.
Ziman, J. 1969. "Information, Communication and Knowledge," *Nature* 224: 318–324.
Zimmerman, David R. 1973. *Rh: The Intimate History of a Disease and Its Conquest*. New York: Macmillan.
Zloczower, Abraham. 1960. "Career Opportunities and Scientific Growth in Nineteenth Century Germany with Special Reference to the Development of Physiology." M.A. thesis, Hebrew University, Jerusalem.
Zuckerman, Harriet. 1977. *Scientific Elite: Nobel Laureates in the United States*. New York: Free Press.

Acknowledgments

I dedicate this book to my wife, Michèle, and my parents, Maurine and Mort, who have given me everything I value. I could not have written this book without their support. And to my brother, Rick, who wanted to know why anyone would bother doing history of science when he could be doing science. Here is an answer.

I must also acknowledge the assistance of several organizations. The idea for a book of this sort began in graduate school, where my research was supported by Danforth and Whiting fellowships. More research and the bulk of the writing were accomplished under a MacArthur Prize Fellowship.

My wife, Michèle, my parents, Jim Atkinson, Timothy Ferris, Michael Ghiselin, Mott Greene, Daniel Kevles, David Rindos, Helen Samuels, and Art Yuwiler have read and commented on all or part of the book. I thank them all. Other people have provided grist for my mill in ways that they may or may not recognize: Algirdas Avizienis, Dalton Delan, Ed Frieman, Scott Gilbert, Allen Kingman, Thomas Kuhn, Jay Last, Don McEachron, David Samuel, Arnold Seid, Cyril Stanley Smith, and Tom van Sant. Many thanks for your friendship, time, interest, and conversation.

Jonas Salk and Art Yuwiler deserve extra thanks for giving me places to work when no one else would, as do James Bonner, the late James Danielli, and David Felten for helping to see that some of the more outrageous theories and experiments reported herein were published. I could not have twisted the lion's tail otherwise. Fred C. Westall deserves special thanks for collaborating on many crazy experiments and for sharing his valuable experience with me.

I must also offer deep thanks to the librarians and archivists who have helped me locate manuscripts, rare books, and photographs. Gerald Geison provided me with an introduction to the Pasteur manuscripts and Marie Laure Prévost guided me through the Pasteur papers in the Bibliothèque Nationale in Paris. Gwyn Macfarlane was kind enough to share his knowledge of the Fleming manuscripts in the British Museum. Gustav Arrhenius put me in contact with various relatives of Svante Arrhenius. Dozens of people have provided me with tips concerning the musical and artistic proclivities of eminent investigators.

Many thanks also to Barbara Robinson for typing and retyping several early drafts, and to Marsha Wise for the index.

Finally, I must thank Angela von der Lippe and Harvard University Press for their willingness to publish such an idiosyncratic work.

Because of these many debts, it is more than usually necessary that I accept full responsibility for all opinions, errors of commission and omission, and editorial matters. I also wish to say, should there be any doubt on the subject, that none of the characters in the book in any way represents any person, living or dead. Though references are made to many real people, the six main characters are entirely fictional.

Index

Agassi, Joseph, 419
Agassiz, Louis, 120
Allen, Garland, 337
Allison, V. D., 146, 186, 421
Alvarez, Luis, 404, 417
Ampère, A. M., 106, 277
Angström, A. J., 242
Arnheim, Rudolf, 331
Arrhenius, Svante, 33, 162, 189, 208–209, 215, 232–239, 242–244, 250, 258–260, 273, 277, 282–290, 345, 355, 365–373, 381–382, 384, 395–396, 401, 408, 415
Austin, James, 133, 140
Avizienis, Algirdas, 416
Avogadro, Amedeo, 212
Axelrod, Julius, 400, 403

Bacon, Francis, 7, 113, 231, 408–409, 412, 414
Baeyer, J. F. A. von, 34, 36, 51–52, 236, 396, 412
Baker, R. A., 276–277
Balard, A. J., 88, 90
Baltimore, David, 383
Bancroft, Wilder, 345, 383
Banting, Frederick, 9, 123, 396
Barber, Bernard, 304
Barcroft, Joseph, 330
Barnes, Barry, 201
Bartlett, Neil, 16
Bates, R. H., 200–201, 418
Beaver, William, 40
Behring, Emil von, 372
Békésy, Georg von, 330, 333, 393
Bell, James F., 123, 346
Ben-David, Joseph, 369
Benzer, Seymour, 386
Bernal, J. D., 33, 88, 110, 328, 352, 354, 367, 412
Bernard, Claude, 56, 84, 113, 115, 132, 136–137, 141, 205, 227, 250, 314, 329, 344, 349–350, 396, 412–413
Berson, Solomon, 337
Bert, Paul, 314
Berthelot, Marcellin, 226–227, 231, 288, 345, 367, 379, 396
Berthollet, Claude Louis, 10–15, 17–26, 29, 31, 42, 47, 58, 87, 89, 107, 133, 151, 189–190, 204, 210–212, 216, 219, 232, 246, 273, 284, 287–288, 344, 357, 365, 368, 373, 381, 384, 386, 388, 417
Berzelius, 212, 242, 285
Besant, Annie, 25, 272
Best, C. H., 396
Beveridge, W. I. B., 140, 332
Billroth, Theodor, 328
Biot, Jean Baptiste, 89–90, 94, 99–100, 102–105, 146, 344, 355, 358, 421
Blalock, J. E., 129–130, 427–428
Blondlot, R., 94
Bohr, Christian, 34, 350
Bohr, Niels, 34, 36, 46, 101, 135, 147, 194, 224, 226, 230, 240, 349–350, 353, 357–358, 375, 413, 415
Boltzmann, Ludwig, 244, 334
Bolyai, Johann, 205–206
Boman, Hans, 175
Bonaparte, Napoleon, 11, 14–15, 20–21, 23, 26, 31, 367, 388
Boring, Edward, 349
Borodin, A. P., 328
Bost, K. L., 428
Boveri, Theodor, 92
Braggs, W. L., 34, 198, 328, 397, 419
Branscomb, Lewis, 384, 403
Bridgeman, Percy W., 349
Bright, Richard, 123
Brillouin, Leon, 165, 299, 352, 425
Brinkman, W. F., 36
Broglie, Louis de, 303, 353, 386
Bronowski, Jacob, 32, 96
Brooks, Chandler, 346–347
Bruce, Robert, 369
Bruner, Jerome, 206

Bunsen, Robert W., 243
Burch, George, 398
Burdon-Sanderson, John, 180, 379
Burnell, Jocelyn Bell, 184
Burnet, Macfarlane, 45, 54, 185, 227, 298, 388, 396, 418
Butlerow, Alexander, 218

Cadeddu, Antonio, 137, 139
Campbell, Donald, 359
Cannon, Walter B., 31, 136, 140, 330
Cantor, Moritz, 345
Carmichael, R. D., 53
Carnelly, Thomas, 369
Carothers, Wallace, 362
Carruthers, Peter, 411, 417
Cavalli-Sforza, L. L., 359–362
Caws, Peter, 85
Chain, Ernest B., 168
Chandrasekhar, Sabrumanian, 10, 32, 162, 332, 386–387, 390
Charcot, Jean-Martin, 123
Chargaff, Erwin, 71, 198, 354, 425
Chevreul, M. E., 25, 107–108
Clausius, Rudolf, 188, 206, 208, 250, 255–256, 285, 287, 290, 367
Cleve, Per T., 242–243
Clough, A. H., 155
Cockroft, John, 95
Cohen, Ernst, 219
Cohen, I. Bernard, 39, 57, 68, 345, 373
Cohen, Robert, 349
Colinvaux, Paul, 378
Colquhoun, D. B., 167, 173
Comte, Auguste, 210–211
Conant, James, 347
Cook, N. D., 129
Cormack, Allan, 400
Craddock, Stuart, 179
Cram, Donald, 333, 386
Crew, F. A. E., 64
Crick, Francis, 30, 36, 52, 70, 75, 78, 95, 116, 164, 198–199, 205, 276, 344, 350, 386, 393, 397, 400, 419
Crombie, A. C., 351
Crookes, William, 183, 272
Curie, Marie and Pierre, 34, 408
Curtin, Dean, 333
Cushing, Harvey, 123
Cuvier, Baron, 93, 211
Cyriax, B., 202
Dainton, Lord, 413
Dalton, John, 187–188, 366
Danielli, James, 62, 129
Darlington, C. D., 92, 329
Darwin, Charles, 6, 45, 47, 132, 194, 206–208, 222, 238, 240, 249, 282, 290, 330, 348, 360, 362, 365, 373, 380, 385, 392–393, 395, 404–405, 414
Davidson, Edward, 260–261, 264–265
Davis, Philip, 334–335
Davy, Humphry, 155, 211, 285, 316–317, 337–338
Debye, Peter, 96, 344, 388, 399–401
Dedijer, Steven, 34
De Forest, Lee, 241, 341
Delbrück, Max, 386, 391, 412
De Kruif, Paul, 260
De Morgan, Augustus, 346
De Vries, Hugo, 246–248
D'Herelle, Felix, 153, 156, 185
Diderot, Denis, 31, 48, 154, 302
Dirac, P. A. M., 26, 332, 387, 419
Disraeli, Benjamin, 134, 180, 409
Dobbs, B. J. T., 355
Dobzhansky, Theodosius, 100, 385
Donnan, F. G., 345, 369
Donohue, Jerry, 198
Doolittle, 427
Drach, Jules, 346
Driesche, Hans, 92, 114, 379, 395
Du Bois-Reymond, Emil, 239, 392
Dubos, René, 92, 100, 116, 194
Duhem, Pierre, 15, 53, 92, 224–225, 299, 314, 332, 344–345, 350, 352, 367
Dumas, J. B. A., 100, 103, 105, 345, 388
Durham, William, 359
Dutrochet, R. J., 248
Dyson, Freeman, 42, 251, 337, 415

Eccles, John, 57, 273, 349
Eddington, Arthur, 349
Edge, David O., 378
Edison, Thomas, 240–241, 341, 363
Edlund, Erik, 243, 282, 284, 373, 405
Ehrenfest, Paul, 204
Ehrlich, Paul, 142, 270, 329, 372
Eiduson, Bernice, 315–316
Eigen, Manfred, 417
Einstein, Albert, 6, 9, 33, 45, 54, 98, 101, 159, 204, 206, 209, 215, 238, 242, 244, 250, 253, 280, 299, 314, 329, 331, 337–338, 345, 348–349, 352–355, 365–366,

373, 385, 387, 392, 404, 408, 410, 412, 419
Eisenberg, Leon, 50
Emerton, Norma, 330
Etzler, John, 337

Faraday, Michael, 132, 209, 316, 329, 365, 414, 418
Feigenbaum, Mitchell, 339
Felten, David, 190
Ferguson, Eugene, 330
Fermi, Enrico, 31, 135, 205, 240, 414
Feyerabend, Paul, 82, 116, 298
Feynman, Richard, 26, 31, 36, 65, 145, 226, 251, 329, 390, 405, 408, 418
Fleck, Ludwig, 201, 375
Fleming, Alexander, 117, 136, 142–153, 155–157, 159–161, 167–174, 176–183, 187–189, 190, 204, 208, 231, 235, 237–238, 240, 254, 273, 291, 295, 300, 303, 329, 348, 364–366, 379–380, 382, 384, 386, 391, 395–396, 412–413, 421
Florey, H. W., 168, 396
Forman, Paul, 305, 349
Foster, Michael, 70
Fourier, Joseph, 368
Fraenkel-Conrat, Heinz, 70
Frank, Philipp, 349
Franklin, Benjamin, 304
Franklin, Rosalind, 198, 330
Fuller, Buckminster, 241
Fulton, Robert, 330

Gajdusek, Carlton, 329
Galdston, Iago, 351
Galois, Evariste, 346
Galton, Francis, 316, 329, 331
Gäth, R., 202
Gaudin, Marc, 218, 220
Gauss, K. F., 205, 277, 334
Geddes, Patrick, 120
Geiger, Hans, 416
Gell-Mann, Murray, 226, 296, 329, 383
Gerard, Ralph, 359–360, 362, 367
Ghiselin, Michael, 347
Gibbs, J. Willard, 206, 303, 346, 369, 395
Giere, Ronald, 378
Glaser, Donald, 396–397
Glashow, Sheldon, 384
Glass, Bentley, 45
Goertzel, V. and M., 315
Goethe, J. W. von, 15, 345, 412
Goldschmidt, Richard, 395, 401, 403
Golomb, Solomon, 26
Good, I. J., 302, 414
Goodfield, June, 97
Goodrich, Edwin, 93
Gould, Stephen Jay, 202, 347, 359, 374
Grafstein, Daniel, 130
Graham, Thomas, 248
Grotthus, C. J. D. von, 284, 288
Gruber, Howard, 330
Guldberg, Cato, 287, 344, 367
Gunning, Willem, 239, 243–244, 371, 405
Guthrie, Leonard G., 315
Gutting, Gary, 86

Haber, Fritz, 385, 388
Hadamard, Jacques, 98, 346, 391
Haden, Francis Seymour, 117
Hadley, Arthur, 393
Hahn, Otto, 396–397, 402–403
Haldane, J. B. S., 34, 280, 350, 353, 385, 388, 395
Haldane, J. S., 34, 194, 350, 392, 400
Hall, Charles Martin, 335
Hamilton, William, 26
Hanson, N. R., 82, 85, 378
Hardin, Garrett, 201, 405
Hardy, Alister, 332, 403
Hare, Ronald, 167–171, 173, 176–177, 185
Harré, Rom, 231
Harris, Errol, 85
Harrison, Anna, 347–348
Hartmann, Fritz, 201
Harwit, Martin, 7, 379–381
Haüy, René Just, 89–90, 355, 358
Heath, Thomas, 345
Heisenberg, Werner, 45, 147, 205, 332, 349, 352, 387, 419
Helmholtz, Hermann von, 25, 121, 222, 240, 243, 259–260, 299, 334, 367
Hempel, C. G., 53
Henle, Jacob, 392
Henri, Victor, 34, 372
Herapath, John, 206–208, 249, 366, 404
Herbst, Curt, 395
Hersch, Reuben, 334–335
Herschel, John, 34, 89, 117–119, 121, 123, 204, 294, 317, 328
Herschel, William, 34, 198, 317, 386
Hertz, Heinrich, 183, 228
Hilbert, David, 332
Hill, A. V., 393

Hindle, Brooke, 330, 337
Hittorf, J. W., 285, 289
Hodgkin, Alan, 178
Hodgkin, Dorothy, 13, 123
Hoffmann, Roald, 32, 211
Hofstadter, Douglas, 335
Holley, Robert, 401, 411
Holton, Gerald, 66, 187, 241, 294, 314, 331, 352, 363, 366
Hopkins, F. G., 372
Horstmann, Friedrich, 246, 252, 344, 367
Hounsfield, Godfrey, 400
Houtermans, Fritz, 121
Hughes, Thomas, 241, 341
Hughes, W. H., 143–144, 146, 172–173
Hull, David, 53, 378
Humboldt, Alexander, 341
Huxley, Julian, 186, 329, 344, 350, 359
Huxley, T. H., 5, 113, 120, 132, 156, 186, 206, 348, 414

Infeld, Leopold, 375, 386, 411

Jacob, François, 101, 223, 350
Jacobi, C. G. J., 132, 334
James, William, 400
Jeans, James, 349
Jerne, Neils, 45, 49, 227
John-Steiner, Vera, 332
Joliot-Curie, Frédéric, 415
Joule, J. P., 385
Judson, Horace, 66

Kac, Mark, 353
Kangro, Hans, 344
Kapitza, Peter, 120
Karnovsky, Manfred, 271–272
Kassler, Jamie, 335
Kaufmann, Walter, 129, 214
Kekulé von Stradonitz, August, 211, 214–218, 236, 242, 277, 282, 298, 317, 357, 375
Keller, Evelyn Fox, 97, 351
Kelvin, Lord, 45, 183, 298
Kendrew, John, 397, 419
Kennedy, Alexander, 341
Kettering, Charles, 97, 134, 139, 341, 409
Kirchhoff, Gustav R., 243, 334
Klein, Félix, 279
Koch, Robert, 392, 395
Koestler, Arthur, 154, 275, 280, 291, 302
Kohlrausch, Friedrich, 244, 282
Kolbe, Hans, 224–225, 337–338
Kovalevskaia, Sofia, 329
Koyre, Alexandre, 349
Krebs, Hans, 34, 35
Krueger, J. M., 271
Kubie, L. S., 314
Kuhn, Thomas, 7, 16, 17, 22, 31, 41, 115, 160, 174, 190, 201–202, 221, 225, 230, 281, 291, 299, 303–304, 343, 345, 353, 357–362, 373–374, 378–379, 413
Kundt, August, 330
Kuznetsov, V. I., 346

Labeyrie, Antoine, 406
Lacepède, B.-G.-E. de, 317
Laënnec, T.-R.-H., 155, 187, 189, 365
Lagrange, J. L., 45, 335
Lakatos, Imre, 225, 357, 378
Land, Edwin, 415
Langmuir, Irving, 84, 94, 390, 416
Lapage, Geoffrey, 330
Laplace, Pierre Simon de, 19, 20, 21, 133, 211–212, 246, 365, 368
Laue, M. T. F. von, 349
Laurent, August, 90, 93, 95, 99, 111, 147, 221, 317, 355, 358
Lavoisier, A. L., 18–20, 138, 288, 395
Lawrence, E. O., 385, 396
Leakey, Louis, 403
Le Bel, Joseph Achille, 209, 216–217, 220–223, 233, 395, 397
Leblanc, Nicholas, 17
Lederberg, Joshua, 97, 115
Lehninger, Albert L., 70
Lehrer, Robert, 175
Leibniz, Gottfried W., 240, 345–346
Lenard, P. E. A., 183
L'Engle, Madeleine, 301
Lenz, Rudolf, 282
Leontief, Vassily, 383
Levarie, Siegmund, 335
Levins, Richard, 350
Lewis, G. N., 288
Lewis, Ralph, 161–162
Lewontin, Richard, 280, 350, 359, 374
Lexell, 198, 204
Liebig, Justus, 34, 222, 248, 316
Lind, H. F., 92
Lipscomb, William N., 277, 332
Lipsky, Eleazar, 425
Lister, Joseph, 123, 146, 180–181, 379
Lobachevsky, Nikolai I., 205
Lockyer, Norman, 47
Lodge, Oliver, 239

Loeb, Jacques, 345, 369
Loewi, Otto, 276, 353, 396
Lorenz, Konrad, 148–149, 189, 329, 332, 391
Loria, Gino, 345
Lucas, Edouard, 26
Lumsden, Charles, 359
Lyell, Charles, 362

Macfarlane, Gwyn, 147, 172–174, 413
Mach, Ernst, 54, 234, 276, 344
Macquer, P. J., 12, 15, 17, 19
Madsen, Thorvald, 372
Mahoney, Michael J., 64, 305
Maier, N. R. F., 226, 419
Maier-Leibnitz, Heinz, 50
Majorana, Ettore, 205
Malthus, Thomas, 93, 186–187, 277
Matthias, Bernd, 227, 276
Maurois, André, 171, 174
Maxwell, J. C., 50, 85, 206, 334, 385, 387, 408
Maxwell, Nicholas, 84, 207, 241, 303
Mayr, Ernst, 347, 350, 352, 420–421
Mazurs, Edward, 272, 274
McClelland, David C., 185
McClintock, Barbara, 71, 97, 330, 351, 403, 408
McConnell, J. V., 7
McCready, Paul, 301
Medawar, Peter, 6, 49, 54, 113, 227, 349, 410–412
Meinong, Alexis, 328
Meitner, Lise, 328, 402–403
Mekler, L. B., 129–130, 427
Melhado, Evan, 91
Mellanby, Kenneth, 403
Mendel, Gregor, 9, 16, 47, 206, 246, 249, 366, 379, 404, 408
Mendeleev, Dmitri, 188, 222, 272, 363, 367
Mendelssohn, Kurt, 206
Merton, Robert, 185
Metchnikoff, Elie, 151–152
Meyer, Lothar, 222, 344
Michelson, Albert, 45, 58, 121, 132, 147, 240, 298, 328
Miller, Arthur I., 330
Miller, Dayton C., 227–228
Miller, Jonathan, 32
Minkowski, Oscar, 117, 136, 140–141, 151, 187, 215, 349
Millikan, Robert, 45, 46, 240, 349, 386
Mirsky, Alfred, 346
Mitscherlich, Eilhard, 89–91, 94, 96, 105, 111–112, 146, 246, 250, 252–256, 344, 355, 358, 421
Mittasch, Alwin, 345
Möbius, P. J., 316
Monod, Jacques, 50, 101, 223, 234, 349–350, 359, 415
Morgan, Thomas Hunt, 337
Morris, Desmond, 148, 403
Morrison, Robert, 336
Morse, Samuel, 330
Morveau, Guyton de, 17
Mulkay, Michael J., 378
Muller, Hans J., 132, 154, 246, 385, 416–417

Nachmansohn, David, 235, 314
Needham, Joseph, 350, 352
Nernst, Walther, 228
Nevanlinna, Rolf, 335
Newton, Isaac, 38, 39, 46, 104, 188, 207, 240, 317, 329, 355, 412
Nicholson, Ben, 92
Nickles, Thomas, 86
Nobel, Alfred, 136
Nollet, J. A., 247–248
Nossal, Gustave, 304
Nye, Mary Jo, 349, 371

Oersted, H. C., 136, 140
Oparin, A. I., 350
Orts, R. J., 427
Osborn, H. F., 154
Ossowska, Maria, 33
Ossowski, Stanislaw, 33
Ostwald, Wilhelm, 25, 33, 34, 189, 206–210, 225, 228, 232–239, 243–244, 255, 258–259, 273, 277, 282, 316, 330, 337, 344–345, 350, 359, 365–367, 371, 382–383, 386, 388, 390, 396

Page, Irvine H., 383
Pappenheimer, J. R., 271
Pardee, Arthur, 222
Partington, J. R., 367
Pasteur, Louis, 6, 10, 68–69, 81, 86–96, 99–100, 102, 103–105, 107–118, 132–139, 143, 146–147, 151, 154, 164, 179–183, 186–190, 194, 204, 208, 213, 216–218, 221, 225, 227, 237–238, 246, 273, 311, 317, 329, 344, 348, 355, 358, 362–367, 379–381, 384–388, 391, 396–397, 401, 408, 412, 417, 421

Pauli, Wolfgang, 206, 415
Pauling, Linus, 7, 10, 59, 135, 139, 199, 205, 238, 264, 277–278, 290, 294, 296, 355, 357, 384, 386, 403, 409, 412, 414
Payne-Gaposchkin, Cecilia, 330
Pearl, Raymond, 119
Penrose, Roger, 26
Perkin, W. H., Jr., 140, 183
Perrin, Jean B., 349, 367
Perutz, Max, 59, 388, 396, 419
Pestalozzi, Johann, 331
Peterson, Sven Otto, 243, 285
Pfaundler, Leopold, 246, 252, 255, 344, 367
Pfeffer, Wilhelm, 248–250, 252–254, 255–256, 265
Pirie, Norman, 395
Planck, Max, 39, 45, 60, 62, 189, 204, 206, 208, 214, 232, 233, 240, 243–244, 258–259, 334, 344–345, 349, 359, 374, 383, 385–386, 418
Platt, W., 276–277
Poincaré, Henri, 15, 53, 84, 217, 224, 238, 240, 277–280, 282, 290, 294, 299, 332, 345, 348, 349, 352, 359, 391
Polanyi, Michael, 84, 97, 115
Polya, George, 62, 83, 414
Popper, Karl, 32, 53, 113, 116, 224, 349, 357, 359, 378
Price, Derek de Solla, 33, 37
Proust, Louis Joseph, 368
Provostaye, M. de la, 90–91, 94, 105, 111–112
Prusiner, Stanley, 190
Pryce, Merlin, 168, 170, 174, 177–178

Rabi, I. I., 135, 393, 407
Ramon y Cajal, Santiago, 92–93, 97, 314, 316, 330, 332, 349, 392, 394–396
Ramsay, William, 345
Raoult, François Marie, 287, 367
Ratliff, Floyd, 333
Rayleigh, Lord, 47, 236, 243, 395–396
Regnault, Henri Victor, 252–253, 256
Reynolds, Joshua, 208
Reynolds, Osborne, 54, 241
Richet, Charles, 54, 117, 136–137, 329, 409–410, 414
Riemann, G. F. B., 132
Rindos, David, 359
Ringer, Sidney, 65, 416
Ritterbush, Philip, 92, 330
Robertson, T. B., 366, 375
Rodley, G. A., 200–201
Roe, Shirley, 330–331
Roebling, John, 337
Roentgen, Wilhelm C., 117, 136, 183, 272, 396–397, 402–403, 416
Rood, Ogden N., 25
Rosenfeld, Leon, 350
Ross, Ronald, 328
Rothenberg, A., 280, 291, 303
Roux, Emile, 138
Roy, Rustum, 395
Rutherford, Ernest, 36, 224–226, 240, 337–338, 385, 393, 396, 416

Sabatier, Paul, 367
Salk, Jonas, 97, 359, 377, 410
Sarton, George, 314, 343, 427
Say, J. B., 368
Schally, Andrew V., 396
Scheele, C. W., 12, 17
Schiller, F. C. S., 81
Schmidt, Carl, 243, 344
Schrödinger, Erwin, 45, 226, 328, 349–350
Schultz, T. W., 403
Schwann, Theodor, 392
Scott, W. M., 178, 186
Seidenberg, Roderick, 379
Selye, Hans, 44
Senarmont, H. H. de, 103, 105
Shaler, Nathaniel, 120
Shepard, Roger, 332
Silverman, D. H. S., 272
Simon, Herb, 383
Simpson, George Gaylord, 331
Smith, Cyril Stanley, 32, 129–130, 183, 339, 427
Snow, C. P., 4, 9, 46, 280, 412, 427
Spencer, Herbert, 350, 359
Sperry, Ambrose, 241, 341
Stahl, Georg E., 199
Stephenson, George, 209
Strassmann, Fritz, 402–403
Sudhoff, Karl, 345
Svoboda, Antonin, 237–238, 416
Sylvester, Joseph, 335
Szent-Györgyi, Albert, 6, 44, 55–56, 65–66, 132, 165, 179, 186, 236, 238, 261–263, 329, 343, 346, 382, 387, 399, 407–408, 412, 414, 416, 418
Szilard, Leo, 41, 135, 382, 386–387, 399, 403–404

Taine, Hippolyte, 210
Taton, René, 106, 172, 182–183, 330, 417

Taylor, D. W., 331
Temin, Howard M., 70, 403
Terroine, Emile, 372
Tesla, Nikola, 241, 341
Thakray, Arnold, 366
Theile, F. K. J., 114
Thomas, Lewis, 50, 114, 134, 139, 155, 403, 415
Thomson, G. P., 54, 113, 396, 410, 415
Thomson, J. J., 34, 35, 41, 53, 65, 154, 226, 240, 385, 392–393, 395
Toulmin, Stephen, 363
Traube, Mauritz, 249, 260
Truchet, Sebastien, 196
Truesdell, Clifford, 6, 341, 346, 352, 376, 403, 409
Twort, Frederick, 154, 156, 185
Tyndall, John, 132, 146, 180, 186

Ulam, Stan, 98, 276, 353

Valéry, Paul, 53
Vallery-Radot, Pasteur, 92–93
Van Deventer, Charles, 216, 240
Van't Hoff, Jacobus Henricus, 68, 189–190, 206, 208–212, 214–224, 226–227, 231, 232–260, 271–273, 277, 282, 287, 289, 316–317, 337, 341, 344–345, 349, 358, 365–367, 369, 371, 382–384, 390, 396–397, 401, 408, 417
Vauquelin, Louis N., 337–338
Virchow, Rudolph, 202

Waage, Peter, 287, 344, 367
Waddington, C. H., 333, 340
Waksman, Selman A., 379
Wald, George, 31, 352
Wallace, A. R., 187, 206, 208, 222, 277
Wallas, Graham, 201, 275
Walpole, Horace, 133
Warburg, Otto, 385
Waterston, John J., 206–208, 243, 249, 366, 379, 404
Watson, James D., 52, 59, 69, 75, 95, 164, 198–199, 205, 329, 385, 386–387, 397, 400, 407, 419
Watt, James, 13, 17
Weaver, Warren, 403, 407
Wechsler, Judith, 333
Weinberg, Alvin, 403
Weiss, Paul, 165
Weisskopf, Victor, 32, 180
Weissman, Gerald, 50
Weizmann, Chaim, 51
Werner, Alfred, 372
Westall, Frederick, 271
Weyl, Herman, 332, 349
Wheatstone, Charles, 330
Wheeler, John, 242, 249, 330, 411
Whewell, William, 210–211
White, L. A., 359
Whitehead, Alfred North, 66, 186, 349
Wiesner, Jerome, 338
Williams, R. R., 388
Williams, William Carlos, 211
Williamson, A. W., 285, 287, 290
Willstätter, Richard, 51, 52, 239, 345
Wilson, C. T. R., 236, 332, 396
Wilson, E. B., 337
Wilson, E. O., 34, 64, 359
Wilson, Mitchell, 42, 135, 280, 313
Wilson, Robert R., 277, 330
Wilson, Tuzo, 219, 241, 346, 372
Wislicenus, Johannes, 218, 222, 271
Wittgenstein, Ludwig, 241, 314
Wollaston, William Hyde, 218
Wood, R. W., 94
Wright, Almroth, 114, 146, 149–151, 186, 238, 395
Wurtz, Adolphe, 211, 216–217, 246, 317, 396

Yang, Chen Ning, 332
Yalow, Rosalind, 337, 382, 403
Young, J. Z., 386
Young, Thomas, 409

Zasloff, Michael, 176
Zloczower, Abraham, 370
Zuckerman, Harriet, 34, 39, 40